刘业翔　院士

1988年刘业翔院士接受人民画报访谈照片

2002年刘业翔院士与学生在一起讨论

2002年刘业翔院士在美国铝业公司技术交流

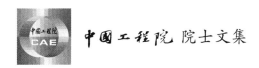

《中国工程院院士文集》总序

2012年暮秋，中国工程院开始组织并陆续出版《中国工程院院士文集》系列丛书。《中国工程院院士文集》收录了院士的传略、学术论著、中外论文及其目录、讲话文稿与科普作品等。其中，既有院士们早年初涉工程科技领域的学术论文，亦有其成为学科领军人物后，学术观点日趋成熟的思想硕果。卷卷文集在手，众多院士数十载辛勤耕耘的学术人生跃然纸上，透过严谨的工程科技论文，院士笑谈宏论的生动形象历历在目。

中国工程院是中国工程科学技术界的最高荣誉性、咨询性学术机构，由院士组成，致力于促进工程科学技术事业的发展。作为工程科学技术方面的领军人物，院士们在各自的研究领域具有极高的学术造诣，为我国工程科技事业发展做出了重大的、创造性的成就和贡献。《中国工程院院士文集》既是院士们一生事业成果的凝炼，也是他们高尚人格情操的写照。工程院出版史上能够留下这样丰富深刻的一笔，余有荣焉。

我向来认为，为中国工程院院士们组织出版院士文集之意义，贵在"真、善、美"三字。他们脚踏实地，放眼未来，自朴实的工程技术升华至引领学术前沿的至高境界，此谓其"真"；他们热爱祖国，提携后进，具有坚定的理想信念和高尚的人格魅力，此谓其"善"；他们治学严谨，著作等身，求真务实，科学创新，此谓其"美"。《中国工程院院士文集》集真、善、美于一体，辩而不华，质而不俚，既有"居高声自远"之澹泊意蕴，又有"大济于苍生"之战略胸怀，斯人斯事，斯情斯志，令人阅后难忘。

读一本文集，犹如阅读一段院士的"攀登"高峰的人生。让我们翻开《中国工程院院士文集》，进入院士们的学术世界。愿后之览者，亦有感于斯文，体味院士们的学术历程。

2012 年 7 月

中国工程院 院士文集

前 言

本文集是从本人与同事及学生们发表的 300 余篇学术论文中精选汇编而成的。文选的时间跨度从 1981 年到 2017 年，按研究工作内容分为八个部分：一、铝电解惰性阳极、惰性可润湿阴极及电解新工艺；二、铝电解工艺优化、自动控制与计算机仿真；三、高温熔盐电解过程电催化；四、湿法冶金电解过程电催化；五、锂离子电池；六、太阳电池；七、学术报告与科技评述；八、其他（氧化铝、TiAl 合金、自蔓延高温合成、传感器、光催化剂、燃料电池电催化、泡沫材料、超细粉体）。本文选可对从事有色金属冶金、新能源材料与器件的科研人员和高校师生提供参考和借鉴。

时光荏苒，白驹过隙，我从事学术研究已经 65 年有余，我将我的终身都奉献给教书育人的事业，奉献给了有色金属冶金事业。于我而言科研从不是一个人的战场，它更像是一代代的科研工作者的接力，更像是一群人分工合作。幸运的是我有一个足够优秀的团队，在我的科研历程中，我和我的团队攻坚克难，团结奋进，不断创新，在铝电解槽在铝电解节能阳极、熔盐电化学、功能电极材料、新能源材料及冶金过程模拟、控制与优化等方面进行了开创性研究与产学研实践，为我国有色金属冶金学科建设、铝电解工业技术进步以及锂离子电池产业发展做出了贡献。在此衷心感谢他们多年来对我的支持和帮助！

本文集是在课题组李劼教授、赖延清教授的统筹协调下进行的，肖劲教授、刘晋教授、张治安教授、田忠良教授、刘芳洋教授、张红亮副教授、蒋良兴副教授、宗传鑫老师一起完成论文的整理工作，非常感谢他们。本文集收录的各篇论文的原发表刊物、发表年代以及署名的合作者，读者可从每篇文章头一页的下注中查到。

回顾我的学术研究道路，虽然艰难，贡献不多，但有深刻体会，借此机会与大家共勉：(1) 科学研究是"学习—实践—再学习—再实践"这一艰难过程的反复与循环，选好专业和专题后，热爱它，钻研它，学而时念之，以最终有所创新，有所成就；(2) 工程科技的研究一定要与现场实际相结合，其目标就是要有实际贡献，为此，要特别重视深入实际，并重视相关基础学科的支撑作用；(3) 学习和实践都要与时俱进，引入新知识、新观点、新理论和新技术，使传统学科优势与新兴学科的新知识相结合，以获得更广阔的发展空间。

最后，对中国工程院提供出版资助，中南大学和中南大学冶金与环境学院对文集出版的支持和帮助深表感谢！

刘业翔

2018年8月

中国工程院 院士文集

目　录

院士传略

» 刘业翔简介 ………………………………………………………… 3
» 刘业翔传略 ………………………………………………………… 4

学术论文

一、铝电解惰性阳极、惰性可润湿阴极及电解新工艺

» 铝电解用 SnO_2 基惰性阳极的研究 ………………………………… 23
» Laboratory Study on TiB_2-based Ceramic Cathode for Aluminium Reduction Cell ………………………………………………………… 29
» Research Progress in TiB_2 Wettable Cathode for Aluminum Reduction ………… 36
» Cup-shaped Functionally Gradient $NiFe_2O_4$-based Cermet Inert Anodes for Aluminum Reduction ………………………………………………… 47
» Alumina Solubility in Na_3AlF_6-K_3AlF_6-AlF_3 Molten Salt System Prospective for Aluminum Electrolysis at Lower Temperature …………………… 55
» Further Research of "Cup-shaped" $NiFe_2O_4$ Based Cermet Inert Anodes for Aluminum Electrolysis ………………………………………………… 63

二、铝电解工艺优化、自动控制与计算机仿真

» Effect of An Alumina Layer at the Electrolyte/Aluminum Interface：A Laboratory Study ………………………………………………… 71
» Inhibition of Anode Effect in Aluminum Electrolysis Process by Anode Dopants：A Laboratory Study ………………………………………………… 80
» 160kA 预焙铝电解槽区域电流效率 ………………………………… 86
» 点式下料铝电解槽氧化铝浓度新型估计模型 ……………………… 91
» Bath Temperature Model for Point-feeding Aluminium Reduction Cells ………… 96
» An Improved Finite Element Model for Electro-magnetic Analysis in

Aluminum Cells ··· 104

三、高温熔盐电解过程电催化

>> Oxygen Overvoltage on SnO_2-based Anodes in NaF-AlF_3-Al_2O_3 Melts Electrocatalytic Effects of Doping Agents ································· 111
>> Progress in Studies of Electrocatalysts and Doped Carbon Anodes in Aluminium Electrolysis Cells ································· 118
>> On the Electrocatalysis of Doped Carbon Anodes in Aluminum Electrolysis ······ 122
>> 高温氯化物熔体氯电极过程的电催化作用 ································· 132
>> Electrocatalysis of the Carbon Anode in Aluminium Electrolysis ················ 137

四、湿法冶金电解过程电催化

>> 锌电解节能惰性阳极的研究——惰性阳极上的析氧超电压 ················ 149
>> 用于Ni和Zn电积的节能形稳阳极（DSA）——实验室研究 ················ 153
>> A Novel Porous Pb-Ag Anode for Energy-saving in Zinc Electrowinning Part Ⅰ: Preparation and Property Tests in Laboratory ································· 159
>> A Novel Porous Pb-Ag Anode for Energy-saving in Zinc Electrowinning Part Ⅱ: Preparation and Pilot Plant Tests of Large Size Anode ································· 174
>> Oxygen Evolution and Corrosion Behaviors of Co-deposited Pb/Pb-MnO_2 Composite Anode for Electrowinning of Nonferrous Metals ················ 185
>> Electrochemical Performance of A Pb/Pb-MnO_2 Composite Anode in Sulfuric Acid Solution Containing Mn^{2+} ································· 198
>> Electrochemical Behaviors of Co-deposited Pb/Pb-MnO_2 Composite Anode in Sulfuric Acid Solution-Tafel and EIS Investigations ································· 212

五、电化学储能材料与器件

>> 天然石墨中嵌/脱锂离子过程的研究 ································· 228
>> Effect of Cooling Modes on Microstructure and Electrochemical Performance of $LiFePO_4$ ································· 235
>> Coating of $LiNi_{1/3}Co_{1/3}Mn_{1/3}O_2$ Cathode Materials with Alumina by Solid State Reaction at Room Temperature ································· 241
>> Synthesis of Nitrogen-containing Hollow Carbon Microspheres by A Modified Template Method as Anodes for Advanced Sodium-ion Batteries ················ 250
>> Confining Selenium in Nitrogen-containing Hierarchical Porous Carbon for High-rate Rechargeable Lithium-selenium Batteries ································· 267
>> A Simple SDS-assisted Self-assembly Method for the Synthesis of Hollow Carbon Nanospheres to Encapsulate Sulfur for Advanced Lithium-sulfur Batteries ········ 280
>> Electrochemical Impedance Spectroscopy Study of A Lithium/Sulfur Battery: Modeling and Analysis of Capacity Fading ································· 290
>> An Electrochemical-thermal Model Based on Dynamic Responses for Lithium Iron Phosphate Battery ································· 302

的高产作用（实验室的数据表明电流效率能提高2%），后来先进铝电解槽都以点式下料技术为特色装备，事实证明，工业上采用点式下料技术以后，普遍都能获得较高的电流效率。

第二个课题的核心技术是电极的电催化作用。最初引起他兴趣和思索的是氯碱工业中正在兴起的电解节能技术，即应用DSA（形稳阳极）的电催化作用可以大大地降低涂层阳极上的析氯过电压。联想到熔盐铝电解炭阳极上同样也有很高的析氧过电压，它是铝电解引起高能耗的重要原因之一。怎样利用电催化作用来解决这个问题呢，他作了深入的学习和思考。须知，铝电解碳阳极上很高的过电压一直是该领域中束手无策的问题，这是因为：第一，形成阳极过电压的理论缺乏，若干假说根据不足；第二，高温熔盐体系至今没有标准参考电极，而且高温强腐蚀环境尚没有任何材料能够制作通用的参比电极；第三，没有测量过电压的可靠技术。他在研究工作中，在铝电解条件下采用了不消耗的SnO_2惰性阳极，由于阳极不消耗，附着在上面的微量掺杂电催化剂不会损失，因此可以进行电化学实验和测量。研究结果得出了在该种条件下可用催化剂掺杂元素的顺序，指出降低铝电解阳极过电压最有效的电催化剂是含有钌（Ru）的化合物（阳极过电压比不加时降低85%）。这篇论文在Electrochimica Acta发表后受到广泛的关注。

《低极距下铝电解的电流效率》是在回国前完成了实验工作回国后写成论文的。在阳极侧部开沟后，极距降低至2.4cm仍可保持高的电流效率。须知，极距降低1mm，就可节电100kW·h，是铝电解节能的重要方向，这一点在此后30年的我国铝电解工业实践中得到了证明。

三、报效祖国硕果累累

1982年，刘业翔从挪威回国，那时候国内的科研条件很差，既缺经费又缺乏基本的实验测试设备，资料欠缺滞后。经过起步期的艰苦努力和探索，特别是在国家自然科学基金多次资助下，刘业翔对电解过程电极上的电催化作用进行了深入系统的研究。

1. 湿法冶金中电积金属用形稳阳极（DSA）和节能阳极开发

在能源和资源日趋紧张的背景下，工业生产过程节能是当务之急。电催化科学研究恰好可在节能方面找到广阔的应用。电催化是使电极与电解质界面上的电荷转移反应得以加速的一种催化作用。20世纪50年代末，出于当时对燃料电池的迫切需要，电催化科学得到了迅速发展。当代电催化科学的研究范围已经远远超过了燃料电池领域，在化学工业、电化学合成、电解工业以及生物电化学等方面的研究与应用都已取得了显著成绩。

刘业翔关注电催化研究，是着眼于如何加速电极上的电化学反应、降低电极电位为目标，直接与节能降耗密切相关，特别是在大能量强电流电解过程中的节能，采用电催化功能电极更会起巨大的作用。电催化科学是与电极材料特性相关联的科学，因此高效稳定的电催化功能电极的研制是电催化科学的重要研究内容。

有色金属冶炼过程中，金属的电积是重要环节而且是一个耗能大项。在电积工艺中阳极析氧造成了能耗、阳极材料腐蚀产物在阴极的共沉积造成了对产品纯度的影响，而且阳极的费用又与阳极材料的选择密切相关。因此，使用电催化析氧性能好、寿命长、成本低的阳极材料对有色冶金工业具有重要的经济意义。随着湿法冶炼工业的发展，对功能优异的阳极材料的需要更为迫切。

金属氧化物涂层阳极，也称形稳阳极，即DSA（Dimensionally Stable Anode），是一种非常重要的电催化功能电极，它的表层材料大致可分为复合涂层、电镀层和热分解涂层三大类，其中热分解成型的钛基金属涂层阳极以其优良的电催化性能，在氯碱工业中获得了广泛的应用，使电解槽的能耗显著降低。金属涂层阳极在工业上成功应用，有力地刺激了这方面研究的发展。其中有色金属电积多采用强酸性电解液，而且阳极过电位高，材料的使用环境相当恶劣，因此阳极材料的选择要求十分苛刻。

湿法冶金上用于含硫酸电解液的阳极，多为铅阳极或铅-银（约1%银）阳极。这种电极较能耐酸腐蚀，易于加工成型，成本低，一般能持续工作数年。但是其缺点也十分明显，首先是铅阳极有很高的析氧过电位，在常用电流密度下过电位可以达到1V，即比理论需要的电能超出很多。这样高的氧过电位会产生无用的热，在锌电积的情况下，还必须安装冷却装置以降低电解液的温度。其次，这种阳极太软，在生产过程中容易弯折变形而引起短路。第三，这种阳极总会使阴极产品遭受铅的污染。因此，采用新型的功能电极材料制作性能更优越的阳极势在必行。

针对上述问题，根据有色金属工业电积过程节能的需要，1983年起，刘业翔率先在国内开展了对各种类型DSA阳极的研究，以找到可用于金属电积的阳极材料，在高酸条件下其表面能存在稳定的电催化剂，并要求在经济上能被工业界接受。普遍调研了各种催化剂的作用，做了大量的实验工作。摸索了不同的基体材料作为电极的可能性，除了Ti以外，还试验了石墨和陶瓷的电极材料。研究结果表明，经过表面涂层处理后的石墨和陶瓷电极，析氧过电位从铅银阳极的0.8V降低到0.2V和0.1V。采用新型DSA代替现行的Pb-Ag阳极，节电最低可达9.6%，最高则接近20%，可见节电潜力巨大。

研究结果在1985年美国TMS年会上报告并收录于《金属电积过程节能》专辑，引起了广泛关注。一直到现在，全世界范围内还有科研团队在进行此类的科研与开发工作。

虽然DSA能够使阳极过电位显著下降，但是一直无法有效解决阳极泥在电极表面的沉积问题和阳极的寿命问题。2005年起，刘业翔开始尝试对现有阳极进行改进，首次提出了多孔阳极的概念，指导学术团队开发出"反重力渗流"铸造技术和设备，制备了多孔Pb-Ag合金阳极，并在湖南株洲冶炼集团股份有限公司进行了持续近2个月的工程化试验，取得了显著的效果：(1) 降低阳极电位约100mV，Zn电积能耗降低90kW·h/t-Zn；(2) 阳极泥生成量减少80%；(3) 阴极锌品质得到提高，产品中Pb含量降低了60%，0号锌合格率100%；(4) 阳极的金属（Pb、Ag）用量只有原来的45%左右，也就是Pb-Ag阳极的投资成本将降低55%。研究结果在2008年全国湿法冶金学术会议和2009年第五届国际湿法冶金会议报道后，引起同行专家的浓厚兴趣与好评，且论文直接被有色金属权威期刊《Hydrometallurgy》主编David M. Muir教授约稿发表。

但是后续的研究发现，Pb基多孔阳极的机械强度和电导率只有传统Pb基平板阳极的25%~30%，使用过程中易折断，阳极电压降较高，这使其成为Pb基多孔阳极工业化应用的主要技术瓶颈。为此，又从结构设计与合金成分方面对多孔阳极进行了改进。创造性地提出了中心为铝合金加强基板的"反三明治"结构多孔Pb基复合阳极新思路。以熔盐化学法成功解决了金属Al与Pb的高强、低阻结合问题后，获得了以金属Al为加强金属基板的轻质、高强"反三明治"结构多孔Pb合金复合阳极。进而又率先将稀土元素引入有色金属电积用Pb合金阳极，对各种Pb基稀土合金的性能进行调研和测试，在使原Pb-Ag（0.8%Ag）阳极中的Ag含量降低了25%的同时，将阳极的力学性能、析氧电催化活性和阳极耐腐蚀性能提高。目前，研究成果正在产业化过程中。

2. 高温熔盐电解电催化炭阳极

从1952年我国第一家电解铝厂抚顺铝厂开工建设至今，我国电解铝工业经历了60多年的发展，已逐步由电解铝的纯进口国变为世界第一大产铝国，电解铝工业在世界上已经具有了举足轻重的地位。但是，令人担忧的是，铝电解是一个高耗能的产业，电解铝生产能耗占整个有色金属工业能耗的80%，电解铝工业总用电量接近全国电力总消费量的5%，电解铝工业的持续发展与国家节能减排政策的矛盾日渐突出。

因此长期以来，铝电解的节能一直很受重视，意义也非常重大。其中，降低铝电解过程槽电压是一个有效的手段。在保证电解过程稳定的基础上降低槽电压需要综合考虑各方面的因素,是一个非常复杂的系统工程,现在工业电解槽上都是以mV为量级来进行槽电压下降节能的考评。现有的技术成熟的铝电解槽槽电压为4.0~4.2V,其中阳极过电压为0.6~0.7V,占总能耗的16%~17%。造成能耗数值巨大,但就是这一项,业内科技与工程人员长期以来不敢去碰它,为什么?因为铝电解过程碳素阳极上过电压问题一直搞不清,自然无有效办法解决,究其原因,一是理论上不清楚,多种学说莫衷一是,又得不到验证;二是测量方法不成熟,材料经不起熔融氟化盐的腐蚀,也没有合适的测量装置;三是测量难度大,例如,要在近1000℃高温下熔融氟化盐电解质中,在强大直流电场下,把探头接触红热的阳极表面,测定其对参比电极的电位变化等。

刘业翔在挪威做过的惰性电极电催化的研究,引导了解决这一问题的思路。对于碳素阳极来讲,可能的降低过电位的方法就是往阳极材料中添加活性物质,实现电催化的效果。而碳素阳极的电催化可不可行,一直存有疑问,因为阳极在电解过程中是消耗性的,电极中的催化剂会进入电解质中,比铝更正电性的杂质会污染产品铝。而惰性阳极与碳素阳极不一样,它本身是没有消耗的。所以碳素阳极电催化的研究,需要有全新的对策和方法。

刘业翔克服了当时面临的很多的困难,采取了一系列的创新性的解决办法,构建了高温炉、高纯刚玉管、铝参比电极、电化学综合测试仪、高速记录仪等装置。最关键的是解决催化剂的选择问题。首先,催化剂必须是由比铝更负电性元素组成的化合物,高温下不分解,且化合物催化剂本身又必须是导电的。同时,催化剂的加入必须能在电极表面形成电化学反应的活性中心,这样才能保证在电流密度比较大的情况下,有更多的带电含氧离子在活性中心上放电。相反,在没有添加剂的情况下电极表面的反应活性中心不够,含氧离子只能在非活性的位置上放电,这就需要额外的能量,从而产生过电位。因此,催化剂的筛选,需要长期的探索和研究。

经过不断地探索,在电催化机理研究和催化剂选择上取得了重大的突破,实验室研究表明合适的催化剂对降低阳极过电位效果非常明显。但要实现工业化的应用,特别是怎么实现催化剂在工业电解槽的阳极中的加入,又成为摆在刘业翔等人面前的一个突出问题。考虑到当时工业电解槽采用的都是自焙阳极,刘业翔和他的学生结合自焙阳极本身的生产工艺,巧妙的通过试验,在阳极糊混料时,把微量的碳酸锂(2~3kg/t-Al)加入到石油焦里面,制成"锂盐阳极糊"。试验结果证明,这种添加方法简单,电极上的电催化效果良好,很好地解决了催化剂工业化应用的问题。

1986年，刘业翔和他指导的第一个研究生肖海明，在西安全国物理化学年会上介绍了他们的碳素阳极电催化研究工作，引起了铝电解工业部门的高度重视。同年受兰州连城铝厂的邀请，以肖海明为主的研究组在铝电解槽现场进行了工业试验，通过往阳极糊中掺入锂盐，即"锂盐阳极糊"，使得阳极过电位降低200mV，试验进行了三个月，取得了非常好的节能效果，引起了当时的中国有色金属总公司重视。在有色总公司的支持下，1989年"锂盐阳极糊"技术开始在全国电解铝系统内推广应用。经过在26个铝厂的电解槽上应用，实现年节电6000万度。"锂盐阳极糊"研究成果在1991年获得有色总公司科技进步一等奖和国家教委科技进步一等奖，1992年被评为国家科技进步一等奖。

此后，为了配合新的预焙阳极工艺（预焙温度高达1200℃），对高温下阳极的电催化行为继续进行了研究，取得了一定的成果，但由于催化剂对产物铝的污染会影响铝的质量，使得高温电催化的实际应用难度较大。不过很有意义的是，这些工作却为后来电解铝惰性阳极的开发和改进起了很好的指导作用。

刘业翔在回国的很长一段时间内，以节能降耗为目标，进行了以开发新型功能电极材料为中心的一系列研究工作，围绕这些方面做了很多的国家级项目和企业项目，连续多次获得国家自然科学基金的资助。1996年，他对过去的工作进行了一次总结归纳，出版了专著《功能电极材料及其应用》，为功能电极材料领域的研究与开发提供了很好的基础参考用书。

此后，他的研究视野更加宽阔，在国家需求和政策的指引下，刘业翔和他的创新团队先后进行了锂离子电池的正、负极材料、超级电容器电极材料以及超级电池电极材料等的开发和研究，又不失时机地往可再生能源和薄膜太阳电池研究领域发展。同时，结合产业需要，以冶金节能和新能源材料及器件为专业方向，刘业翔及其团队开展了一系列产学研工作，获得了持续稳定的发展。

四、开发高科技，走向产学研

1985年以后，身处社会主义市场经济的大环境下，刘业翔深深地感觉到科研工作不能仅仅停留在实验室层面，科研成果的价值也不能由发表几篇论文来衡量。科技工作者一定要为增强国家竞争力、提高人民生活水平服务。改革开放总设计师邓小平同志提出了"发展高科技，实现产业化"的号召，这引起了他深深的共鸣和反思。高科技产品是我国参与国际竞争的重要支柱，因此，结合自身的特点，开发节能新技术以及新能源产品逐渐成为新时期刘业翔科研工作的重心。并且从一开始，刘业翔就确立了科研工作必须密切结合市场需要，坚定的走产业化道路的指导思想。随着研究队伍的壮大和领域的不断拓展，刘业翔及其学术团队在冶金节能及新能源领域开展了一系列卓有成效的产学研结合的工作。

1. 铝电解控制系统的开发与产业化

旧中国根本没有"铝工业"。直到新中国成立后的1954年，我国在前苏联156项建设项目的帮助下，建起了山东铝厂（生产氧化铝）和抚顺铝厂（生产电解铝），这才有了"铝工业"。可那个"工业"离现代两个字实在差得太远了，耗电量大、污染严重不说，产量还低得很——当时最大铝电解槽的容量才45kA，1960年后我国发展到60kA、80kA，就再也上不去了，当时（1965年）国外已经到了160kA甚至200kA。电解槽容量（电流强度kA值）的大小关系到单槽铝产能的多少，它是现代铝电解技术水平的标志。1962年抚顺铝厂设计和建造了国内最大的80kA自焙电解槽，后来又做过135kA预焙电解槽，但都因生产不稳定，效果差，技术不过关而告失败。

为什么我国大容量铝电解槽上不去？主要原因是，在电解过程中铝电解槽内有非常强的磁场，直流电流越大，引起的磁场影响越大，当年电流才45kA时刘业翔在铝电解槽旁实习时，铝饭盒里的铁勺子都是立着的，工人干活用的铁锹常被粘在槽沿钢板上。而且槽中的熔融铝液跟着磁场有规律地旋转波动。直流电要上到160kA，"磁场"问题不解决，不仅能耗高，而且铝的再氧化损失大，就不能达到提高产量和降低能耗的目的。除了磁场外，还有"温度场"、"流体场"等一系列技术关键我国没有掌握。买国外的技术吧，人家是天价，开口就要几千万美金。但是，再贵也得买啊。1984年，国家花2亿美元从日本引进了一套160kA的铝电解槽系列。随后，在消化日本的技术时，遇到了难题，其他的硬件较好解决，就是电解槽的自动控制软件看不懂。学冶金的技术人员不懂计算机，学计算机的人员又不懂铝电解，因此，这些控制软件被认为是"天书"一部，相当长时间没法解决。

为了解决技术自主开发的问题，刘业翔同他的博士研究生李劼进行了艰苦的科研攻关。那是1986年，23岁的李劼考回母校，师从刘业翔教授，第一年基础课上完后，李劼就按导师的指引，一头扎进了国家"七五"攻关项目"大型预焙铝电解槽自适应控制技术的开发应用"，这一干就是20年。

研究期间，刘业翔和李劼发现，要消化日本的技术，关键是吃透电解槽的控制系统和软件。李劼在导师指导下自学了计算机和程序语言等系列知识，这对于本来一直从事轻金属冶金专业的他来说，是要克服很大困难的。经过三年的努力，到硕士毕业的时候，终于掌握了日本的控制系统和软件，并在刘业翔指导下继续攻读博士学位,继续攻克先进的控制技术。在刘业翔指导和支持下,李劼在获得博士学位前后,还往控制系统的相关的硬件发展,他组织了行业内的工程技术人员,开始了电解槽控制柜的研究和开发工作。并以刘业翔的名字命名,发起组成了湖南中大业翔科技有限公司,主要从事铝电解技术和自控技术的研发、设计、制造和推广应用,并开发出新的大型的"中国制造"的电解槽自控技术和装备。

业翔公司成立初期,市场竞争非常激烈。当时国内的工业自控公司专业性强,市场占有率高,各铝厂都有固定的设备供应公司。因此业翔公司的产品出来后,销售面临着很大的困难,主要是恶性降价竞争,如果公司的产品卖1万元一台,对方就降价至9000元,公司的产品降到9000元,对方就降价至8000元。当公司的发展生死攸关的时候,李劼毅然决定,产品先无偿地提供给企业试用,待该铝厂取得明显效果后作为这项技术先进性的样板,再向其他铝厂推广。正是这一个重大决策,为以后业翔公司产品的推广应用奠定了牢固的基础。后来企业反馈的结果表明,业翔公司产品完全能满足目前国内铝电解行业的需要,起到了很好的高产节能示范作用。

经过二十多年的努力,业翔公司研发的铝电解槽槽控箱技术和设备已广泛应用于国内外100余家电解铝企业,为我国近年来铝电解工业实现大幅节能减排并促进我国大型槽炼铝成套技术走出国门发挥了重要作用。同时,刘业翔和李劼团队还承担了铝电解节能国家重大项目和课题的研究,取得了丰硕的成果。如针对大型铝电解槽寿命较低的问题,创造性地提出了铝电解槽"电、磁、热、流、力"多物理场(多场)的耦合仿真新方法,开发出铝电解新型电极材料等。之后,又趁热打铁开发出铝电解智能控制技术,使我国在这一领域的研究与应用水平,从落后于国外20年到目前达到国际领先水平。经2003年中国有色金属工业协会对88家应用企业统计并组织成果鉴定,平均吨铝直流电耗从13800~14000kW·h降低到13200~13500kW·h(同期国际先进水平),总计年节电10.5亿千瓦时、年减排PFC约1470t(等效CO_2减排1000万吨)。成果获2004年度国家科技进步二等奖和2003年度国家重点新产品称号。

2. 新能源材料的开发与产业化

锂离子电池不仅是目前手机、笔记本电脑等现代数码产品中应用最广泛的电池,而且大容量、大功率的锂离子电池也是目前最具有发展前景的动力电池,它是新材料、新能源和新能源汽车中的重要项目,在我国战略性新兴产业中具有重要的地位。

刘业翔学术团队自1996年就开始了锂离子电池正极材料的研发。最开始的时候主要是研究锰酸锂材料,但一次偶然的机会,改变了整个团队的研究方向和后续的发展轨迹。那是在1998年,刘业翔团队成员胡国荣博士后参加武汉一家公司举办的关于锂电池的技术交流报告会。与会者大部分是一些企业家。会议间隙,作为会场唯一的博士,他被众多企业家围了起来。一位广州的老板问他在研究什么,胡国荣说自己在研究锰酸锂电池材料,对方问他为什么不研究钴酸锂。"当时日本已经在批量生产钴酸锂电池材料,已经八九年了,但我从大学研究与创新的角度考虑,就不愿做钴酸锂。"胡国荣刚说完,就被那位老板批评了一通。他说"大学净搞一些闭门造车的东西,研究锰酸锂,谁知道多少年以后才能用?现在我们很多企业都在等米下锅,国内买不到钴酸锂,要买日本的,一吨上百万,实际成本只有三四十万"。那时锂电池很贵,1200元一块,一部手机几万块,国内做电池的企业很想做锂电,但目前只能从日本进口。

那位广州老板语重心长的话使胡国荣深受刺激。他回去向导师刘业翔讲了自己的想法,刘业翔欣然同意他研究钴酸锂,并说"为什么不研究钴酸锂?这是核心材料,钴酸锂本身就不成熟,我们应该赶快研究,解决国民经济的急需"。半年的时间,胡国荣就拿出一个样品。跟日本的样品做了对比,发现比较接近,大家都很兴奋,就对外发布信息。因为这个是很新的项目,很多企业找上门来。在那之前,教授们参与企业,更多的是技术转让模式。学校拿到的技术服务费很少。而以技术入股办企业结果就不一样。20世纪90年代末期,高校以技术入股办企业蔚然成风。刘业翔和胡国荣最后选定了上海一家投资公司,最后经过深入的洽谈,成立了湖南瑞翔新材料有限公司,协定中刘业翔和胡国荣团队的技术占25%的股份,投资公司占75%,中南大学也持有其中很小一部分股份。胡国荣作为技术负责人,出任公司副总裁,专管技术。

可是,"实验室小杯子做出的东西,到了生产层面,如何控制质量,谈何容易"。七八个月之后,初期1500万元的投资耗尽了,产品仍然不能达到要求。但刘业翔觉得当时的研究思路应该没有问题,一直鼓励胡国荣坚定信念。最后,胡国荣在一次去云南出差的时候,在宾馆里冷静思考了一个星期,拿出两套方案。按照新方案生产的产品拿到检测机构检验,半个月后,接到了检测机构的电话。对方质疑这是否是直接买的日本的样品,根本不可能是国内公司研制出来的。前两个月检测的时候还是垃圾,现在怎么这么好呀。胡国荣心里一阵狂喜,技术攻克后,瑞翔的市场一下子就打开了。

从2001年10月份开始,由于市场需求太大,瑞翔生产能力已经严重不足,它们开始大力拓展生产线。当时日本的原材料120万元一吨,在瑞翔产品出来后，不得不降到了 80 万元，后来瑞翔批量生产，价格降到了 49 万元，价格上比日本便宜 35%。2008 年，瑞翔销售额达到了 6.8 亿元。美国、日本、韩国和国内天津力神等国际国内非常重要的电池生产厂家都成为了瑞翔的客户，而瑞翔的产品也受到了用户的高度评价。

为了更好地把科学研究和国家需求相结合，便于技术的转移，以"先进电池材料教育部工程中心"为依托，2009 年刘业翔团队又成立了"业翔先进电池技术转移中心"，作为学校和市场沟通的桥梁，既进行前沿的基础研究和技术开发，同时又注重高端人才的培养。校内研究团队、技术转移中心和校外学科性公司的良性互动，使得研究成果可以很快由样品变为小批量的产品，然后再转向大规模工业生产，刘业翔及其研究团队正在探索这样一条科学的高效的产学研结合的道路。

除此之外，刘业翔团队在新能源材料的研究与开发中，还平行进行着铝电解惰性阳极，CIGS（铜铟镓硒）薄膜太阳电池，新型动力电池、超级电池与超级电容器，有色金属电积用新型 Pb 合金阳极及铅蓄电池轻型板栅材料的研发。

五、还想讲的话

刘业翔从教 58 年，弹指一挥间，回顾自己的学术生涯，他还想讲几点自己的体会与希望。

1. 人生箴言

（1）要培养坚强的意志。完成伟大的事业不仅在于志向和体力，而在于坚忍不拔的毅力。一个人必须百折不挠的经过一番刻苦奋斗，才会在向往的目标上有所成就，挫折是人生的财富。要培养坚强、执著不服输的精神，学会勇敢地面对生活中的各种打击和挫折，锲而不舍、百折不挠地朝着自己的人生理想奋进。

（2）终生学习，与时俱进。"吾生有涯，而知无涯"，何况现在是"知识爆炸"的时代，知识老化加速，社会变化急剧，任何人都不可能一劳永逸地拥有足够的知识，所以需要终身学习与时俱进。同时有了好想法，更需要努力去实现。

（3）对国家对人民有高度的责任感。饮水思源，知恩图报，这既是良好的品德，也是支持自己努力工作的动力和鞭策。

2. 寄语青年

一是要有求识的渴望。善于获取知识，具有较宽广的知识面；二是要有提出问题、发现问题和解决问题的能力，特别是要有实践能力，能把好的想法转化成现实的能力；三是搞科研除了要有从事科学研究的基本素质外，还要有创新激情，要有幻想、梦想和联想，多想出智慧；四是一定脚踏实地、实事求是、不畏艰难、勇于攀登；五是要有善于团结、合作、沟通、利他的团队精神。

3. 寄望未来，为节能减排、开发新能源再作贡献

（1）惰性阳极取得成功。

惰性阳极，是指那些在目前通用的冰晶石-氧化铝熔盐电解中不消耗或微量消耗的阳极，它是铝业界的梦想。半个多世纪以来，惰性阳极的研究和开发工作时断时续，是全球铝行业及相关领域科技人员一直努力实现的目标。

用惰性阳极取代碳素阳极有以下主要优点：一是省去了数量巨大的优质炭的消耗（1t 铝需要 0.4t 炭阳极，同时减少了大量的二氧化碳排放），除去了污染较重的炭阳极生产制造工厂。二是铝电解生产排放的是氧气，而不是二氧化碳，因此可以实现清洁生产。三是由于免除了炭阳极周期性更换（平均每天换 1~2 块，属高温作业），避免电解槽槽罩的频繁开启造成热损失，保持电解槽生产平稳和热平衡的稳定，便于完全自控。采用惰性阳极的电解槽还可以进一步的节能。有人怀疑，阳极析氧和阳极析出二氧化碳电位前者高于后者将近 1V，是否能实现节能。这可以通过精确的减少和保持极距解决，由于阳极是不消耗的，其形状能保持稳定，可以控制较低的极距，这方面的节能可以比常规的电解槽降低 20% 左右。

目前还没有应用惰性阳极的铝厂，其生产成本尚为未知数。但是从这几个主要方面推算，铝生产成本大致与采用预焙阳极的相当，甚或稍低，投资费用也会降低，加上环境因素，采用惰性阳极的革命性意义不言自明。正因为如此，刘业翔和同事们数十年来对惰性阳极的研究从没有终止过。在国家 863 计划的大力支持下，刘业翔课题组联合中国铝业公司研究院，惰性阳极研究已经由实验室研究开始走向工程化试验，电解持续时间已经超过美国报道的同类试验时间，目前试验已初现曙光。

（2）薄膜 CIGS 太阳电池的广泛应用。

与第一代的硅材料太阳电池相比，第二代的 CIGS（铜铟镓硒）薄膜太阳电池成本更低，柔性耐久，能与建筑材料很好的结合，正成为目前研究的热点。刘业翔学术团队较早开始了可再生能源领域的研究，发挥冶金学科所擅长的知识基础和技术手段，特别是电化学科学技术。采用非真空的电沉积为主的方法进行了 CIGS 薄膜太阳电池的制备、结构与性能的研究。在理论基础和工艺技术两方面都取得了重要进展，同时还着手薄膜电池器件的制

作，进一步提高转化率，降低成本，简化制备流程，为产业化作准备。在国家发展可再生能源政策的支持下，目前正与若干科研院所和高科技企业联合承担"863"项目，开展产业研合作，希望在3~5年内能研制和试用成功并得到广泛应用。

（3）新型动力电池（超级电池）电动车广泛通行。

超级电池就是把铅酸电池和超级电容器相关电极耦合在一起，产生出具有新功能的动力电池。它同时具备铅酸电池和超级电容器的功能，产生的功率大、使用寿命长、充电时间短、制造成本低，是今后的高品质动力电池之一。

对铅酸电池的功过应当重新评价，人们较多的关注铅的污染及引起的环境危害。在当前的技术和政策的指引下，其危害性是可以控制的并可把它降至最小。而铅酸电池的优点是技术成熟，生产成本低，安全可靠。经过技术改造，例如，改进板栅的结构提高其性能，减轻其重量，增大其功率密度并升级为超级电池，可以为低成本的动力电池开辟一条新的途径。

超级电池的研究和开发成功，能使古老的铅酸电池重新焕发青春，意义重大。刘业翔及其课题组成员正在研发新型的板栅材料，力争先期制成的新型铅酸电池能符合当前国家对电动自行车用电池的要求（电池组的重量低于10kg），使占市场份额百分之九十的铅酸电池自行车能继续奔驰。

（4）科技新才茁壮成长，创新团队在竞争中崭露头角。

几十年的科研和教学工作经验使他明白，培养优秀的科技人员是事业成功的核心问题。特别是近些年来国家对拥有大批创新人才的殷切期望，刘业翔希望自己的团队在教学和科研发展中不断壮大，做到：1）注意挑选优秀的学生早期（例如大三学生）进入团队，及早参与科研课题，从实践中得到科研团队的培养和锻炼，及早感受科研团队积极努力向上的氛围，并在科研实践中受到初步训练；2）强化以研究生为主体的学术活动，激发大家的创新意识和博采众长的良好习惯；3）尽可能创造条件让学生参与国内和国际的重要学术活动，使他们尽早获得开阔的视野，多向国内外优秀的团队学习；4）定期进行总结和考核，多予鼓励、启发与奋发向上的动力。此前，已有众多本团队的学生在事业上取得了显著的成绩，为国家做出了重要贡献，这样的培养道路，希望继续坚持和发扬光大。

一、铝电解惰性阳极、惰性可润湿阴极及电解新工艺

铝电解用 SnO_2 基惰性阳极的研究[*]

摘要 所研制的电极在电解温度下的电阻率为 $(0.6~1.4)\times10^{-3}\Omega\cdot cm$,比碳素阳极低得多,腐蚀速率为 $(0.5~3)\times10^{-3}g/(cm^2\cdot h)$。在 $0.6~1.0A/cm^2$ 的范围内析氧超电压为 $0.1~0.2V$,小型电解实验的电流效率为 $86\%~88\%$。本文还讨论了 SnO_2 基惰性阳极在工业上应用的前景及经济性问题。

近年来,关于铝电解用惰性阳极的研究尽管取得了很大进展和许多有意义的成果[1~5],但是还没有达到工业实用的水平,到目前为止也未看到工业应用的实例。究其原因,仅从技术角度看,惰性阳极目前存在的主要问题首先是耐腐蚀性还不够强,使用寿命还不够长,其次是制造适合工业上应用的大型电极有一定困难,Fe-Ni 基惰性阳极的导电性较差,也存在电极大型化的问题,其耐腐蚀性也未见得达到令人满意的程度。

本文的目的是介绍我们关于 SnO_2 基惰性阳极实验室研究的结果,并对若干技术经济问题进行讨论。

1 SnO_2 基惰性阳极的实验室研究结果

1.1 材料的选择和电极制备工艺

在众多的可以作为惰性阳极的材料中,我们认为 SnO_2 基材料更有希望,据 K. Grjotheim[6] 的资料,SnO_2 在冰晶石-氧化铝熔体中的溶解度约为 0.01%,有适当的添加剂的 SnO_2 基电极在电解温度下的电阻率和碳素电极接近,在冰晶石-氧化铝熔体中的析氧超电压也很低,其线膨胀系数大约在 $(4~5)\times10^{-6}/℃$ 之间,相当于刚玉的一半,属于低膨胀陶瓷一类。SnO_2 材料的热导率也比较大,约为 $29.3W/(m\cdot K)$,属于高导热性陶瓷,因此,SnO_2 烧结体抗热震性良好。相对来说 SnO_2 有充足的来源,所以我们把它作为主要的研究对象。

SnO_2 基电极的制备工艺如下:将添加有 Sb_2O_3、CuO、MnO_2、ZnO、Cr_2O_3、Fe_2O_3 等添加剂的 SnO_2 粉料磨细、混匀,在 $98~196MPa$ 的压力下冷压成型,然后进行烧结。在 500℃ 以下发生黏结剂的挥发,到 1200~1400℃ 便发生一系列物理化学变化,使体积收缩,表观密度、电导率和机械强度同时增加。SnO_2 电极烧结时的失重曲线和烧结温度制度如图 1、图 2 所示,其外形如图 3 所示。

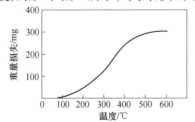

图 1 SnO_2 基电极烧结时失重曲线

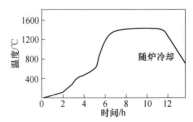

图 2 SnO_2 基电极烧结温度曲线

---

[*] 本文合作者:王化章、肖海明。原发表于《中南矿冶学院学报》,1988,19(6):636-642。

1.2 各种添加剂对电极物理性质的影响

纯 SnO_2 不仅成型困难而且电阻率很大，因此研究各种添加剂对电极性质的影响，以期获得性能最佳的电极组成便是十分重要的了。在前人[1~8]研究的基础上，我们挑选了 6 种添加剂 Sb_2O_3、CuO、MnO_2、ZnO、Cr_2O_3、Fe_2O_3 用正交试验设计，配制了 27 种不同配方的电极，对其室温电阻率、高温电阻率、表观密度、真密度、气孔率、抗压强度等物理性质进行了系统的测定，通过方差分析判明了各种添加剂对电极的各种性质的影响，择优选出了数种综合性能较好的配方，结果表明：使电阻率降低的有效成分是 Sb_2O_3。而 ZnO 只对降低低温电阻率有效，Fe_2O_3 和 Cr_2O_3 对降低电阻率特别不利，当它们的含量超过 0.1% 时，就使电阻率急剧升高，CuO 和 MnO_2 虽然对降低电阻率没有直接的贡献，但它们能促进烧结，显著降低试样的显气孔率，提高表观密度，间接地对降低电阻率有利，所以它们对电阻率的影响呈现一个凸形的曲线，有一个极大值存在，同时显气孔率的降低和表观密度的提高，对提高抗腐蚀性也是有利的。ZnO 和 Fe_2O_3 对降低显气孔率，提高抗压强度有显著的影响，它们的含量和显气孔率的关系存在着一个极小值，和抗压强度的关系存在着一个极大值，因此加很少量的 Fe_2O_3 也许是很有利的。

实验测定的电阻率-温度特性曲线如图 4 所示，和理论曲线的特点是一致的，其室温和高温的伏安特性分别示于图 5a 和 b。

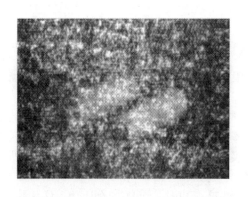

图 3　SnO_2 基电极外形图　　　　图 4　SnO_2 基电极的电阻率-温度特性

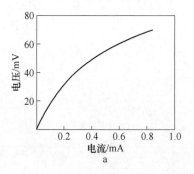
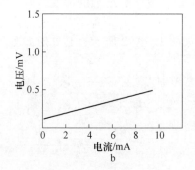

图 5　室温和高温度的伏安特性

a—室温下（9℃）的伏安特性；b—950℃下的伏安特性

1.3 SnO$_2$基电极的耐腐蚀性

惰性阳极的耐腐蚀性不仅关系到阳极的使用寿命，还关系到所生产铝的纯度，尽管 SnO$_2$ 在冰晶石熔体中的溶解度很小，但是当有铝存在时，腐蚀速率则大大加剧。实验首先研究了鉴定电极被腐蚀程度的方法，传统的称重法固然简便，但渗入电极的电解质往往不易清洗干净造成误差，而精确的体积法则可避免这种误差。前期研究结果[8]表明，影响电极耐腐蚀性的因素主要是电极本身的组成和使用条件。根据我们所优选的电极组成，在不通电的情况下，且坩埚中无铝时腐蚀速率为 $1.53×10^{-4}$ g/(cm^2·h)，有铝时的腐蚀速率为 $9.815×10^{-3}$ g/(cm^2·h)，在通电的情况下电极的腐蚀速率和阳极电流密度有关，实验结果如表 1 所示。

表 1 阳极电流密度和腐蚀速率关系

（电解质组成：2.5NaF·AlF$_3$+3%CaF$_2$+4%MgF$_2$+5%Al$_2$O$_3$；

电解温度：945~950℃；实验时间 12h；极距 4~5cm）

阳极电流密度/A·cm^{-2}	0	0.53	0.8	0.97	1.53	2.05
腐蚀速率/g·(cm^2·h)$^{-1}$	13×10^{-3}	3.85×10^{-3}	2.04×10^{-3}	0.6×10^{-3}	10.3×10^{-3}	16.8×10^{-3}

当电流密度接近 1A/cm^2 时腐蚀速率最小约为 $0.6×10^{-3}$ g/(cm^2·h)。

根据已有的实验资料，可以认为 SnO$_2$ 基被腐蚀的原因，除了机械破损之外在正常情况下主要由于下面的反应：

$$\frac{4}{3}Al + SnO_2 \longrightarrow \frac{2}{3}Al_2O_3 + Sn \tag{1}$$

$$\Delta G_T^\ominus = -128200 + 0.4T$$

当 $T=1273$K 时，$\Delta G_{1273}^\ominus = -534258.3$J

平衡常数 $K = 8.39×10^{21}$

$$2AlF + SnO_2 \longrightarrow \frac{2}{3}Al_2O_3 + \frac{2}{3}AlF_3 + Sn \tag{2}$$

$$\Delta G_T^\ominus = -122360 + 22.83T$$

当 $T=1273$K 时，$\Delta G_{1273}^\ominus = -3903568.36$J

平衡常数 $K = 1.04×10^{16}$

$$4Na + SnO_2 \longrightarrow 2Na_2O + Sn \tag{3}$$

$$\Delta G_T^\ominus = -61900 + 18T$$

当 $T=1273$K 时，$\Delta G_{1273}^\ominus = -163117.4$J

平衡常数 $K = 4.94×10^6$

由式（3）所生成的 Na$_2$O 和电解质中的 AlF$_3$ 会发生交互反应。

$$2Na_2O + \frac{4}{3}AlF_3 \longrightarrow \frac{2}{3}Al_2O_3 + 4NaF \tag{4}$$

$$\Delta G_T^\ominus = -38840 - 91.97T$$

当 $T=298$K 时，$\Delta G_{298}^\ominus = 47831.48$J

当 $T=1273$K 时，$\Delta G_{1273}^\ominus = -327362.98$J

平衡常数 $K = 2.71×10^{13}$

反应式（3）和式（4）可以综合写为

$$4Na + \frac{4}{3}AlF_3 + SnO_2 \longrightarrow \frac{2}{3}Al_2O_3 + 4NaF + Sn \qquad (5)$$

反应所生成的 Sn 使金属铝受到污染。

从以上的热力学计算可以看出反应式（1）~式（4）向右进行趋势都很大，很彻底，据此可以对 SnO_2 基电极在冰晶石-氧化铝熔体中被腐蚀的过程作这样的描述：以各种形式进入电解质中的 Al、Al^+ 和 Na 通过扩散和对流转移到阳极区，当它们与阳极接触时便发生上述的反应，使电极遭受腐蚀，随着电解质的循环，这一过程将不断地进行下去，但是这一过程受到各种因素的制约：电极被阳极极化是防止 SnO_2 被还原的有利因素，而阳极气体析出所造成的湍流由于加速了传质过程，促进阳极的腐蚀，这是不利因素；当电流密度小时，前者起主导作用，电流密度大时，后者起主导作用。因此，腐蚀速率随电流密度的变化出现一个极小值，达到极小值的电流密度和电解质的运动状态有关，并不是固定不变的，可以预言随着电极配置方式和电解槽结构形式的改变，最佳电流密度也会改变。

1.4 SnO_2 基电极在冰晶石氧化铝熔体中的电化学特性

在冰晶石-氧化铝熔体中采用惰性阳极时的电解反应为

$$Al_2O_3 \longrightarrow 2Al + \frac{3}{2}O_2$$

在 1000℃下其可逆分解电压为 2.17V，我们前期采用了多种电化学测量方法对 SnO_2 基电极的阳极过程进行了研究[9]，在 1010℃下的稳态极化曲线如图 6 所示。根据此极化曲线可以算出阳极超电压在 $0.6~1.0A/cm^2$ 范围时为 $0.1~0.2V$。

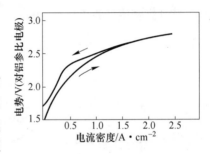

图 6　SnO_2 基惰性阳极稳态极化曲线

1.5 使用 SnO_2 基惰性阳极的实验室电解实验[10]

实验所用电解质组成为 $2.5NaF·AlF_3$，$4\%MgF_2$，$3\%CaF_2$，$3\%LiF$，$10\%Al_2O_3$，电解电流 4A 左右，电解温度为 1000℃ 左右，用电解槽内铝的增量来确定电流效率。结果表明，电流效率一般为 86%~88%，而在同样条件下碳阳极的电流效率可达 90%，由于影响电流效率的因素是多方面的[11]，我们认为只要条件适当同样可以获得高的电流效率。

1.6 所研制的 SnO_2 基惰性阳极性质一览

为了和目前使用的碳阳极比较，碳阳极的有关性质也一并列出，从表 2 所列数据可看出惰性阳极的高温电阻率和析氧超电压、气孔率、抗压强度等方面都优于碳阳极，不足的方面是室温电阻率还较高，耐腐蚀性需要增强，其使用寿命也还需要进一步考核。

表 2　SnO_2 基惰性阳极和碳阳极性质一览表

测量项目	SnO_2 基惰性阳极	预焙碳阳极	自焙阳极烧结体
高温电阻率/$\Omega·cm$	$(0.6~0.4)\times 10^{-3}$	$(6.0~6.5)\times 10^{-3}$	$(8.0~8.5)\times 10^{-3}$
显气孔率/%	3~7	25~38	32
表观密度/$g·cm^{-3}$	6.2~6.46	1.5~1.6	1.4~1.55

续表2

测量项目	SnO₂基惰性阳极	预焙碳阳极	自焙阳极烧结体
真密度/g·cm⁻³	6.3~7.1	2.02	1.8~1.9
抗压强度/MPa	30~50	>30	>27
腐蚀速率/g·(cm²·h)⁻¹	(0.5~3)×10⁻³	—	—
阳极超电压/V	0.1~0.2	0.45~0.55	0.45~0.55
电流效率/%	86~88	85~92	85~92

2 讨论

（1）从目前的研究水平看SnO_2基惰性阳极技术上的可行性，作为铝电解用惰性阳极最少要满足4个条件：

1）良好的导电性，在电解温度下的电阻率应当和碳阳极相近，室温电阻率也不过大。

2）足够长的使用寿命，6个月以上。

3）不需要另外处理就能生产出工业纯铝。

4）阳极的制造容易工业化，成本不高。

从表2可见，第一条已不成问题，我们所制惰性阳极，在电解温度下的电阻率低于碳素阳极，而第二条和第三条是联系在一起的，也就是说，电极的耐腐蚀性不够，不仅使用寿命短，所生产的铝也会被严重污染；反之耐腐蚀性强，不仅使用寿命长，所生产的铝纯度就高。

（2）SnO_2基惰性阳极的经济性和实用前景。从技术上讲足够的耐腐蚀性和良好的导电性是使之能够实用的关键，那么在经济上是否合理？为了定量地进行分析我们提出一个经济性指标，用符号E表示，它的定义如下：

$$E = \frac{每千克 SnO_2 阳极所生产的铝的价值}{每千克 SnO_2 阳极的费用}$$

从公式看提高经济性指标E的方法是：

1）尽量降低SnO_2电极的材料费和加工费，所谈SnO_2的材料费是化学纯试剂的价钱，如果工业生产大批量使用，则会显著降低；电极的加工费也是伸缩性较大的数值，它随着加工工艺的改进，是很有潜力可挖的。

2）增强耐腐蚀性，延长电极的使用寿命从理论上讲只要所生产铝的纯度合乎要求，则阳极可以一直使用下去，可是实际上由于机械原因或热物理原因的破坏，或者尺寸的改变使电流密度发生显著变化都要更换阳极，因此，总有一定的使用寿命。对此，提高电极的耐腐性和机械强度，以及抗震性都是重要的。

3）设计合理的阳极形状以提高电流负荷d值。

至于提高铝价和电流效率其作用是有限的，提高电流负荷和延长使用寿命对于提高经济性指标来说，其作用是等价的，但是提高d值相对来说比较容易，而成倍地延长寿命则较为困难，以各种形状的电极为例，来说明形状和电流负荷的关系，空心的比实心的好，在壁厚不变的情况下，直径大较直径小好，盘状又较圆筒状好，而以板状，以垂直面工作的电极形状最好。

参 考 文 献

[1] U S Pat, 3 960 678.
[2] U S Pat, 4 379 033.
[3] U S Pat, 3 713 550.
[4] Ger Offen, 2 446 314.
[5] Billehaug K et al. Aluminium, 1981, 57：146.
[6] Grjotheim K et al. Aluminium Electrolysis—Fundamentals of the Hall-Heroult Process, 2nd ed, Dusseldorf：Aluminium Verlag, 1982.
[7] 王化章, 刘业翔. 铝电解用惰性阳极的物理性质. 第三届全国铝电解学术会议论文集, 1984.
[8] 蔡祺风, 刘业翔. 轻金属, 1986（9）：38.
[9] 肖海明, 刘业翔. 有色金属（国际版）, 1986（4）：57.
[10] 黄永忠, 等. 冰晶石体系中氧化铝电解时电流效率的研究. 有色金属学会第一届轻金属学术年会论文集, 1986.
[11] Jarrett N. Journal of Metals, 1984（4）：71.

Laboratory Study on TiB$_2$-based Ceramic Cathode for Aluminium Reduction Cell*

Abstract The experimental results on bench scale of TiB$_2$-based ceramic catbode used in aluminum cells were described including manufacture process of this cathode and electrolysis test in laboratory cell. The results show that the process of preparing electrode consisting of cold press and sintering is reasonable and economic, the electrode produced according to this technique was of good conductivity, mechanical strength and good wettability by aluminum, but the corrosion rate of the electrode during electrolysis is rather high as far as now. Adoption of the cathode in aluminum reduction cells would be possible to save energy to a great extent due to the shorten A-C distance, The corrosion mechanism is also discussed.

1 Introduction

Saving energy is not only requirement of society but also of aluminium industry itself. In the market economy as a material aluminium have to compete with other materials such plastics as well as between the aluminium producer. Therefore low-energy, low-cost cell is a vital condition of developping aluminium industry. Analysis of the voltage distribution in a existing aluminum cell operating at less than 50% power efficiency shows there is significant room for saving energy in voltage drop in electrolyte, being 38 percent of total cell voltage, If the anode cathode distance (ACD) is decreased from 4.45cm to 1.91cm, the cell voltage would be decreased from 4.64 volt to 3.64 volt, consequently the D.C. energy consumption would be decreased from 15.19kW·h/kg-Al to 11.92kW·h/kg-Al, implying saving energy 3.27kW·h for each kilogram aluminum, that is considered as significant. However it is very difficult to operate at such low ACD in a traditional Hall-Herault cell, resulting from the upheave and fluctuation of molten aluminium pad, even though many efforts have been done to adjust or compensate magnetic fields both by altering bus location and magnetic improvement devices. Therefore three major problems must be solved before the ACD can be shortened to take this advantage. The first problem is the periodic shorting that occurs between the molten aluminium pad and anode with increasing frequency and longer time as the ACD is reduced. Meanwhile the direct contact of the anode gas with the cathode aluminium has become possible. This two reasons would cause the decrease in current efficiency. The second most important problem to be solved is the maintain of heat balance, the lower of ACD imply the decrease in energy into the cell, in order to maintain operating temperature of electrolyte as the interpolar space is reduced, either the current must be increased or the cell insulation must be enbanced to reduce heat loss. The third problem is the supply of dissolved alumina must be adequated by suitable feeding device and

---

* Copartner: Wang Huazhang, Wang Huiling, Wang Xiangmin. Reprinted from Light Metals, 1994:505-511.

program, or to make sure the volume of electrolyte flow through the interpolar spacing, replacing depleted electrolyte in dissolved alumina by fresh electrolyte saturated alumina. In order to solve these problems, developing the novel electrode materials, designing new construction cell and set up the optimum operating parameters, all of these must be done.

As a novel cathode material the following requirements must be met:

(1) Good conductivity.

(2) High corrosion resistance to fluoride melts and molten aluminium.

(3) Good wettability by molten alumnium.

(4) Enough mechanical strength.

(5) Good resistance against thermal shock.

(6) The price is cheap.

Since the early 50's the study on this area has started, through the extensive investigation, the carbide, boride and nitride of transient metals so called Refractory Hard Metal (RHM) materials particular titanium diboride is considered as cathode material to be hopeful. K. Billehaug and H. A. Øye has made comprehensive review references published before 1980[1,2] in which the material selection, manufacture process as well as service way has been reported, since then the development of titanium diboride cathode has got great progress, recently a review of RHM cathode development has made by Curits J, McMinn[3]. In order to overcome the defects such as the brittle and sensitive to thermal shock and improve other physical properties, the cathode material is made from TiB_2-based ceramic with TiC, SiC, CrB_2. AlN, BN etc. or cermet with Fe, Co, Ni. Meanwhile many efforts were made to develop coating composite being consists of TiB_2 graphite and various resins. In this area the achievement of TiB_2 coating cathode used into 105kA VSS cells by Martin Marietta company which was published on the 1984, 1985 AIME annual meeting respectively is remarkable and inspiring[4,5]. The U. S. patent by William M. Buchta and Larry G. Boxall et al were also published[6,7]. Since then Yexiang Liu, Xianan Liao et al. improved the formulation and processes of coating, followed by conducted industrial test on the 50kA HSS cells in several aluminium plants of China and obtained affirmative save energy results[8,9]. There are many documents on TiB_2-based ceramic materials such as the Pacific North-West Laboratory supported in financial by U. S. Department of Energy has reported their experimental results on TiB_2-based material used for aluminum cell cathode in July 1988[10], and the reynolds metals company published their experimental results on TiB_2-graphite cathode for pilot reduction cell[11].

However, up to now the reduction cell operating in short interpolar distance using TiB_2-based cathode have not yet empoyled in industry. The cause resulting in this situation come from two factors, one of them is economy, the price of raw material TiB_2 powder with high purity is rather high, additional complex manufacture process led to raise the cost of TiB_2 cathode, the other is technique reason, the main problems are corrosion resistance relating to service time and cell structure, that is to say that cell design would be suitable low interelectrode spacing operation. Present work aim at developing various composites based on titanium diboride, that produced by our technique with low cost, and employ fabrication process to be industralization easily, as well as test the various performance during the electrolysis process to provide the basis for application in industry.

2 Preparration and Performance Test of Electrode

2.1 Raw material

TiB_2 powder was supplied from the Hefei aluminum plant. The composition was (wt%): Ti 60.74, B 29.67, C 2.47, O 1.82, Fe 0.19, particle size 3.4~9.66μm. TiC powder was supplied from Zhuzhou carbide alloy plant. The composition was (wt%): Ti 78.85, C 19.36, O 0.60, average particle size 3μm. SiC was supplied from Shanghai abrasive disic plant, SiC content don't less than 98%. AlN was supplied from the Central South University of Technology, powder metallurgy Institute. The Co, Fe, Ni with chemical reagent grade was purchased from market.

2.2 Preparation of an electrode

First the TiB_2 powder were passed a −200 mesh sieve and mixed with other component and a binding agent according to certain proportion, then pressed in a mould at 2000~3000kg/cm^2 to the shape of cylinder of 12mm diameter and 30mm height. After then, the pressed "green" samples were subject to sintering at the temperature as high as 1800~2000℃ for 10 hours in the reduction atmosphere, finally cool slowly to the room temperature. Some physical properties of the electrodes obtained were listed in Table 1.

Table 1 The composition and some properties of TiB_2-based composite electrode

Electrode No.	Composition of electrode /wt%	Apparent density /g·cm^{-3}	Porosity /%	Resistivity (room temp.) /μΩ·cm	Compressive strength /MPa
Y1	TiB_2(90% pure)	2.963	13.62	2.51	73.6
X1	TiB_2(75% pure)	2.261	26.02	65.41	3.9
Y5	40% TiB_2 + 60% TiC	3.909	4.18	2.51	159.3
Y8	50% TiB_2 + 50% SiC	2.701	9.07	435.0	71.05
Y15	50% TiB_2 + 10% TiC + 40% SiC	2.729	9.22	39.0	70.1
Y16	40% TiB_2 + 30% TiC + 30% SiC	3.074	8.29	10.88	112.6
Z4	TiB_2 + Fe	3.053	7.33	8.93	43.3

2.3 The test of oxidation resistance of the electrode at high temperature

Put the sample in a corundum crucible without cover which was placed in a resistor furnace. Heated it from room temperature to 1000℃ and maintain it at this temperature for 4 hours, followed by taking it out and cool to the ambient temperature. It is found that the color of the sample surface become yellow, furthermore its hardness raised. The analysis result by X-ray diffraction indicate that the layer of yellow material was titanium dioxide. So the TiB_2-based sample exposed to the oxidation atmosphere at high temperature would be subjected to oxidation therefore the weight gain by 2%~3.7% generally.

3 Electrolysis Test in a Laboratory Cell

3.1 Experimental facilities and procedure

The experimental facilities is shown in Fig. 1.

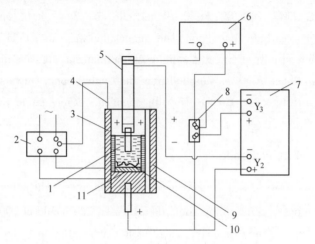

Fig. 1 Arrangement of electrolysis device in laboratory

1—Electrolyte; 2—Temperature controller; 3—TiB$_2$ electrode; 4—Pt-Rh thermocouple;
5—Stainless steel pipe; 6—DC power; 7—X-Y recorder; 8—Standard resistor ($R_0 = 0.001\Omega$);
9—Resistor furnace; 10—Aluminum; 11—Corundum crucible

The graphite crucible of 5.2cm inside diameter, 10cm height containing electrolyte serve as anode, in which a sintered alumina crucible of such diameter as inside diameter of graphite crucible and only 2.0cm height was placed on the bottom serve as collector of aluminum electrolyzed. The TiB$_2$ cathode was supported in a stainless steel pipe used as a conductor. Formulated electrolyte was charged in a graphite crucible that placed in a furnace then heating it until all of electrolyte is fused and reached 960℃ maintaining it at this temperature for several minutes, insert the TiB$_2$ cathode in a fused electrolyte. The depth of immersion was adjusted by clip on the furnace lid. The electrolysis current was normally kept at 8A throughout an experiment. The duration of an experiment was 2.5~8 hours. A moderate amount of Al$_2$O$_3$ and 5g of cryolite were added into electrolyte every hour to compensate for evaporation loss. The current and cell voltage were recorded by X-Y recorder, and the Counter-Electromotive Force was determined by means of breakdown current method near the end of electrolysis. When an experiment was finished the TiB$_2$ cathode was taken off and cool to the room temperature for examination by scanning electron microscope. At the same time the current efficiency was calculated by the weight of aluminum produced during experiment.

3.2 Results and discussion

For each type of TiB$_2$ based electrode at least two times test were carried out with the same electrolyte with molar ratio of 2.4, and containing 5% MgF$_2$. The content of Al$_2$O$_3$ was kept saturation. The results of electrolysis test was shown in Table 2.

Table 2 The Results of electrolysis test

Electrode No.	Composition of electrode /wt%	Cell voltage /V	CMEF /V	Duration /h	Current efficiency /%	Energy comsumption /kW·h·(kg-Al)$^{-1}$	Corrosion rate /g·(cm^3·h)$^{-1}$
Y5	40% TiB$_2$ + 60% TiC	3.04	1.41	4	75.43	12.01	4.2 × 10^{-3}
Y8	50% TiB$_2$ + 50% SiC	3.23	1.46	4	—	—	—
Y13	60% TiB$_2$ + 10% TiC + 30% SiC	3.74	1.72	2.5	—	—	—
Y15	50% TiB$_2$ + 10% TiC + 40% SiC	3.14	1.42	4.5	—	—	—
Y16	40% TiB$_2$ + 30% TiC + 30% SiC	3.22	1.62	3.5	—	—	—
Z1	TiB$_2$ + C	2.75	1.56	4	69.74	11.75	7.1 × 10^{-3}
Z12	TiB$_2$ + TiC + Ni	2.90	1.48	8	76.73	11.26	9.84 × 10^{-3}
Z16	TiB$_2$ + TiC + AlN	3.55	1.58	6	72.86	14.52	4.35 × 10^{-3}

The cell was operated at 8A with cathode current density of 1.13~1.17A/cm^2, and anode current density of 0.1A/cm^2. Therefore the cell voltage and the CMEF was also lower. The cell voltage increases when the anode current density is increased. In general the current efficiency is rather low. The reason led to this result have two factors, one of them is mechanical loss result from good wettability of the TiB$_2$-based electrode by molten aluminum therefore a part of aluminium adhere to the surface of TiB$_2$-based electrode is very difficult to take off entirely. The other is higher loss of reoxidation of aluminium solved in electrolyte because of lower interpolar distance. In spite of these unfavorable factors the unit energy consumption calculated based on data in Table 2 is considerable low, we can still see that the energy saving effect have a great attraction. If extend the electrolysis duration and manage to raise the current efficiency the energy saving effect become even more considerable.

The corrosion rate of an electrode was determined based upon the content of titanium in metal and electrolyte. The value of corrosion rate was normally 10^{-3} order of magnitude, the unit being g/(cm^2·h) and the value is still considered as very great.

The content of titanium in molten aluminum is much higher than the solubility of TiB$_2$ in aluminium and the more deep investigation and study is needed to find out the cause.

In order to get an insight into the behavior of TiB$_2$-based cathode during electrolysis and its wettability by molten aluminium, the TiB$_2$-based cathode electrolyzed were mounted and examined by scanning electron microscopy. The chemical composition of these samples were determined by electron microprobe analysis. The SEM picture of longitudinal cross sections of TiB$_2$-based cathode electrolyzed were shown in Fig. 2. From Fig. 2 it can be seen that a layer of aluminum > 750μm thickness adhering tightly on the surface of the cathode indicates the good wettability, and a part of aluminum has penetrated into inside of the cathode, and full fill the pore exist in cathode internal. From the images of titanium and silicon distribution it can be also seen that a small quantity of titanium and silicon appears in the aluminum layer near the cathode surface that shows the components of TiB$_2$ and SiC has been solved into aluminum to some extent. The sodium come from electrolyte entered also into the internal porosity of the cathode.

It is possible that the electrolyte always penetrates into the cathode at the beginning of electrolysis before aluminum layer formed. It's also possible that the sodium deposited on the cath-

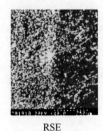 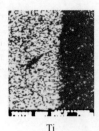   

      RSE   Ti   Si   Al   Na

Fig. 2　SEM photograph of TiB_2 + TiC + SiC electrode after electrolysis

ode dissolved into aluminum layer then transfer to the cathode internal. This sort of penetration of aluminum and sodium was likely a major causes led to destroy of electrode. Of course the study of mechanism of the destruction in detail is needed.

4　Conclusions

(1) Cold pressing and sintering is an economic and reasonable process for fabrication of TiB_2-based electrode which have a good conductivity and mechanical strength except the bigger porosity.

(2) The TiB_2-based electrode prepared by using this process have a good wettability molten by aluminium and can be operated at lower interpolar distance to get saving energy effect.

(3) The addition of a large amount of TiC makes the properties and performance of electrode great improvement. The experiments demonstrate that the better composition was 40% TiB_2 and 60% TiC.

(4) The aluminum and sodium always penetrate towards internal of electrode and the components of electrode such TiB_2, TiC, SiC tend to dissolve in aluminum. So that the corrosion rate was still rather high, the emphasis of more work should focus on the diminution of corrosion rate and prolong the service time of the electrode.

References

[1] K. Billehaug, H. A. Øye. Inert Cathodes for Aluminum Electrolysis in Hall-Heroult Cells(I). Aluminum, 1980, 56(10):642.

[2] K. Billehaug, H. A. Øye. Inert Cathodes for Aluminum Electrolysis in Hall-Heroult Cells(II). Aluminum, 1980, 56(11):713.

[3] Curtis J. McMinn. A Review of RHM Cathode Development (Paper presented at the 120th AIME Annual Meeting, San Diego, California, 1 March 1992):419.

[4] Larry G. Boxall, Arthur V. Cooke. Use of TiB_2 Cathode Material: Application and Benefits in Conventional VSS Cells (Paper presented at the 113th AIME Annual Meeting, Los Angeles, California, 28 February 1984):573.

[5] Arthur V. Cooke, W. Mark Buchta. Use of TiB_2 Cathode Material: Demonstrated Energy Conservation in VSS Cells (Paper presented at the 114th AIME Annual Meeting, February 1985):545.

[6] William M. Buchta, Baltimore Md. Process for Manufacture of Refractory Hard Metal Containing Plates for Aluminium Cell'. U. S. patent 4,582,533, Apr. 15,1986.

[7] Larry G. Boxall, et al. Aluminum Cell Cathode Coating Method, U. S. patent 4,466,996, Aug. 21,1984.

[8] Xian-An Liao, Ye-Xiang Liu. Some Important Performances of TiB$_2$ Coated Cathode Carbon Block (Paper presented at the 119th AIME Annual Meeting, Anaheim, California, 18 February 1990):409.

[9] Ye-Xiang Liu, et al. Observation on the Operating of TiB$_2$-Coated Cathode Reduction Cells (Paper presented at the 120th AIME Annual Meeting, San Diego, California, 1 March, 1992):427.

[10] C. H. Schilling. Laboratory Testing of TiB$_2$-Based Cathodes for Electrolytic Production of Alumimium (Report, Prepared for the U. S. Department of Energy Under Contract DE-Acoh-76RLOI830, July 1988).

[11] T. R. Alcorn, D. V. Stewart and A. T. Iabereaux. Pilot Reduction Cell Operation Using TiB$_2$-G Cathodes (Paper presented at the 119th AIME Annual Meeting, Anaheim, California, 18 February 1990):413.

Research Progress in TiB$_2$ Wettable Cathode for Aluminum Reduction*

Abstract Titanium diboride wettable cathodes are regarded as ideal for aluminum reduction because of their excellent wettability with molten aluminum. The TiB$_2$ inert wettable cathode materials for aluminum reduction may be divided into three groups: pure TiB$_2$ ceramic cathode, TiB$_2$ composite cathode, and TiB$_2$ coating. This paper briefly describes international research progress on TiB$_2$ inert wettable cathodes as well as problems faced, and concentrates on the activities of Central South University, Changsha, China, in researching the ambient temperature solidified TiB$_2$ cathode coating. At the same time, the results of the coating applied in many aluminum smelters in China are presented, and the pattern of cathode surface of 160kA cells coated with the ambient-temperature-solidified TiB$_2$ cathode coating after one year operation is discussed in comparison with the normal cells.

1 Introduction

With its high energy consumption and emissions, the aluminum reduction industry faces an immense challenge as energy savings and emission reductions have become matters of global urgency. New technology has reduced specific energy consumption to some extent. However, the industry still requires 13000~15000kW·h/t aluminum, compared to the theoretical consumption of 6330kW·h/t[1] required for the electrochemical decomposition of alumina (with carbon anode). Thus, the energy efficiency is less than 50%.

A large fraction of the energy consumed in the reduction cell is expended as IR drop in the gap between the anode and cathode (ACD)[2]. If one centimeter of ACD can be reduced, about 1800kW·h/t Al is saved in the current commercial electrolyte system. Therefore, decreasing ACD is an effective approach for saving energy.

Because carbon material is employed as the cathode for the present aluminum reduction cell, a substantial pool of metal (about 20cm) is needed on the surface of carbon cathode. Although the aluminum pool to a certain extent protects the cathode bottom, the electromagnetic forces create movements and standing waves in the aluminum and the aluminum/electrolyte interface. To avoid shorting between the metal and anode, the ACD must be kept a safe 4cm to 6cm in production. So, as for the present aluminum reduction cell with a carbon cathode, the potential of saving energy is very limited. However, an a inert wettable cathode, which is wettable by molten aluminum and does not require a high aluminum pool, can form a stable aluminum layer of 3mm to 5mm. By reducing the enormous disturbance of the magnetic field, the ACD can be reduced without adverse effects on current efficiency. Therefore, the voltage drop between the anode and cathode is reduced, and significant power savings may result.

---

* Copartner: Li Jie, Lü Xiaojun, Lai Yanqing, Li Qingyu. Reprinted from JOM, 2008, 60(8): 32-37.

Inert wettable cathode material must exhibit the following characterisitics: good electrical conductivity, good resistance to corrosion by high temperature and molten fluoride salt, wettability by molten aluminum, satisfactory mechanical strength and resistance to cracking due to thermal or chemical forces, capability of being produced and fabricated into required shapes, low cost and broad source of raw material[3].

Since the 1950s inert wettable cathode material has been used by international aluminum organizations. Extensive research has the refractory metal (RHM) materials to be the candidate material of cathodes for Hall-Heroult process aluminum reduction cells. Among them, TiB_2 shows the greatest potential. According to published comprehensive literature reviews[2,3], TiB_2 inert wettable cathode materials for aluminum reduction are divided into three groups: TiB_2 ceramic material, TiB_2 composite material and TiB_2 coating. This paper presents a review of research into the TiB_2 inert wettable cathode, and introduces the research and development of the ambient temperature solidified TiB_2 cathode coating studied in Central South University (CSU), as well as its industrial application.

2 Research Progress in TiB_2 Wettable Cathode Materials

2.1 TiB_2 ceramic cathode material

Titanium diboride is a very strong covalence compound with poor sintering capability due to its low diffusion coefficient. Only through heating can a densely sintered TiB_2 material with good mechanical properties be obtained. Generally, the preparation techniques of TiB_2 ceramic material have direct hot pressing and pressureless sintering or cold pressing following by high-temperature sintering. The density of hot-pressed TiB_2 ceramic cathode material is high, closer to theoretical density, and the relative density is higher than 95%. Patent and literature report that hot-pressed TiB_2 materials had better properties than those processed by other methods[4,5]. For TiB_2 pressureless sintering, temperature in excess of 2000℃ are often necessary due to the high melting point of TiB_2.

Some techniques have been developed to decrease sintering temperature. According to the literature, cold-pressed and sintered parts of high density at 98% to 99.5% of the theoretical density can be produced from TiB_2 powders with high specific surface of $3\sim15m^2/g$[6]. A combination of C and Cr, Fe, Ni, Co, CrC, B_4C, or WC can be used as a sintering aid to produce TiB_2 parts with a final density of greater than 95% of the theoretical density. R. Gonzales achieved a high-density TiB_2 material prepared with a hot-pressed technique and sintering aids[7].

Hot-pressing, is expensive, however, and does not allow the net-shape fabrication. The cost of cold-pressed sintering is relative low and net-shape fabrication is possible to a certain extent, but a higher sintering temperature is needed.

The main drawback of TiB_2 ceramic material is the poor resistance to thermal shock. J. R. Payne from kaiser Aluminum and chemical corporation even claims that, as a general rule, TiB_2 material should not be subjected to a temperature gradient greater than 200℃ or the life of material will shorten[8]. In addition, joining to the substrate has proved to be a formidable task. The literature[9-11] introduced some ways of joining and bonding to the matrix, but this was never entirely successful.

2.2 TiB$_2$ composite cathode material

Titanium diboride composite cathode material was investigated to remedy the brittleness of pure TiB$_2$ ceramic material and improve the mechanical properties and the resistance to thermal shock. The components vary, but most of them are other RHM or non-oxide ceramic. Carbide and borides such as TiC, ZrC and ZrB$_2$ were mixed with TiB$_2$ since they have similar properties. The mechanical and sintering properties of TiB$_2$ material are greatly improved by these additions. Among them, TiB$_2$-TiC composite material shows excellent resistance to thermal shock[4,5,12]. Comparing to TiB$_2$, the chemical properties of carbides are not as stable as TiB$_2$, and their electrical resistivities are also high. In the literature[14], some nitrides such as Si$_3$N$_4$, BN or AlN have been used to form the composites with TiB$_2$. Generally, the mechanical properties and chemical stability are improved by these non-conductive components, but at the expense of the degraded electrical conductivity. Reynolds[13] developed TiB-AlN-Al and TiB$_2$-AlN composite material, and mainly tested the properties of TiB$_2$-AlN. These composite materials may decrease requirements TiB$_2$ powder purity, but their electrical resistivities are increased. In addition, a breakage problem remained with the composite materials. H. Lu[4] studied TiB$_2$-WSi$_2$ composite material, and concluded that the addition of WSi$_2$ improved evidently sintering properties of TiB$_2$, but the resistance to corrosion in electrolyte was reduced because of high solubility of WSi$_2$ in electrolyte.

Another important component material is carbon. H. A. Øye[15] et al carried out a serial study of TiB$_2$-C composites as the cathode in the Hall-Heroult cell. The composition of TiB$_2$-C cathode is 0%~50% TiB$_2$, 33%~83% anthracite, and 17% binder. The specimens were heated to 1250℃ and placed in a vessel packed with carbonaceous powders. Their compressive strength were 14.4~47MPa, and less sodium was present on TiB$_2$ areas than on carbon areas in specimens after electrolysis. The literatures[16,17] reported that wetting of TiB$_2$-C composite by liquid aluminum occurred after an immersion time of 6 hours, and discussed the chemical stability of TiB$_2$ particles in molten aluminum and bath as well as the cathodic process. The results showed that Al$_2$O$_3$ dissolved in bath enhanced dissolution of TiB$_2$. The cathodic overvoltage reduced with the increase of TiB$_2$ content, and a continuous aluminum layer was formed on the surface of cathode.

Two 70kA pre-baked aluminum reduction cells at the Kaiser Mead Smelter were retrofitted and operated with mushroom-shaped TiB$_2$-graphite cathode elements for 4 to 5 months to evaluate the technical and economical benefits[18]. The major problem revealed in two tests was breakage of the TiB$_2$-graphite cathode elements, this problem prevented the energy consumption targets to be fully achieved.

In conclusion, TiB$_2$ composite cathode material is considered to have the greatest potential for use in drained aluminum reduction cell. It not only overcomes the sintering difficulty of TiB$_2$ ceramic cathode material, but also reduces the cost. However, its properties, especially resistance to cracking and corrosion as well as electrical conductivity, must be further improved[19,20].

2.3 TiB$_2$ coating

A TiB$_2$ coating can be applied on the substrate of inert wettable cathode materials. The substrate materials contain carbon, nickel, molybdenum and steel, etc., but when considering the TiB$_2$ expansion coefficient, carbon is preferred. Various TiB$_2$ coating techniques are listed in Fig. 1.

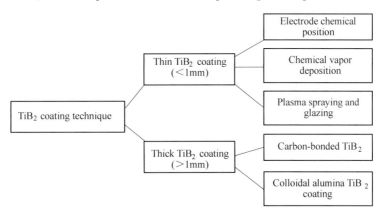

Fig. 1 Titanium diboride cathode coating techniques

H. Pierson[21] et al studied TiB$_2$ coatings deposited on a graphite substrate, and the thickness was greater than 700μm. A. Becker and J. Blanks[22] prepared graphite pipes that were chemically vapor deposited with TiB$_2$, and experimentally evaluated the performance of materials on the laboratory scale cell. The results showed that the coating had a short lofe time because of the grain boundary attack and subsequent pull-out. U. Fastner[23] et al. investigated the electrochemical deposition coating on steel and molybdenum substrates, respectively, which was deposited using a NaCl-KCl-NaF-KBF$_4$-K$_2$TiF$_6$ electrolyte at 600℃. The results showed that the pulse sequences and the current densities used significantly affected the homogeneity of the layers deposited, the crystal size, texture and other physical properties such as electrical and thermal conductivity. Coatings were also deposited from the Na$_3$AlF$_6$-Al$_4$B$_2$O$_9$-TiO$_2$ and Na$_3$AlF$_6$-Al$_4$B$_2$O$_9$-CaTiO$_3$ systems on graphite, glassy carbon, nickel and tungsten cathodes[24]. The coating thickness of the deposited TiB$_2$ was 15~20μm (deposition rate:50μm/h).

H. Lu et al[25] adopted a plasma spraying technique to prepare TiB$_2$-MoSi$_2$ coating and studied the effect of process parameters such as spray distance and additive MoSi$_2$ content on the properties of the coating.

Katharina Seitz and Frank Hiltmann[26,27] fabricated a vacuum plasma spraying TiB$_2$ coating. The coating layers was wetted by molten aluminum and showed a good adherence to the carbon substrate materials. Its electrical resistivity is lower than that of the carbon substrate. The advantage of plasma spraying technique is quick deposition speed, but the single phase TiB$_2$ coating is typically porous because of the difficulties with melting the starting powders. The porosity on the coatings may cause penetration by aluminum and the electrolyte, which leads to deterioration of the substrate material.

The coating techniques described are confined to a laboratory study and the coating was thin. The coating susceptibility to the penetration of aluminum and electrolyte, and adherence to the substrate need to be improved.

In recent years, coatings of carbon-bonded TiB_2 and colloidal alumina TiB_2 coating have been extensively investigated, and numerous industrial tests conducted. Moltech's TINOR TiB_2-colloidal alumina coating was tested in eight smelters worldwide. Titanium and boron analyses in the tapped aluminum found that the coating lasted about three years. Preliminary results showed a dramatic reduction in sodium uptake by the carbon cathode. H. A. Øye[28] studied the properties of colloidal multi-layer alumina-bonded TiB_2 coating. The thermal expansion was somewhat higher than that of carbon materials, the electrical conductivity was of the same order as graphite, and porosity of the coating was about 30%. Other properties were tested including abrasion resistance, wettability by liquid aluminum, resistance to sodium penetration and bonding to the substrate. Dissolution of TiB_2 by aluminum was approximately 0.3mm/year. Patents[29-32] claimed different preparations of TiB_2/colloidal alumina carbon cathode coatings in Hall-Heroult and drained cells. The results showed that the mechanical properties and electrical resistivity of the coating increased with increasing particle size of the TiB_2; decreasing the particle size of the colloidal alumina increased the flexural strength and decreased the electrical resistivity of the coating. The coating comprised several layers: the protective layer, the aluminum-wettable layer, the anchorage layer, and the substrate layer.

Comalco[33] developed a TiB_2 cathode coating in drained cells. The coating was a mixture of TiB_2 and carbon produced by baking out a mixture of TiB_2 and a carbonaceous cement. The coating permitted good wetting of the Al and close anode-cathode distance in the drained cells. Coating life appeared to be a problem as the Al metal pad apparently reacts with the carbon portion of the cement to produce Al_4C_3. Thus the carbon portion of the coating slowly disappeared, simultaneously releasing the TiB_2 particles.

M. O. Ibrahiem[34,35] et al studied the effect of the binder type, mixing methods, and coating method on properties of TiB_2 coating such as the chemical stability and wetting. When the TiB_2 content of the coating was greater than 50%, the coating showed excellent wettability by molten aluminum. After 48 hours, a penetration depth of 166μm was reached through the open pores and grain boundaries. Some Al_4C_3 and Al_2O_3 formed at the aluminum coating interface as well as in the aluminum pool.

As compared to coatings applied by plasma spraying, electrochemical deposition and chemical vapor deposition, the thickness of carbon-bonded coating and colloidal alumina coating may increase, with a thickness of 3~8mm. However, as the thickness of the coating increases it is difficult to cracking or a reduction of other properties. The hurdle of the short lifetime of coating must still be overcome.

3 Ambient Temperature Solidified TiB$_2$ Cathode Coating

Since the 1980s, extensive researches have been done on the inert wettable cathode by team by Liu Yexiang at Central South University. At the early stage of this research, H. Wang[36] et al. studied many kinds of composites of TiB_2 matrix such as TiB_2-TiC, TiB_2-SiC, TiB_2-SiC-TiC, TiB_2-AlN-TiC and so on. The specimens were sintered at 1600~2000℃ by the cold press method, and their porosity was 4%~10% with a higher corrosion rate of the electrode. Industrial experiments on carbon-bonded TiB$_2$ cathode were done by Liu Yexiang[37] et al on nine sets of 75kA söderberg cells at the Lianchen aluminum smelter. The cathodes were solidified and hensed in atmosphere at less than 300℃ for 40~50 hours, followed by carbonized coating using the heat generated by the preheating and start-up. Compared with the contrast cell, the testing cell was more stable, and its cell bottom was cleaner, current efficiency higher, the voltage drop of the cell bottom lower, and its consumption of energy reduced 200kW·h/t Al. The amount of NaF added to each cell also reduced 300kg. Xianan Liao[38] et al did the industrial tests on carbon-bonded TiB$_2$ cathode at the Hefei aluminium plant in 1995. They employed 28 sets 60kA söderberg cells, and another 12 identical cells as contrast cells. Similarly, results indicated that TiB$_2$ cathodes offered improved current efficiency, the voltage drop of the cell bottom and the distribution of the cathode current. Before 2000, this smelter experienced a two-month production halt due to a power shortage, and during the restart the physical appearance and chemical property of the cells had changed significantly. However, there was no failure in the 28 testing cells until 2000 and they still operated regularly, with pot lifes of 6 years, the contrast cells, however, had deteriorated gradually.

Söderberg cells always start up with liquid aluminum baking, which makes TiB$_2$ coating carbonized quickly, resulting in the deterioration of bonding between TiB$_2$ and the carbon substrate. This limits the coating in such areas as wear resistance and extension of pot life. In addition, a heat source is generally needed in order to solidify at high temperature before preheating and start up of cells. This will lead to high cost, operational difficulties and extended time for solidification, which become one of the main barriers of application of TiB$_2$ cathode coating.

To resolve these problems, Centrol South University developed concept of ambient temperature solidified TiB$_2$ cathode coating (ATC) in 2001, With the support of the National High-Tech Research and Development Program of China and the State Basic Research Development Program of China. This process means the coating can solidify at ambient temperature with no need to increase the heat source. That is, after overlaying coating, only presetting at ambient temperature for 24 hours, preheating, start-up and production of the cell can begin according to the normal procedure.

Q. Li[39] investigated the ambient temperature solidification and found excellent properties of TiB$_2$ cathode coating such as resistance to sodium expansion, resistance to thermal shock, lottle electrical resistance, high compressive strength and strong bond strength. After eight quick cooling and heating circle, the coating still showed great strength. At the same time, after 12 hours of anode and cathode inversed analogue experiment, the coating was perfect and the surface wettable by aluminum.

Then at the end of 2001, the ambient temperature solidified technique of TiB$_2$ was put to an industrial test in four 160kA prebaked aluminum electrolysis cells in Guangxi Branch, Aluminum Corporation of China Limited. The 2~3mm thickness of TiB$_2$ coating paste of with 60 weight percent TiB$_2$ was smoothly painted on the pretreated surface of carbon cathode lining. The cell with painted coating was kept at room temperature for 24 hours so that the cathode coating was fully solidified, then it was carbonized with coke resistive preheating. The results indicated that this test was very successful because of the convenience of preparation, low cost, good solidification and carbonization, startup stability, reduction of the penetration of sodium and electrolyte, neatness of the bottom of cell, decreave of volt drop in the bottom of cell, and, nerease in current efficiency.

In the 160kA cell painted with ATC, 700kg of Na_2CO_3 can be saved during commissioning of a new cathode lining. After startup, the CR stays in the normal range, proving that the coating can really slow down the rate of sodium penetration and reduce the rate of sodium penetration into cell lining. The surface appearance of the cathode in 160kA pre-baked contrast cell and the 160kA cell with TiB_2 coating after 1 year of operation are shown in Figs. 2 and 3, respectively. As shown in the figures, in the contrast cell without TiB_2 coating, the cathode surface was obviously uneven, there were many more cracks on it and the shape change was much greater. The coated cathode surface (with 60% TiB_2 content in coating), however, was still flat and no obvious cracks and the shape change was small. With the same results as tested in a laboratory, the TiB_2 cathode coating showed that it would resist shape change of cathode lining in the primary aluminum production process.

Fig. 2　A cathode surface without TiB_2 coating in 160kA prebaked cell after one year of operation

Fig. 3　A cathode surface with TiB_2 coating in 160kA prebaked cell after one year of operation

In general, the lifespan of aluminum electrolysis cells is determined the deterioration of cathode lining until it is finally broken. The TiB_2 cathode coating could slow down or prevent the deterioration, reduce the lining shape change, thus prolong the life of cells.

Bo Ye[40] and Xiaojun Lu[41] et al. added dimensional stable particles to the coating, and found that it can improve the mechanical properties of coating and increase the thickness (12~15mm) of the coating (Fig. 4). During industrial tests in 75kA prebaked cells, little loss of TiB_2 coating with ambient solidification resulted, at 1.22% on average every month with a maximum of 4.60%, compared with the contrast cell, the volt drop at the bottom of cell decreased by nearly 10mV. On the anaphase of the test, the reduction was 50mV on average and the average current efficiency increased by 2.06%. Based on the described technology, Lü[41] optimized further the properties of TiB_2 coating by selecting the grain size of materials, kinds of carbonaceous additive agents, and enhancement agents, with a result of greatly omproveed conductivity, mechanical properties, and resistance to sodium expansion of TiB_2 coating.

There are six patents on ATC, which is used by many companies such as the Guangxi branch of Chalco, Shanxi Guanlu Aluminum Co., Ltd., Zhengzhou Longxiang Aluminum Co., Ltd., Jiaozuo Wangfang Aluminum Co., Ltd., Yu Gang Long Quan Aluminum Co., Ltd., Sichuan

Fig. 4　A Sample of ambient temperature solidified thick TiB$_2$ coating (12~15mm)

Fig. 5　The cathode surface of large-scale prebaked cell with ambient temperature solidified TiB$_2$ coating material

Qimingxing Aluminum Co., Ltd., Qingtongxia Aluminum Co., Ltd. and Shenhuo Aluminum Co., Ltd. (see Fig. 5).

Q. Li[42] and J. Fang[43] et al. also studied TiB$_2$/C cathode composite material sintered at 1000℃, which is the temperature requirement of wettable cathode used in aluminum reduction. They found that kinds of carbon in TiB$_2$ cathode composites have a great influence on the property of resistance to sodium penetration. They proposed that the materials containing sodium be added to the cathode in order to enhance the property of resistance to sodium penetration. That is, sodium is added into the cathode materials beforehand to reduce the differential concentration between surface and interior of cathode early during electrolysis in order to reduce the penetration of sodium into the cathode and sodium induced expansion. Based on the idea of premixed sodium, Q. Li[43], and J. Fang[44] et al. propsed a waste cathode lining which contained sodium and electrolyte, which are used as the raw materials for TiB$_2$/C cathode composites. They found that during electrolysis the expansion rate can be slowed, avoiding, to a certain extent, the failure of cathode materials resulting from rapid expansion. At the same time, selecting waste cathode lining as a component of TiB$_2$/C cathode composites may provide a new way of reusing the waste lining and protecting environment.

As for the TiB$_2$ cathode material in the drained cell, efforts are under way to overcome the difficulties of technology in order to develop the drained cell successfully.

4　Conclusion

Study on the inert wettable TiB$_2$ cathode has been ongoing for several decades, and industrial tests have obtained some fruits. However, more efforts are needed to transform these fruits to massive industrial application, especially for application in drained cells. Many problems will need to be solved in future studies for wettable cathode material. The first challenge is improving the sintering characterisitics and mechanical property of TiB$_2$ materials while reducing the corrosion rate, electrical resistivity and cost, while prolonging the life of material.

Next, it is generally believed that the smaller the contact angle, the better the wetting property between inert wettable TiB$_2$ cathode and melt aluminum. Yet to be determined, however, is the optimal contact angel required for aluminum electrolysis or the drained cell. If a quantitative value can be given, this will demonstrate the direction for the development of TiB$_2$ cathode.

Finally, the order of wetting (Aluminum vs. salt) will need to be addressed in future studies as it may influence the long-term stability of the wettable cathode material.

5 Acknowledgements

The authors are grateful for the financial support of the National Basic Research Program of China (2005CB623703).

References

[1] Zhongyu YANG, editor, Light Metals Metallurgy (in Chinese). Beijing: Metallurgical Press of Industry, 2006:179-180.

[2] K. Billehaug. Inert Cathodes in Aluminum Reduction in Hall-Heroult Cell. Aluminum, 1980, 54(2): 642-718.

[3] Yexiang LIU. Research Progress of Inert Anode and Wettable Cathode for Aluminum Reduction (in Chinese). Light metals, 2001, 10(5):26-29.

[4] S. K. Das, P. A. Foster, G. J. Hildeman. Electrolytic Production of Aluminum Using a Composite Cathode. U. S. Patent 4308114, 1981.

[5] L. G. Boxall, A. V. Cooke. Use of TiB$_2$ Cathode Material Application and Benefits in Conventional VSS Cells. Light Metals, 1984, 5(3):573-588.

[6] A. V. Cooke, W. M. Buchta. Use of TiB$_2$ Cathode Material: Demonstrated Energy Conservation in VSS Cells. Light Metals, 1985, 10(4):545-566.

[7] R. Gonzales, M. Barandika, D. Ona. New binder phase for the consolidation of TiB$_2$ hardmetals. Matl. Sci. and Eng. A, 1996, 216(6):185-192.

[8] J. R. Payne. Bonding of Refractory Hard Metal. U. S. Patent 4093524, 1978.

[9] H. I. Kaplan. Refractory Surfaces for Alumina Reduction Cell Cathodes and Methods for Providing Such Surfaces. U. S. Patent Defensive Publication T993002, 1980.

[10] H. I. Kaplan. Cathodes for Alumina Reduction Cells. U. S. Patent 433813, 1982.

[11] Curtis J McMinn. A Review of RHM Cathode Development. Light Metals, 1991, 20(7):419-425.

[12] H. Zhang, V. de Nora, J. A. Sekhar. Materials Used in the Hall-Heroult Cell for Aluminum Production. Light Metals, 1994, 26(8):412-415.

[13] N. E. Richards, et al. Electrolytically Conductive Cermet Compositions. U. S. Patent 3328280, 1967.

[14] Huimin LU, Huanqing HAN, Ruixin MA, et al. Titanium Diboride and Wolfram Silicide Composite Used as Aluminum Reduction Inert Cathode Materials. Light Metals, 2006:687-690.

[15] H. A. Øye. Sodium and Bath Penetration Into TiB$_2$-Carbon Cathodes During Laboratory Aluminum Reduction. Light Metals, 1992, 17(3):773-778.

[16] Martin Dionne, Gilles L Esperance and Amir Mirtchi. Wetting of TiB$_2$-Carbon Material Composite. Light Metals, 1999, 22(7):389-394.

[17] S. C. Raj, M. Skyllas-Kazacos. Electrochemical Studies on Wettability of Sintered TiB$_2$ Electrodes in Aluminium Reduction. Electrochimica Acta, 1993, 5(38):663-669.

[18] A. Tabereaux, J. Brown, I. Eldridge, et al. The Operational Performance of 70Ka Prebake Cells Retrofitted with TiB$_2$-G Cathode Elements. B J Welch, eds. , Aluminium Smelting Conference. Queenstown, New Zea-

land,1998,12(16):257-264.

[19] Jilai XUE, Qingsheng LIU, Wenli OU. Sodium Expansion in Carbon/TiB_2 Cathodes During Aluminum Reduction. Light Metals,2007:1061-1066.

[20] Yaowu WANG, Naixiang FENG, Jing YOU, et al. Study on Expansion of TiB_2/C Compound Cathode and Sodium Penetration During Reduction. Light Metals,2007:1067-1070.

[21] H. O. Pierson, A. W. Mullendore. Thick Boride Coating by Chemical Vapor Deposition. Thin Solid Film, 1982,95(2):99-104.

[22] A. J. Becker, J. H. Blanks. TiB_2-coating Cathodes for Aluminum Smelting Cells. Thin Solid Films,1984, 119(7):241-246.

[23] U. Fastner, T. Steck, A. Pascual, et al. Electrochemical Deposition of TiB_2 in High Temperature Molten Salts. Journal of Alloys and Compounds,2007,31(6):1-4.

[24] S. V. Devyatkin, G. Kaptay. Chemical and Electrochemical Behavior of Titanium Diboride in Cryolite-Alumina Melt and in Molten Aluminum. Journal of Solid State Chemistry,2000,154(7):107-109.

[25] Huimin LU, Wellton JIA, Ruixin MA, et al. Titanium diboride and molybdenu silicide composite coating on cathode carbon blocks in aluminum reduction cells by atmospheric plasma spraying. Light Metals, 2005:785-788.

[26] Katharina Seitz, Frank Hiltmann, Titanium Diboride Plasma Coating Carbon Materials Part Ⅰ:Coating Process and Microstructure. Light Metals,1998,3(12):379-383.

[27] Katharina Seitz, Frank Hiltmann. Titanium Diboride Plasma Coating Carbon Materials Part Ⅱ:Characterization. Light Metals,1998,3(12):385-390.

[28] H. A. Øye, V. de Nora, J. J. Duruz, et al. Properties of a Colloidal Alumina-Bonded TiB_2 Coating on Cathode Carbon Materials. Light Metals,1997,5(12):279-286.

[29] Jean-Paul Huni, K. The, A. A. Mirtchi, et al. Refractory Coating for Components of an Aluminum Reduction Cell. U. S. Patent 0046605A1,2001.

[30] J. A. Sekhar, Jean-Jacques Duruz, J. J. Liu. Slurry and Method for Producing Refractory Boride Bodies and Coatings for Use in Aluminum Electrowinning Cells. U. S. Patent 6783655B2,2004.

[31] T. T. Nguyen, Jean-Jacques Duruz, V. de Nora. Dense Refractory Material for Use at High Temperatures. U. S. Patent 0224220A1,2003.

[32] J A. Sekhgar, V de Nora, J Liu, et al. TiB_2/colloidal Alumina Carbon Cathode Coating in Hall-Heroult and Drained Cells. Light Metals,1998,10 (14):605~615.

[33] G. D. Brown, et al. TiB_2 Coated Aluminum Reduction Cells:Status and Future Direction of Coated Cells in Comalco. B. J. James,eds. ,Proceedings of the 6th anstralian alminum smelting workshop,1998.

[34] M. O. Ibrahiem, T. Foosnæs, H. A. Øye. Stability of TiB_2 C Composite Coatings. Light Metals, 2006: 691-696.

[35] M. O. Ibrahiem, T. Foosnæs, H. A. Øye. Chemical Stability of Pitch-based TiB_2 C Coatings on Carbon Cathodes. Light Metals,2007:1041-1046.

[36] WANG Huazhang,et al. Titanium Diboride Based Ceramic for Aluminum Reduction(in Chinese),Light Metals,1993:26-31.

[37] Yexiang LIU, Xianan LIAO, Fuling TANG, et al. Observation on the Operating of TiB_2-Coated Cathode Reduction Cells. Light Metals,1992:427-429.

[38] Xianan LIAO, Yongzhong HUANG, Yexiang LIU. Potline-scale application of TiB_2 oating in Hefei aluminium & carbon plant. Light Metals,1998:685~688.

[39] Qingyu Li. Development and Industrial Application of Wettable Inert TiB_2 Cathodic Composite Coating for Aluminum Reduction. Ph. D. thesis,Central South University,2003.

[40] Bo YE. Preparation and Properties Study of Wettable Thick TiB_2 Cathode Coating for Drained Aluminum

Reduction Cell. M. D. thesis, Central South University, 2005.

[41] Xiaojun Lü. Study on the Electrical Conductivity, Compressive Strength and Resistance to Sodium Penetration of TiB_2-C Composite Cathode Coating. M. D. thesis, Central South University, 2006.

[42] Qingyu LI, et al. The TiB_2-carbon Composites Sintered at Moderate Low Temperature Used as Wettable Cathode for Aluminum Reduction(in Chinese), Journal of Central South University of Technology, 2003, 34(1):24-26.

[43] Jing FANG. Properties study and preparation of wettable inert TiB_2/C composite cathode material for aluminum reduction. M. D. thesis, Central South University, 2004.

[44] Qingyu LI, Yanqing LAI, Jie LI, et al. The Effect of Sodium-Containing Additives on the Sodium-Penetration Resistance of TiB_2/C Composite Cathode in Aluminum Reduction. Light Metals, 2005:789-791.

Cup-shaped Functionally Gradient NiFe₂O₄-based Cermet Inert Anodes for Aluminum Re duc tion*

Abstract Application of inert anode and wettable cathode technology for aluminum reduction will result in significant energy and environmental benefits, so it has become a research focus for several decades. The candidate as inert anode concentrates on oxide ceramic, cermet and alloy. This paper reviewes briefly their research progress and presents the achievements of Central South University, Changsha, China, in researching $NiFe_2O_4$-based cermet inert anode, which includes the preparation and optimization of material performance, the joint between the cermet anode and metallic bar, as well as the results of electrolysis testing for large inert anode group. At the same time, the problems for $NiFe_2O_4$-based cermet inert anode faced are discussed.

1 Introduction

Though great progress has been made in the primary aluminum production, the industry still requires 13000~15000kW·h/t aluminum and its energy efficiency is less than 50%. Therefore, the aluminum industry faces the immense challenge of energy savings and emission reductions. The implementation of new technology based on inert electrodes could result in significant energy, cost, productivity and environmental benefits, and it has become a research focus for several decades[1].

To be successfully implemented, an inert anode must offer: low solubility in fluoride melts containing dissolved aluminum, high electronic conductivity, high resistance to anode oxygen, little contamination of aluminum produced, ability to be fabricated in large shapes, stable electrical connection, adequate mechanical strength, relatively low cost and ready availability[2]. Despite intensive research efforts, no materials have yet been found that meet all these strict requirements. Most investigations of candidate inert anodes materials have been focused on ceramics, metals and cermets[3]. Cermets are a class of materials in which both metallic and ceramic components are present. Ideally, these could have the desirable properties of metals as well as those of ceramics. $NiFe_2O_4$-based cermets is one of the most promising materials as inert anode for aluminum electrolysis[4]. This paper presents a review of their research progress and especially describe the achievements of Central South University in developing $NiFe_2O_4$-based cermet inert anode supported by the National Basic Research program of China and Development Program of China since July 2001.

2 Research Progess in Inert Anode Materials

2.1 Ceramic anodes

For resistance to chemical attack by pure oxygen at 960℃, the choice of a fully oxidized materi-

---

* Copartner: Tian zhongliang, Lai Yanqing, Li Zhiyou, Li Jie, Zhou Kecao. Reprinted from JOM, 2009, 61(5):34-38.

al is appealing. Tin oxide and spinels such as $Ni_xFe_{3-x}O_4$ were considered potential inert anode materials, owing to their putative chemical inertness[5]. Cell tests using anodes with composition (in mass fraction) of 96%SnO_2-2%Sb_2O_3-2%CuO demonstrated good results, low-temperature and low-ratio electrolyte trials yielded erosion rate of nearly 20mm/a[6]. However, their electrical conductivity, electrical connection capability to a metal bus, thermo-mechanical properties, and expanding the inert anode make them a poor choice for application.

2.2 Metal anodes

Metal or alloys have several advantages compared to ceramic materials, including being easy to fabricate, non-brittle, good conductors, and providing good electrical connection. Metal-based anodes were studied by Argonne National Laboratory, Northwest Aluminum Technologies, and mainly Moltech in recent years. The Argonne National Laboratory anodes consist of Cu-Al alloy, and the tests of 10~100A at low temperature demonstrated goode results[7]. Northwest Aluminum Technologies carried out the tests on a 200A scale with the Cu-Ni-Fe alloy anode. Laboratory tests obtained current efficiencies of between 60% and 78%[8]. R. V. Kaenel[9] reported the results of Ni-Fe alloy anode on a 25kA pilot cell scale from Moltech. The dissolution rate of anode was 2.1mm/a, the impurities in the metal aluminium recovered at the cathode was lower than 1000ppm, and the current efficiency was above 90%.

However, metals are unstable in the presence of oxygen at high temperatures, it is difficult but crucial that the anode will be covered by a coherent, relatively thin oxide layer and self-repairing oxide layer. At the same time, a low-melting fluoride electrolyte containing KF or K_3AlF_6 is used which needs a new type cathode replacing the current cathode material to meat the requirement of penetration resistance.

2.3 Cermet anodes

Cermet anodes consist of a ceramic phase and a metal phase. Cermets are attractive since they combine the advantages of ceramic (desirable for their chemical inertness) and metals (desirable for their high electrical conductivity and mechanical properties). The ceramic phase was general ferrite, such as $Ni_xFe_{3-x}O_4$, $Ni_yFe_{1-y}O$, $NiFe_2O_4$ + NiO, $ZnFe_2O_4$ + ZnO and $NiFe_{2y}Zn_zO_{(3y+x+z\pm\delta)}$. The metallic phase was metal or alloy, such as Cu, Cu-Ni, Ni-Fe, Cu-Cr and Cu-Ag. For example, the 17%Cu + 83% (51.7%NiO, 48.3%Fe_2O_3) material had a conductivity of nearly 90S/cm, and displayed excellent corrosion-resistant properties in laboratory cells[10,11]. Alcoa conducted, with the support by U. S. Department of Energy, considerable work about this kind of material and determine to research further. However, operational difficulties developed throughout the tests of 6kA pilot cell scale due to breakage of the anode conductor stems, cracking and breakage of the cermet anodes, unequal anode current distribution, and alumina muck build-up in the cell[11].

3 Cup-shaped Functionally Gradient $NiFe_2O_4$-based Cermet Inert Anodes

To solve above-mentioned problems of $NiFe_2O_4$ based cermet inert anodes, the research group of

Central South University, Changsha, China, put forward a research program according to the design idea of metal/ceramics functionally gradient composite materials and obtain the "cup-shaped" $NiFe_2O_4$-based cermet inert anode. The metal/ceramics composite anodes' composition, organization, and structure are changed continually so as to lessen the intermediate coefficient of thermal expansion, and further improve their chemical and physical properties, such as electrical conductivity and chemical inertness.

3.1 Optimization of material composite

To be used as the inert anodes for aluminum electrolysis, cermets should be fabricated without the spillage or asymmetric distribution of metallic phase, and they should be densified. With these objectives, the sintering performance of several kinds of $NiFe_2O_4$-based cermets containing metallic phase Cu, Ni and Cu-Ni were studied. The results showed that Ni-$NiFe_2O_4$ cermet with symmetrical distribution of metallic phase and the relative density of 96.85% could be obtained under controlled atmosphere.

The corrosion resistance of $NiFe_2O_4$-based cermet anode to Na_3AlF_6-Al_2O_3 melts is affected not only by the ceramic matrix but also by the metallic phase. Though the cermet including 17% Cu and 83% ceramic (51.7% NiO + 48.3% Fe_2O_3) was chosen and tested by Alcoa, the report failed to exhibit the substantial proof to explain why the excess amount of NiO in the cermet was 18%[12,13]. To determine the content of NiO in ceramic matrix, performances including mechanical property, electrical conductivity and corrosion resistance were studied. The results were listed Table 1.

Table 1 Effect of excess NiO content on the ceramic performance

Excess NiO content in ceramic phase/%	Relative density/%	Electrical conductivity at 1000℃/S · cm^{-1}	Solubility in melts/%	
			Ni	Fe
0	93.04	2.10	0.0085	0.0695
10	92.96	1.17	0.0101	0.0602
20	91.27	1.27	0.0106	0.0528
30	88.33	0.34	0.0112	0.048
40	84.96	0.23	0.0126	0.042

From the data listed in Table 1, the relative density and the electrical conductivity will recede with the content of NiO in ceramic phase increasing. From this, a conclusion could be drawn that the addition of NiO impedes the improvement of these performances. However, it must be noted that the relative density of material will affect its electrical conductivity under the same condition. If NiO-$NiFe_2O_4$ ceramics with the same relative density can be obtained, the electrical conductivity may increase. The effect of excess NiO content in cermaic phase on the electrical conductivity of Ni-NiO-$NiFe_2O_4$ cermets containing metal Ni 5% (Table 2) might prove this.

Table 2 Electrical conductivity of Ni-NiO-NiFe$_2$O$_4$ cermets at various temperature

Excess NiO content in ceramic phase/%	Relative dendsity/%	Electrical conductivity at various temperature/S · cm^{-1}							
		300℃	400℃	500℃	600℃	700℃	800℃	900℃	960℃
0	96.50	7.36	10.87	14.63	12.82	14.03	15.38	18.95	22.73
10	97.99	10.21	13.33	17.4	21.91	25.59	29.51	31.99	34.47
20	93.65	27.53	30.22	35.33	39.99	42.53	46.37	49.99	52.2
30	95.15	8.504	10.95	14.36	17.45	21.5	21.15	27.21	29.3
40	93.30	9.207	11.54	14.52	17.69	18.75	20.64	23.2	25.46

As a kind of inert anode materials in aluminum reduction, its corrosion resistance to Na$_3$AlF$_6$-Al$_2$O$_3$ melt is the most important. From Table 1, the solubility of Fe and Ni in the bath from NiO-NiFe$_2$O$_4$ ceramics are inversely related to each other. The solubility of Ni increases but Fe solubility and overall solubility of NiO-NiFe$_2$O$_4$ ceramics decreases with increasesing NiO content. The results (Table 3) from the electrolysis testing for Ni-NiO-NiFe$_2$O$_4$ cermet inert anodes containing metal Ni 5% also shown that the addition of NiO in Ni-NiO-NiFe$_2$O$_4$ cermets could improve the corrosion resistance, but amount added affected this property to a limited extent.

Generally, the material containing 17% metal Ni and the excess 10% of NiO in cermaic phase was selected and was further tested for larger current intensity.

Table 3 Steady-state concentration of impurities Ni and Fe in the bath during electrolysis

Excess NiO content in ceramic phase/%	Relative density/%	Steady-state concentration of impurities/%	
		Ni	Fe
0	96.15	0.0093	0.018
10	98.11	0.0085	0.0077
20	93.35	0.0094	0.0096
30	94.96	0.0099	0.0068
40	93.79	0.0080	0.0077

3.2 Preparation

To develop the functionally gradient large-size "cup-shaped" cermet inert anodes, the research group of Central South University introduced the preparation flow-sheet of 17%Ni-(10%NiO-NiFe$_2$O$_4$) cermet inert anodes as shown in Fig. 1.

During milling, organic solvent and dispersant were used rather than water because water may cause oxidation of the metal particles. The green samples were sintered at 1350℃ for 4 hours in an atmosphere of efficaciously controlled oxygen partial pressure and the desired cup-shaped 17% Ni-(10%NiO-NiFe$_2$O$_4$) cermet samples (Fig. 2a)

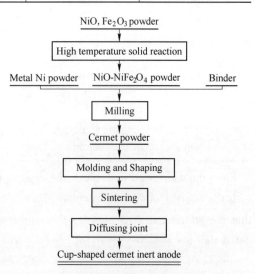

Fig. 1 A preparation flow-sheet of 17%Ni-(10%NiO-NiFe$_2$O$_4$) cermet inert anodes

could be obtained.

To attain the firm juncture between the cermet and metallic bar, some kind of transition material and a new joining techniques were adopted. Thus, the cup-shaped $NiFe_2O_4$ based cermet inert anode were prepared approximately 100mm in diameter and 120mm in height and a relative density of samples was above 93.00% (Fig. 2b). And this kind of jointing structure was proved to be firm and reliable by suspending it from the industrial cell of 160kA.

Fig. 2   Photos of large $NiFe_2O_4$ based cermet inert anode

3.3 Electrolysis test of "cup-shaped" cermet inert anode

To determine if electrolysis testing with the large-size anode yielded results similar to those from the laboratory test with smaller anodes, the operating performance of a 100mm diameter "cup-shaped" inert anode of cermet with 17% metal Ni and 83% ceramic ($10NiO$-$90NiFe_2O_4$) was evaluated in a laboratory electrolysis cell.

The results indicated that no major operational difficulties were encountered during electrolysis is which lasted for 101.5 hours and the inert anode exhibited good general performances (Fig. 3). The steady-state average concentration of impurity Ni in the bath was close to the solubility; however, Fe concentration was lower than its solubility. The content of the main contamination for aluminum produced was Ni 0.1288%, Fe 1.0074%. The corrosion rate of inert anode under electrolysis conditions based on the content of impurity Ni in metal aluminum recovered at cathode was approximately 8.51mm/y.

Fig. 3   A Cermet inert anode after testing for over 100 chours

During testing, an interesting phenomenon was observed that the gas bubbles, almost like a froth with 0.5~1.0mm diameter, evolved at the anode surface. The most value of voltage fluctuations is 49mV caused by the bubbles(Fig. 4), which is smaller than 135mV caused by the graphite anode[14].

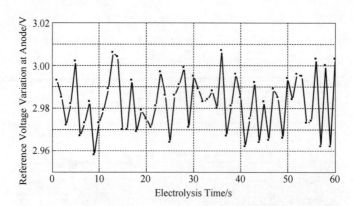

Fig. 4 Voltage fluctuation due to bubble formation on anode surface

3.4 Electrolysis test of anode group

Based on the achievement of previous works, six large-size cup-shaped cemet inert anodes with 17%Ni and 83% (10NiO-90NiFe$_2$O$_4$) were organized and were conducted during the electrolysis tesitng for 28 days (Fig. 5). Though the anodes didn't crack and invalidate, the joint between the cermet and metal bar was fixed during testing, the current efficiency was low at about 40% and the erosion rate of anodes to Na$_3$AlF$_6$-Al$_2$O$_3$ melt under the operational condition could not meet the requirements. How to improve the corrosion resistance and the current efficiency should be emphasized in the following studies.

Fig. 5 An anode group after electrolysis for 28 days

3.5 Distribution of impurity ions in bath during electrolysis

An interesting phenomenon, in which the concentrations of impurities in electrolyte samples taken from positions near the anode and near the cathode were very different, was observed after electrolysis test with NiFe$_2$O$_4$-based cermets. This phenomenon was also observed by other researchers[10,15]. To learn the phenomenon, a so-called upright electrolysis cell was designed and the electrolysis test was carried out at different current densities. The results approved the concentration gradient and uneven distribution between the anodic area and cathodic area of impurities Ni and Fe corroded into the bath from cermet inert anodes during electrolysis (Fig. 6), which might cause the severely uneven distributions of impurities in the frozen bath and the existance of a 1~2mm thick "skim layer" around the Al cathode cooled down after electrolysis tests as reported in the previous studies.

4 Conclusions

Study on the inert anodes has been ongoing for several decades, with greater attention paid to the materials of metals and cermets than others recently. However, the greatest challenge is to

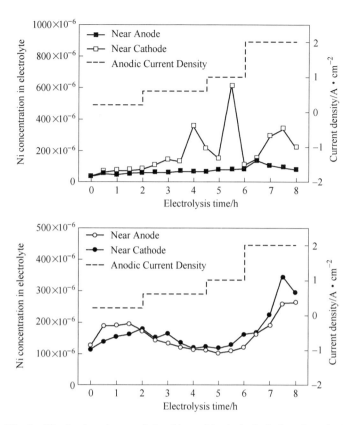

Fig. 6 Distributing characteristic of impurities in bath during electrolysis

improve its corrosion resistance to Na_3AlF_6-Al_2O_3 melts.

According to the design idea of metal/ceramics functionally gradient composite materials, the functionally gradient large size "cup-shaped" $NiFe_2O_4$ based cermet inert anodes were developed and exhibited good general performance. Some problems, including the electrical fueled joint of the cermet and metallic bar at high temperature, cracking and breakage of cermets during electrolysis were solved. However, more efforts such as improving the corrosion resistance for largescale testing and the electrical conductivity should be made to achieve success.

5 Acknowledgements

The authors are grateful for the financial support of the National Basic Research Program of China (2005CB623703) and the National High-Tech Research and Development Program of China(2008AA030503).

References

[1] J. Keniry. JOM,2001,53(5):43-47.
[2] B. J. Welch, et al. JOM,2001,53(2):13-18.
[3] R. P. Pawlek. Light Metals 2004,ed. A. T. Tabereaux (Warreudale,PA:TMS,2004):283-287.
[4] E. Olsen, et al. Journal of Applied Electrochemistry,1999,29(3):293-299.
[5] J. H. Yang, et al. Light Metals 1993,ed. S. K. Das (Warreudale,PA:TMS,1993):493-495.
[6] A. M. Vecchio-sadus, et al. Light Metals 1996,ed. W. Hale (Warreudale,PA:TMS,1996):259-265.

[7] J. H. Yang, et al. Light Metals 2006, ed. T. J. Galloway (Warreudale, PA: TMS, 2006): 421-424.

[8] T. R. Beck, et al. U. S. patent 6419812 (16 July, 2002).

[9] R. V. Kaenel, et al. Light Metals 2006, ed. T. J. Galloway (Warreudale, PA: TMS, 2006): 397-402.

[10] E. Olsen, et al. Light Metals 1996, ed. W. Hale (Warreudale, PA: TMS, 1996): 249-257.

[11] T. R. Alcom, et al. Light Metals 1993, ed. S. K. Das (Warreudale, PA: TMS, 1993): 433-443.

[12] G. R. Tarcy, Light Metals 1986, ed. R. E. Miller (Warrendale, P A: TMS, 1986): 309-320.

[13] E. Olsen, et al. Journal of Applied Electrochemistry, 1999, 29 (3): 301-311.

[14] R. D. Peterson, et al. Light Metals 1990, ed. C. M. Bickert (Warreudale, PA: TMS, 1990): 385-393.

[15] P. Chin, et al. Canadian Metallurgical Quarterly, 1996, 35(1): 61-68.

Alumina Solubility in Na_3AlF_6-K_3AlF_6-AlF_3 Molten Salt System Prospective for Aluminum Electrolysis at Lower Temperature*

Abstract The alumina solubility in the title system within the composition range of KR $\{m(K_3AlF_6)/[m(K_3AlF_6)+m(Na_3AlF_6)]\}$ 10%~50%, a ternary Na_3AlF_6-K_3AlF_6-AlF_3 molten system with 23%~29% (mass fraction) AlF_3 was investigated by measuring the mass loss of a rotating sintered corundum disc. And the following empirical equation was derived when superheat degree was no more than 60℃: $w(Al_2O_3)_{sat} = A \times (T/1000)^B$, where $A = -1.85774 + 26.754234 w(AlF_3)^{-0.3683} - 0.00783KR^{2.363} + 0.010266KR^{2.3048} + 0.7902w(AlF_3)^{0.00652}$, $B = 112.4625 - 53.2567w(AlF_3)^{0.4236} + 5.1079w(AlF_3)^{0.9241} + 0.01542w(AlF_3)^{1.3540}$. Considering both higher alumina solubility and not too high superheat degree are required, alumina solubility of different compositions at not the same temperature but the same superheat degree was studied, which will be more industrial helpful for selecting prospective compositions. The results show that the composition deserved to be further tested in lower temperature cells is 10%~30% KR and 23%~26% (mass fraction) AlF_3.

Key words Alumina solubility, Lower temperature electrolyte, Na_3AlF_6-K_3AlF_6-AlF_3 molten salt system, Aluminum electrolysis

1 Introduction

Interest in low melting baths has emerged again since last decades and the interest in inert anodes has become the re-search focus in aluminum reduction industry. On the one hand, lowering electrolysis temperature can be expected to reduce energy and carbon materials consumption and to prolong cells life. This is also the most important reason that researchers have been committed to lowering electrolysis temperature for aluminum reduction since Hall-Héroult method was patented. On the other hand, previous studies have shown that the application of low melting baths could provide a relative better service environment for inert anodes and the corrosion rate could be markedly alleviated[1,2].

Previous researches about low melting baths can be basically divided into NaF-AlF_3 system with low molar ratio(CR) and KF-AlF_3 system. However, weak alumina solubility is the main problem for NaF-AlF_3 system, declining from 10% at the cryolite composition to 3% at the eutectic[3]. It is hard to be applied in current cells due to deposition of alumina which will hinder the further operations. While system KF-AlF_3 has better alumina solubility and much wider range of low-temperature liquid compositions, the penetration of potassium to carbon lattice will result in the wear of carbon cathode[4]. However, it should be noted that the developmentin

---

* Copartner: Li Jie, Yuan Changfu, Tian Zhongliang, Wang Jiawei, Lai Yanqing. Reprinted from CHEM. RES. CHINESE UNIVERSITIES. 2012,28(1):142-146.

celldesignandcell-related materials make it possible that the operation at higher superheat degree and higher potassium content is allowed. Therefore, system $NaF-AlF_3$ with low CR added with bearable content of K_3AlF_6 is considered to be of some interest for aluminum electrolysis at lower temperatures.

Part of $NaF-KF-AlF_3$ system's phase diagram, as shown in Fig. 1, has been derived by Danielik et al.[5] and Barton et al.[6] through thermodynamic calculation and thermo-analysis method, respectively. In addition, Wang et al.[7] and Huang et al.[8]

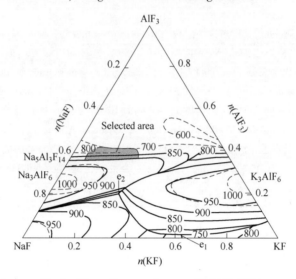

Fig. 1 Phase diagram of system $NaF-KF-AlF_3$

Solid line—tested and calculated by ref. [5]; dashed line—tested by ref. [6];
selected area—location of experimental points; e_1, e_2—eutectic point.

also tested the liquidus temperature and electrical conductivity of part of this system. However, alumina solubility which is recognized as one of the most important properties for aluminum electrolyte has rarely been investigated before except several compositions tested by Meng et al.[9] and Zhou et al.[10] using EDTA titration method to measure the solubility of α-Al_2O_3 powder which is still not enough to give a roadmap for industrial application.

In this work, our intention is to offer guidance for selecting low melting bath compositions in ternary system Na_3AlF_6-K_3AlF_6-AlF_3 from a perspective of alumina solubility. It is recognized that a frozen ledge must be maintained to prevent lining materials from direct attack of fluoride molten salt unless new container materials have been developed for freezeless operation, which means the electrolyte temperature must be kept rather close to its liquidus temperature. Therefore, it is more practically meaningful to evaluate alumina solubility of different compositions at different temperatures but the same superheatdegree.

Alumina solubility can be tested by visual methods[11,12], quenching techniques[13], disc-test method[3,14,15], EDTA titration method[9,10] and LECO oxygen analyzer method[16]. Alumina solubility in some systems like pure Na_3AlF_6 and Na_3AlF_6-AlF_3-additives has been systematically tested[3,15] and certain compositions of $KF-AlF_3$ have also been tested[16]. Higher alumina solubility will be obtained by the former two methods due to oversaturation, and lower alumina solubility by LECO method due to incomplete reduction of oxide. Therefore, simple and accurate disc-test method was chosen in this work.

2 Experimental

Based on the goal of electrolysis temperature below 920℃, a certain superheat degree should also be taken into account, and the liquidus temperature of investigated compositions should be in a range of about 750~900℃. Meanwhile, the content of potassium cryolite is not expected to be too high. Considering these reasons, the composition range of melts tested is: $m(K_3AlF_6)/[m(K_3AlF_6)+m(Na_3AlF_6)]$ (defined as KR in this paper) 10%~50% and AlF_3 23%~29% (without special explanation, % denotes mass fraction all through in this paper), as illustrated in Fig. 1.

The experimental set-up and procedure were based mainly on the work of Solheim et al.[3] and Skybakmoen et al.[14]. Alumina solubility was determined by measuring the mass loss of a rotating sintered corundum disc. The chemicals used were reagent potassium cryolite, reagent sodium cryolite and sublimed AlF_3 (>99.5%) and all the raw materials were dried before using to remove the moisture in vacuum at 120℃ for 24h. The corundum disc, containing a minimum of 99.8% $\alpha\text{-}Al_2O_3$ (purchased from Wulian New Ceramics Co., Ltd., Shanghai, China), was fastened to a rotating stainless steel shaft and submerged into the melt, as shown in Fig. 2. A total of 1200g of melt was kept in an argon atmosphere in a graphite crucible covered with a graphite lid. The liquidus temperature of $Na_3AlF_6\text{-}K_3AlF_6\text{-}AlF_3$ within the tested range was determined by the following empirical equation which had been derived by our group:

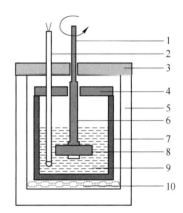

Fig. 2 Experimental arrangement of alumina solubility test

1—Stainless steel shaft; 2—TC(Pt/Pt10Rh); 3—Furnace lid; 4—Corundum lid; 5—Furnace; 6—Protective sleeve; 7—Graphite crucible; 8—Sintered corundum disc; 9—Bath; 10—Alumina powder.

$$T = 1003.5 + 0.081w(AlF_3)^{2.316} + 5.87KR^{0.657} - 0.024w(AlF_3)^{2.220}KR^{1.140} + 0.035w(AlF_3)^{2.170}KR^{1.084} \quad (1)$$

In this work, the experimental temperature was set based on the superheat degree plus the liquidus temperature of melt calculated from Eq. (1). The temperature was monitored during the experiment with a Pt/Pt10Rh thermocouple. The uncertainty of temperature was about ±1℃.

The test began when the disc of alumina was immerged into the melt. The disc was rotated at a speed of 340r/min. Having been rotated for enough time, the disc was lifted above the melt and cooled with the furnace. And then the disc was cleaned by leaching in a hot aqueous solution of $AlCl_3$ which can dissolve the remaining fluoride salts. The alumina saturation was then calculated from the mass loss of the disc.

3 Results and Discussion

3.1 Test of method

Since the dissolution rate increases when the agitation becomes fierce, the dissolution process is mostly considered to be controlled by mass transport through the electrolyte boundary layer adjacent to the disc surface. The necessary time to achieve equilibrium can be estimated by Levich equation[17]

$$t = -\frac{1.61Vv^{1/6}}{AD^{2/3}\omega^{1/2}}\ln(1 - w/w^*) \tag{2}$$

where V is the volume of the melt, v is the kinematic viscosity, A is the surface area of the disc, D is the diffusion coefficient of the alumina, ω is the angular velocity of the disc, and w and w^* are mass fractions of alumina in the bulk of the melt and at the disc surface, respectively. The following equation can be obtained upon simplification:

$$w = a(1 - e^{bt}) \tag{3}$$

where $a = w^*$ and $b = -[(1.61Vv^{1/6})/(AD^{2/3}\omega^{1/2})]$.

Alumina solubility data of the molten bath with 25% AlF_3 and 10% KR (liquidus temperature tested by Wang et al.[7] is 875℃) at three different superheat degrees can be well fitted by Eq. (3) (Fig. 3). The values of a and b in Eq. (3) were not expected to vary much without great change in experimental conditions. Based on these experimental results, a contact time of 6h between the disc and the melt was chosen for subsequent experiments in order to ensure saturation.

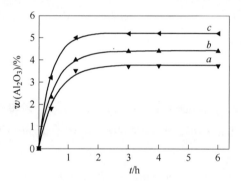

Fig. 3 Time-alumina solubility curves at different superheat degrees in melt $w(AlF_3) = 25\%, KR = 10\%$.
Superheat degree: a—20℃; b—40℃; c—60℃

The sources of error can result from these aspects as follows: (1) initial alumina concentration introduced by chemicals, moisture adsorption and so on; (2) incomplete removal of salt from corundum disc after testing; (3) total mass loss of melt during a run due to vapor loss and some penetration into the pores of the graphite crucible. The former two factors would cause experimental results to be lower than actual results, while the last would cause opposite effect. However, these influences were not considered significantly and may offset mutually. Furthermore, using a larger amount of melt can further reduce experimental error. And this is the reason that 1200 g of melt was adopted, which is more than 200 and 600 g adopted by Skybakmoen et al.[14] and Yang et al.[16], respectively.

In order to ensure the reliability of the results, the solubility of alumina in the melt containing 47.35%KF+52.65%AlF_3 at 700℃ was tested. The value obtained is 4.88%, which is very close to the results of 4.70% (obtained by Kryukovsky et al.[18]) and 5.04% (obtained by Yang et al.[16]). Duplicated runs agreed to an error of ±0.20% Al_2O_3, which is within an estimated uncertainty of ±0.30%.

3.2 Effect of melt composition under same superheat degree

It is well known that what kind of effect of each component added, like AlF_3 and K_3AlF_6, has on the alumina solubility under the same temperature from a lot of previous work. As discussed above, liquidus temperature, alumina solubility and superheat degree should be all taken into consideration. For example, it had been recognized that alumina solubility would rise by adding K_3AlF_6 into Na_3AlF_6-based system at the same temperature, but meanwhile, the liquidus temperature would decrease, too, which would result in undesirable higher superheat degree. So it might be more practical helpful to figure out the effect of melt composition on solubility under the same superheat degree than under the same temperature from an industrial perspective.

In consideration of this, the typical effect of melt composition on the alumina solubility under the same superheat degree is illustrated in Fig. 4.

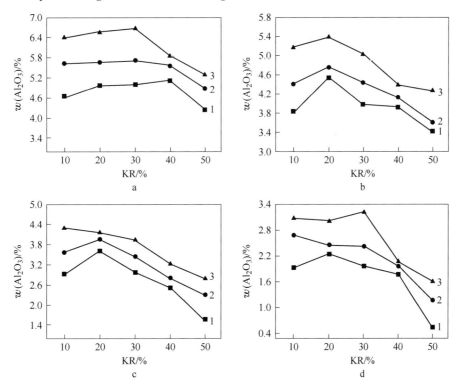

Fig. 4 Effect of KR on alumina solubility under each superheat degree with different AlF_3 mass fractions

a—$w(AlF_3) = 23\%$; b—$w(AlF_3) = 25\%$; c—$w(AlF_3) = 27\%$; d—$w(AlF_3) = 29\%$

Superheat degree: 1—20℃; 2—40℃; 3—60℃

The results indicate that (1) alumina solubility decreases with $w(AlF_3)$ increasing by horizontal comparison of figures, which is in accordance with previous work in Na_3AlF_6-AlF_3 and KF-AlF_3 systems; (2) under each superheat degree, with different mass fractions of AlF_3, the effect of KR on alumina solubility is not the same. With increase of KR from 10% to 50% under the same superheat degree and $w(AlF_3)$, the alumina solubility mostly appear to increase at first and then decrease, or sometimes increases all through except the case in which $w(AlF_3)$ is 29% and superheat degree is 60℃. This complication results from our unique method of plotting. This phenomena can be only approximately unders-tandable by realizing that the experimental temperature of each point is different, increasing in KR will cause decline of liquidustemperature[7], which means testing temperature would become lower because superheat degree is pre-determined. Then alumina solubility will decrease by this point[16]. However, increase in KR, meanwhile, has a beneficial effect on alumina dissolution because of the increase of $w(K_3AlF_6)$[16,19]. Then the trend of each line is decided by what effect taken advantage of; (3) basically, the content of AlF_3 has greater influence on alumina solubility than temperature on the premise that too high superheat degree is not allowed. When superheat degree is increased by as much as 40℃, alumina solubility increases by no more than 1.7%, while $w(AlF_3)$ is increased from 23% to 29%, alumina solubility decreases by about 3.0%.

3.3 Experimental data and empirical equation

The data (Table 1) were well fitted to the following equation within the composition ranges of 10% ~ 50% KR and 23% ~ 29% (mass fraction) AlF_3 under superheat degree no more than 60℃:

$$w(Al_2O_3)_{sat} = A \times (T/1000)^B \quad (4)$$

where $A = -1.85774 + 26.754234w(AlF_3)^{-0.3683} - 0.00783KR^{2.363} + 0.010266KR^{2.3048} + 0.7902w(AlF_3)^{0.00652}$

$B = 112.4625 - 53.2567w(AlF_3)^{0.4236} + 5.1079w(AlF_3)^{0.9241} + 0.01542w(AlF_3)^{1.3540}$

Correlation coefficient of this empirical equation was 0.994 and maximum deviation was found to be 0.47% (Table 1). A wider average deviation at bigger superheat degree was found, which resulted from the deterioration in stability of experimental data as superheat degree decreases.

Table 1 Experimental (Expt.) and calculated (Calcd.) alumina solubility (%, mass fraction) of investigated compositions of Na_3AlF_6-K_3AlF_6-AlF_3 salt system and their differences (Δ) at different superheat degree

$w(AlF_3)$/%	KR/%	T^*/℃	20℃			40℃			60℃		
			Expt.	Calcd.	Δ	Expt.	Calcd.	Δ	Expt.	Calcd.	Δ
23	10	895	4.64	4.84	0.20	5.64	5.41	-0.23	6.4	6.03	-0.37
23	20	888	4.96	5.06	0.10	5.67	5.66	-0.01	6.57	6.31	-0.26
23	30	871	5.00	5.08	0.08	5.73	5.70	-0.03	6.67	6.36	-0.31
23	40	846	5.12	4.83	-0.29	5.57	5.43	-0.14	5.87	6.09	0.22
23	50	815	4.24	4.32	0.08	4.87	4.88	0.01	5.29	5.49	0.20
25	10	875	3.82	4.02	0.20	4.40	4.54	0.14	5.17	5.12	-0.05
25	20	868	4.55	4.21	-0.34	4.78	4.75	-0.03	5.38	5.35	-0.03
25	30	847	3.97	4.10	0.13	4.43	4.64	0.21	5.04	5.26	0.22
25	40	817	3.94	3.73	-0.21	4.12	4.25	0.13	4.39	4.83	0.44
25	50	780	3.42	3.15	-0.27	3.60	3.61	0.01	4.26	4.11	-0.15
27	10	853	2.91	3.13	0.22	3.58	3.60	0.02	4.29	4.12	-0.17

Continued Table 1

$w(AlF_3)$ /%	KR/%	T^*/℃	20℃			40℃			60℃		
			Expt.	Calcd.	Δ	Expt.	Calcd.	Δ	Expt.	Calcd.	Δ
27	20	845	3.62	3.24	-0.38	3.95	3.73	-0.22	4.14	4.26	0.12
27	30	820	2.97	3.02	0.05	3.46	3.49	0.03	3.94	4.02	0.08
27	40	784	2.52	2.56	0.04	2.79	2.97	0.18	3.23	3.45	0.22
27	50	739	1.53	1.96	0.43	2.29	2.29	0.00	2.78	2.68	-0.1
29	10	828	1.93	2.21	0.28	2.69	2.60	-0.09	3.08	3.05	-0.03
29	20	818	2.25	2.24	-0.01	2.46	2.64	0.18	3.01	3.10	0.09
29	30	789	1.97	1.96	-0.01	2.44	2.32	-0.12	3.22	2.75	-0.47
29	40	747	1.78	1.50	-0.28	1.97	1.79	-0.18	2.07	2.13	0.06
29	50	692	0.53	0.97	0.44	1.18	1.18	0.00	1.61	1.43	-0.18

* Liquidus temperature calculated by Eq. (1).

3.4 Equal alumina solubility map

By combining Eq. (1) with Eq. (4), equal alumina solubility map as well as isothermal diagram can be drawn in together as shown in Fig. 5. An alumina mass fraction of 2% ~ 4% is kept in most modern Hall-Héroult cells[20]. In consideration of the "state of the art" of the feeding system and automatic control technology, only these melt compositions with a minimum of 4% solubility are of some interest for industrial aluminum reduction at lower temperatures to avoid problems produced by insufficient alumina solubility.

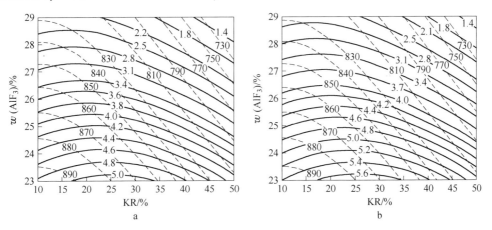

Fig. 5 Isothermal curves (dashed line) and equivalent alumina solubility
curves (solid line) calculated *via* Eq. (1) and Eq. (4)
Superheat degree: a—20℃; b—40℃

Therefore, based on the required alumina solubility and electrolysis temperature, it is recommended that the composition deserved to be further tested in cells is 10% ~ 30% KR and 23% ~ 26% AlF$_3$.

4 Conclusions

Based on the research objective of electrowinning aluminum at lower temperatures than those of nowadays reduction cells, published ternary phase diagram of Na$_3$AlF$_6$-K$_3$AlF$_6$-AlF$_3$ molten salt system, and the pros and cons of potassium cryolite, the alumina solubility within a composition range of 10% ~ 50% KR and 23% ~ 29% AlF_3, was selectively investigated by measuring the mass loss of a rotating sintered corundum disc submerged in the melt. The results are as follows: (1) evaluation of alumina solubility of different compositions under the same superheat degree rather than the same temperature is a more fare way from a industrial perspective. However, this will also make it a little more complicated to be explained because testing temperature changes with composition change. Under the same superheat degree, alumina solubility will decrease with the increase of AlF_3, but will not increase all through with KR increasing, often reaches a peak in between, around 10% ~ 30%; (2) all the experimental data are well fitted to an empirical equation with a correlation coefficient of 0.994. Combined with empirical equation describing the relation between liquidus temperature and composition, equal alumina solubility map and isothermal diagram have been drawn in together, which can somehow guide further testing work of prospective compositions of this salt system in cells at lower operational temperatures. It is recommended, based on this work, that the composition deserved to be further tested in cells can be focused on 10% ~ 30% KR and 23% ~ 26% AlF_3.

References

[1] Thonstad J., Rolseth S., Trans. Inst. Min. Metall., Sect. C, 2005, 114, 188.

[2] Wang J. W., Lai Y. Q, Tian Z. L., Li J., Liu Y. X., Proc. TMS-AIME, Light Met., 2007, 525.

[3] Solheim A., Rolseth S., Skybakmoen E., Støen L., Sterten Å., Støre T., Proc. TMS-AIME, Light Met., 1995, 452.

[4] Grjotheim K., Welch B. J., Aluminum Smelter Technology—A Pure and Applied Approach, 2nd Ed., Aluminum-Verlag, Düsseldorf, 1988, 52.

[5] Danielik V., Gabèová J., J. Therm. Anal. Calorim., 2004, 76, 763.

[6] Barton C. J., Bratcher L. M., Grimes W. R.; Ed.: Thoma R. E., Phase Diagrams of Nuclear Reactor Materials., US AT. Energy Comm., Oak Ridge, 1959, 32.

[7] Wang J. W., Lai Y. Q., Tian Z. L., Li J., Liu Y. X., Proc. TMS-AIME, Light Met., 2008, 513.

[8] Huang Y. G., Lai Y. Q., Tian Z. L., Li J., Liu Y. X., Li Q., Proc. TMS-AIME, Light Met., 2008, 519.

[9] Meng Q., Kan S., Lu S., Ding H., Zhang X., Chinese Journal of Rare Metals, 2010, 34, 905.

[10] Zhou C., Ma S., Li G., Sheng J., Non Ferrous Metals, 1998, 50, 81.

[11] Fenerty A., Hollingshead E. A., J. Electrochem. Soc., 1960, 107, 993.

[12] Xiao L., George L., Wills S. F., Valerie A., Proc. TMS-AIME, Light Met., 1994, 359.

[13] Foster P. A. Jr., J. Am. Ceram. Soc., 1975, 58, 288.

[14] Skybakmoen E., Solheim A., Sterten Å., Metall. Mater. Trans. B, 1997, 28B, 82.

[15] Skybakmoen E., Solheim A., Sterten Å., Proc. TMS-AIME, Light Met., 1990, 317.

[16] Yang J., Graczyk D. G., Wunsch C., Hryn J. N., Proc. TMS-AIME, Light Met., 2007, 537.

[17] Levich V. G., Physicochemical Hydrodynamics, Prentice-Hall, Inc., Englewood Cliffs, New Jersey, 1962, 172.

[18] Kryukovsky V. A., Frolov A. V., Tkatcheva O. Y., Redkin A. A., Zaikov Y. P., Khokhlov V. A., Apisarov A. P., Proc. TMS-AIME, Light Met., 2006, 411.

[19] Robert E., Oslen J. E., Danek V., Tixon E., Ostvold T., Gilbert B., J. Phys. Chem. B, 1997, 101, 9447.

[20] Grjotheim K., Kvande H., Introduction to Aluminum Electrolysis. Understanding the Hall-Héroult Process, 2nd Ed., Aluminum-Verlag, Düsseldorf, 1993, 48.

Further Research of "Cup-shaped" NiFe$_2$O$_4$ Based Cermet Inert Anodes for Aluminum Electrolysis*

Abstract The new aluminum electrolysis technology based on inert electrode has been received much concern for several decades because of the environment and energy advantages. The key to realize this technique is the inert anode. This paper presents the recent developments of China in NiFe$_2$O$_4$-based cermet inert anode, which includes the optimization of material performance, the joint between the cermet inert anode and metallic bar, as well as the results of 20kA pilot testing for a large-size inert anode group. At the same time, the problems for NiFe$_2$O$_4$-based cermet inert anode faced are also discussed.

1 Introduction

Though great progresses have been made in the primary aluminum production, the industry still is confronted with great challenge as emission reductions and energy savings. The application of inert anode(non-consumable) instead of the traditional consumable carbon anode will avoid the generation of the greenhouse gas CO$_2$ and also eliminate perfluorocarbon(PFC) by products and other polluting emissions such as PAHs[1]. It has become an urgent goal in Hall-Héroult electrolysis cells for the production of aluminum since the middle of 1980s, in views of technical, commercial and environmental perspectives[2,3].

To be successfully implemented, an inert anode for aluminum electrolysis must include the following requirements[4]: (ⅰ) strong corrosion resistance and low corrosion rate, (ⅱ) good electronic conductivity, (ⅲ) not to contaminate the metal aluminum produced to any significant degree, (ⅳ) thermally stable up to electrolysis temperature as well as exhibit adequate resistance to thermal shock and (Ⅴ) relatively low cost and ready availability. Despite intensive research efforts, no materials have yet been found that meet all these strict requirements.

After many years' study, the materials of inert anode for aluminum electrolysis were focused on alloy[5,6] and cermet[1,7]. Among them, NiFe$_2$O$_4$ based cermets is one of the most promising materials as inert anode for aluminum electrolysis[7]. This paper presents the recent developments in NiFe$_2$O$_4$-based cermet inert anode since 2009 in China, which includes the optimization of material performance, the joint between the cermet anode and metallic bar, as well as the results of 20kA pilot testing for a large size inert anode group. Of course, all these progresses are based on previous works[7,8] including the preparation of material and the electrolysis testing and are obtained under the fund of the National High-Tech Research and Development Program of China.

---

* Copartner: Zhongliang Tian, Yanqing Lai, Zhiyou Li, Dengpeng Chai, Jie Li. Reprinted from JOM. 66(11):2229-2234.

2 Previous Progresses and Problems

Since July 2001, much research work has been done in $NiFe_2O_4$-based cermet inert anode for aluminum electrolysis, and many progresses were also achieved, which mainly included: (i) the material composite was optimized and the inert anode with excess 10% of NiO in ceramic phase was determined for further researching, (ii) a functionally gradient large-size "cup-shaped" $NiFe_2O_4$-based cermet inert anode was designed and its preparation flow-sheet was also founded, (iii) the results from 4kA scale-testing for 17%Ni/($NiFe_2O_4$-10NiO) cermet inert anode shown that the anodes didn't crack and invalidate, the joint between the cermet and metal bar was fixedness during testing.

Despite of these progresses, the corrosion rate of anodes to Na_3AlF_6-Al_2O_3 melt under the operational condition is too fast to meet the requirements, and the current efficiency of the cell is as low as about 40%. The density and mechanical performance of the large-size $NiFe_2O_4$-based cermet inert anode must be promoted. The joint configuration between the cermet anode and metal bar also needs improving to meet the requirement of good stability while it is energized.

In order to solve these problems above-mentioned, the material composite and the preparing technique were optimized furthermore to improve its comprehensive performance. Also, the low temperature electrolyte of Na_3AlF_6-K_3AlF_6-AlF_3 was developed to better its service environment. Based on these achievements, a 20kA scale electrolysis testing was conducted for over 100 days.

3 Promotion of Cup-shaped $NiFe_2O_4$-based Cermet Inert Anode

3.1 Optimization of $NiFe_2O_4$-based cermet inert anode

Based on previous works, $NiFe_2O_4$-10NiO was selected as ceramic phase of inert anode. The effect of sintering atmosphere and pressure on the relative density of ceramic matrix $NiFe_2O_4$-10NiO was studied, and the appropriate technology was determined that nitrogen atmosphere should be adopted at the pressure of 0.3~0.7atm, and the range which the partial pressure of oxygen should be controlled is between 50×10^{-6} and 200×10^{-6}. To enhance its conductivity and mechanical property, the powder of Cu-Ni alloy, which has coating structure, was used to replace Cu powder or Ni powder or the powder of them prepared by the mechanical mixing. At the same time, some rare earth oxides such as Yb_2O_3、Y_2O_3 and CeO_2 were added to strengthen the grain boundary structure and improve the density of $NiFe_2O_4$-based cermet inert anode.

Thus, not only the density of $NiFe_2O_4$-based cermet inert anode was promoted, but also the penetration phenomenon of the electrolyte was not reduced enormously after electrolysis testing in Na_3AlF_6-Al_2O_3 melt at 960℃ and its performance of anti-corrosion was strengthened. In addition, the selective dissolution of metal phase was restrained because of the adoption of the powder of Cu-Ni alloy coating structure and rare earth oxides.

3.2 Preparation of large-size inert anode

Through the spray drying process of raw material powder, near net shaping of inert anode blocks, degreasing and sintering process of high density inert anode, a small batch of cermet inert anode was completed and the cup-shaped $NiFe_2O_4$-based cermet inert anode were prepared, which its size was approximately 110mm in diameter and 150mm in height (Fig. 1a) and the metal phase Cu-Ni distributed uniformly in the ceramic phase $NiFe_2O_4$-10NiO (Fig. 1b). The relative density of anodes is above 97.00%, the electrical conductivity at 900℃ is more than 60s/cm, the density distribution difference of each other is less than 2% (Table 1).

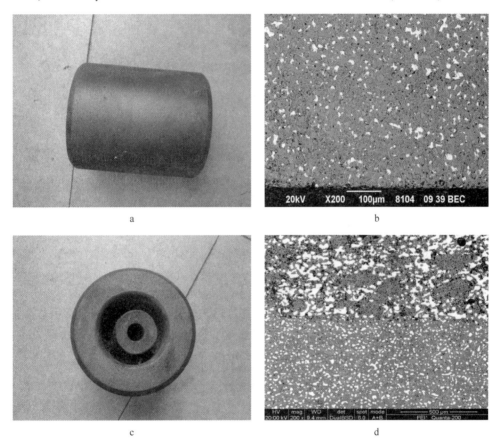

Fig. 1 Large-size $NiFe_2O_4$-based cermet inert anode

a—Photo of cermet inert anode sintered; b—SEM of inert anode;

c—Photo of inert anode jointed; d—SEM of junction between anode and transition material

Table 1 Comprehensive performance of large-size $NiFe_2O_4$-based cermet inert anode

Relative density/%	>97
Bending strength at room temperature/MPa	159~176
Residual flexural strength at 300℃ temperature difference/MPa	95~120
Electrical conductivity/S·cm^{-1} (at 900℃)	>60
Density consistency (density distribution difference)/%	≤2
Rate of finished products (degreasing-sintering)/%	>92

According to the design idea of metal/ceramics functionally gradient composite materials, the electrical jointing structure between $NiFe_2O_4$-based cermet and metallic bar was improved, and some kind of transition material and a new jointing technique were adopted. The diffusion of the substrate each other improves the adhesion strength of the interface, and the strengthen of joint is over 25MPa, its electrical conductivity at 900℃ is over 150S/cm. This kind of jointing structure was proved to be firm and reliable by suspending it from the industrial cell of 160kA and by testing in the pilot cell later, and can meet the requires. From its SEM(Fig. 1d), the bonding interface is close connected.

4 Low Temperature Electrolysis of $NiFe_2O_4$-based Cermet Inert Anode

4.1 Low temperature electrolyte of Na_3AlF_6-K_3AlF_6-AlF_3

From previous works, a relative better service environment for $NiFe_2O_4$-based cermet inert anode could be provided and its corrosion rate could be markedly alleviated if the electrolyte with low liquidus temperature and high concentration of Al_2O_3 is adopted during electrolysis. It is well known that the alumina Solubility is more in Na_3AlF_6-K_3AlF_6-AlF_3 Molten than in Na_3AlF_6-AlF_3 Molten[9,10].

To offer basic data for the development of low temperature electrolyte for aluminum electrolysis based on Na_3AlF_6-K_3AlF_6-AlF_3 system, the physical chemistry performances were measured, including the first eutectic temperature, the solubility and dissolving velocity of Al_2O_3, the electrical conductivity of Na_3AlF_6-K_3AlF_6-AlF_3 system within the composition range of K_3AlF_6/(K_3AlF_6+Na_3AlF_6)0 wt.% ~ 50 wt.% and AlF_3 0 wt.% ~ 30 wt.%. These basic data were obtained, and the relationship among the liquids temperature, electrical conductivity and saturated concentration of Al_2O_3 with the electrolyte composition was founded.

4.2 Electrolysis testing of $NiFe_2O_4$-based cermet inert anode

The effects of the content of K_3AlF_6 and AlF_3, and overheating temperature of Na_3AlF_6-K_3AlF_6-AlF_3 low temperature electrolyte on the corrosion rate of $NiFe_2O_4$-based cermet inert anode were investigated. Compared with the traditional Na_3AlF_6-AlF_3 electrolyte at 960℃, the use of low temperature bath of Na_3AlF_6-K_3AlF_6-AlF_3 system can improve the service environment and reduce the corrosion rate of $NiFe_2O_4$-based cermet inert anode. For example, if the effect of potassium on the cathode expansion is not considered, the corrosion rate of Cu/($NiFe_2O_4$-10NiO) cermet inert anode in low temperature electrolyte containing much K_3AlF_6 is about 0.85cm/a, which is deduced from the lab electrolysis tests. However, when the bath is replaced by Na_3AlF_6-AlF_3 electrolyte at 960℃, the corresponding corrosion rate is 5.27cm/a.

Upon the above studies, the composition of Na_3AlF_6-K_3AlF_6-AlF_3 system appropriating for $NiFe_2O_4$-based cermet inert anode was optimized. It was suggested for 20kA pilot test to adopt the suitable composition of electrolyte: K_3AlF_6/(K_3AlF_6+Na_3AlF_6)18 wt.% ~ 20 wt.%, superfluous AlF_3 concentration 24 wt.% ~ 26 wt.%, Al_2O_3 concentration >4.0 wt.%. The results from 500 hours electrolysis testing of (Cu-Ni)/(NiFe$_2$O$_4$-10NiO) cermet inert anode at the anode current density of 0.95A/cm^2 show that the corrosion rate of the anode is less than 1.5cm/a deduced from the electrolysis test, when the electrolyte containing the above-mentioned composition is adopted (Fig. 2).

Fig. 2 (Cu-Ni)/(NiFe$_2$O$_4$-10NiO) cermet inert anode after 500h testing
a—Photo of inert anode; b—SEM of the anode bottom

5 20kA Scale Electrolysis Testing

5.1 Cell design

A theoretical model cell based NiFe$_2$O$_4$-based cermet inert anode was created (Fig. 3), and its physical fields including electric, magnetic, thermal, flow and stress field were calculated on the platform CFX10.0 by the finite element method. The effect of inert anode configuration on these physical fields was analyzed to get a cell structure with better physical field distribution. The flow and fluctuation of the melt was also studied on the integrity of the melt movement. Additionally, the simulation model of inert anode gas movement and its stirred electrolyte flow was created to study the influence of technical and anode structure parameters.

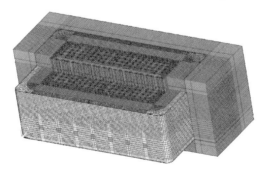

Fig. 3 Electromagnetic finite element model of cell

According to the results of 20kA pilot cell physical field simulation and taking into account the simplicity of replacement of inert anode damaged, the cell structure was optimized further (Fig. 4). The size of cell was 4260mm×1910mm, the side of which was made of SiC, and the cathode was set for 6 groups with double-cathode steel sticks.

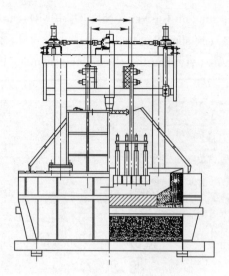

Fig. 4 Side view of cell

5.2 20kA Pilot testing

To start up the cell, carbon anodes and the mixture of petroleum coke grain were utilized as heating source in the pilot cell. After operating some time stably, carbon anodes were replaced by inert anode groups pre-heated, which were organized by sixteen large-size cup-shaped $NiFe_2O_4$-based cermet inert anodes and protected by a coating material.

When the replace of $NiFe_2O_4$-based cermet inert anode groups was finished, the concentration of Al_2O_3 in the bath was close to 4.0%, the cell voltage is about 7.20V. After a while, the cell voltage increased quickly to 8.03V, and then decreased to around 7.38V by using conventional effect vanishing bar and adding some Al_2O_3 to cell. These phenomena, which includes the voltage alteration, disorder and peak appearing in anode current distribution, is very similar with that of the anode effect which occurs usually in the traditional cell with carbon anode.

To avoid this phenomenon appear again, the concentration of Al_2O_3 in the electrolyte was increased and it was about 5.0 wt.% ~ 6.0 wt.%, the cell voltage changed in the range of 7.4 ~ 7.6V, and the cell current was 18.5kA. Though the cell has worked over 100 days, the current distribution was very uneven in the anode groups during inert anode working (Fig. 5). Some of the inert anodes were dropped into the cell during electrolysis (Fig. 6). The external morphology of the dropped inert anode at 900℃ remained the same as that of a new one. It must be mentioned that the phenomenon of inert anode dropping is caused not by the jointing structure between anode and metallic bar, but not by the fragile of electric conduct rod. And this confirmed further that the firm and reliable of jointing structure between $NiFe_2O_4$-based cermet anode and metallic bar.

And from the photos after pilot (Fig. 7a), $NiFe_2O_4$-based cermet inert anodes exhibit good anti-corrosion to Na_3AlF_6-K_3AlF_6-AlF_3 melt and the expanding phenomenon disappeared.

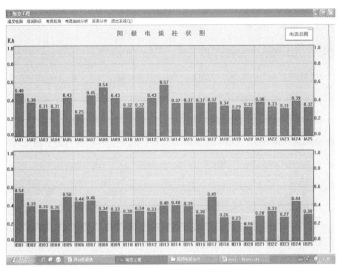

Fig. 5 Real-time monitoring of anode current distribution in operation

Fig. 6 Profile of inert anode dropped during electrolysis

Judging from the bottom of anode fracture section, it included the interface layer, the interim layer and the compact layer. The thickness of interim layer became gradually thinner from the centre of bottom to the side anode section (Fig. 7b).

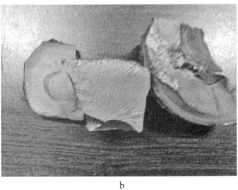

a b

Fig. 7 Photo inert anodes after electrolysis

a—Inert anode groups after electrolysis; b—Profile of inert anode section

6 Conclusion

The results from 20kA scale pilot over 100 days shown that the comprehensive performance of $NiFe_2O_4$-based cermet inert anodes was improved by optimizing the material composite and the preparing technique. At the same time, the low temperature electrolyte of Na_3AlF_6-K_3AlF_6-AlF_3 was developed to better its service environment.

However, uneven anode current distribution could lead to some part of anode current overload, and the anode electric conduct rod was so fragile that the breakage was concentrated at the junction between the electric conduct rod and the anode bar. Of course, the corrosion resistance, thermal shock resistance of cermet inert anode should be improved further.

7 Acknowledgements

The authors are grateful for the financial support of the state natural science fund(51222403) and the National High-Tech Research and Development Program of China(2008AA030503).

References

[1] R. P. Pawlek. Inert Anodes: an Update[A]. Light Metals 2014, ed. J. Grandfield (Hoboken, NJ: TMS, 2014), 1309-1313.

[2] J. D. Weyand. Manufacturing Processes Used for the Production of Inert Anodes[A]. Light Metals 1986, ed. R. E. Miller(Warreudale PA: TMS, 1986), 321-339.

[3] B. J. Welch. Inert Anodes-the Status of the Materials Science, the Opportunities They Present and the Challenges that Need Resolving before Commercial Implementation[A]. Light Metals 2009, ed. G. Bearne (Warreudale PA: TMS, 2009), 971-978.

[4] E. Olsen, J. Thonstad. Nickel ferrite as inert anodes in aluminium electrolysis: Part I Material fabrication and preliminary testing[J]. Journal of Applied Electrochemistry, 1999, 29(3): 293-299.

[5] S. Helle, M. Pedron, B. Assouli, B. Davis, D. Guay, L. Roué. Structure and high-temperature oxidation behaviour of Cu-Ni-Fe alloys prepared by high-energy ball milling for application as inert anodes in aluminium electrolysis[J]. Corrosion Science, 2010, 52: 3348-3355.

[6] I. Gallino, M. E. Kassner, R. Busch. Oxidation and Corrosion of Highly Alloyed Cu-Fe-Ni as Inert Anode Material for Aluminum Electrowinning in As-Cast and Homogenized Conditions[J]. Corrosion Science, 2012, 63, 293-303.

[7] Z. L. Tian, Y. Q. Lai, J. Li, Z. Y. Li, K. C. Zhou, Y. X. Liu. Cup-Shaped Functionally Gradient $NiFe_2O_4$ Based Cermet Inert Anode for Aluminum Reduction[J]. JOM, 2009, 61(5): 34-38.

[8] W. X. Li, G. Zhang, J. Li, Y. Q. Lai. $NiFe_2O_4$-Based Cermet Inert Anodes for Aluminum Electrolysis[J]. JOM, 2009, 61(5): 39-43.

[9] E. Robert, J. E. Olsen, V. Danek, E. Tixon, T. Østvold, B. Gilbert. Structure and Thermodynamics of Alkali Fluoride-Aluminum Fluoride-Alumina Melts, Vapour Pressure, Solubility, and Raman Spectroscopic Studies[J]. J. Phys. Chem. B, 1997, 101: 9447-9457.

[10] J. H. Yang, J. N. Hryn, B. R. Davis, A. Roy, G. K. Krumdick, J. A. Pomykala Jr.. New Opportunities for Aluminum Electrolysis with Metal Anodes in a Low Temperature Electrolyte System[A]. Light Metals 2004, ed. A. T. Tabereaux(Charlotte, NC: TMS, 2004), 321-326.

Effect of An Alumina Layer at the Electrolyte/Aluminum Interface: A Laboratory Study*

Abstract Alumina particles can accumulate at the interface between liquid aluminum and cryolite-alumina melts, due to the high interfacial tension (450mN/m). Theoretical calculations indicate that alumina spheres of sizes up to 7.5mm diameter can rest at the interface. Alumina which was carefully added to the system, was found to accumulate at the interface, as evidenced by chemical analysis and by microscopy.

The rate of oxidation of dissolved metal when CO_2 was passed over the melt, decreased strongly when excess alumina was added. Correspondingly, current efficiency of a laboratory electrolytic cell increased in the presence of an alumina layer at the interface. The concentration of dissolved metal in the bulk of the melt decreased, and a concentration gradient was set up within the alumina layer. Evidently, mass transfer of dissolved metal was impeded by the presence of the layer.

Settling of alumina at the bath/metal interface probably plays a part in the mechanism of alumina dissolution in commercial aluminum cells.

1 Introduction

Alumina is fed to aluminum electrolysis cells in the form of a fine powder. It was shown previously[1] that only a minor proportion of the alumina feed dissolves immediately after an addition has been made. Part of the undissolved alumina will settle in the cell. Since the density of alumina ($\sim 3.75 g/cm^3$) is far higher than that of the electrolyte bath ($\sim 2.05 g/cm^3$) and of the metal ($\sim 2.30 g/cm^3$) the alumina will have a tendency to sink to the bottom of the cell and form a socalled bottom sludge underneath the metal pad. However, due to the strong interfacial tension between metal and bath ($\sim 450 mN/m$) alumina particles which sink through the bath, may come to rest at the metal/bath interface and remain there until they dissolve.

Schadinger[2] observed that when a large excess of alumina was added to a crucible containing cryolite and aluminum, the alumina accumulated above the metal. The melt would then become transparent as opposed to the foggy appearance in the absence of the alumina layer. The metal reoxidation reaction with CO_2 did practically stop when excess alumina was added. The alumina layer then inhibited the dissolution of metal into the melt.

Kent[3] suggested that a thin layer of alumina sludge could form on the metal surface in commercial cells. This layer would make the interface be very still and probably inhibit the dissolution of aluminum into the bath, thus having a beneficial effect on the current efficiency.

---

* Copartner: Jomar Thonstad. Reprinted from Light Metals, 1981: 303-312.

Kozmin et al[4] established by chemical analysis that excess alumina added to a laboratory crucible accumulated near the metal surface. The alumina content in this region was around 25% (by weight). Samples taken from commercial cells showed in some cases an enrichment of alumina near the metal surface, e. g. ~16% Al_2O_3 near the metal as against 4% Al_2O_3 in the bulk.

Accumulation of undissolved alumina at the metal/electrolyte interface is of interest not only with regard to current efficiency, but even more so in connection with alumina feeding. Compared to alumina present in the bottom sludge, an "alumina sludge" floating on top of the metal will be in a far more favorable position for dissolution in the bath.

The purpose of the present work was to examine more closely the conditions for formation of an alumina layer at the metal/bath interface and to study its effect on the rate of reoxidation of metal and on the current efficiency.

2 Maximum Particle Size at the Interface

From everyday life it is well known that small objects with higher density than water can float on water, due to the surface tension. Likewise, in a system with two liquid phases, heavy particles can rest at the interface between the two liquids. When a solid particle with a density higher than those of either liquid sinks through the upper liquid towards the interface, the following two things can happen:

(1) The interfacial forces balance the gravitational force and any other force (dynamic), and the particle settles at the interface.

(2) If the gravitational force is the stronger, the particle will make its way through the interface and sink to the bottom.

Theoretical equations have been derived[5] which relate particle size (spheres) to the densities and interfacial tensions for the three phases involved. For the present system data are scarce concerning the interfacial tension between solid alumina and liquid cryolite. therefore, a simplified treatment presented by Maru et al[6] was adopted, whereby the interfacial tensions between the solid and the two liquids are set equal and then neglected.

The shape of the interface with a spherical particle resting at it is show in Fig. 1.

The vertical component of the interfacial force (F) acting upwards will be

$$F = 2\pi x_c \sigma_{23} \sin\phi_c \qquad (1)$$

Where σ_{23} is the interfacial tensions between the two liquids. The meaning of the other symbols are apparent from Fig. 1. The gravitational force can be calculated when density differences are know, taking into account the fact that the sphere rests partly below the level of the lower liquid.

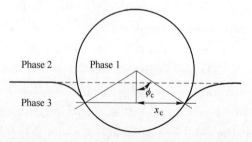

Fig. 1 A spherical particle (phase 1 (alumina)) resting at the interface between two liquids (phase 2 (cryolite), phase 3 (alumina)). Radii of contact-x_c, angle-ϕ_c

The resulting equations are complex. However, Maru et al[6] have show that a simplified equation can be derived to find the critical radii of the sphere (R_{crit}), i. e. the largest size that can rest at the interface,

$$R_{crit} = \frac{1.27}{\sqrt{\frac{\Delta\rho_{23} \cdot g}{\sigma_{23}}\left(\frac{\Delta\rho_{12}}{\Delta\rho_{31}} - 1\right)}} \quad (2)$$

where $\Delta\rho$ symbolizes the density differences between the phases indicated (see Fig. 1). By introducing relevant data for the cryolite-alumina(sat) system at 1000℃ [7,8], we obtain,

$$R_{crit} = 7.5mm$$

It can be shown that the magnitude of R will increase somewhat with decreasing alumina density and by AlF_3, CaF_2 or MgF_2 additions to the melt[9]. The real value of R_{crit} will be lowered when there are waves at the interface and by the impact of a moving particle. Nevertheless, considering that the particle size of commercial alumina is below 0.15mm, it is clear that single alumina particles will tend to settle at the metal-bath interface. Only various kinds of agglomerates of alumina and pieces of the top crust are able to penetrate the interface.

This conclusion is of particular interest in connection with alumina feeding systems such as point feeders, which apply frequent feedings without breaking much of the top crust. The alumina dissolution immediately after feeding may be limited by heat transfer[10]. It is likely that part of the alumina added will sink towards the metal surface. Insofar no lumps and agglomerates are formed, some of the alumina may settle at the interface and be carried away to be dissolved in other parts of the cell.

3 Evidence of Alumina Layer

Excess alumina was added batch wise to crucibles containing aluminum and cryolite-alumina melt at 1000℃. The melt was usually stirred and then left to settle for some time before cooling. The length of this period(hours) had no detectable influence on the results. The amount of alumina added was varied.

The crucibles were cut longitudinally and the area above the metal was examined. If the melt had been exposed to oxidizing gases (CO_2 or air), the alumina layer could easily be distinguished as a grey zone, while the rest of the section would be almost white, as shown in Fig. 2. The greyness is indicative of the presence of dissolved metal in the molten state[7]. As shown in the following, measurements of limiting current support the conclusion that the grey zone was identical to the aluminum layer.

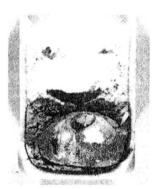

Fig. 2 Longitudinal section of sintered alumina crucible (45mm diameter) containing soudified aluminum and a cryolite-alumina(sat) mixture to which excess alumina was added(0.62g/cm^2). The layer of excess alumina is distinguishable as a grey zone(see text)

Because of the curvature of the aluminum surface, the alumina apparently was not distributed evenly across the surface. For large additions of alumina the upper surface of the alumina layer tended to be horizontal, as may be seen in Fig. 2. The thickness of the layer would then vary greatly from the center to the sides.

Samples of the alumina layer (the grey zone) were crushed and treated with 30% aluminum chloride solution for the convectional extraction analysis. Alumina contents of around 25%. Al_2O_3 were found, which is in agreement with the data given by Kozmin et al[4]. The bulk of the melt contained about 15%. Al_2O_3, which is close to the saturation concentration at 1020℃[7].

A sludge prepared by stirring cryolite-alumina melt with excess alumina will contain will contain about 40 wt% Al_2O_3[10]. In comparison the present alumina layer which is formed by free settling, must be very loosely packed. Since about half of the alumina will be in solution, the volume fraction of solid alumina is as low as 0.08%.

Samples of the layer were also examined on a scanning election microscope. The fluoride phase was extracted from the surface by etching with aluminum chloride solution. The layer appeared to be inhomogeneous, exhibiting areas with rather dense packing as well as voids, as shown in Fig. 3.

The present results together with the earlier works by Schadinger[2] and by Kozmin et al[4] show that excess alumina which is added in such a way that no lumps are formed, will settle at the melt/metal interface. The alumina layer formed will have a very open structure.

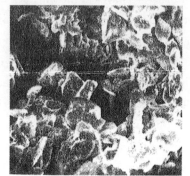

Fig. 3 Scanning electron micrograp(×1000) of alumina layer near the aluminum surface. The fluorides have been extracted from the surface layer

4 The Alumina Layer and the Reoxidation Reaction

The loss in current efficiently in aluminum electrolysis is mainly due to the reoxidation of aluminum by the anode gas,

$$Al + \frac{3}{2}CO_2 = \frac{1}{2}Al_2O_3 + \frac{3}{2}CO \tag{3}$$

The reactants are assumed to be dissolved aluminum and gaseous or dissolved CO_2.

As shown in the preceding paper[11], the rate of formation of CO(rf_{CO}) was studied by passing CO_2 over a cryolite-alumina melt in contact with aluminum, measuring the CO content in the exit gas. Changes in the concentration of dissolved metal in the bulk of the melt were followed by recording the limiting current for anodic oxidation of dissolved metal. The limiting current was measured with a platinum electrode at 1400mV positive to aluminum, and it was designated i_{1400}. Experimental details are given in the preceding paper[11].

The results[11] indicated that the reaction is governed by slow mass transfer at boundary layers at the aluminum/bath and bath/gas interfaces. In this work the metal/bath interface is of particular interest because the presence of an alumina layer at that interface would be expected to impede the mass transfer of dissolved metal and thereby lower the rf_{CO}.

Experiments were conducted with continuous recording of rf_{CO} and i_{1400}, while alumina was added batchwise through a tube. The size of the additions were normally 0.76g Al_2O_3 corresponding to 0.1g Al_2O_3 per cm^2 cross sectional area of the crucible.

As shown in Fig. 4 the rf_{CO} decreased gradually by each addition, and a new stable value was attained after 10~20 minutes. For the first two additions the average decrease in rf_{CO} was 47.6% (+2.6% standard deviation) per 0.1g Al_2O_3/cm^2 added.

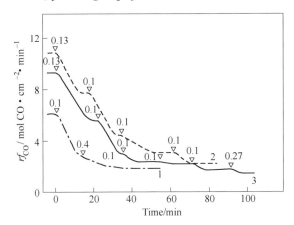

Fig. 4 Changes in rf_{CO} by addition of excess alumina to cryolite-alumina (sat) melts in contact with aluminum. Amounts(g) added per cm^2 cross sectional area are indicated on the curves for the three individual experiments shown, 1—1000℃, 2,3—1020℃

In Fig. 5 the rf_{CO} and the i_{1400} are plotted versus the amounts of alumina added. The i_{1400} decreased even more strongly than the rf_{CO}. The curves tend to level off when about 0.4g Al_2O_3/cm^2 had been added. These effects are demonstrated even more clearly in Fig. 6, where changes in rf_{CO} and i_{1400} are given relative to the initial values before alumina was added.

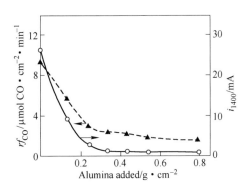

Fig. 5 The rf_{CO} and i_{1400} as a function of the amount of excess alumina added at 1020℃

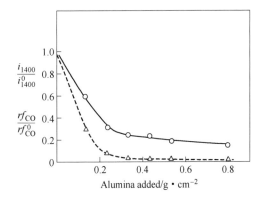

Fig. 6 Changes in rf_{CO}(○) and i_{1400}(△) by addition of excess alumina relative to the initial values (before alumina was added)

In accordance with the preceding work[11] it was found that when moving the platinum electrode to different positions in the melt, the i_{1400} did not change noticeably. Consequently, it did not appear to be any concentration gradient in the bulk of the melt with respect to dissolved metal. However, when the electrode was lowered into the alumina layer, the i_{1400} exhibited a steady increase as the metal surface was approached, The two curves shown in Fig. 7 were obtained in the same experiment, but at different positions. They indicate that the thickness of the alumina layer is variable, as suggested above. Evi-

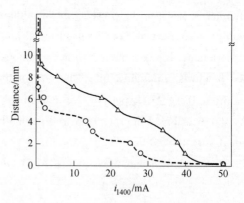

Fig. 7 Changes in i_{1400} as the platinum electrode is lowered is lowered into the alumina layer. Amount of added alumina:0.62g/cm². The distance refers to the distance above the metal surface at the position of the electrode

dently there is a gradient with respect to dissolved metal within the alumina layer. Since the fraction of solid material is so low(0.08%) the mass transfer within the layer probably occurs by convective diffusion.

5 Effect of Mechanical Stirring

The experiments described above were all conducted without any kind of forced convection within the cell. The influence of mechanical stirring was tested using a propeller-shaped stirrer made of boron nitride of 19mm diameter.

When the stirrer was located in the melt 15mm above the metal, the rf_{CO} decreased in a similar manner as without stirring, although the numerical values were higher, as shown in Fig. 8. However, when the stirrer was immersed into the metal, the effect of alumina additions on the rf_{CO} was very slight. It is then likely that the alumina added was swept towards the walls of the crucible by the moving metal.

Provided that the loss in current efficiency (CE) is caused exclusively by the reoxidation reaction, the CE can be expressed as

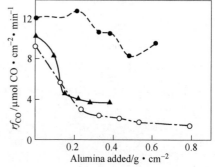

Fig. 8 Effect of mechanical stirring on the rf_{CO} as a function of the amount of excess alumina added
1—No stirring; 2—Stirring in the melt(125r/min); 3—Stirring in the metal(210r/min)

$$CE = \frac{k_1 i - k_2 rf_{CO}}{k_1 i} = 1 - \frac{k_1 rf_{CO}}{k_2 i} \tag{4}$$

Where the constants k_1 and k_2 refer to Faraday's law. As shown in the preceding chapter the rf_{CO} was diminished in the presence of an alumina layer at the bath/metal interface. A beneficial effect on the CE is then to be expected.

Measurements of CE were carried out in the cell shown in Fig. 9. The cell was contained in an alumina crucible of 45mm inner diameter. Electrical contact to the aluminum cathode (F) was provided by a small graphite rod (G) protruding through a hole in the bottom of the crucible. The height of the metal was in average 1.5cm and the interpolar distance was 4.3cm. The electrolyte consisted of cryolite +3.8%AlF_3+10%Al_2O_3. The total current was 5A, the current densities being 0.7A/cm^2 at the anode and 0.3A/cm^2 at the cathode.

The CE was determined by weighing the metal before and after the experiment and relating it to Faraday' law. Certain precautions had to be taken to minimize metal losses before ad after the electrolysis. The cell was brought to the operating temperature without any aluminum present, and the piece of aluminum prepared for the cathode was preheated to around 400℃, and then added. The experiment was started about 15min. later when the temperature was equilibrated. At the end of the experiment the cell was taken out of the furnace to cool rapidly. The cell was crushed and the piece of aluminum was collected and dipped into a NaCl-KCl melt to remove any adhering fluorides. The clean aluminum lump was then weighed.

In experiments where excess alumina was added, a first addition of 0.06g/cm^2 was made about 15 minutes after the beginning of the experiment. This addition was repeated at 20 min. intervals during the two hour runs on order to replenish the melt and the alumina layer. Inspection of crucibles after the experiments indicated that the final thickness of the alumina layer was about 4mm.

As observed previously in similar cells[12], the cell voltage increased during electrolysis from e.g. 3.2V initially to 3.9V after two hours. The additions of alumina slowed down the voltage increase.

The reproducibility of the CE data was fairly good. At 980℃ and without excess alumina added the CE was 81.3%±2% standard deviation. With excess alumina the CE was raised to 84.7%±2.5%. The temperature dependence of the CE is show in Fig. 10. The result obtained in presence of excess alumina were consistently higher than those without. As shown the observed temperature dependence is in excellent agreement with data given by Abramov et al[13], based on a similar experimental technique.

The improvement of the CE due to excess alumina at e.g. 980℃ was from 81.3% to 84.7%. The loss due to reoxidation then decreased from 18.7% to 15.3% and the rate for the reoxidation reaction thus was diminished by 18%. Comparison with the data on rf_{CO} is hampered by the fact that the thickness of the alumina layer during the CE measurements varied due to additions and consumption. Taking the average amount present in the layer to be 40 percent of the total amount added, this would correspond to approximately 0.1g Al_2O_3/cm^2 in average over the time of electrolysis. The calculated decrease in the reoxidation loss of 18% should then be compared with the corresponding value for the reoxidation experiments, being 47.6%.

Attempts to explain this difference can only be tentative. A rather strong concentration gradient will probably be set up in the alumina layer during electrolysis. The melt within the layer will be enriched with respect to NaF[7] and the metal solubility will consequently increase, having a negtive influence on CE.

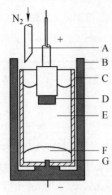
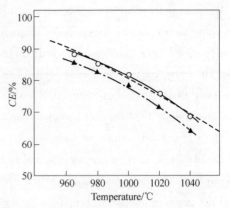

Fig. 9 Laboratory electrolytic cell for current efficiency determination
A—Alumina charging tube and inert gas inlet;
B—Graphite crucible; C—Alumina crucible;
D—Anode with alumina shield; E—Electrolyte;
F—Aluminum cathode; G—Graphite rod, cathode contact

Fig. 10 Current efficiency (CE) as a function of temperature
——— Excess alumina added;
—·—·— No alumina added;
- - - - - Abramov et al[13]

The practical implication of these results for the operation of commercial aluminum cells is not clear. The beneficial effect on CE by an aluminum layer which was suggested by Kent[3], has been established on a laboratory scale, but it is not known to what extent such layers can persist in commercial cells over any length of time.

6 Acknowledgement

A fellowship granted by The Norwegian Agency for International Development to one of the authors (Y. L.) is gratefully acknowledged. The authors are indebted to Messrs. S. Rolesth, R. Ødegård, and A. Solheim for assistance with certain parts of the experiments. The apparatus for the reoxidation studies was constructed by S. Rolseth.

References

[1] J. Thonstad. Semicontinuous Determination of the Concentration of Alumina in the Electrolyte of Aluminum Cells. Met. Trans., 1997, 8B(3):125-130.

[2] R. Schadinger. The Reaction between Carbon Dioxide and Metal fog in Aluminum Electrolysis. (Italian) Alluminio, 1953, 22:691-697.

[3] J. H. Kent. The Attainment of High Current Efficiency on Aluminum Reduction Furnace by a Cold Running Technique. J. Metals, 1970, 22:30-36.

[4] G. D. Kozmin, P. V. polyakov, A. V. Sysoyev, A. M. Dsyplakov, V. A. Kyyukovskiy. Dissolution of Alumina in the Electrolyte of an Aluminum Reduction Cell. Sov. J. Non-Ferrous Met., 1976, 17(7): 31-32.

[5] H. M. Princen. The Equilibrium Shape of Interfaces, Drops and Bubbles, Rigid and Deformable particles at Interfaces. in Surface and Colloid Science, Egon Matijevic ed.; John Wiley & Sons, 1969.

[6] H. C. Maru, D. T, Wasan, R. C. Kintner. Behavior of a Rigid Sphere at a Liquid-Liquid Interface. Chem. Engineering Science, 1971, 26:1615-1628.

[7] K. Grjotheim, C. Krohn, M. Malinovsky, K. Malinovsky, K. Matiasovsky, J. Thonstad, Aluminium Electrol-

ysis, Aluminium Verlag GmbH, Dvsseldorf, 1977.

[8] E. W. Dewing, P. Desclaux. Interfacial Tension of Aluminium in Cryolite Melts Saturated with Alumina. PP. 30-34 in Symposium on Molten Salt Electrolysisis in Metal Production. Grenoble, France, Sept. 1977.

[9] Yexiang Liu, J. Thonstad, to be published.

[10] J. Thonstad, P. Johansen, W. Kristensen. Some Properties of Alumina Sludge. paper presented at 109th AIME Annual Meeting, Las Vegas, Nevada, Feb. 1980. light Metals, 1980:227-239.

[11] S. Rolseth, J. Thonstad, W. kristensen. On the Mechanism of the Reoxidation Reaction in Aluminum Electrolysis. Paper to be presented at 110th AIME Annual Meeting, Chicago, Feb. 1981.

[12] C. Castellano, D. Bratland, K. Grjotheim, T. MiIftvoglu, J. Thonstad. Current Efficiency Measurements in Laboratory Aluminium Cells IV: Depletion of Alumina and Cell Voltage. Can Metall. Q., 1979, 18(1): 13-18.

[13] G. A. Abramov, M. M. Vetyukov, I. P. Gupalo, L. N. Lozhkin, A. A. Kostyukov. Theoretical Principles of the Electrometallurgy of Aluminium (in Russian), Metallurgizdat, Moscow, 1953:319-325.

Inhibition of Anode Effect in Aluminum Electrolysis Process by Anode Dopants: A Laboratory Study[*]

Abstract The critical current densities (ccd) of carbon anodes doped with lithium, lead, chromium. The doped anodes possess ccds 2 to 4 times higher than undoped ones. None AE were observed during a rather long time electrolysis in the melts containing 1wt% Al_2O_3, while AE occurred on the undoped anodes under much lower c. d. It showed that these dopants inhibited AE to an obvious extent. Mechanism concerned was also discussed, and suggested that the formation of CF_x which result in occurrence of AE was inhabited by these dopants.

1 Introduction

Frequent anode effect will cause the operation of reduction cell unstable, and decreased the current efficiency. It seems that anode effect is surely undesirable today. The modern advanced reduction cell installed point feeding devices and well controlled by computer should reduce the anode effect to a minimum.

The occurrence of anode effect depends largely on the anode materials, and mainly on the surface property of anode. Modification of electrode surface which is an effective approach to improve the properties of electrode material is now extensively applied in the fields of electrochemistry and electrochemical engineering[1,2].

As for the carbon anode in aluminium electrolysis which could be subjected to surficial modification would also be improved its properties. We considered that the modification of carbon electrode is a worthy noted field to research.

Chemical modified the electrode will change the chemical composition and texture of the electrode both in the bulk and on the surface. One of the chemical modification methods is doping the carbon electrode with some chemical salts. After doping treatment, the carbon electrode poses some new performances. Recently, we presented a paper[3] dealing with the beneficial effect of the doped carbon anodes, which reduced the anodic overvoltage obviously in the aluminum electrolysis process. Our preliminary explanation is that the amount of active sites on the electrodes surface was increased by the doping treatment, and increased the electrode reaction rate thereby, so reduced the anodic overvoltage.

Along the lines of above mentioned thought, we attempted to research the anode effect by chemical modified electrode. It is well known that the occurrence of anode effect is resulted from the decreasing of alumina content in the bath to a certain extent, as well as the increasing of anodic potential[4]. During the case, CF_x which has a very low surface energy was formed on

---

[*] Copartner: Xiao Haiming and Xion Guangchen. Reprinted from Light Metals, 1991, 2:489-494.

the anode surface and the anode was not wetted by the melt. In consequence, anode effect occurred[5].

2 Experimental

Spectroscopic grade graphite rods with 10mm diameter were employed as anode material. The following analytical grade agents were the dopants: LiF, Pb(NO$_3$)$_2$, CrCl$_3$, NaCl and PbCl$_2$. The doping treatment was as follow: The cleaned graphite rods were degassed under vacuum for 30 minutes, then dipping with 5wt% dopant solution for 12 hours and dried at 110℃ for 30 minutes. The last step was heat treatment at 400℃ in air stream for 4 hours. After these treatments, dopants were distributed homogeneously on the surface and in the bulk of electrode. The same kind of graphite rods without doping were also employed in our experiments for comparison.

Anodic critical current density was measured by potential sweep technique. The so-called critical current density(ccd) is that the maximum current density which is attained before the normal anode reaction is superseded by the anode effect.

Using two electrodes system in the measurement, the counter electrode was a graphite crucible containing 150g cryolite-alumina melt. The doped graphite rod which shielded with BN tube was the anode. The apparent working area of all the anode was 0.78~0.79cm^2.

Each kind of doped electrodes was prepared three pieces, and each piece of electrodes measured repeatedly three times at least. ccd was the average value of measurements.

3 Results and Discussion

3.1 The ccd of doped anodes

The ccd of doped anodes was obvious higher than that of undoped anodes. Fig. 1 shows the potential sweep curves of CrCl$_3$ doped graphite anode and undoped ones.

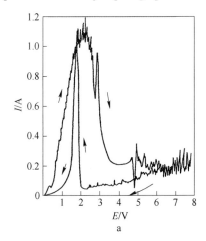
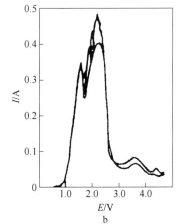

Fig. 1 Potential Sweep curves of graphite anodes
(3NaF·AlF$_3$-2wt%Al$_2$O$_3$ melt, 1025℃, v=300mV/s)
a—Doped with CrCl$_3$; b—Undoped

Making a comparison between a and b in Fig. 1, it can be seen that the peak point current of doped anode was increased two and three fold, the sharp of peak and the irreversible extent were also on the surface of doped electrode.

In addition, the typical current-Potential curves of some doped anodes (Shown in Fig. 2) are also interesting.

It is shown in Fig. 2 that the ccd of doped anodes was two to four fold higher than undoped ones. The results listed in Table 1 supported our above mentioned thought, modified treatment improved the performance of electrode and thus inhibited the occurrence of anode effect.

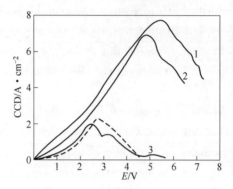

Fig. 2 Current-Potential curves diagram of some doped anodes
(2.7NaF · AlF$_3$-1wt% Al$_2$O$_3$ melt, 1000±5℃)
1—Doped with Pb(NO$_3$)$_2$; 2—Doped with CrCl$_3$; 3—Undoped

Table 1 ccd of various doped anodes
(2.7NaF · AlF$_3$-1wt%Al$_2$O$_3$ melt, 1000±5℃)

ccd	Doping agent							
/A · cm^{-2}	None	CrCl$_3$	Pb(NO$_3$)$_2$	RuCl$_3$	LiCl	NaCl	PbCl$_2$	LiF
	2.08	7.41	8.85	9.17	6.64	4.52	8.15	5.8

3.2 The effect of alumina content on the ccd of doped anodes

It is well known that alumina content in the bath is an important variable affecting the occurrence of anode effect. We examined the effect of alumina content on the ccd of doped anodes, and plotted the I-V curves of various doped anodes with 0.5wt% to 5wt% alumina content. There was no considerable ccd difference could be found when alumina content was below 0.5wt%. The ccd of doped anodes began to rise obviously as alumina content was increasing, and showed the ability of inhibition to anode effect. The higher alumina content the more effective inhibitive ability could be seen. Fig. 3 shows these typical results.

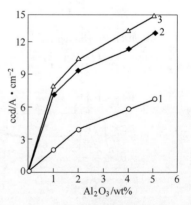

Fig. 3 ccd of some doped anodes vesus alumina content in 2.7NaF · AlF$_3$ melt 1000±5℃. 1—doped with Pb(NO$_3$)$_2$; 2—doped with LiF; 3—undoped

3.3 The lasting period of time by inhibited anode effect

During electrolysis the lasting period of time by inhibited anode effect should be governed by the distribution of dopant in the doped electrode. We examined the dopant distribution in the doped electrode by X-ray spectra and SEM X-650 (Hitachi). Here we take the electrode doped with $CrCl_3$ for an example. Fig. 4 is a BEI photograph of cross-section of $CrCl_3$ doped electrode. The feature of BE is that the back electrons yield is increasing with the increasing of atomic number of an element the more bright is the image shown in the photograph.

The atomic number of carbon is 6 and that of Cr is 24, so the bright spots in Fig. 4 are chromium and dark ones are carbon. Fig. 5 shows $K\alpha$-ray scanning photograph of chromium.

It showed in Fig. 5, that there were quite many peaks of $CrK\alpha$-ray sweep curves represented chromium content.

Fig. 4 BEI photograph of $CrCl_3$ deposed electrode (100×)

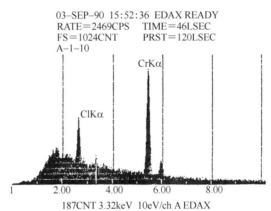

Fig. 5 $CrK\alpha$-ray screen photograph of $CrCl_3$ doped graphite electrode

It also can be seen from Fig. 4 and Fig. 5 that the dopant is homogenous distributed in the electrode. It was expected that the dopant in the electrode would act as an inhibitor to restrain anode effect during the electrolysis process until the doped electrode was consumed entirely. Checking the lasting time of an inhabited AE has been carried out. Both electrodes doped with $Pb(NO_3)_2$ and $CrCl_3$ were subjected to electrolysis in the cryolite melts containing 1wt% Al_2O_3. The current densities conducted to the doped anodes were 30% ~ 70% higher than that of undoped graphite electrode. In consequence, none anode effect occurred in spite of conducting high current densities on the electrodes and electrolysis lasted thirty minutes. The current passed through the doped anode has not declined under the constant potential. Fig. 6 showed the results recorded on the X-Y recorder.

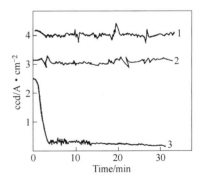

Fig. 6 Lasting period of time in inhibited AE by doped anodes under high c. d
1—doped with $Pb(NO_3)_2$, E_{cell} = 6.0 volts;
2—doped with $CrCl_3$, E_{cell} = 4.5 volts;
3—undoped, E_{cell} = 5.0 volts

3.4 The possible mechanism related to inhibited anode effect by the dopants

In spite of several viewpoints relevant to the mechanism of anode effect have been proposed, the viewpoint of CF_x formation can express our experimental results satisfyingly. The fluorine-containing ions began to discharge on the carbon anode when oxygen-containing ions depleted in the vicinity of anode surface, and followed the CF_x formation. Due to CF_x has a very low surface energy and its isolating property, once the carbon anode covered with CF_x, the later was not wetted by the melt. Then the gas evolved from anode would adherent and accumulated there growing bigger bubbles, blocking the path of electricity passing and resulted in rising the current densities on the unoccupied places, thus, increasing the anodic potential and producing more CF_x until onset of anode effect. It may be said that the occurrence of anode effect depends upon the forming amount of CF_x.

The role of dopant could be explained below. As fluorine-containing ions began to discharge at the anode the dopant ions reacted with the discharged fluorine to form metallic fluorine MF_x more easily than to form CF_x, where M is a metal such as lithium, lead and chromium. Thus, limited the occurrence of anode effect. Thermodynamic analysis unsupported this explanation. The free energies of formation of PbF_2, LiF and CrF_3 were larger than that of CF_x (Listed in Table 2). Therefore, the ccd were increased when dopants presented.

Table 2 The free energies of formation reactions of some fluorides

Possible reaction	$\Delta G_{1273K}/\text{kcal} \cdot \text{mol}^{-1}$
$C+2F_2 = CF_4$	+178.97
$Pb+F_2 = PbF_2$	-128.90
$Li+\frac{1}{2}F_2 = LiF$	-117.57
$Na+\frac{1}{2}F_2 = NaF$	-102.61
$Cr+\frac{3}{2}F_2 = CrF_3$	-203.75

The formation of a ternary conducting layer CF_xM_y at the interface is an another possible explanation. D. Devilliers and co-worker[6] proposed above suggestion recently. They have studied the CF_x flim.

Which was formed at the surface of carbon anode in the production of fluorine. Modified carbon electrodes have been empty in their investigation. They concluded that the formation of a ternary conducting layer CF_xM_y at the interface improved the performances of carbon anode and thereby the electrolytic process. The conductivity of some of CF_xM_y was higher than that of graphite. The surface compound on the modified carbon electrode was thought to enhance electron transfer at the interface and to improve the wettability of the electrode by the melt.

As for the influence of alumina content on the ccd, the possible explanation is as follow. It was mainly the oxygen-containing ions discharged on the anode when alumina content in electrolyte was rich enough. Discharged oxygen reacted with carbon to form C_xO under high current density (e.g. industrial c.d.) and leading to a higher anode overvoltage. More active sites on the anode surface would be provided by the dopant which presented as metallic oxides. These sites promoted to a higher rate of anodic reaction and thus reduced anodic overvoltage to some extent. Under a higher anodic potential the fluorine-containing ions would discharge and combine the metallic ions to form MF_x dominantly. As alumina content was very low, e. g. 0. 5wt% in the bath, the oxygen-containing ions were serious depletion in the vicinity of anode surface, fluorine-containing ions discharged dominantly, the amount of CF_x formation was much more than that of MF_x formation. Thus, the ccd was not so high and leading to onset of anode effect. When alumina content was around 1wt%, both oxygen and fluorine-containing ions discharge to gather. C_xO and CF_x were formed simultaneously. In the case of dopant presence, fluorine would combine the M to form MF_x. The later has better conductivity and wettability to the melt, resulted in ccd increased and anode effect inhibited.

It is clearly that the dopants containing lead or chromium were harmful to the alumina electrolysis process. We just took these dopants as an example to manifest their function in sustraining anode effect. From the practical viewpoint, the dopants containing lithium or sodium etc. are beneficial but it is needed to investigate the modified carbon electrode further and meticulously.

References

[1] R. W. Murray. Chemically modified electrodes, in electroanalytical chemistry, a series of advances, Vol 13 (ed. by A. J. Bard) Marcel Dekker, New York, 1984:191-368.

[2] E. Barendrecht. Chemically and physically modified electrodes: some new developments. Journal of Applied Electrochemistry, 1990, 20:175-185.

[3] Yexiang Liu, Himing Xiao. A New Approach to Reduce the Anodic Overvoltage in the Hall-Heroult Process. Light Metals, 1989:185-275.

[4] K. Grjotheim, et al. Aluminium Electrolysis-Fundationals of the Hall-heroult Process, 2nd Ed. Dussedldoff: Aluminium-Verlag, 1982.

[5] M. Chemla, D. Devilliers. Study of CF_x Passivating Layers on Carbon Electrodes in Relation to Fluorine Production. Proceedings of the Joint International Symposium on Molten Salts (1987), The Electrochem. Soc. Inc. 1987, Hawaii, 546-556.

[6] D. Devilliers, et al. Polish Carbon Electrodes for Improvingthe Fluorine Production Process. Journal of Applied Electrochemistry, 1990, 220:91-96.

160kA 预焙铝电解槽区域电流效率*

摘 要 本文论述了区域电流效率的意义,利用电流效率综合机理模型和区域参数估计模型计算出了 160kA 预焙槽的区域电流效率。结果表明,160kA 预焙槽中各区域的电流效率很不均匀,最大值与最小值相差 20% 以上,阳极底掌下的区域电流效率相差可达 4% 以上。采用区域电流效率的观点,分析了某厂 160kA 预焙铝电解槽电流效率的特点,认为低电流效率区域的存在和面积偏大导致电解槽平均电流效率偏低,并由此提出了提高电流效率的途径。

铝电解槽中各部位的工作状况有所差别,大型铝电解槽尤其如此。目前所说的电流效率通常为槽子在一段时间内各部位电流效率的平均值,即空间和时间的均值。这个平均值对于考查某台槽子的工作情况既方便又实用,但存在某些不足,它掩盖了槽中各处特征的差异,无法了解槽中某一区域或某块阳极的工作状态,难以挖掘槽子的潜力。目前的研究表明,同一槽中各阳极下的极距可能相差 1~2cm,熔体流速相差 10cm/s 以上,电解质温度也相差 8~10℃,各阳极底掌下电流效率相差可达 10% 左右[1]。鉴于这种情况,分区域来讨论大型铝电解槽的电流效率是很有必要的,这将有利于分析电解槽内部工作状态,充分发挥槽内各处的最大效益,为进一步寻找某些电解槽电流效率低的原因,寻求提高电流效率的途径打下了基础。

1 区域的划分

对于一台预焙槽的区域划分有一定的任意性,一般根据问题的需要来划分。区域划分越细,就越能细致地分辨槽中各处的工作情况,但却会增加许多工作量,对于实际问题的解决也不是十分必要;区域划分太大,又达不到分析解决问题的目的。因此,本文的区域划分遵循下列原则:(1) 区域内磁流条件相差不大;(2) 区域内部是连续的,跨度尽可能小;(3) 计算机处理方便;(4) 对铝电解生产操作条件的优化及提高电流效率有利。

本研究把某厂 160kA 预焙阳极电解槽划分成 9 个区域,相邻的 3 个阳极为 1 个区域,24 个阳极分成 8 个区域,电解槽边部为 1 个区,见图 1。

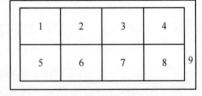

图 1 160kA 预焙槽区域划分示意图

2 区域电流效率的计算

2.1 计算方法

电解槽中铝的损失主要受传质过程的影响,而对于在整个传质过程中哪一阶段是

---

* 本文合作者:曾水平。原发表于《中国有色金属学报》,2000,2:274-277。

律速阶段，一些研究者有不同的观点，因而导致有不同计算电流效率的机理模型建立[2~4]。

这里采用充分考虑整个传质过程影响的综合机理模型[5,6]，此模型表达为：

$$\left.\begin{array}{l} E_c = 100 - 219 d_{en}^{-1} D_{me}^{0.67} \mu^{-0.5} u_e^{-0.83} d^{-0.17} \rho^{1.5} c_{Al}^* (1-f) \\ f = c_m / c_m^* \\ D_{me} = D_m (\sigma_r / \sigma)^{0.5} \end{array}\right\} \quad (1)$$

式中，E_c 为电流效率的百分数，%；d_{en} 为阴极电流密度，kA/m^2；D_{me} 为校正界面张力时的等效扩散系数；μ 为电解质黏度，$Pa \cdot s$；u_e 为电解质相对铝液的平均流速，m/s；d 为极距，m；ρ 为电解质密度，kg/m^3；c_{Al}^* 为电解质中 Al 的饱和浓度；f 为金属浓度比例系数；c_m 为电解质主体的金属浓度；c_m^* 为电解质中金属的饱和浓度；D_m 为扩散系数；σ_r 为对应于扩散系数 D_m 时的界面张力；σ 为铝和电解质界面张力。

各区域的温度采用区域平均温度的估计模型计算，对于 160kA 电解槽实测期间的温度估计模型为[6]：

$$T = 965 + 0.65 \bar{d}(\bar{d} - 3) \quad (2)$$

式中，T 为温度，K；$\bar{d}$ 为电解质中平均的电流密度，A/m^2。换阳极时，此区域的温度需特殊处理。

各区域的极距采用区域平均极距的估计模型计算。对于 160kA 电解槽各区域的极距估计模型为[6]：

$$L_i = 0.81/I_i \quad (3)$$

式中，L_i 为区域的平均极距，m；I_i 为区域流过的电流，kA。

各区域的界面张力平均值估计模型为：

$$\sigma_界 = 550 - 110 d_c \quad (4)$$

式中，$\sigma_界$ 为铝和电解质界面张力，$10^{-3} N/m$；d_c 为区域平均阴极电流密度，A/cm^2。

比例系数 f 值的估计模型为：

$$f = 0.5 - 0.5 d_a \quad (5)$$

式中，d_a 为区域阳极平均电流密度，A/cm^2。

计算过程所用到的物理化学参数参考文献[7]的取值。

2.2 计算结果

利用式（1）~式（5），并结合铝电解槽中物理场的计算模型，计算了各区域电流分布均匀时 160kA 电解槽各区域的电流效率，计算结果见图 2。

图 2 中"10"对应的电流效率为整个电解槽的平均电流效率计算值。由于区域电流效率的测定比较困难，必须在对应区域的阳极上直接钻孔取 CO_2 气体进行分析。鉴于条件所限，我们只校验了平均电流效率，结果表明对于一周内的平均电流效率计算值与实测值的误差小于 0.6%。

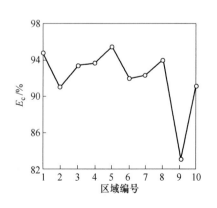

图 2 电流效率与电解槽中区域的关系图

3 讨论

3.1 160kA 预焙槽电流效率的特征

从上面的计算结果可知，电解槽中各区域的电流效率是不相同的，最大值与最小值相差20%以上，阳极底掌下的区域电流效率相差可达4%以上。1、5 和 8 区域属高电效区域；2、3、4、6 和 7 区域属中等电效区域；9 区域即边部区域属低电效区域，这与该区域的磁流条件和电流分布是一致的。

某厂 160kA 预焙槽电流效率除具有铝电解生产的一般规律外，还具有下列特征。

（1）槽周边（阳极底掌投影外围）的电流效率比槽平均电流效率低15%左右，主要原因是这个区域电流密度小，熔体流速大。

（2）各阳极底掌下的局部工作状况很不一致，区域电流效率相差不等，相差最大者达4%以上。这主要是由于母线配置造成槽内各处磁场条件不同[8]，以及各阳极导电情况不同所致。

（3）通过计算可知，换阳极所引起的平均电流效率损失超过1%，这是因为新阳极在很长时间达不到额定电流值，大大增加了熔体内水平电流，而导致熔体流速增加。另外换阳极的区域垂直电流密度大大减少。

（4）在相同操作条件下，并非所有阳极导电严格相等，电流效率最高。这是由于槽子设计造成各处磁场条件不同，而不均等的电流分布能在一定程度上补偿磁场不均匀的影响。

（5）槽内有的区域电流效率高达96%，但槽平均电流效率却低于90%，这说明低电流效率的区域占面积较大，槽子的设计和操作方面有待改进。

3.2 提高电流效率的途径

提高电流效率的途径有过许多一般性论述[7,9]，本文针对 160kA 槽的实践和区域电流效率的计算结果进行了讨论。

（1）缩小铝液镜面的表面积：区域电流效率计算表明，电流效率最低的区域是周边区域。这个区域电流效率比阳极底掌下区域电流效率低10%～15%，而目前电解槽的设计和操作使这一区域的面积占阳极表面总面积的15%～20%，这是使槽平均电流效率低的一个重要原因。据粗略估计，若能使这个区域面积缩小一半，平均电流效率可提高1%～2%，160kA 槽子工作时，炉帮最薄处恰好在铝液与电解质界面附近，这显然是不合理的。改进的办法有：

1）槽子设计时采用窄炉面操作，减少阳极至槽壁空间，从而减小阴极表面。

2）槽子设计和大修时，重新考虑保温材料的布置，使电解质/铝液界面处炉帮增厚，使铝液表面积尽可能小，这可以通过电解槽热平衡的精确计算和内衬优化设计来实现。

（2）减少伸腿长度：研究表明，金属-电解质界面处的伸腿厚度对电流效率有较大的影响[10]。目前 160kA 预焙槽伸腿长度一般在 20cm 左右，伸腿处还有一些软沉淀的存在。铝液中水平电流有较大的值，前面曾经指出，炉帮伸腿一般在阳极底掌投影的边缘最合适。伸腿的长短主要由电解槽的热场来决定，欲使伸腿位置最佳，除可以在操作过程中人工消减伸腿外，主要应通过改进电解槽的热场设计来实现。如果伸腿能从目前状况改进到伸腿长度为5cm以内，可使熔体流速降低15%~20%，电流效率可相应提高1%左右。

（3）优化母线配置：阳极底掌下不同区域电流效率差别的主要原因之一是磁场条件不一致，改善磁场条件的手段是改造母线设计。某厂160kA预焙槽的母线配置存在一定问题，这是导致设计电流效率不高（87.5%）的主要原因之一。如果能把现行160kA槽中高电流效率区域的磁场条件强加于全槽，据粗略估计电流效率可提高2%~3%。

（4）阳极设置高度的最佳化：目前认为阳极设置高度对生产有一定的影响，但阳极本身有自我调节功能。当某一阳极消耗过快时，局部极距增加会使该阳极导通电流减少，其消耗速度自动下降；而当某种原因导致某一阳极消耗速度减慢时，局部极距减小会使该阳极导通电流增加，其消耗速度也随之增加。这样似乎不用强调阳极设置高度的严格性，生产过程也不用人工干预，阳极本身的自调节功能使生产正常进行，然而正是阳极在自调节过程中引起电流效率的损失。

阳极设置高度的最佳化有过专门报道[11]，它基本是从阳极预热和导电性方向来论述，没有从电流效率的角度去分析，得出的结论是安装新阳极的底掌应高于工作阳极底掌0.5~1cm。从电流效率的角度分析，阳极应尽快达到其额定电流，达到额定电流的时间越短，电流效率损失越小。这样在有条件时阳极的安装应分两次完成，第一次安装高度低于工作阳极底掌2cm左右，待导电达额定电流时，进行第二次安装，从而缩短新阳极不正常导电的时间，这样可以使换阳极引起的电流效率损失降低20%左右。

（5）阳极个别调整：电解过程中，由于某种原因如接触不好、阳极质量差异、阳极长包等，导致某一阳极工作不正常，长时间偏离额定电流值，有时需要人工处理，有时它能自己恢复正常工作。这种情况在实际生产过程中一般靠阳极的自调节性能恢复正常工作。由于阳极的自调节功能有限，会引起电流效率的损失。我们认为，出现这种情况时，在适当处理使阳极故障消失后，应采用调整个别阳极高度的方法，使其迅速达电流额定值，从而提高电流效率。当然，调整个别阳极必须是在有条件的厂房才能实现，并且需精确控制，否则调整个别阳极会有不良后果。

（6）延长高电流效率区域的工作时间：电解槽各区域的正常工作时间，就是对应区域阳极的正常导电时间。阳极的导电性能主要由阳极本身的温度和质量决定，实际生产过程中换上去的冷阳极在15h内不能正常导电[9]，由此造成电流效率损失。如果在换阳极时用充分预热好的新阳极，则完全可以使其在1h内正常导电，这将大大改善槽内熔体的电流分布，对提高本区域的电流效率和相邻区域的电流效率都有利。另外，在没有条件预热阳极时，为了保证更高电流效率区域工作时间，可以考虑在高效区需换新极时，先把低效区的工作阳极移植到高效区，把冷的新阳极换在低效区。当各区域电流效率差别较大时，这在经济上是合算的，但此方案需经用工业实践来证明其可行性。

4 结论

（1）本文以电解槽中溶解金属传质过程的综合机理模型为基础，结合实际情况，计算了电解槽中各区域的电流效率，首次对大型铝电解槽采用区域电流效率的观点进行了分析和讨论。

（2）现场测量和计算了某厂160kA预焙铝电解槽电流效率，给出了铝电解槽中各区域电流效率的分布图。

（3）针对某厂 160kA 预焙铝电解槽电流效率和电流分布的特点，分析了电流效率低的原因，提出了提高电流效率的方法。

参 考 文 献

[1] Alcorn T R, Mcminn C J, Tabereaux A T. Current efficiency in aluminium electrolysis by anode gas analysis [J]. Light Metals, 1988: 683-695.

[2] Dorreen M M R, Hyland M M, Welch B J. Current efficiency studies in a laboratory aluminium cell using the oxygen balance method [J]. Light Metals, 1998: 483-489.

[3] Lillebuen B, Mellerud T H. Current efficiency and alumina concentration [J]. Light Metals, 1985: 637-645.

[4] Dewing E. Loss of current efficiency in aluminium electrolysis cell [J]. Metallurgical Transaction B, 1991, 22B: 177-182.

[5] Zeng Shuiping, Liu Yexiang, Mei Chi. Mathematical model for continuous detection of current efficiency in aluminum production [J]. Trans Nonferrous Met Soc China, 1998, 8 (4): 683-687.

[6] 曾水平. 铝电解槽电磁场计算及电流效率连续监测的研究 [D]. 长沙：中南工业大学，1996.

[7] Grjotheim K, Krohn C, Malinovsky M, et al. Aluminum electrolysis (2nd Edition) [M]. Aluminum-verlag. Dusseldorf, 1982.

[8] 曾水平，蔡祺凤，梅炽，等. 铝电解槽内磁场的计算 [J]. 中国有色金属学报，1995，5（1）：34-38.

[9] Qiu Zhouxian. Aluminum Electrolysis in Prebake Cell. Beijing: Metallurgy Industry Press, 1988, 6.

[10] Torstein Haarberg T, Solheim A, Johansen S T, et al. Effect of anodic gas release on current efficiency in hall heroult cells [J]. Light Metals, 1998: 475-481.

[11] Utigard T. Optimum of anode setting height [J]. Light Metals, 1991: 273-280.

点式下料铝电解槽氧化铝浓度新型估计模型*

摘 要 采用基于机理分析获得的状态空间模型和非线性系统的自适应推广的 Kalman 滤波算法实现了对铝电解槽 Al_2O_3 浓度的直接估计。用虚拟噪声补偿了简化模型包含的时变误差和过程中的时变噪声。用实测数据证明了模型与算法的有效性。

当前国际上先进的控制系统是通过对与 Al_2O_3 浓度有一定对应关系的参数进行在线估计来间接获取 Al_2O_3 浓度信息的，并以此为基础构成 Al_2O_3 度的自适应控制系统。一般采用输入-输出模型，未考虑模型误差及过程噪声的时变性。

为了改进 Al_2O_3 浓度的估计模型，我们以采用将连续下料控制策略的 160kA 中心点式下料预焙槽为测试槽，采用机理分析与现代时间序列分析的建模理论相结合的方法，建立了基于非线性系统自适应推广的 Kalman 滤波算法的铝电解槽 Al_2O_3 浓度估计模型。该模型可实现对 Al_2O_3 浓度这一状态参数的直接估计。采用了虚拟噪声补偿技术。用实测数据证明了模型的可行性和先进性。

1 非线性系统的自适应推广的 Kalman 滤波算法

文中仅考虑带非线性观测模型的纯量系统

$$x_{k+1} = a \cdot x_k + b \cdot u_k + w_k \tag{1}$$

$$y_{k+1} = h(x_{k+1}, k+1) + v_k \tag{2}$$

式中，x_k 为状态变量；y_k 为观测变量；u_k 为输入变量；a 和 b 为常数；h 为可微函数；w_k 和 v_k 为统计特性（即均值和方差）未知、时变且相互独立的高斯白噪声。将 h 线性化的常规方法是将其在状态 x_{k+1} 的一步预报值 $x_{k+1|k}$ 处展开成 Taylor 级数，但是当一步预报误差较大时有导致滤波发散的可能。假设在 $k \sim k+1$ 的采样间隔内 $\partial h/\partial x$ 似不变，故采用一阶差分将观测模型线性化得

$$z_{k+1} = \frac{\partial h}{\partial \hat{x}_k} x_{k+1} + S_{k+1} + \eta_{k+1} \tag{3}$$

其中

$$z_{k+1} = y_{k+1} - y_k \tag{4}$$

$$S_{k+1} = -\frac{\partial h}{\partial \hat{x}_k} \cdot \hat{x}_k \tag{5}$$

$$\eta_{k+1} = v_{k+1} - v_k + \xi_{k+1} \tag{6}$$

式中，z_{k+1} 成为新的观测变量；ξ_{k+1} 用来补偿模型线性化引入的时变误差，故 η_{k+1} 成为虚拟观测噪声；显然，它也带有未知时变统计。至此，非线性系统（1）、（2）的状态估计问题转化为带未知时变噪声统计的线性系统（1）、（3）的自适应滤波问题。容易推导出其 Kalman 滤波算法为：

---

* 本文合作者：李劼、黄永忠、王化章。原发表于《中国有色金属学报》，1993, 24（3）：25-28。

$$\hat{x}_{k+1} = \hat{x}_{k+1|k} + K_{k+1} \cdot \varepsilon_{k+1} \tag{7}$$

$$\hat{x}_{k+1|k} = a \cdot \hat{x}_k + b \cdot u_k + \hat{q}_k \tag{8}$$

$$\varepsilon_{k+1} = z_{k+1} - \frac{\partial h}{\partial \hat{x}_k} \hat{x}_{k+1|k} - S_{k+1} - \hat{r}_k \tag{9}$$

$$K_{k+1} = P_{k+1|k} \left(\frac{\partial h}{\partial \hat{x}_k} \right) \left[P_{k+1|k} \times \left(\frac{\partial h}{\partial \hat{x}_k} \right)^2 + \hat{R}_k \right]^{-1} \tag{10}$$

$$P_{k+1|k} = a^2 P_k + \hat{Q}_k \tag{11}$$

$$P_{k+1} = \left(1 - K_{k+1} \frac{\partial h}{\partial \hat{x}_k} \right) P_{k+1|k} \tag{12}$$

式中，$\hat{q}_k$ 和 $\hat{Q}_k$ 为时变噪声 w_k 的均值和方差估值；$\hat{r}_k$ 和 $\hat{R}_k$ 为时变噪声 η_k 的均值和方差估值。采用按极大后验原理和指数加权渐消记忆法推导的时变噪声统计估值器[1,2]，可得次优无偏后验估值：

$$\hat{q}_{k+1} = (1 - d_k)\hat{q}_k + d_k(\hat{x}_{k+1} - a\hat{x}_k - u_k) \tag{13}$$

$$\hat{Q}_{k+1} = (1 - d_k)Q_k + d_k(K_{k+1}^2 \varepsilon_{k+1}^2 + P_{k+1} - a^2 P_k) \tag{14}$$

$$\hat{r}_{k+1} = (1 - d_k)\hat{r}_k + d_k\left(z_{k+1} - \frac{\partial h}{\partial \hat{x}_k} x_{k+1} - S_{k+1}\right) \tag{15}$$

$$\hat{R}_{k+1} = (1 - d_k)R_k + d_k\left[\varepsilon_{k+1}^2 - P_{k+1|k}\left(\frac{\partial h}{\partial \hat{x}_k}\right)^2\right] \tag{16}$$

式中，$d_k = (1-b)/(1-b^{k+1})$，b 为遗忘因子，$0<b<1$。

通过适当地选取遗忘因子 b 并给定初值（x_0, P_0, q_0, Q_0, r_0, R_0），按递推方式进行上面的自适应推广的 Kalman 滤波，可得状态 x 的估计。并因采用了虚拟噪声补偿，故能有效地抑制滤波的发散[1]。

2　Al_2O_3 浓度的非线性估计模型的建立

2.1　观测模型

以表观槽电阻增量（记为 z_k）作为观测变量，综合国内外以及我们所作的一些理论与实验研究结果[3,4]，推导了一个具有式（3）形式的观测模型（机理模型）。重写式（3）如下：

$$z_{k+1} = \frac{\partial h}{\partial \hat{x}_k} x_{k+1} + S_{k+1} + \eta_{k+1} \tag{3}$$

式中，z_{k+1} 为表观槽电阻增量（即变化速率）；x_{k+1} 为 Al_2O_3 浓度；S_{k+1} 的意义与式（5）中的相同；η_{k+1} 的意义与式（6）中的相同，即它不仅包含有 z_{k+1} 的观测噪声，而且包含有理论模型与实际过程的偏差，因此，设其为带有未知时变统计的白噪声；$\partial h/\partial x k$ 为表观槽电阻（$\mu\Omega$）对 Al_2O_3 浓度（wt.%）的偏导数。取有关工艺参数为常态值（如电解质温度为 955℃，极距为 4.5cm，经过修正的阳极电流密度为 0.664A/cm^2，以及电解质组成为 CR=2.7，CaF_2=5%），可推导出 $\partial h/\partial x_k$ 与 x_k 的关系为

$$\frac{\partial h}{\partial \hat{x}_k} = -0.638\hat{x}_k^{-1} - 0.0706(\hat{x}_k^{1.5} - 1.226\hat{x}_k + 0.331\hat{x}_x^{0.5}) +$$

$$[0.105 + 0.0854(1 - 0.008\hat{x}_k)(1 - 0.002\hat{x}_k)^{-2.5}$$
$$1.425(1 - 0.0121\hat{x}_k - 0.0122\hat{x}_k^2) \times (\hat{x}_k - 0.347)^{-2}] \times \exp(0.0207\hat{x}_k) \quad (17)$$

2.2 状态模型

Al$_2$O$_3$ 浓度的状态模型可用类似于式（1）的形式简单地表达，即

$$x_{k+1} = x_k + u_k + w_k \quad (18)$$

此处，带有未知时变统计的白噪声 w_k 是用来补偿由估算得到的输入变量 u_k 中所包含的时变误差。u_k 为 $k \sim k+1$ 采样间隔内增加的 Al$_2$O$_3$ 浓度，估计式为

$$u_k = \frac{UF_k - UE_k}{m_b} \times 100\% \quad (19)$$

式中，m_b 为电解质质量，kg；设定其为常数；UE_k 是 $k \sim k+1$ 采样间隔时电解消耗的 Al$_2$O$_3$ 量，kg，估算式为

$$UE_k = 1.079 \times 10^{-2} CE \cdot T \cdot I_k \quad (20)$$

式中，CE 为电流效率，按一段时间内的出铝量确定；T 为采样间隔，min；I_k 为 $k \sim k+1$ 采样间隔内的平均电流强度，kA；UF_k 为 $k \sim k+1$ 采样间隔内溶入电解质中的 Al$_2$O$_3$ 量，kg，它与同一采样间隔内加入电解质中的 Al$_2$O$_3$ 量（记为 UD_k）存在区别，因为加入的 Al$_2$O$_3$ 需要一定的溶解时间。我们的研究表明，UD_k 与 UF_k 的关系可近似地用一个一阶惯性型低通数字滤波器的输入与输出的关系来模拟，即

$$UF_k = (1 - T/T_f)UF_{k-1} + T/T_f \times UD_k \quad (21)$$

式中，T 仍为采样间隔；T_f 为溶解时间常数，T_f 与电解槽工艺条件及物料的溶解性能有关，我们用试验确定其为一个合适的常数（$T_f \approx 10$min）。

UD_k 按 $k \sim k+1$ 采样间隔内打壳下料器的下料次数乘以每次的设定下料量近似计算

$$UD_k = NB \cdot W_f \quad (22)$$

式中，UB_k 为下料次数；W_f 为下料设定定量。

3 保证滤波收敛和提高浓度估计精度的措施

3.1 表观槽电阻的预滤波处理

$$R_n = (V_n - \varepsilon)/I_n \quad (23)$$

式中，V_n、I_n 分别为时刻 n 处的槽电压、系列电流采样值；R_n 为对应时刻的表观槽电阻采样值；ε 为代表表观反电动势的设定常数。依据我们针对测试槽所作的理论与试验研究，取定 $\varepsilon = 1.60$。

R_n 中包含着由铝液波动、阳极气体排放、系列电流波动、打壳下料时的机械搅动等引起的快时变噪声。在我们的测试槽中，由铝液波动引起的噪声的方差最大，其频率成分常在 0.01~0.04 之间。因为方差较大的快时变噪声的存在，所以若不预先进行低通滤波处理，将导致浓度估计发散。因此我们以三阶低通 Butterworth 模拟滤波器为原型，按双线性变换原理[6]设计的三阶低通数字滤波器对 R_n 进行滤波。该滤波器的性能是，通带上限频率（记为 f_p）处的衰减为 3dB；取阻带起始频率为两倍的 f_p，则阻带内最小衰减 $\geq$18dB，转移函数表达式为

$$H(z) = (1 + z^{-1})^3/\{[(c+1) - (c-1) \cdot z^{-1}] \times [(c^2 + c + 1) -$$

$$2(c^2-1)z^{-1}+(c^2-c+1)z^{-2}]\} \tag{24}$$

式中，z 代表传递函数的复变量；常数 c 由通带上限频率 f_p 及采样间隔确定

$$c = \cot(\pi f_p T) \tag{25}$$

由于 Al_2O_3 浓度相对而言为慢时变参数，因此取浓度估计所需的采样间隔为 2min。根据香农采样定理，槽电阻的采样间隔需小于 60s。在本试验中，我们采用了 $T=1×10s$ 的采样间隔，并选用 $f_p=1/400Hz$ 作为式（24）的滤波器的通带上限频率。试验表明，滤波性能良好。再按浓度估计的采样间隔（2min）对滤波器的输出序列进行抽样后，计算出一阶差分作为浓度估计模型的观测量（Z_k）。

3.2 表观槽电阻异常时的处理

为避免表观槽电阻的异常导致浓度估计的发散，在预处理算法中设立了一个判断表观槽电阻异常变化的环节。在出现异常时暂停自适应 Kalman 滤波，而仅用状态模型即式（18）进行浓度的预报，在异常状态消失后，恢复滤波估计。

3.3 特定控制策略的采用

国际上一些先进的铝电解槽计算机控制系统采用了一些富有成效的提高浓度间接估计精度的控制策略。本文提出的模型中，重要的控制策略有以下几点。

（1）Al_2O_3 浓度工作区设置在表观槽电阻对浓度变化反应敏感（即 $\partial h/\partial x$ 较大）的低浓度区。对于我们的测试槽，合适的浓度工作区为 1.8%～3.5%。

（2）为保证浓度估计精度，必须保持输入信号 u_k 有足够强度，为此须定期或交替安排"欠量"下料与"过量"下料。浓度控制的目标是维持浓度的变化范围不超过设定的工作区。尤其是在浓度估值高于上限时，需立即安排"欠量"下料，使浓度估值返回工作区，因为表观槽电阻增量对浓度的偏导数（$\partial h/\partial x$）在 4% 浓度附近存在一个零点，而且浓度高于 4.4% 后将会因沉淀导致槽底电压降升高，造成浓度的观测模型显著偏离实际过程。

（3）为了校验浓度估计值，尤其是当估计值频繁地高于工作区上限时，需定期停止下料，等待阳极效应，利用浓度向阳极效应发生的临界浓度趋近时所引起的表观槽电阻快速上升来校验浓度估值。对受测槽型的大量测试表明，在一定工艺条件下，阳极效应发生的临界浓度一般在 1.0%～1.2% 的狭窄范围内，因此当临近阳极效应引起的表观槽电阻跃升使浓度估值达到甚至低于临界浓度时，便取临界值作为新的滤波估计起点，并迅速使下料转为"过量"下料，以避免阳极效应真正发生。

（4）在进行浓度估计的同时，按质量平衡计算方式计算"参考浓度" C_k。其计算式为

$$C_{k+1} = C_k + u_k \tag{26}$$

式中，u_k 同于状态模型即式（18）中的输入变量。比较浓度估值 $\hat{x}_k$ 与"参考浓度" C_k，当发现恒向偏差时，先用上述定期安排的阳极效应等待过程来校验浓度估值，然后通过修正下料器的参数 W_f 值（参见式（22））来修正由 W_f 偏离实际值所引起的输入变量 u_k 的计算误差，以提高其后的浓度估计精度。

4 模型的验证

用某厂 160kA 中间点式下料预焙槽作为测试槽，用自制的铝电解槽工艺参数微机

采集系统[5]获得大量实测数据，然后对本文介绍的模型与算法的可行性进行了离线验证。图1给出的是其中一次试验的部分结果，实测值与模型估计值吻合良好。

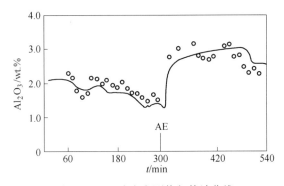

图1　Al_2O_3 浓度实测值与估计曲线

——估计曲线；○—实测值；AE—阳极效应

5　结论

从机理分析出发可建立铝电解槽的 Al_2O_3 浓度估计模型。采用虚拟噪声补偿技术可补偿简化的模型所包含的时变误差和过程中的时变噪声。用非线性系统的自适应推广的 Kalman 滤波算法并配用一些特定的预处理算法和控制策略可实现对 Al_2O_3 浓度的实时估计。

参 考 文 献

[1] 邓自立，王建国. 自动化学报，1987，9，13（5）：375-379.
[2] 邓自立，王建国. 信息与控制，1988，1：1-4.
[3] Cell Voltage. In：The 7th International Course on Process Metallurgy of Aluminum,（挪威），1988.
[4] 李劼，等. 第五届全国铝电解学术会议论文集，1992.
[5] 李劼，黄永忠，刘业翔. 轻金属，1992，9：21-26.
[6] 宗孔德，胡广书. 数字信号处理. 北京：清华大学出版社，1988，6.

Bath Temperature Model for Point-feeding Aluminium Reduction Cells[*]

Abstract To solve the problem of heat balance(HB) supervision & control of point-feeding aluminium reduction cells, an adaptive predicting estimation model(APEM) of dynamically balanced bath temperature(DBBT) is established, which co-operates with a temperature measuring device, the later employing the method of temperature dynamic measurement (TDM) and operating in intermittent way. Since it can realize quantitive description of the relationship between the output-DBBT and the inputs-heating power and equivalent feeding rate, of a HB system, APEM can be used in design of a HB supervision & control system of a point-feeding cell. Its effectiveness has been tested by using the data measured on a 160kA point-feeding prebaked anode aluminium reduction cell.

1 Introduction

It is well known that good maintenance of heat balance(HB) of cells is a key to the attainment of good technological and economic targets, therefore, realizing on-line supervision and control of HB is a long-standing problem that aluminium metallurgists have hoped to solve. As the bath is a strongly corrosive melt, conventional contact-type sensors are unable to operate continuously for a long time, and so, they are only used in manual measurements of short-term tests. Contactless-type sensors are unavailable because the surface of a cell is covered by crust, and the crust thickness is variable. Nowadays, there is a kind of indirect measurement in use, which measures cell shell temperature or side wall heat flux. However because the delay-time of its respose can reach one hour to several hours and its respose is disturbed by variations of ledge thickness, it can only be used for the quantitive analysis of medium-long term changing tendency of HB. At present, in some advanced control systems a kind of HB control model has been used which is based on energy balance calculation, but obviously this is a kind of "open loop" control method of poor accuracy. To solve the difficult problem-HB supervision & control, our lab has carried on studies on the method of temperature dynamic measurement(TDM) and has reaped first fruits in lab[1], The device of TDM can automatically insert a thermocouple into bath, using the dynamic varying curve of the potential signal obtained in the transient contacting process to calculate the bath temperature corresponding to the steady state of the potential signal. In this way, thermocouple-bath contacting time can be reduced, and so, the service life (measuring times) can be extended. But the weakness of the method is that the service life of a thermocouple is still restricted by the material of its productive shell, and, therefore, the method

---

[*] In collaboration with Li Jie, Huang Yongzhong, Wang Huazhang, Han Nu, Yang Xingrong. Reprinted from Transaction of Nonferrous Metals of China, 1994, 4(1).

may have industrial use value only when the measuring frequency is restricted in the range of once per several hours. To make up for this weakness and to realize real-time and accurate supervision and control of the HB states of point-feeding aluminium reduction cells, we have established a bath temperature model which co-operates with a temperature measuring device, the later employing TDM and operating in intermittent way.

2 Analysis and Testing of HB Characteristics

Because crust thickness and ledge thickness vary with changing bath temperature, aluminium reduction cells possess quite strong self-balance ability. In addition, techniques of semicontinuous point-feeding and cell voltage autoregulation have been widely used, therefore, with average-time concept, the HB system of a cell can be regarded as "a dynamic balance process", the slow-time varying tendency (the smoothed curve) of a varying curve of bath temperature as "dynamically balanced bath temperature" (DBBT), and DBBT as the supervision & control parameter of HB.

The first input variable which influences HB is "heating power" (denoted by Q), which is calculated with:

$$Q = A_p - A_r - A_e \tag{1}$$

where A_p denotes electric power inputted into the cell. It is calculated from on-line sampling values of cell voltage and line current:

$$A_p = V_c \cdot I \tag{2}$$

where V_c and I denote cell voltage (V) and line current (kA), respectively.

A_r denotes electrochemical reaction power (kW) required by the electrolytic process. By using thermodynamic data (at 1223K)[2], an equation for approximate calculation of A_r can be derived from the electrochemical reaction equation, that is:

$$A_r = (0.441 + 1.131r) \cdot I \tag{3}$$

where I still denotes line current (kA); r denotes current efficiency, which is assumed to be a constant, but can be modified when data from aluminium tapping are available.

A_e denotes the power (kW) dissipated in metered external resistances (outside the HB system). By giving the value of the external resistances $R_e(\mu\Omega)$, A_e becomes a function of line current I(kA):

$$A_e = R_e \cdot I^2 / 1000 \tag{4}$$

The second input variable which influences HB is material feeding rate. In general, it is alumina feeding rate calculated and ordered by the process control system, but its accurate calculation becomes difficult when manual operations (anode setting, aluminium tapping) occur which can cause additional feedings and additional heat dissipation. Nevertheless, to maintain the continuity of HB supervision & control, we take the influence of a manual operation on HB as being equivalent to an increase of alumina feeding rate. This is realized on the basis of thermodynamic data[2] and our field tests, and the equivalency can be dealt with automatically by the process control system. As for electrolyte additions, due to long addition period and small addition amount, we also, on the basis of thermodynamic data[2], take their influences on HB as being equivalent to increases of alumina feeding rate. As a result, the second input variable can be defined as "equivalent feeding rate" of aluminium. Now, the HB system of a point-feeding cell has been simplified as a double input-single output (DISO) system, the inputs being heating power (denoted by Q) and equivalent feeding rate (denoted by F_{eq}), the output being DBBT (denoted by T_b).

To test HB characteristics of point-feeding cells, we chose a 160kA point-feeding prebaked anode aluminium reduction cell in normal operation as a test cell, and used a self-made data sampling system[3] to sample cell voltage, line current, bath temperature, etc. Raw bath temperature signal was obtained with a NiCr-NiSi thermocouple that was protected by a silicon nitride tube. To eliminate high frequency disturbances in the temperature sampling tunnel, lowpass filters were set in both hardware and software of the sampling system, and quite fast sampling rate (1~10s) was used to sample primary bath temperature signal.

In order to carry on system identification experiments, we designed the input variables according to pseudo-random binary sequence (PRBS), and this was realized by adjusting the anode-cathode distance (ACD) and the time interval of feeding.

It has been determined by our tests that, under normal conditions, the settling time and the time-constant of the HB system of the test cell are about 60min and 15min, respectively, hence, bit width of PRBS and sampling period for DBBT estimation were all set as 12min, and the lowpass-filtered average of primary bath temperature signal in every sampling period (12min) was taken as the measured value of DBBT in the corresponding period.

3 Adaptive Predicting Estimation Model (APEM) of DBBT

3.1 Identification of CAR model

A controlled auto-regression (CAR) model has been used to describe the HB system of a cell. By using a linear difference equation, the CAR model is expressed by

$$y(k) = \sum_{i=1}^{n_0} a_i y(k-i) + \sum_{i=0}^{n_1} b_i u_1(k-i) + \sum_{i=0}^{n_2} c_i u_2(k-i) + d + e(k) \qquad (5)$$

where $y(k)$ denotes the output variable, here representing DBBT; $u_1(k)$ and $u_2(k)$ denote the two input variables (i.e. the controlled terms); d denotes the model bias (here temporarily assumed as a constant); $e(k)$ denotes a Gaussian white noise term with zero mean; n_0, n_1 and n_2 are the orders of auto-regressive part, and of the two controlled terms, respectively. Suppose in a HB system the base output corresponding with the base inputs ($Q = Q_0, F_{eq} = F_0$) is equal to T_0 (i.e. $T_b = T_0$). In order to enhance model identification accuracy, the sampled sequences of input and output variables (denoted by $Q(k), F_{eq}(k)$ and $T_b(k)$, respectively) should go through a disposal of subtracting their base values. Moreover, $Q(k)$ should go through a disposal of reducing by suitable times. Hence, the input and output variables of the CAR model become:

$$u_1(k) = [Q(k) - Q_0]/100 \qquad (6)$$
$$u_2(k) = F_{eq}(k) - F_0 \qquad (7)$$
$$y(k) = T_b(k) - T_0 \qquad (8)$$

where equivalent feeding rate $F_{eq}(k)$ is represented by the equivalent feeding times in the sampling period $(k-1) \sim k$; $Q(k)$ is calculated with eqs. (1) ~ (4), but cell voltage V_c and ling current I used in those equation should be replaced by their average values in the period $(k-1) \sim k$ (denoted by $V_e(k)$ and $I(k)$). Then, the following equation can be derived from eqs. (1) ~ (4):

$$Q(k) = V_c(k) \cdot I(k) - (0.441 + 1.131r) \times I(k) - R_e \cdot I(k)^2/1000(\text{kW}) \quad (9)$$

The base values Q_0, F_0 and T_0 are approximately substituted by the assembly averages of $Q(k)$, $F_{eq}(k)$ and $T_b(k)$, respectively, with a CAR model auto-identification algorithm[4] based on recursive least-square method and the parsimony principle, the determination of suitable model order and time-delay, as well as the fitting of model parameters can be carried out automatically by computer. Finally, a parsimony-parameter CAR model can be obtained.

As for out test cell, DBBT sampling period (T_s) has been set as 12min. The base values: T_0 = 958℃, Q_0 = 400kW, F_0 = 2times/12minutes, have been given on the basis of statistical results. The CAR model identification results are:

$$y(k) = a_1 y(k-1) + a_2 y(k-2) + b_0 u_1(k) + b_1 u_1(k-1) +$$
$$b_2 u_1(k-2) + c_0 u_2(k) + c_1 u_1(k-1) + d + e(k) \quad (10)$$

where

$$[a_1, a_2, b_0, b_1, b_2, c_0, c_1, d] =$$
$$[1.075, -0.199, 3.694, -1.330, 0.276, -0.124, -0.204, -0.174]$$

3.2 Multistep predicting estimation (P. E.) of DBBT and P. E. errors

Now that CAR model has been obtained, by setting the white noise term $e(k) = 0$, the model can be used to estimate DBBT in the way of multistep prediction. In order to keep generality, here we still use the general equation, (eq. (5)), in the following discussion, therefore, the following P. E. equation can be acquired:

$$\hat{y}(k) = \sum_{i=1}^{n_0} a_i \hat{y}(k-i) + \sum_{i=0}^{n_1} b_i u_1(k-i) + \sum_{i=0}^{n_2} c_i u_2(k-i) + d \quad (11)$$

where $\hat{y}(k-i)$, $i = 0, 1, \cdots, n_0$, denote the P. E. values of $y(k-i)$. Obviously, when $\hat{y}(k-i)$, $i = 0, 1, \cdots, n_0$, on the right side of eq. (11) are replaced with known $y(k-i)$, eq. (11) becomes the case of one-step P. E., otherwise, it is the case of multistep P. E.. When multistep P. E. is carried on, deviations between P. E. values and actual values will gradually increase due to measurement noise, environmental disturbances, time varying characteristics of the process, and error accumulation caused by recursive calculation, therefore, available P. E. steps are limited. As to our test cell, by using eq. (10) for recursive P. E., it has been found that, under basically normal conditions, P. E. deviations are generally within ±2.5℃ as long as recursive P. E. steps do not go beyond 20. Hence, the average sampling interval of TDM (temperature dynamic measurement) can be set as 4h, i. e. every 4h or so, it is necessary to use observations of TDM to reset starting points of P. E. as well as to provide innovations for adaptive modification of model parameters.

As the sampling interval of TDM is by far larger than the sampling period of DBBT, observations of TDM are not enough to modify all the parameters in the CAR model. Our studies have shown that the majority of P. E. errors can be attributed to variations of model bias d. For example, when a variation of the overall heat transfer coefficient of the cell results in the base value of DBBT (corresponding to the base inputs Q_0 and F_0) varying from T_0 to $T_0+\Delta T_0$, the model bias will vary from d to $d+[1-\sum_{i=1}^{n_0} a_i]$ under the assumption that in the CAR model all parameters except d do not vary.

3.3 Adaptive modification of the CAR model bias d

With the assumptions that TDM sampling interval is m times DBBT sampling period and the estimate of d at point k is $\hat{d}(k)$, the estimate $\hat{y}(k-m)$ of $y(k+m)$ can be obtained with the CAR model (eq. (11)) calculating from $k+1$ to $k+m$ in the way of recursive prediction, that is:

$$\hat{y}(k+m) = \sum_{i=1}^{n_0} a_i \hat{y}(k+m-i) + \sum_{i=1}^{n_1} b_i u_1(k+m-i) + \sum_{i=0}^{n_2} c_i u_2(k+m-i) + \hat{d}(k) \tag{12}$$

To derive a modification algorithm, we use several assumptions. First, it is assumed that TDM gives a measured value at point $k+m$, which, after being subtracted the base value T_0, is denoted by $T_d(k+m)$ and use as the observation of $y(k+m)$. Second, the observation noise term (denoted by $\xi(k+m)$) is assumed to be an independent Gaussian noise term with zero mean, i.e.

$$T_d(k+m) = y(k+m) + \xi(k+m) \tag{13}$$

Third, variations of model bias d are assumed to fit the generalized random-walk model, i.e.

$$d(k+m) = d(k) + w(k+m) \tag{14}$$

where $w(k+m)$ is assumed to be an independent Gaussian white noise term with zero mean.

With these assumptions and steady state Kalman filtering principle, an algorithm for adaptive modification of the model bias d can be derived:

$$\hat{d}(k+m) = \left[\frac{T_d(k+m) - \hat{y}(k+m)}{K_m}\right] \times \hat{d}(k) + K_t \tag{15}$$

where K_t is a modification factor ($K_t<1$), which has the meaning of steady-state Kalman filter gain; $T_d(k+m)$ is the observation given by TDM at point $k+m$; $\hat{y}(k+m)$ is the P.E. value given by the CAR model at point $k+m$; K_m is calculated with the following recursive equation:

$$K_m = \sum_{i=1}^{n_0} a_i K_{m-i} + 1, \text{if } m < i$$

Then $$K_{m-i} = 0 \tag{16}$$

where n_0 is the sub-order of the auto-regressive part in the CAR model.

3.4 Resetting of starting points of P. E.

After being used to modify the model bias d, $T_d(k+m)$ is also used to reset starting points $\hat{y}(k+m-i)$, $i=0,1\cdots n_0$, which will be used in P.E. of DBBT during next interval $(k+m-i) \sim (k+2m)$. Because the estimate of d in the interval $(k+1) \sim (k+m)$ has been modified at point $k+m$ from $\hat{d}(k)$ to $\hat{d}(k)$, the first modified values (denoted by $\hat{y}_1(k+m-i)$) of $\hat{y}(k+m-i)$ can be obtained with eq. (11)

$$\hat{y}_1(k+m-i) = \hat{y}(k+m-i) + K_{m-i} \cdot [\hat{d}(k+m) - \hat{d}(k)] \quad (17)$$

where K_{m-i} is still calculated with the same recursive algorithm as eq. (16).

Because at point k and point $k+m$ the innovations provided by the observations of TDM are:

$$\varepsilon(k) = T_d(k) - \hat{y}_1(k) \quad (18)$$

$$\varepsilon(k+m) = T_d(k+m) - \hat{y}_1(k+m) \quad (19)$$

the innovations at $k+m-i$ may be approximately calculated with

$$\varepsilon(k+m-i) = \varepsilon(k) + (m-i)/m \times |\varepsilon(k+m) - \varepsilon(k)|, 0 < i < m \quad (20)$$

Now, by imitating the steady-state Kalman filtering, the second modified value (denoted by $\hat{y}_2(k+m-i)$) of $\hat{y}(k+m-i)$ can be obtained:

$$\hat{y}_2(k+m-i) = \alpha \cdot \varepsilon(k+m-i)\hat{y}_1(k+m-i) \quad (21)$$

where the modification factor $\alpha (\alpha<1)$ may be determined as a suitable constant by tests, but, when such manual operating as anode setting, manual feeding, etc, or anode effects take place during $k \sim (k+m)$ due to poor reliability of $\hat{y}(k+m)$, $\alpha = 1$ should be set, let the resetting of the starting points of P.E. depends entirely on the observations of TDM. By replacing k, $\hat{y}(k-i)$ and d in the CAR model (eq. (11)) with $k+m-i$, $\hat{y}_2(k+m-i)$ and $\alpha(k+m)$, respectively, the resetting of the starting points of P.E. is realized.

3.5 General constitution of APEM

From the above discussion, we can sum up the general constitution of APEM in Fig. 1.

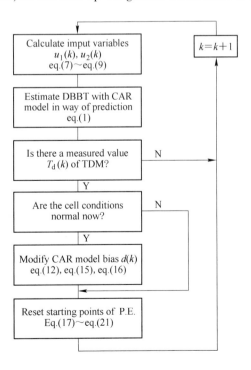

Fig. 1 General constitution of APEM

4 Verification of APEM

Since we have not carried on the testing of TDM together with that of APEM, we chose 4h as the sampling interval and sampled the primary temperature values from those sampled by the data sampling system, so as to simulate observations of TDM, and then, to verify APEM.

By off-line testing, it has been shown that APEM can solve the problem of DBBT estimation as long as cell conditions are basically normal (i. e. except anode effects and about 1h after an anode effect, as well as seriously abnormal cell conditions). Generally, estimation deviations are within ± 3℃. Fig. 2 shows a testing example.

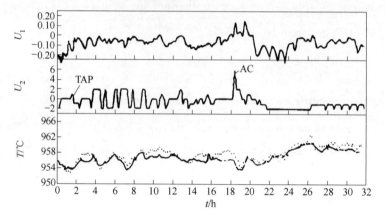

Fig. 2 Estimated curve and measured values of DBBT

u_1—the input variable depending on heating power $Q(k)$; u_2—the input variable depending on equivalent feeding rate $F_{eq}(k)$; ×—measured temperatures (℃) from simulated TDM; datted line—measured temperatures (℃) of DBBT; solid lines estimated curve of DBBT from APEM; TAP—aluminium tapping; AC—anode setting; Bath composition: CR ≈ 2.7%, CaF_2 ≈ 5; Sampling period for primary signals = 1s; Sampling period for DBBT estimation = 12min

Finally we should point out that: ①in actual use, the estimation accuracy of APEM will depend, to a great extent, on the measurement accuracy of TDM; ②APEM does not demand that the sampling interval (m) of TDM is constant, hence, in order to keep observations of TDM being as close as possible to the actual values of DBBT, and then, to enhance the estimation accuracy of APEM, measurements of TDM can be arranged at the moments that the input variables are quite stable; ③with average-time concept, it can be concluded that, as long as the mean of observation errors of TDM is equal to zero, the unbiassedness of DBBT estimation can be ensured.

5 Conclusions

(1) An adaptive predicting estimation model (APEM) of dynamically balanced bath temperature (DBBT) has been established, which co-operates with a temperature measuring device, the later employing the method of temperature dynamic measurement (TDM) and operating in intermittent way, APEM consists of a DISO CAR model an algorithm for adaptive modification of the CAR model bias, and an algorithm for resetting of P. E. (predicting estimation) starting points.

(2) By using the data measured on a test cell, it has been verified that APEM possesses quite satisfactory estimation accuracy. The values of bath temperature measured intermittently (once per several hours) by TDM can be used to modify the slow-timevarying model bias, and therefore, APEM has the ability to be adaptive to the slow timevarying process of the characteristics of the heat balance (HB) of a cell.

(3) Having has realized quantitative description of the relationship between the output-DBBT and the inputs-heating power and equivalent feeding rate, of the HB system, APEM can be used in the design or a HB supervision & control system. The combination of the parameter estimation method with TDM can become an effective way to solve the problem of HB supervision & control of point-feeding aluminium reduction cells.

References

[1] Wang Huazhang, Liu Yexiang, et al. CN 91106724. 8. 1991.
[2] Qiu Zhuxian. Physicochemistry of Alumimium Metallurgy, Shanghai: Science and Technology Press of Shanghai, 1985.
[3] Li Jie, Huang Yongzhong, Liu Yexiang. Light Metals, 1992(9):21-26.
[4] Deng Zili, Ge Yixin. Modern Times series Analysis and Appication. Beijing, Knowledge Press, 1989:32-40.

An Improved Finite Element Model for Electro-magnetic Analysis in Aluminum Cells*

Abstract This article presents the use of an improved finite element model to calculate the static electro-magnetic field for an aluminum reduction cell. Consisting of three solid cells and their surrounding bus bars, the model can evaluate the nonuniformity of the current distribution in the inside conductors and bus bar system, and couple the current into the sequential magnetic analysis through a conversion routine. Voltage potential distribution in the molten aluminum was investigated based on one industrial 320kA aluminum cell with two designed bus bar arrangements. Characteristics of magnetic components' distributions were also given.

1 Introduction

The high-amperage aluminum reduction cells with capacity over 300kA have been used widely not only in the new built aluminum plants, but also for the retrofitting of the old ones. In the worldwide, there is a trend to develop extremely large commercial electrolysis cells toward the level of 400kA, 500kA and even more. As is well known, AP 50 prototype pots were set up for the approval of the structure design, control system and economic performance[1].

Magnetic compensation is an important factor to be considered in the design process for high-amperage cells. Interest in this subject is for stabilizing the shape of the bath-metal interface and controlling the velocity of the molten liquids. Numerical calculations such as with finite element method (FEM)[2,3] or boundary element method[4] are performed to solve the Maxwell equations with presence of ferromagnetic material. Work on the electro-magnetic field has been done that focused on building the complete and complex electro-magnetic mathematic model[5], designing new bus bar arrangements especially for very large electrolysis cells[6], and measurements on newly used industrial cells for the validation purpose[7].

Currently high-amperage cells are arranged side by side in the potline, which allows current to come into anodes through side risers. The current out from the upstream side of the cell can be directed to the next anode risers under the bottom or through passing around the end side of the cell. The current out from the downstream can be connected to the next anode risers as close as possible. The bus bar sizes have to be adjusted to achieve "resisitivity balance" in the whole circuit and make sure that current flowing out in the collector bars can be evenly distributed. All is done to serve for reducing the magnetic field.

From the realistic point of view, all conductors inside and outside the cell should be connected and defined as an electric integrity to get the current and voltage distribution, of which this kind may quite differ from that of supposed based on symmetrical structures. Meanwhile it is

---

* Copartner: Li Jie, Liu Wei, Lai Yanqing. Reprinted from JOM, 2008, 60(2): 58-61.

needed to consider how this change occurring in the electric field will affect the magnetic field. Despite the fact that the magnetic distribution is caused by both the inside conductors and bus circuit, the question is how much each one will contribute to the magnetic field alone. The answer is meant to choose the most effective bus bar arrangement for the magnetic compensation. If the static Maxwell equations are solved numerically by finite element method, one has to deal with the problem of building the continual meshes, where solid conductors, lining, steel shell, open air domain are involved. Fortunately bus bars can be represented by source elements to supply current and modeled separately. These questions will be studied in this investigation with constructing three cells and their surrounding bus bars and by FEM.

2　Development of Finite Element Model

The fundamental physical laws applicable to the electric field are the Ohm's and Coulomb's Laws which govern the electric current distribution. To the magnetic field are Ampere's Law and the magnetic permeability definition which govern the magnetic field distribution. In our electric model the inside conductors and bus circuit are connected by electric constrained boundary conditions. The contact phenomenon exists between the carbon block and the steel bar. It is considered by the means of using the definition of the electric contact and magnetic contact for the electro-magnetic solution. The solution procedure is seen as time independent or static.

Before the development of the computer model, assumptions should be made for simplifying the model. The complex shell and cradles were replaced by a simple rectangular box. The superstructure didn't contribute to the magnetic filed and therefore were omitted from the model. Infinite boundary was defined at the open boundary of air. A given ledge profile existed surrounding the liquids and the interface between the bath and metal was flat.

The ANSYS code is employed to build the geometry model, assign material properties, mesh the model with all hexahedrons and then get electric and magnetic finite element models. They are both object to forces and boundaries. Five anode risers of the upstream cell are applied with given value of potline current. Another five of the downstream cell are applied with zero electric potential. Magnetic scalar potential is applied on nodes at surrounding surfaces of air.

The flow procedure of the electro-magnetic sequential coupled simulation is as follows. First the electric model is built and solved by the electric scalar potential to get the current distribution. Then this model will be converted to the corresponding magnetic model by changing element types, materials and boundary conditions, which is solved by the magnetic scalar potential. Current of the conductive parts in the cell can be stored in the result file and shared by the magnetic solution procedure. The bus bars are simply treated as conductive two-dimetinal(2-D) wire-bars in the electric model. As for the magnetic field they must be precisely transferred to current source elements according to the carrying current, the flowing direction and the cross sections of every bus bar. Fig. 1 is the electric model with three solid cells and the bus bar circuits.

The origin of the coordinate system is selected near the bottom of Fig. 1 located at the bottom of cathode lining and in the center of the cell with the upward direction as x axis, the direction to the upstream cell as y axis and the outward direction as z axis.

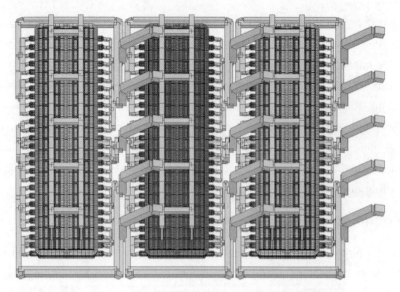

Fig. 1 Schematic of the electric model with three solid cells and the bus bar circuit
(see the attached figures)

3 Local Bus Bar Designs for Compared Analysis

In fact commercial aluminum cells are running at 320kA with one set of designed bus bars. In our computer model the bus bar arrangements have been redesigned that is quite different from the original. One bus bar configuration is finally set up as the basic setting. Two local bus adjustments differ near one corner at the upstream side (Fig. 2). They are ready for the detailed analysis of the impact of the bus bar changing on the electric and magnetic field. The solid structure and dimensions of the model cell are the same as the commercial 320kA electrolysis cell.

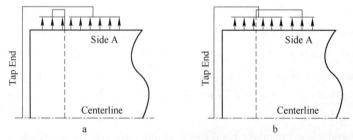

Fig. 2 Two local bus bar configurations (a & b) near the corner at the upstream side
(see the attached figures)

4 Analysis of the Electro-magnetic Field

According to the steps mentioned above, the static electro-magnetic field was calculated. The voltage solution in the cathode and the metal, current distribution in the steel bars and magnetic flux in the middle of the metal were of interest.

4.1 Voltage distribution

Cathode voltage drop (CVD) was calculated by the electric contact model. The CVD was about

0.290V with semi-graphitic blocks used in this case. Contact phenomenon can be responsible of an ohmic drop of about 100mV. Predicted isopotential curves concentrates near the bar outlet and this kind of distribution agrees well with Reference[8].

Molten aluminum has a very large electric conductance but it still goes through about ten milli-voltage drops, which will cause horizontal current flowing in the metal. Examples of two local bus bar configurations show that changing of the bus bars' grouping and location caused different voltage distributions in the metal. The voltage potential was higher where steel bars are connected by six flexes to direct current passing around the end side of the cell than where steel bars are connected by three flexes to carry current going under the bottom of the cell. The redistribution of voltage is influenced by the bus bars' equivalent resistivity, which is dependant on its paths and cross sections. So ideally optimized bus bar configuration will make voltage potential nearly even in sides of the metal, but there is no doubt that the voltage is higher in the middle than in the side. To direct current effectively from the metal vertically into the block, the cathode structure with full-length collector bars is preferred.

4.2 Current distribution

Fig. 3 shows how much current steels bars could take out in two local bus bar configurations. Variations of current were seen at the upstream side, which correspond to the resulted voltage distribution mentioned above. This is largely due to the change of the local bus bars' grouping and localization.

Also found in Fig. 3, there were seven cathode bars collecting current over 6500A at the positive side, which are wired to the circuit going beneath the cell. The current was much smoothly and uniformly distributed in the negative side than in the positive side. It's suggested that the cathode bus should be positioned under the cell to get the current uniformity. In two examples the 50 : 50 split for two sides was not reached, however the currents flowing out of the negative side were only about 1% more than the positive side.

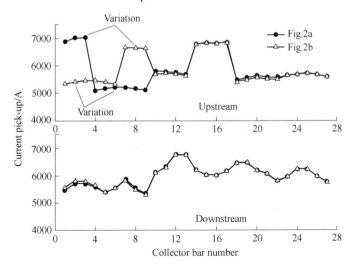

Fig. 3 Current distributions in the steel bars

(see the attached figures)

4.3 Magnetic flux distribution

Magnetic flux dependent on the non-uniformity of the current distribution was calculated. In both examples, distribution of the horizontal magnetic component Bx in the middle of the metal was antisymmetric along the longitudinal direction of the centerline of the cell with its value ranged from -138 Gauss to 169 Gauss. And that of By was also antisymmetric along the transversal direction with its value ranged from -23 Gauss to 23 Gauss. It is seen that local bus bars' adjustments don't have any effects on the horizontal magnetic field.

However, the vertical component was influenced indeed. In the Fig. 2a bus pattern, the maximum and minimum values of Bz were 30 Gauss and -25 Gauss respectively with the average being 7 Gauss while in the Fig. 2b bus pattern, the maximum and minimum values of Bz were 28 Gauss and -24 Gauss respectively with the average being 8 Gauss. In both patterns max/min values of Bz were localized at two corners of the metal near the upstream side. Bz was relatively large in the area between the first and second risers (from the coordinate origin toward the x axis) for the latter bus pattern, which was caused by current increasing in the three-groped collector bars and decreasing in the second risers. This fact indicates that the adjusted bus circuit can cause current redistribution in the risers and meanwhile risers are important contributors to the magnetic field.

4.4 Characteristic of the magnetic field

Generally, discussion on the magnetic field is based on the final calculated results without dealing with inside conductors and outside bus circuits separately. It's necessary to know how much the bus bar network can cause the magnetic field and how it will compensate the magnetic pattern introduced by the inside currents passing in the anodes, bath, aluminum, cathode and steel bars. The second bus configuration is for this kind of purpose.

The horizontal component Bx induced by the inside currents was antisymmetric along the length and ranged from -65 Gauss to 67 Gauss while that caused by the bus network had the similar distribution with min/max values being -75 Gauss to 117 Gauss. Concerning By, the distributions were much different. The perfect antisymmetric pattern occurred in the former case with max/min values existing at the end compared to the latter case where max/min values appear at the sides. Two horizontal parts of the magnetic field would finally seem to superpose to strengthen themselves.

Inside conductors caused the vertical component Bz to form two diagonal peaks and troughs at four corners (Fig. 4a). Horizontal current out of collector bars was the main reason and it also determined the basic distribution pattern for the whole magnetic field. Bz by the bus network is shown in Fig. 4b. This kind of distribution would assist in strengthening the basic pattern at the upstream side while compensating it at the downstream side. It is suggested that some amount of current should be arranged to go around the end side of the cell to reduce the magnetic field.

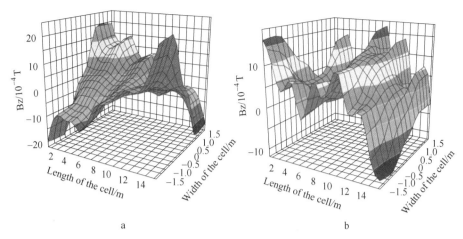

Fig. 4 Vertical magnetic components induced by inside currents(a) and bus circuits(b)
(see the attached figures)

5 Discussion

The upstream and downstream solid cells were considered in our model. So designed tests were computed to conform that they were actually involved in the magnetic solution procedure. To design large aluminum cells, the length/width ratio, confining the number of collector bars and their interval spaces is important to the MHD stability since currents in bars can induce the basic magnetic field. Another worry is that assumptions and applied boundary conditions on the model might cause effects on the solution precision but it could be very limited.

6 Conclusions

The finite element model of the electro-magnetic field is improved by building the contact mechanisms between steel bars and cathode blocks to make sure that voltage drops is properly distributed in the metal and the cathode. The model including three solid cells and the bus circuits is used to take into account impacts of bus bar adjustments on the electric and magnetic distribution. It is concluded that it's important to balance resisitivity of bus bars in the whole circuit by means of directing current under the bottom of the cell as possible and meanwhile having to put some amount of current around the end side of the cell for the magnetic compensation.

7 Acknowledgements

The authors are grateful for the financial support of National Natural Science Foundation of China(No. 50374081 & No. 60634020). The authors thank Zhengzhou Research Institute of Aluminum Corporation of China Ltd and Henan Zhongfu Industrial Co. ,Ltd for sharing their experiences on cell design and operation.

References

[1] C. Vanvoren et al. Light Metals, ed. J. L. Anjier. New Orleans, LA:TMS, 2001:221-226.

[2] J. Li et al. Acta Metallurgica Sinica. English Letters,2006,19(2):105-116.
[3] Z. W Yan et al. Electric Machines and Control,2005,9(4):326-329(in Chinese).
[4] Y. Sun et al. Acta Metallurgica Sinica,2001,37(3):332-336(in Chinese).
[5] G. V. Arkhipov. JOM,2006,58(2):54-56.
[6] M. Dupuis. Light Metals,ed. H. Kvande. San Francisco,CA:TMS,2005:449-454.
[7] C. W. Jiang. Ph. D. thesis. Central South University,Changsha,China,2003(in Chinese).
[8] R. Beeler. Light Metals,ed. P. N. Crepeau. San Diego,CA:TMS,2003:241-245.

Oxygen Overvoltage on SnO$_2$-based Anodes in NaF-AlF$_3$-Al$_2$O$_3$ Melts Electrocatalytic Effects of Doping Agents[*]

Abstract Oxygen overvoltage on sintered anodes made of SnO$_2$+2wt.%Sb$_2$O$_3$+2wt.%CuO was measured in 2.7 NaF-AlF$_3$-Al$_2$O$_3$(sat) melts at 1000℃. Prior to use the sintered electrodes were dipped into solutions containing doping agents. Doping with Ru, Fe and Cr showed a marked electrocatalytic effect. Ru was the most effective, and compared with undoped material the coefficients for the initial Tafel line decreased from $a=0.11$ to 0.070V and from $b=0.065$ to 0.035V/dec. Steeper slopes appeared at high cds, being 0.14V/dec for Ru, but the overvoltage was as low as 0.15V even at 4A/cm^2.

1 Introduction

Electrolytic decomposition of alumina dissolved in NaF-AlF$_3$ melts,

$$Al_2O_3 = 2Al + 3/2O_2 \qquad (1)$$

requires the use of inert anodes. Such materials have so far not been developed commercially. The conventional Hall-Heroult process for aluminium production uses consumable carbon anodes, and the anode product is CO$_2$,

$$Al_2O_3 + 3/2C = 2Al + 3/2CO_2 \qquad (2)$$

The reversible *emf* of this process at 1000℃ is -1.169V as against -2.196V for (1)[1]. The large difference in *emf* between the two processes is partly offset by higher overvoltage on carbon(0.4~0.6V) than on inert anodes(0.1~0.15V) at normal cds(~0.7A/cm^2). Inert anodes can probably be operated at a substantially shorter interpolar distance than carbon anodes, and the total cell voltage has been estimated to be lowered by 20% by introduction of inert anodes in aluminium cells[2]. For this and other reasons the study of inert anodes for aluminium electrolysis is of considerable interest, if suitable anode materials can be found.

Early work in this field was concentrated on the use of anodes made of metal(Cu) or carbon flushed with gas(CH$_4$)[3]. The use of massive oxide anodes was first reported by Belyaev and Studentsov[4,5] in the 1930's. Later works are found in the patent literature, as reviewed by Billehaug and Øye[6]. Little is known about the electrochemical behaviour of oxide electrodes in fused salts, whereas many studies have been made in aqueous systems[7]. Overvoltage data for oxide electrodes in cryolite-alumina melts(Na$_3$AlF$_6$-Al$_2$O$_3$) have not been reported. In this system the oxygen overvoltage on platinum is about 0.15V at 1A/cm^2 and the process appears to

---

[*] Copartner: J. Thonstad. Reprinted from Electrochem. Acta., 1983, 28(1):113-116.

be charge transfer controlled[8].

Several oxide materials have been suggested for use as anodes in aluminium electrolysis[3,6]. For the present work SnO_2-based materials were chosen because of good electrical conductivity, ease of preparation and acceptable corrosion resistance[9]. It is known from aqueous systems that doping agents may have a marked influence on anodic overvoltage[10], so the effects of such agents were tested also in the present case.

2 Experimental

2.1 Preparation of anodes

The basic composition of the anodes was 96wt.% SnO_2+2wt.% Sb_2O_3+2wt.% CuO, as recommended by Alder[9]. This material has low electrical resistivity(0.0034Ω at 1000℃) compared to pure SnO_2.

Pre-dried chemically pure oxides were passed through a 200 mesh sieve and mixed with a binding agent-camphor or water-and then pressed in a mould at 2.200~2.500kg/cm² to the shape of cylinders of 13mm dia. and 8~12mm height. The pressed "green" samples were heated from room temperature to 1250℃ during 5h and held at this temperature for 5h and then cooled slowly to room temperature. During sintering the apparent density increased from 4.3~4.4g/cm³ to 6.3~6.5g/cm³, as compared to the calculated theoretical density of 6.9g/cm³.

A simple corrosion test based on weight loss indicated that the corrosion in a 2.7NaF-AlF$_3$-10wt.% Al_2O_3 melt(molar ratio NaF/AlF_3 = 2.7) at 980℃ in the absence of aluminium was very low. With aluminium present during polarization measurements the corrosion rate was of the order of 0.04g/(cm²·h). Alder[9] reported a rate of 0.054g/(cm²·h) in the absence of aluminium. The corrosion rate tends to increase when aluminium is present because aluminium is soluble in the electrolyte to a certain extent(~0.1 wt.% Al[3]), and the dissolved metal can reduce the oxide materials. When the anode is polarized, the dissolved metal will be oxidized anodically or by the oxygen gas evolved. Anodic polarization or sparging with oxygen has been shown to protect such anodes against attack by dissolved metal[9].

Attempts to blend the oxides of the doping agents with the anode raw materials prior to pressing were not successful, due to impaired mechanical strength of the "green" pressed samples. Doping was then accomplished by dipping the sintered anodes into a solution of 10g/L of one of the following salts, $FeCl_3$, $CrCl_3$, $MnCl_2$, $CoCl_2$, $NiCl_2$, and $RuCl_3$. Alcohol served as solvent except for $RuCl_3$ which was dissolved in 20% HCl in water. After drying at 150℃ the samples were heat treated for 4~5h at 350~420℃ in air. During this treatment the sorbed chlorides were converted to oxides.

Electrical contact to the electrodes was accomplished by twisting a platinum wire around the electrode in a groove made at the upper end of the electrodes, as indicated in Fig.1. Conducting platinum cement was applied to secure good contact.

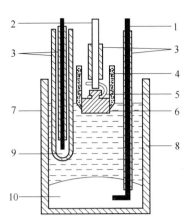

Fig. 1　Experimental cell

1—Mo wires; 2—steel rod, 3—sintered alumina tubes; 4—boron nitride sheath;
5—Pt wire; 6—SnO$_2$-based anode, 13mm dia.; 7—alumina crucible, 45mm i. d.;
8—2. 7 NaF-AlF$_3$-Al$_2$O$_3$(sat) melt; 9—aluminium reference electrode; 10—aluminium counter electrode

2.2　Experimental cell

The electrolyte was made up of hand-picked Greenland cryolite(Na_3AlF_6), sublimed AlF$_3$ and technical grade alumina to the composition 2. 7NaF-AlF$_3$-10 wt. % Al$_2$O$_3$. The melt was contained in a sintered alumina crucible as shown in Fig. 1. Super purity aluminium(99. 998%Al) was used for the counter and reference electrodes. The latter was housed in a sintered alumina tube, furnished with a hole for contact with the electrolyte. The open circuit potential difference between the two aluminium electrodes was within 5~10mV. The cell was kept under nitrogen atmosphere in an electrically heated furnace at 1000 ±2℃.

2.3　Measurement technique

Steady state polarization curves were recorded on an x-y recorded by using a potential step generator and a potentiostat. Rates of 3~6steps/min and 10~20mV per step were found to be suitable for increasing as well as decreasing potential. The ohmic resistance between the anode and the reference electrode was determined with a pulse technique, using a pulse generator and a storage oscilloscope. A part from measurements made before and after each run, current pulses were also superimposed on the polarization current in order to measure resistance during gas evolution. The resistance increased slightly(<10%) with increasing cd, i. e. increasing gas evolution rate. Visual inspection of the anode showed that gas escaped as tiny bubbles evenly distributed around the anode, as opposed to the large bubbles formed on carbon anodes. Depending on the positioning of the electrodes, the ohmic resistance ranged from 0. 16 to 0. 30Ω with 0. 003~0. 01Ω standard deviation. Accurate resistance measurements were essential since the current went as high as 5A.

3　Results and Discussion

The open circuit potential of oxide electrodes can in some cases be used to evaluate their stability[11]. The open circuit potential of the present cell ranged from 1.60 to 1.72V. In comparison the standard *emf* at 1000℃ of the reactions

$$2Al + 3/2SnO_2 = Al_2O_3 + 3/2Sn \qquad (3)$$

$$2Al + 3CuO = Al_2O_3 + 3Cu \qquad (4)$$

$$2Al + 3Cu_2O = Al_2O_3 + 6Cu \qquad (5)$$

are 1.36[12], 1.97[1], and 1.84[1] V respectively. This indicates that the electrodes did not act as reversible SnO_2/Sn electrodes, contrary to the behaviour of pure SnO_2 electrodes[12]. The observed open circuit potential was probably rather a mixed potential for the couples involved. The doping agents seemed to have a certain influence on the open circuit potential, the values being 1.68V for undoped and Mn-doped electrodes, 1.60~1.62V of those doped with Ru, Fe, Cr, Ni, and 1.72V for Co. The significance for these variations is not clear.

The open circuit potential may also have been affected by the presence of dissolved aluminium in the melt. By anodic polarization a limiting current was observed before the beginning of oxygen evolution. The limiting current is probably caused by anodic oxidation of dissolved metal[3]. This process will occur particularly at low *cds*, since at high rates of oxygen evolution the dissolved metal may be oxidized by the oxygen gas. The hysteresis observed between ascending and descending polarization curves shown in Fig. 2 is probably mainly due to these effects of the dissolved metal. In general the descending curves were considered to be the most reliable, and they gave reasonable potential readings for oxygen evolution down to $0.03 A/cm^2$.

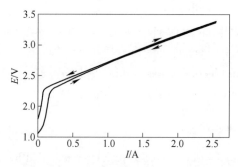

Fig. 2 Typical steady anodic polarization curve for an SnO_2-based anode vs an aluminium reference electrode measured by slow potential steps (3 steps/min, 10mV per step) with increasing (→) and decreasing (←) potential. The ohmic drop is included

To obtain the anodic overvoltage (η) the polarization data (anode-reference) were corrected for the ohmic drop, and the calculated reversible *emf* for reaction (1), (2.196V) was subtracted. Typical curves are presented in Fig. 3 as semi-logarithmic plots of η vs log i. The curve representing the undoped sample showed only one linear section followed by a marked upward curvature at higher *cds*. The doped samples showed up to three straight sections, as appears from the plots. Repeated measurements with the same electrodes doped with Fe and Ru showed no consistent time dependent changes, indicating that the electrocatalytic behaviour of these electrodes did not change during the time span of these experiments (3~4h). Long time tests were not conducted. However, the long time stability of the electrocatalytic effects may be questioned since there has been reported[3] a slight solubility in molten cryolite for most of the oxides of the doping agents used. In the case of ruthenium, volatile higher oxides (RuO_3, RuO_4) may form in oxygen atmosphere at such high temperatures[7].

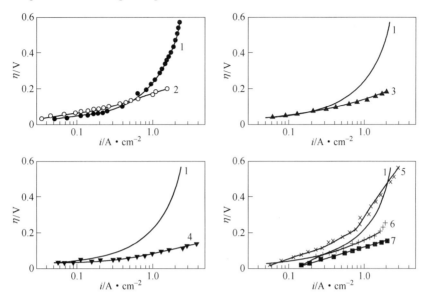

Fig. 3 Semi-logarithmic plots of anodic overvoltage on SnO_2-based anodes which were not doped(1) or doped with Mn(2), Fe(3), Ru(4), Ni(5), Co(6), Cr(7)
For curves with more than straight segment, the data for each segment are listed consecutively

Kinetic data derived from the selected curves are given in Table 1 in the form of Tafel coefficients ($\eta = a + b\lg i$) and extrapolated exchange cds (i_0). There was considerable scatter of the data, and since only 3~6 runs were made in each case, reliable limits of error cannot be given for each set of data. The overall standard deviation was about 15%, and in some cases the scatter seemed to be even larger. More extensive measurements would be needed to identify the sources of error and to limit the uncertainty so as to get more precise data. Possible sources of error are inhomogeneity of the electrode materials, drift in the potential of the reference electrode (~10mV), anodic oxidation of dissolved aluminium at low cds, and fluctuations in resistance and wetted surface area at high cds.

Table 1 Tafel coefficients (a, b) and exchange cds (i_0) for oxygen evolution on SnO_2-based anodes in 2.7NaF-AlF_3-Al_2O_3(sat) melts at 1000℃

Doping agent	a/V	b/V	i_0/A·cm^{-2}	Range/A·cm^{-2}
None	0.11	0.065	0.023	0.05~0.30
Mn	0.13	0.070	0.013	0.02~0.18
	0.17	0.13	0.04	0.25~1.80
Fe	0.11	0.070	0.022	0.06~0.45
	0.12	0.12	0.01	0.55~2.00
Ru	0.070	0.035	0.001	0.085~0.50
	0.080	0.093	0.14	0.60~4.0
Co	0.16	0.16	0.01	0.18~0.43

Continued Table 1

Doping agent	a/V	b/V	i_0/A·cm^{-2}	Range/A·cm^{-2}
Co	0.17	0.19	0.13	0.55~1.65
Cr	0.12	0.11	0.09	0.15~0.70
	0.12	0.17	0.18	0.85~2.0
Ni	0.19	0.14	0.035	0.06~0.30
	0.26	0.26	0.10	0.40~0.80
	0.29	0.59	0.32	0.90~3.00

It is evident from the data in Table 1 as well as from Fig. 3 that some of the doping agents have a marked electrocatalytic effect on oxygen overvoltage on SnO_2-based electrodes. In particular Ru have a strong positive effect, the overvoltage at $1A/cm^2$ being only about one third of that for undoped electrodes. Also Fe and Cr are beneficial. In the cd range commonly used in fused salt electrolysis, (0.5~$1A/cm^2$) the effectiveness of the doping agents can be ranked as follows,

$$Ru > Fe \approx Cr > Co \approx Mn > Ni$$

Doping with Mn and Ni had a negative influence at low medium cds.

It is well known that RuO_2 has a beneficial effect on oxygen overvoltage also in aqueous solutions[7]. The overvoltage is higher in aqueous solutions than in the present system, and the reaction mechanism is obviously different in the absence of water. The electrocatalytic effect may be due to changes in stoichiometry, valence, crystal structure, nature of conduction, surface states, etc.[7,13-20]. Generation of more active sites will be one simple explanation. Most of the doped anodes gave semi-logarithmic plots with two different slopes. Such behaviour may be due to a change of rate-determining step, a change in surface coverage, a change of reaction mechanism, or change of oxidation states at the electrode surface[7].

In a previous study of oxygen overvoltage on platinum in cryolite-alumina melts at 1000℃[8] two slopes were observed on semi-logarithmic plots, i.e. 0.083 and 0.245V/dec. These values were shown to agree well with the expected slopes for a two-step discharge reaction

$$``O^{2-}" \longrightarrow O_{(ad)} + 2e \tag{6}$$

$$``O^{2-}" + O_{(ad)} \longrightarrow O_{2(ad)} + 2e \tag{7}$$

which should be $2.3RT/3F(=0.084V)$ and $2.3RT/F(=0.25V)$ if the charge transfer coefficient $\alpha=0.5$ and the second step (7) is rate-determining at low cds ($2.3RT/3F$) and the first step (6) at high cds ($2.3RT/F$). The "O^{2-}" ion is assumed to be a dissociation product of Al-O-F complexes[3].

The slopes given in Table 1 evidently depend upon the doping agents. The lowest slope is lower than $2.3RT/3F$ and the highest is higher than $2.3RT/F$. A further discussion of the reaction mechanism on the basis of the present results is hardly warranted.

References

[1] JANAF Thermochemical Tables, 2nd edn., U.S. Dept. Commerce, Nat. Bur. Stand., Washington, 1971.
[2] N. Jarrelt. AI Che Symp. Ser., 1981, 11(204).

[3] K. Grjotheim, C. Krohn, M. Malinovsky, K. Matjasovsky and J. Thonstad, Aluminium Electrolysis. 2nd edn. Aluminium-Verlag, Düsseldorf, 1982.
[4] A. I. Belyaev and Ya. E. Studentsov, Lēgk Metally, 1937, 6(3): 17.
[5] A. I. Belyaev, Lēgk Metally, 1938, 7(1): 7.
[6] K. Billehaug and H. A. Øye, Aluminium 57, 146(1981); ibid., 1981, 57: 228.
[7] S. Trasatti and G. Lodi, in Electrodes of Conductive Metallic Oxides, (Edited by S. Trasatti). Part A, p. 301; Part B, D. 522. Elsevier, Amsterdam, 1981.
[8] J. Thonstad, Electrochim. Acta, 1968, 13: 449.
[9] H. Alder, U. S. Pat. 1976, 3(960): 678; 1976, 3(974).
[10] V. De Nora, Diaphragm Cell Chlorine Prod. Proc. Symp., p. 15. Sot. Chem. Ind., London, 1977.
[11] B. V. Tilak and N. L. Weinberg, AIChE Symp. Ser., 1981, 77(204): 60.
[12] M. Rolin and A. Ducouret, Bull. Soc. chim. Fr., 1964, 790: 794.
[13] Ya. M. Kolotyrkin, V. V. Losev. D. M. Shub and E. Roginskaya, Soviet Electrochem., 1979, 15: 245.
[14] R. S. Yeo, J. Orehotsky, W. Visscher and S. Srinivasan, J. Electrochem. Soc., 1981, 128: 1900.
[15] L. D. Burke and J. F. Healy, J. electroanal. Chem., 1981, 124: 327.
[16] M. H. Miles, E. A. Klaus, B. P. Gunn, J. R. Locker and W. E. Serafin, Electrochim. Acta, 1978, 23: 521.
[17] D. Galizziolo, F. Tantardini and S. Trasatti, J. appl. Electrochem, 1974, 4: 57.
[18] L. I. Krishtalik, Electrochim. Acta, 1981, 26: 329.
[19] S. Yamada, Y. Matsumoto and E. Sate, Denki Kagoku, 1981, 269.
[20] M. Morita, C. lnakura and H. Tamura, Electrochim. Acta, 1977, 22: 325.

Progress in Studies of Electrocatalysts and Doped Carbon Anodes in Aluminium Electrolysis Cells[*]

Abstract The second generation electrocatalysts with little or without lithium salt have been studied in laboratory. The promising electrocatalysts which can be used on prebaked anodes in Hall-Heroult process can reduce anodic overvoltage (η_a) up to 200mV, and those which can be used on Söderberg anodes can reduce η_a up to 148mV. Some arguments about the lithium salt-containing anode paste are discussed.

1 Introduction

Aluminium electrolysis industry consumes huge amount of electric energy. It is now facing urgent aspect-the rising of energy prices. Electrocatalysis is a hopeful way to save energy. We[1-3] have been studying the electrocatalytic effects of doped carbon anodes in Hall-Heroult process since 1985. After tested in industrial cells, 2000 HSS cells have used lithium carbonate containing anode paste in their production in China so far. This technique being employed by many smelters is not an accidental phenomenum. It has really shown the merit of saving energy. This paper will report progress related to our studies on the doped carbon anodes and offer explainations to some questions.

2 Progress Related to Studies on Second Generation Electrocatalysis

We started the studies on the second generation electrocatalysts after the lithium carbonate doped anode paste found application in industry. The aims of these studies are as follows.

(1) To reduce the cost of electrocatalyst. The comsumption of Li_2CO_3 is around 3~4kg/t-Al, oweing to the high price of lithium carbonate (more rhan 30000 yuan/t), the economical efficiency is not so good. It is necessary to find a type of electrocatalysts containing little or no Li_2CO_3, and the additives should be no harm to the metal quality and the cell operation.

(2) To improve the electrocatalysts introducing method. The previous mixing method was not satisfactory, because it was not really a doping method.

(3) To seek for a type of electrocatalysts used in prebaked anode which should maintain electrocatalytically active after baking at 1250℃.

A number of new type electrocatalysts were selected from fifty dopants by these studies. They are suitable for both Söderberg anodes and prebaked anodes. Some of these electrocatalysts are listed in Tables 1 and 2.

---

[*] Copartner: Wang Xiangmin, Huang Yongzhong, Wang Huazhang, Yang Jianhong. Reprinted from Transaction of Nonferrous Metals of China, 1994(2):92.

Table 1 Results of electrocatalysts used for Söderberg anode

Electrocatalysts	Overvoltage reduced[1]/mV
K-Ca Salts	148
Li Salts	147
Li-Mg Salts	80
Li-Mg-Ca Salts	74
Mg-Al Salts	68
Mg-Fe Salts	54

[1] Comparison was made between doped anodes and undoped ones at 0.8A/cm² anode current density.

Table 2 Results of electrocatalysts used for prebaked anode

Electrocatalysts	overvoltage reduced[1]/mV
Ba-Fe Salts	208
Mg-Al Salts	170
K-Ca Salts	150
Li Salts	8

[1] as Table 1.

The experimental procedures presented elsewhere[1-5]. The possible mechanism of electrocatalytic effect have also been discussed. We suggested that the active sites which transformed from dopants were formed at the anode surface. Through the preparation of electrodes, active sites should possess electronic conductivity and huge surface area. Which are beneficial to promote the main anodic reaction. That is to say, these sites would accelerate the electrons exchange velocity of oxygen evolution reaction (CO_2 formed), thereby decrease the excessive energy consumption, thus decrease the anodic overvoltages, However, the dopants selection and the electrodes preparation need technical know-how.

3 Explanation Concerning Differences in Experiments and Mesurements

Feng, Naixiang et al[6,7] argued that no measurable effect of the lithium carbonate addition was found on the anodic overvoltage or on the rate of the anodic reaction. However, in our studies, we can obtain the measurable η_a reduction almost every time when testing the lithium salt doped anode in laboratory cells. The main differences were in the preparation of electrodes, and the experimental conditions and methods employed. The dopants should be distributed homogenously in the electrode, and possess huge surface area after special treatment. However, the preparation method is also a knowhow at the moment. We are willing to provide some well prepared electrode samples to those who are interested in the measurements of η_a on their request. We believe that the can get quite the same measuring results as we did.

"Doping" is one of the chemical modified electrode methods through which the modified electrode possess many new functions, such as electrocatalysis, selectivity and stability and so on. Using our know-how, an isolate porcelain thin piece exhibited electronic conductivity after doping and heat-treatment.

We agreed that the graphite crucible which contained the cryolite-alumina melt should be shielded by alumina tube, otherwise not only horizontal current would disturb the measurements, but also the composition of melts would be changed.

It should be pointed out that the measurement of η_a on an operating reduction cell is very difficult and is hard to obtain reproduceable results. The reasons are as follows: ① The operating anodes on an industrial cell are far from electrochemically steady state; ② The real current density changes with time and different measuring locations on the anode, especially with the fluctuation of current supply. As is well known, the η_a varies with current densities; ③ The η_a measuring method is not very good until now. Among existing methods W. Haupin's is the best one, which was employed in our in-site measurement. The measured η_a data were instant. It was required to keep the experimental conditions and the measuring technique as constant and strict as possible. An expert group invite by CNNC made the η_a measurement in industrial cells in Shandon Aluminium Smelter and Liancheng Aluminium Smelter in 1989 for judging the experiments and tests both in our laboratory and these two smelters. They obtained effective measurement results.

4 Evaluation Related to Energy Saving Effect Using Lithium Salt-containing Anode Paste

Until now, many aluminum smelters still continue employing lithium salt-containing anode paste in their cells, and it is one of the major energy saving measures, But, Sen[7] questioned that there was no effect on reducing the cell voltage, because the average cell voltage was not decreased. However, the data cited in Sen's paper showed the energy saving effects in the aluminium smelters which use lithium salt-containing paste. According to the following well known equation, one can calculate the energy consumption W and make the comparison:

$$W = 2980 V_{Cd}/CE, kW \cdot h/t\text{-Al}$$

where 2980 is constant, related to electrochemical equivalent of aluminium; V_{Cd} is the average cell voltage(V) and CE is current efficiency(%).

This expression demonstrates that the energy consumption can be reduced by either lowering the average cell voltage or increasing the current efficiency. Thus, Shandon Aluminium Smelter obtained the annual energy reductions 81, 203, 232 and 292kW · h/t-Al respectively from 1987 to 1991. Hushun Alunminium Smelter was more effective and obtained annual energy reduction 231, 300, 245 and 330kW · h/t-Al respectively from 1987 to 1991. Table 3 listed some parameters of one of the potlines(No. 100) of Hushun Aluminium Smelter.

Sen evaluated the energy saving results merely by average cell voltages and did not pay attention to the increase of CE. As a matter of fact, As a matter of fact, the reduced anodic overvoltages were in several decades to 200mV, and they are hard to detect by a voltmeter which is mounted on the reduction cell and with a 0.2V precision. Besides, the cell voltages are controlled and regulated by computer in existing smelters, and maintained in a certain region. Once there is a η_a reduced in a reduction cell, it would be compensated by an enlarged anode-cathode distance(ACD). As is well known, an greater ACD is beneficial to the CE. As Table 3 showed was increased annually.

Table 3 Some parameters of reduction cells of potline No. 100 in Hushun Aluninium Smelter

Parameters	Years				
	1987	1988	1989	1990	1991 Jan. ~ Jun.
Ave. Cell Voltage/V	4.367	4.359	4.377	4.367	4.370
CE/%	88.91	89.43	90.24	89.68	90.27
D. C. Energy Consumption /kW·h·(t-Al)$^{-1}$	14759	14528	14459	14514	14429
Li$_2$CO$_3$ Containing in Bath(wt.)/%	—	—	2.83	2.75	2.70
Energy Saving/kW·h·(t-Al)$^{-1}$		231	300	245	330

Electrocatalysis and functional electrode materials are attracting more and more attention because they can lead to energy reductions in many technological fields including electrometallurgy. We will continue our work in this direction.

References

[1] Liu Y X, Thonstad J. Electrochim Acta, 1983, 28(1):113-116.
[2] 严大洲, 刘业翔, 肖海明. 中南矿冶学院报, 1989, 20(5):505-511.
[3] Liu Y X, Xiao H M, Xiong G G. Light Metals, 1991(Warrendale, PA. TMS), 489-494.
[4] Liu Y X, Xiao H M, Chan Z M. Light Metals, 1989(Warrendale, PA TMS), 275-280.
[5] Liu Y X, Wang X M, Huang Y Z, Wang H Z. Light Metals, 1993(Warrendale,. PA. TMS), 599-601.
[6] Feng. Naixiang et al. Carbon, 1991, 29(1):39-41.
[7] 沈时英. 轻金属. 1993(12):26-28.

On the Electrocatalysis of Doped Carbon Anodes in Aluminum Electrolysis*

A literature review is given on the field of electrocatalysis of the carbon anode reaction in aluminium eletrolysis. An experimental study was conducted on the anodic overvoltage of graphite anodes with an additive consisting of Mg-Al oxide. Results obtained in saturated cryolite-alumina melts at 1010℃ showed a certain electrocatalytic effect, the reduction in anodic overvoltage at 0.8A/cm^2 being 35mV in average.

The performance of the carbon anode in aluminum electrolysis cells is an important factor with respect to carbon consumption, current efficiency, metal quality and energy consumption. Therefore, much attention has been paid to anode problems, and a number of studies have been carried out. Research regarding the effect of additives (impurities) on the electrical conductivity, oxidation resistance and anode consumption has been reviewed in a monograph[1].

Since 1980, several studies have been reported on the electrocatalytic effects of carbon anodes doped with additives. The research has mostly been focused on reducing the anodic overvoltage(η_a) and thereby possibly reducing the energy consumption.

Electrocatalysis in aluminum electrolysis is an interesting and promising way of saving energy. In the present work, a literature review on the field of anodic electrocatalysis is given, and careful measurements were made of anodic overvoltage of graphite anodes with an additive (spinel type compound $MgAl_2O_4$-MgO).

1 Literature Review

1.1 The effect of various additives on anodes performance

Since the concept "electrocatalysis" was first introduced in the 1960s[2], great progress has been made in this field, especially in the chlor-alkali industry, fuel cells, photoelectrocatalysis, organic synthesis, etc. In the following, literature data on the effect of additives on the anodic overvoltage (η_a) in aluminum electrolysis is reported.

Thonstad and Hove[3] studied the effect of some additives to carbon anodes on the anodic overvoltage(η_a). Amounts of 0.4% ~ 1.3% were added to the coke. The results showed that η_a decreased slightly by the addition of Fe_2O_3 and Na_2CO_3. While H_3BO_3 led to a slightly higher overvoltage, the differences being of the order of 5mV.

Braunworth et al.[4] suggested AlF_3 as an anode additive that could reduce the excess carbon consumption, based on tests in horizontal stud S derberg cells. This has also been supported by

---

* Copartner: J. Thonstad and J H Yang. Reprinted from Aluminum, 1996, 72(11):836.

laboratory tests[5].

Liu and Thonstad[6] showed that for oxygen evolution on SnO_2-based anodes in cryolite-alumina melts at 1000℃, η_a was lowered by adding dopants to the anode. The electrocatalytic effectiveness was ranked as follows:

$$Ru > Fe, Cr > Co, Mn > Ni$$

At a current density of $1A/cm^2$, η_a for a Ru doped anode was only about 1/3 of that of an undoped electrode. Liu[7] proposed to apply some additives to the carbon anode to improve its electrochemical activity so as to reduce the η_a. Several research works have since been carried out in China. A number of electrocatalysts have been proposed and tested. Liu and Xiao[7-9] tested the dopants $CrCl_3$, $MnCl_2$, $CoCl_2$, $NiCl_2$, $RuCl_2$, Li_2CO_3 and NaCl. In cryolite-alumina melts at 1000℃, the η_a of anodes doped with $CrCl_3$, Li_2CO_3 and $RuCl_2$ could be reduced by 275, 181 and 148mV, respectively at $0.85A/cm^2$ as compared to an undoped anode. Tests were later carried out in industrial cells. Yao et al.[10] carried out similar experiments. An addition of 1.5wt.% Li_2CO_3 was made to Söderberg anodes, and η_a was reduced by 380mV at $0.96A/cm^2$ (the original η_a calue was not given). However, Feng et al.[11,12] claimed that test with 1.3wt.% Li_2CO_3 addition showed no measurable effect on the anodic overvoltage. Yu[13] found that Na_2CO_3, MgF_2 and NH_4VO_3 possessed some catalytic effect on the anode reaction, while HBO_3, $Na_2B_4O_7$ and Li_2CO_3 acted as inhibitors.

Duan et al.[14] presented results of a test in a 100A laboratory cell. S derberg type anodes containing 0.3wt.%~0.5wt.% Li_2CO_3 showed the best electrocatolytic effect, i.e. at $0.88A/cm^2$, η_a was reduced from originally 444mV to 285mV. The experiments were conducted at 1000℃ in melts containing 0.5wt.% Li_2CO_3 and CR = 2.9 (molar NaF/AlF_3 ratio). Yan[15] studied the effect of additives of rare earth elements (Ce, Y, Nd and Pr) or a mixture of these elements on the η_a. In the range of industrial current densities ($0.5~1.0A/cm^2$), η_a was reduced by 220~340mV, e.g. the a of undoped graphite was 585mV at 1000℃ and $1.04A/cm^2$, compared to 245mV after doping with yttrium salt.

Li[16] investigated the use of Li_2CO_3, Ba-Li, Li-Na-Fe and Mg-Fe complex salts as anode dopants. The results showed that, compared with the η_a value (454mV at $0.85A/cm^2$) for undoped anodes, the η_a of Söderberg type anodes was reduced by 261mV for Ba-Li salt addition, 198mV for Li-Na-Fe, 125mV for Mg-Fe and 147mV for Li_2CO_3 addition. For prebaked type anodes, the following reduction in η_a was found: Li_2CO_3 22mV, Mg-Fe salt 64mV, and Ba-Li salt 66mV. These results indicate that the elevtrocatalytic effect was diminished when the doped carbon anode was baked at a higher temperature (1200℃). Liu[17] studied the electrocatalytic effect of additives of complex salts such as Ca-Mg, Li-Ca-Mg salts etc. the η_a of an undoped Söderberg type anodes was 440mV at $0.85A/cm^2$, being reduced by 213~275mV after doping. The catalytic activity of the anode decreased after being baked at 1250℃, and η_a was reduced by only 88~130mV (the original η_a value of prebake type undoped anode was 446mV).

Hu[18] tested the electrocatalytic activity of carbon anodes doped with complex salts, such as Fe, Ca, K-Ca, K-Ca-Fe, K-Ca-Mg-Fe. The doped anodes contained 0.2 wt.% complex salts and were baked at 1250℃. The η_a of an undoped prebaked type anode was 638mV at $0.85A/cm^2$, being reduced by 99159mV after doping. The authors mentioned above[15-18] also tried to interpret the anodic reaction mechanism. Haarberg et al.[19] presented result related to industrial type carbon anodes with Fe_2O_3 additive. The η_a of a carbon anode was reduced by more than 100mV from an initial calue of 0.8V at $1A/cm^2$ Liu et al.[20] used several additives separately as dopants in carbon anodes. Samples baked at 1200℃ still exhibited electrocatalytic activity e.g. Ba-Fe salt lowered η_a by 200mV, Mg-Al salt 170mV, and Ca-K salt 150mV (at 1000℃, saturated cryolite alumina melt and $0.8A/cm^2$, the η_a of undoped prebaked type anode being 590mV). It was suggested that perovskite and spinel compounds formed from these dopants at high temperature could lead to the formation of active sites at the electrode surface, promoting the electron transfer so as to accelerate the anode reaction rate and reduce the η_a. The main experimental results obtained were summarized[21,22] and are listed in Table 1.

Table 1 Effect of electrocatalysts added to Söderberg and prebaked anodes at $0.8A/cm^2$ (difference between undoped and doped anodes)

Anode type	Electrocatalysts	η_a, reduced by mV	Ref.
Söderberg anodes	K-Ca salts	148	[22,24]
	Li_2CO_3	147	[20,35]
	Li-Mg slats	80	[20]
	Li-Mg-Ca salts	74	[20]
	Mg-Al salts	68	[35]
	Mg-Fe salts	54	[20]
Prebaked anodes	Ba-Fe salts	208	[24]
	Mg-Al salts	170	[24]
	K-Ca salts	150	[24]
	Li_2CO_3	8	[20]

In addition to the research on anodic overvoltage (η_a), there are many works in the literature dealing with other properties of carbon anodes containing additives. A few relevant papers will be mentioned here.

Liu et al.[23] and Qiu et al.[24,25] studied the effect of dopants on the anode effect and critical current density (ccd) for the initiation of the anode effect[25,26] of carbon anodes in aluminum electrolysis. Liu's result showed that doping with Pb, Cr and Ru salts increased the ccd and inhibited the occurrence of the anode effect. Qiu's experiment showed that carbon paste containing lithium salt showed better wettability by the bath, higher ccd and lower η_a.

Kuang[27] and Solli[28] tested the effect of impurities of V, S, Al, Fe, Ni, Li and Na on the carbon consumption in aluminum electrolysis. The results showed that the compounds Na_2CO_3, NiO, Fe_2O_3, AlF_3, $LiCl$, V_2O_5 increased the electrolytic consumption of the anodes, while Al_2O_3 showed a weak catalytic effect and sulphur behaved as an inhibitor. However, sulphur increased the dusting of the anode.

The effect of lithium salt additives and anode paste containing lithium salt has been discussed in China since 1987[11-13,29-34]. The main arguments were as follow:

(1) The η_a value which appeared in some works[7,10] was not correct.

(2) Shen and Li[29-31] maintained that the reduction in η_a by 380mV reported in paper[10] was unlikely.

(3) Feng and Yu[11-13,32] did not find any measurable effect on η_a by Li_2CO_3 addition.

(4) The good performance of industrial cells employing paste containing lithium salt was not due to reductions in η_a[33,34], but resulted from the lithium salt which was transferred to the bath, improving the properties of the bath, and thereby the electrolysis process.

As a result of these discussions, some conclusions were given[22,35]:

(1) The calculation of η_a was not correct at the very beginning of these studies, but it was corrected later on. The experimental results, i.e. the differences of η_a obtained by comparing doped and undoped anodes were correct.

(2) The choice of carbon materials, the preparation of electrodes and the measurement techniques were rather different among various authors. This could be the cause of the discrepancies.

(3) Industrial practice in cells using lithium bath showed that anodes made from paste containing lithium salt performed better than normal carbon anodes, as demonstrated in Table 2.

(4) Electrocatalysis in high temperature melts is quite a new field which can lead to effective energy savings. More work is needed, in particular to explain discrepancies.

Table 2 Main technical and economical parameters of some Chinese aluminum smelters using anode paste containing lithium salt. CE current efficiency

Smelters	Potline current /kA	Av. Cell voltage /V	CE				Anode	
			CE /%	Increase /%	kW·h/t-Al	LiF in bath /%	Paste consump /kg·t^{-1}	Energy saving /kW·h·t^{-1}
Liancheng①	75.2	4.41	90.2	1.1	14580	3.40	493	152
Shandong	63.1	4.32	89.9	1.3	14330	2.58	536	150
Fushun	60.6	4.37	89.7	0.5	14510	2.87	529	85
Baotou①	62.5	4.34	89.6	0.6	14450	2.51	528	60
Lanjiang	50.6	4.29	90.3	0.9	14160	2.03	517	156

① The parameters listed here are results compared with direct addition of Li_2CO_3 to the bath.

1.2 Anode paste containing lithium salt

Since 1987, industrial tests of carbon anodes containing Li_2CO_3 have been carried out in China. Cooperation was established between the Central-South University of Technology and the Liancheng Aluminium smelter[36,37], and between the North-East Institute of Technology and the Shandong Smelter[38] to perform such tests for nine month on two potlines in each smelter. Good results were obtained, the energy saving in the Shandong Smelter being 460kW·h/t-Al and at the Liancheng Aluminum Smelter 305kW·h/t-Al. Since then this technique of doping the anode paste with lithium salt has spread to many aluminum smelters in China.

In 1989, China National Nonferrous Metals Corporation (CNNC) organized an evaluation and appraisal of this technique by Chinese specialists in the field of aluminum electrolysis. It was concluded that the average cell voltage of the test potlines was reduced by 50 ~ 63mV, the current efficiency was increased by 0.54%, and the overall energy savings were 246kW · h/t-Al (Shandong) and 152kW · h/t-Al (Liancheng), respectively. The energy savings were obvious, and the economical benefit was also good. Some main technical and economical parameters of several aluminum smelters which use paste containing lithium salt are listed in Tables 2 and 3. Normally 0.3% ~ 0.8% (wt) Li_2CO_3 is added to the paste.

Table 3 Some parameters of a potline in the Fushun Aluminum Smelter

Parameters	1987	1988	1989	1990	1991(Jan. ~ Jun.)
Av. Cell voltage/V	4.37	4.36	4.38	4.37	4.37
Current efficiency/%	88.9	89.4	90.2	89.7	90.3
Energy consumption/kW · h · (t-Al)$^{-1}$	14760	14530	14460	14510	14430
Li_2CO_3 in bath(wt)/%			2.83	2.75	2.70
Energy saving/kW · h · (t-Al)$^{-1}$		231	300	245	330

By the end of 1991, there were 2000 horizonal stud S derberg (HSS) cells belonging to 16 aluminum smelters in China which employed anode paste containing lithium salts, and 50 million kW · h electric power per year was saved (according to the statistical data provided by CNNC). Owing to outstanding word on energy saving, Liu and his co-workers and other contributors were awarded the first prize of the National Science and Technical Progress Prize of 1992.

2 Experimental

Preparation of doped and undoped graphite electrodes:

In order to have a uniform anode material, spectrographic pure graphite rods were chosen. The electrodes were made with the shape of a cone, as shown in Fig. 1a, in order to facilitate the escape of the gas bubbles.

Doping was accomplished by dipping the cleaned and dried graphite electrodes into a 100mL aqueous solution with 0.02mol MgO+0.01mol Al(NO$_3$)$_3$ · 9H$_2$O+5mL HNO$_3$+10mL ethyl alcohol for 24 hours. Before dipping, the graphite rods were held under vacuum in a bottle which was connected to a water ejection pump for half-an-hour, then a valve to a funnel was opened and the solution was allowed to flow slowly into the bottle until the graphite rods were totally immersed, and pumping was continued for another half-an-hour. After drying at 110℃, the impregnated rods were heat treated at 480 ~ 500℃ in air for 2 hours. Electrodes from the same graphite rod were divided into two parts, one was doped as described above, the other was undoped. Identical measurements were made with the two parts in order to keep the conditions as similar as possible.

3 Experimental Technique

The measurements were performed with one undoped and one doped graphite electrode simultaneously, as shown in Fig. 1b. A galvanostat gave a constant current, and an X-Y recorder recorded the potential between the two working electrode. The ohmic drop between the anode and the reference electrode was determined with a pulse technique, using a current interruptor switch (Model 800 IR measurement system, Elctrosynthesis Company. Switching time: 0.2μs, duration of current interruption 10μs) and a storage digital oscilloscope (LeCroy Scopestation 140). A dummy cell with a resistor (0.3 ohms) and a capacitors (24.7 microfarad) in parallel was used to test if the IR measuring system and the calculation method. The electrolyte was a cryolite-alumina saturated melt at 1010℃, and hand-picked natural grade alumina was used in the measurement.

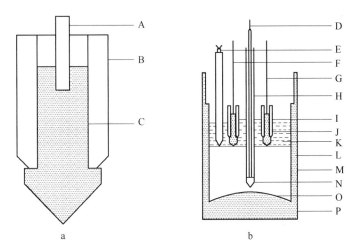

Fig. 1 The shape of the graphite electrodes

A—3mm steel rod; B—Sintered alumina sheath; C—Graphite anode; Laboratory cell;
D—Reference electrode; E—Pt-Pt 10% Rh thermocouple; F, G—3mm steel rod; H—Reference
electrode in sintered alumina closed end tube; I, J—Working anodes (one doped, the other undoped.);
K—Cryolite-alumina saturated melt; L—Alumina crucible; M—Graphite crucible;
N, O—99.99% pure Al; P—Graphite plug going through a hole in the alumina crucible

Pre-electrolysis was performed with an auxiliary anode consisting of a 10mm diameter graphite rod dipping 10mm into the bath, and applying 0.6A for 0.25h. Subsequently the aluminum reference electrode and the anodes were lowered into the melt and kept for 40 minutes, before the experiment was started. The cell current was decreased in steps of 0.2A from 1.9A, and a steady state potential was normally reached after 40s. The anodic potential corrected for the ohmic drop in the electrolyte was derived as follows: After the ohmic drops were obtained from the current-interruptor potential-time curves recorded by the oscilloscope, a linear regression of the ohmic drops at different currents was made to get the electrolyte resistance between the working electrode and the reference electrode. Subsequently the polarization potential of the anode was calculated by subtracting the ohmic voltage drop from the measured potential recorded by the X-Y recorded. At $1A/cm^2$ the ohmic drop was typically 380mV.

4 Results and Discussion

4.1 Electrocatalytic effect of doped graphite electrodes

Measurements were conducted with nine sets of undoped and doped graphite electrodes with the same dopants and melt composition, respectively. A plot of anodic potential corrected for the ohmic voltage drop versus log current density is shown in Fig. 2. It appears that the reproducibility of the anodic potential was rather poor. However, the difference between undoped and doped graphite anode is obvious. By subtracting the reversible potential (1.164, 1010℃ [1]) a regression analysis gave the following Tafel equation ($\eta = a + b \lg i$) for undoped graphite anodes,

$$\eta = 0.599(\pm 0.033) + 0.177(\pm 0.0124)\lg i \qquad (1)$$

and for doped graphite anodes,

$$\eta = 0.565(\pm 0.028) + 0.188(\pm 0.0106)\lg i \qquad (2)$$

where the values inside the paranthesis indicate the standard deviation for the respective Tafel constants. At $0.8 A/cm^2$, the potential difference between undoped and doped electrodes are about $-35 mV$ in the favor of the doped specimens.

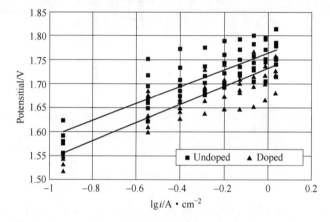

Fig. 2 Potential of undoped and doped graphite anodes versus $\lg i$

Extrapolation of the equations above yielded the exchange current density i_0 of doped graphite, being $9.88\ 10^{-4}\ A/cm^2$, as against $4.13 \times 10^{-4} A/cm^2$ for the undoped. As seen from Eqs. (1) and (2), there is not a large difference in the Tafel slopes (b), which may indicate that the electrocatalytic effect is related to an increase of the active surface to an increase of the active surface area of the doped graphite anode during electrolysis.

Fig. 3 shows a plot similar to Fig. 2 where the uppermost and lowermost sets of data in Fig. 2 were removed, the reason being that these two runs gave abnormal shapes. In this case the Tafel relations were as follows:

for undoped graphite anode

$$\eta = 0.593(\pm 0.016) + 0.183(\pm 0.007)\lg i \qquad (3)$$

and for doped graphite anodes

$$\eta = 0.563(\pm 0.014) + 0.197(\pm 0.006)\lg i \qquad (4)$$

It is noticed that the Tafel coefficients showed little change compared to Eqs. (1) and (2), while the standard deviation was improved.

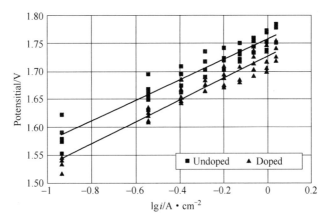

Fig. 3 Potential of undoped and doped graphite anodes versus lg$i$ where two abnormal curves have been removed(see Fig. 2)

4.2 Analysis of the amount of dopants in the graphite electrodes

SEM studies were attempted in order to determine the distribution of dopants inside the graphite electrodes. However, the concentrations of the elements (Mg or Al) were below the detection limit. Analysis by ICP(Inductively Coupled Plasma Emission Spectrometer) of the ash after incinerating the carbon showed a content corresponding to about 300×10^{-6} Mg in the doped graphite electrodes. If we assume that all the pores of the graphite were filled with solution during doping, and that all the dopant was taken up by the anode, the maximum amount of Mg in the graphite would have been 800×10^{-6} according to the porosity of the graphite and concentration of the solution.

The authors express their gratitude towards the National Natural Science Foundation of China and the Norwegian Research Council and the Norwegian aluminum industry for financial support.

References

[1] K. Grjotheim, C. Krohn, M. Malinovsky, K. Matiasovsky, J. Thonstad, Aluminiumveriag, Dusseldorf, 1982: 392-394.

[2] J. OM bockris and Z. S. Minevski, Electrocatalysis: Past, Present and Future, Electrichim. Acta, 1994, 39: 1471-1479.

[3] J. Thonstad and E. Hove, On the Anodic Overvoltage in Aluminum Electrolysis, Can. J. Chem., 1964, 42: 1542-1550.

[4] V. A. Braunworth, J. A. Brown, E. A. Hollingshead and P. J. Rhedey, The Addition of Aluminum Fluoride to Soderberg Paste, Light Metals, 1975: 325-337.

[5] M. Sorlie, Z. Kuang, J. Thonstad, Gas Reactivity and Electrolytic Consumption of Aluminum Cell Anode with Aluminum Fluoride Additions, Proc. 21st Conference on Carbon, Buffalo, NY, USA, 1993: 677-678.

[6] Y. X. Liu and J. Thonstad, Oxygen Overvoltage on SnO_2-based Anodes in NaF-AlF$_3$-Al$_2$O$_3$ Melts. Electrocatalytic Effects of Doping Agents, Electrochim. Acta, 1983, 28(1): 113-116.

[7] H. Xiao and Y. Liu, Study on the Electrocatalytic Activity of Doped Carbon Anodes in Aluminum Electrolysis, Proc. 6th Meeting of Physico-Chemistry of Metallurgical Processes, Xian, China(in Chinese), 1986.

[8] Y. Liu and H. Xiao, A Study on the Electrocatalytic Activity of Doped Carbon Anodes in Crylite-Alumina Melts, Proc. Joint International Symp. MoltenSalt. The Electrochemical Soc. Inc. Honolulu, Hawaii, USA, 1987:744-750.

[9] H. Xiao and Y. Liu, Studies on the Electrocatalytic Activity of Doped Carbon Anodes in Aluminum Electrolysis, J. Cent. South Inst. Min. Metall. (in Chinese), 1988, 19(3): 241-248.

[10] G. Yao, Z. Qiu, J. S. Zhang, Z. L. Zhang and J. J. Li, The Electrocatalysis of Lithium Carbonate in Anodic Reaction in Hall-Heroult Process, Light Metals(in Chinese), 1988(12):22-26.

[11] N. Feng, M. Zhang and Z. Qiu, The Effect of Additives on the Anodic Overvoltage of Carbon Anodes in Aluminum Electrolysis, Light Metals(in Chinese), 1989(7):26-30.

[12] N. Feng, M. Zhang, K. Grjothem and H. Kvande, Influence of Lithium Carbonate Aluminum electrolysis Cell, Carbon, 1991, 29(1):39-42.

[13] Z. Yu, G. Li and R. Cao, The Effect of Some Additives on the Overvoltage of Carbon Anode in Aluminum Electrolysis, Non-ferrous Metal(in Chinese), 1987, 39(4):59-63.

[14] C. C. Duan, G. A. Xia. M. -D. Zhang and X. Zhai, Study on the Relationship Between Lithium Salt Addition and Overvoltage of Carbon Anode in Aluminum Electrolysis, Light. Metals (in Chinese), 1990 (6):27-30.

[15] D. Yan, Y. Liu and H. Xiao, Studies on the Carbon Anode with Dopants for Saving Energy in Aluminum Electrolysis, J. Cent. -South Inst. Min. Metall. (in Chinese), 1989, 20(5):505-511.

[16] Z. Li, Master Degree Thesis, Central South University of Technology(CSUT), Changsha, China 1990.

[17] H. Liu, Master Degree Thesis, CSUT, 1991.

[18] W. Hu, Master Degree Thesis, CSUT, 1992.

[19] G. M. Haaberg, L. N. Solli and A. Sterten, Electrochemical Studies of the Anode Reaction on Carbon in NaF-AlF$_3$-Al$_2$O$_3$ Melts, Light Metals 1994(TMS. Warrendale. PA, USA):227-231.

[20] Y. Liu, X. Wang, Y. Huang, J. -H. Yang and H. Wang, New Type Electrocatalysts for Energy Saving in Aluminum Electrolysis, Light Metals 1994(TMS. Warrendale. PA, USA):247-251.

[21] Y. Liu, X. Wang, New Progress of Studies on Lithium Salt Containing Carbon Anode Paste, Light Meltals (in Chinese), 1994(5):28-30.

[22] Y. Liu, X. Wang, Y. Huang, H. Wang and J. -H. Yang, Progress in Studies of Electrocatalysts and Doped Carbon Anodes in Aluminum Electrolysis Cells. Trans. Nonferrous Metals Soc., China 1994, 4 (2): 92-94.

[23] Y. Liu, H. Xiao and G. Xiong. The Inhibition of Anode Effect in Aluminum Electrolysis Progress by Anode Dopants: A Laboratory Study, Light Metals 1991(TMS. Warrendale. PA, USA):489-494.

[24] Z. Qiu, T. Xun, Y. Yue, K. Yao, K. Grijotheim, H. A. Qye and H. Kvande, Carbon Anode with Lithium Salt Addition, Light Metals 1995(TMS. Warrendale. PA, USA): 749-751.

[25] Z. Qiu, Y. Yue, R. Ma and B. Tian, Aluminum Electrolysis with Lithium Salt-Containing Carbon Anode, A-luminium. 1995, 71(3):343-345.

[26] Z. Qiu and Y. Yue, The Study of Critical Current Densities on Lithium Salt Containing. Carbon Anode, Light Metals(in Chinese), 1994(5):31-33.

[27] Z. Kuang, On the Consumption of Carbon Anodes in Aluminum Electrolysis, Dr. Theses, Norwegian Inst. Technology, Trodheim, 1994:46-62.

[28] L. N. Solli, Carbon Anodes in Aluminum Electrolysis Cells. Factors Affecting Anode Potential and Carbon Consumption. Dr. Theses, Norwegian Inst, Technology, Trondheim, 1994:71-92.

[29] S. Y. Shen. On Measurement of Anodic Overvoltage in Aluminum Electrolysis by Using Reference Electrode, Light Metals(in Chinese), 1989(10):24-29.

[30] D. Y. Li, On Laboratory Measurement of Anodic Overvoltage in Aluminum Electrolysis, Light Metals(in

Chinese),1990(6):31-35.

[31] S. Y. Shen, On the Lithium Salt Paste in the Aluminum Electrolysis Industry of China, Non-Ferrous Metals (in Chinese),1991(3):41-43.

[32] D. X. Li. X. N. Wang, The Laboratory Measurement on Overvoltage of Anode Containing Lithium Salt in Aluminum Electrolysis, Proceedings of a Special Conference on Lithium Salt Paste(in Chinese), Shenyang, China, November, 1992.

[33] S. Shen, The Practical Effect of Anodic Overvoltage in the Aluminum Industrial Cells, Light Metals(in Chinese),1993(12):31-35.

[34] S. Sun, On the Practical Effectiveness of Overvoltage Reduced by Lithium Salt Anode Paste Used in Aluminum Electrolysis cells, Light Metals(in Chinese),1993(12):31-35.

[35] J. H. Yang, Y. Liu, X. Wang, Y. Huang and H. Wang, Discussion of views and Divergences on Electrocatalysis of Lithium Salt Paste in Aluminum Electrolysis Cells, J. Cent. South Inst. Min. Metall(in Chinese) 1994,25(3):326-332.

[36] Y. Liu, H. Xiao. X. Liu and Z. Chen, A New Approach to Reduce the Anodic Overvoltage in the Hall-Heroult Progress. Light Metals 1989(TMS. Warrendale. PA,USA):275-280.

[37] Y. Liu, X. Wang, Y. Huang and H. Wang. A New Field to Reduce Energy in Hall-Heroult Progress-The Research and Application of an Anode Paste Containing Lithium Salt, Light Metals, 1993 (TMS. Warrendale. PA, USA):599-601.

[38] Z. Qiu. The Application of Lithium Salt Carbon Anode Paste in Aluminum Electrolysis, Non-ferrous Metals (in Chinese),1991(1):22-25.

高温氯化物熔体氯电极过程的电催化作用*

摘要 以石墨为基底,采用热分解工艺制备氧化物涂层电极;用慢速线性电位扫描技术结合改进的断电测试技术,对过渡族金属氧化物,镧系稀土氧化物及其复氧化物电极,在700℃高温氯化物熔体中的析氯电催化性能进行测试。结果发现,高温下 Pr, Tb 和 Tm 等一些稀土氧化物和复合氧化物对氯析出显示出高的催化活性;与石墨电极相比,0.6A/cm² 的工业电流密度下可降低析氯过电位 80~110mV,证实了高温氯电极过程存在电催化作用。依过电位降低的大小,实验获得镧系稀土氧化物析氯电催化活性的顺序为:Pr≥Tb>Tm>Dy≥La≥Y≥Sm>Yb≥Nd。

高温熔盐体系电催化活性的影响因素复杂,实验和测试难度较大,特别是由于高温腐蚀性以及制备方法或工艺的原因,许多在低温下原具有催化活性的物质(电极材料)在高温下却失去催化活性或失稳,给高温电催化研究带来许多不利因素。Uchida 等曾对金属氧化物在 $NaAlCl_4$(175℃) 和 LiCl-KCl(450℃) 熔体中的阳极材料进行了研究[1-7],发现 Ru_2O,Rh_2O_3 和 IrO_2 等贵金属氧化物是析氯电催化很好的物质。但由于 Uchida 等人所采用氯化物熔体的温度较低,而电化学冶金工业所采用的熔盐体系的温度普遍高于此温度,因此有必要研究高温(>650℃)氯化物熔体氯电极过程的电催化作用。

1 实验方法

1.1 实验装置及电极制备

实验装置如图 1 所示。电解槽采用直径 90mm×100mm 的透明圆形石英电解槽,外置直径 100mm×250mm 的石英套筒;电解质采用等摩尔的 KCl-NaCl 混合盐,电解前,120℃烘干 24h 以上,在 700℃下,2mA/cm² 的电流密度预电解 20~24h。

氧化物电极采用热分解工艺制备。基底为直径 6mm 的光谱纯石墨棒(上海碳素厂生产),使用前,磨平,在盐酸中煮沸,超声波清洗,高温焙烧去除挥发组分,丙酮除油,蒸馏水洗涤,烘干,用高温胶封装在石英玻璃管中。电极设计成 L 形结构,使气泡易于逸出,减少了溶液的 IR 降压和气泡放出引起的气膜电阻的波动。

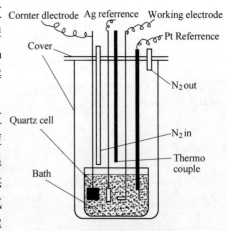

图 1 实验装置示意图

---

* 本文合作者:谢中。原发表于《金属学报》,1999, 35 (3):296-299。

1.2 电化学研究方法

电化学测试采用三电极体系，熔盐体系无通用参比电极。采用 Ag/AgCl 和裸 Pt 丝两种参比。AgCl 参比组成为 AgCl∶(NaCl+KCl)(1∶1mol) = 1∶9。验证实验表明，Ag/AgCl 参比采用石墨作为隔膜材料，改善参比电极的稳定性，8h 内不对称势小于 5mV，裸 Pt 作为准参比具有很好的稳定性，12h 不对称电势小于 10mV；两种参比结合使用提高了阳极过电位测定的可靠性。

熔盐体系气体电极反应欧姆压降的准确扣除或补偿是高温电催化研究的关键。实验中阳极过电位的测定采用 M273 恒电流/电位仪，断电时间为 3μs；记录用先进的 Lecroy LS140 型数字存储式示波器，其采用速率达 100Ms/s，频带为 100MHz（10^{-8}s），可以捕捉断电流瞬间电位变化及衰减过程。

1.3 实验过程及仪器

实验在 N_2 气下进行，温度控制在 (700±2)℃。电解质溶化后，插入石墨棒，采用 DH1716 型直流电源，在 $3mA/cm^2$ 的电流密度预电解 1~2h，除去吸附的水和氧。预电解完毕，换插研究电极，平衡 5min，进行电化学测试；测试过程中，每次保持工作的位置和插入深度一致。电化学测试仪器为 EG & G PAR 公司的 Model273 恒电流/电位仪，由 M270 电化学研究软件通过计算机控制和记录。采用慢速线性电位扫描技术（扫描速率为 5mV/s）结合断电流法绘制消除欧姆降压的极化曲线。实验以光谱纯石墨作阳极进行验证。结果表明所采用的测试装置可行，电化学测试方法可靠。计算得石墨阳极析氯的 Tafel 斜率为 182mV，交换电流密度为 $0.0308A/cm^2$。

2 实验结果与讨论

2.1 过渡族金属氧化物的高温析氯行为

实验对（1）Fe、Co、Ni、Cr、Mo、Sn 和 Ru 的氧化物；（2）Co+Mg、Co+Fe、Co+Ni、Ni+Fe 和 Fe+Mg（$M_2M'O_4$ 尖晶石型结构），$Co_{0.75}Fe_{2.25}O_4$ 和 $Ni_{0.85}Fe_{2.15}O_4$（$A_xFe_{3-x}O_4$ 铁氧体结构）及 Sn+Sb 的复合氧化物在 700℃ 高温 KCl-NaCl 熔盐中析氯过程进行测试。测定各涂层电极的阳极极化曲线与相同电流密度下石墨电极的曲线进行比较，依据过电位的高低来表征氧化物电极的电催化活性[8]。结果表明，贵金属 Ru 和金属氧化物及大多数复氧化物在高温下未发现对氯析出具有明显的催化作用；相反，一些氧化物如 Mn 和 Mo 涂层电极的表面电阻增大，电极在极化过程中处于钝化状态；仅铁氧体 $Co_{0.75}Fe_{2.25}O_4$ 复合氧化物显示出一定的析氯催化活性，其稳定极化曲线如图 2 所示，$0.6A/cm^2$ 的工业电流密度下可降低析氯过电位 80mV；计算得 Tafel 斜率 137mV；交换电流密度分别为 $0.041A/cm^2$。Tafel 斜率较石墨电极的低，而交换电流密度高

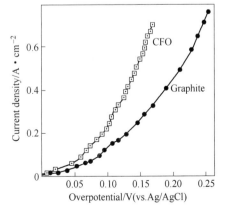

图 2 $Co_{0.75}Fe_{2.25}O_4$ 氧化物与石墨电极的极化曲线

于石墨电极的值，表明了氧化物具有电催化活性。

过渡族元素具有特殊的 d 电子结构和 d 空轨道，具备了作为电催化剂的电荷因素。但实验结果表明在高温氯化物熔体中这些金属氧化物对 CER 的电催化作用不明显，和在水溶液中的行为相比有较大的差异，尤其贵金属 Ru 低温下的析氯催化活性在 700℃以上的高温氯化物熔体中基本消失。说明高温熔盐体系的特殊性和复杂性。其原因可能是热分解制备的氧化物膜在高温电解质中易于被侵蚀或发生溶解，而失去活性。$Co_{0.75}Fe_{2.25}O_4$ 显示一定的催化活性究其原因可能是制备的复氧化物产生协同效应。复氧化物的种类繁多，是电催化剂选择研究的一个重要方向。

2.2 稀土氧化物的高温析氯性能

稀土由于电子组态和 4f 电子的运动特性是他们具有很多特殊的理化性能，成为探找新型高新技术材料（包括电催化剂）的重要研究对象。5d 空轨道提供了良好的电子转移轨道，可作为催化作用的电子"转移站"，在化学催化和低温电催化中作为催化剂以初露端倪[9]；形成氧化物时，正离子外层 d 和 s 电子的空态（d^0s^0）形成交叠的导带，具有半导体的特性，在高温下具有良好的电学性能，所有这些都使稀土元素氧化物作为高温析氯电催化剂成为可能。

实验对 La、Ce、Pr、Nd、Sm、Eu、Gd、Tb、Dy、Er、Tm、Yb 和 Y 等 13 种稀土氧化物涂层电极在 700℃下，KCl-NaCl 熔体中阳极析氯的电极曲线进行了测定，结果列于图 3。

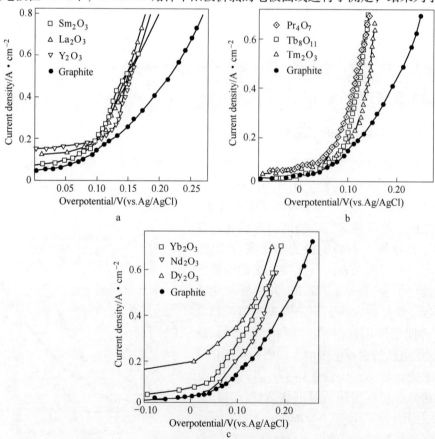

图 3 稀土氧化物涂层电极的极化曲线

a—Sm, La, Y oxides; b—Pr, Tb, Tm oxides; c—Yb, Nd, Dy oxides

在 0.6mA/cm² 的电流密度下，石墨和各种稀土氧化物电极的过电位及降低值列于表1。根据过电位测试结果，可将稀土氧化物电极分为三类。析氯催化活性显著的有 Pr、Tb 和 Tm，相同条件下与石墨电极相比较，降低过电位达 80~100mV；Yb 和 Nd 过电位降低在 50~60mV 左右；Ce、Gd、Eu 和 Er 涂层电极的表面电阻升高或发生钝化现象。由表1的数据，依过电位降低的大小可进一步获得单一稀土氧化物电极析氯电催化活性的排序为：Pr≥Tb≥Tm>Dy≥La≥Y≥Sm>Yb≥Nd。

表1 700℃下稀土氧化物电极的过电位（电流密度 0.6A/cm²），Tafel 斜率 b 和交换电流密度 i_0

Rare earth oxide	Overpotential /mV	Overpotential lower/mV	b /mV	a /mV	i_0 10^2/A·cm^{-2}	Current density range /A·cm^{-2}
La	158	-78	126	186	3.35	0.15~0.60
Ce	293	57	—	—	—	—
Pr	135	-101	114	157	4.20	0.10~0.70
Nd	185	-51	151	212	3.92	0.05~0.70
Sm	161	-75	121	182	3.13	0.10~0.60
Gd	521	285	—	—	—	—
Tb	138	-98	120	164	4.30	0.05~0.35
Dy	157	-79	316	238	17.6	0.20~0.70
Tm	150	-86	158	215	4.34	0.05~0.35
Yb	178	-58	184	209	7.31	0.07~0.70
Y	160	-76	122	191	2.72	0.20~0.40
Graphite	236	0	182	275	3.08	0.05~0.70

将具有电催化活性的稀土氧化物的极化曲线转化为 Tafel 关系，计算得相应电流密度范围内 Tafel 斜率 b，截距 a 和交换电流密度 i_0 列于表1；根据电催化性能的判断标准，催化性能良好的电催化剂通常具有高的交换电流密度和低的 Tafel 斜率。交流电流密度和 Tafel 斜率所显示的稀土氧化物催化活性与极化曲线测得的过电位结果基本一致。少数例外可能是高温实验误差所致。

相同电极制备条件下（消除表面结构因素的影响），各种稀土氧化物的析氯特性存在差别。可能是由于稀土元素原子内部未充满的4f电子层在其氧化物的能带结构中形成了特殊的4f能级；4f能级的位置决定了稀土氧化物的半导体性能，进而影响活性中心的表面态。表面态不同，使吸附态中间产物处于不同的能量状态，其催化活性是不同的，这反映出不同稀土氧化物析氯催化活性的差异。

2.3 稀土符合氧化物高温析氯性能

对 La 和（La+Pr+Nd+Sm+Eu）分别与 Co、Ni、Mn 和 Cr 合成的钙铁矿型复合氧化物，稀土 Yb 和 Y 与 Cr 合成的石榴石型复氧化物，及混合稀土分别与 Ru 形成混合氧化物电极的析氯性能进行了测定，发现 Ru 与混合稀土氧化物电极产生较好的协同效应（图4），可降低析氯过电位 110mV；Tafel 斜率等于 133mV，交换电流密度 $9.17×10^{-2}$ A/cm²。这种行为可能与钌氧化物的性质有关，钌氧化物电极对植被热分解温度较为敏感。$RuCl_3$ 具有析氯活性的热分解温度较低，而稀土氧化物显示催化活性的热分解温度较高，两者混合时可能产生协同反应，改善了半导体氧化物的性能，显示了高的析氯电催化活性。

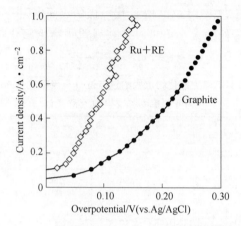

图 4　稀土与钌复合化合物涂层电极极化曲线

3　结论

稀土 Pr、Tb、Tm 的氧化物及（RE+Ru）形成的复合氧化物在高温氯化物熔体中显示出高的析氯电催化活性。稀土氧化物的电催化活性排序为：$Pr \geqslant Tb > Tm > Dy \geqslant La \geqslant Y \geqslant Sm > Yb \geqslant Nd$。

过渡金属氧化物及复合氧化物仅铁氧体 $Co_{0.75}Fe_{2.25}O_4$ 复合氧化物显示出一定的催化活性。

参 考 文 献

［1］Uchida I, Laitincn H A, Ken-ichiro. Proc of 1st Int Symp on Moiten Salt Chem and Technol, Japan, 1983, 65, 219, 223, 227.

［2］Uchida I, NiKi K, Laitinen H A. J Electrochem Soc, 1978, 125：1759.

［3］Uchida I, Urushibada H, Akahoshi H, Toshima S. J Electrochem Soc, 1980, 127：757, 995；1981, 128：2351.

［4］Uchida I, Toshima S. J Appl Electrochem, 1979, 126：647；1982, 129：115.

［5］Uchida I, Urushibada H, Toshima S. J Electroanal Chem, 1979, 96：45.

［6］Uchida I, Asano H, Toshima S. J Electroanal Chem, 1978, 93：221；1980, 107：115；1981, 124：165.

［7］Trasatti S, Lodi G. In：Trasatti S ed, Electrodes of Conductive Metallic Oxides, Part B, Amsterdam：Elsevier Scientific Pub. Co. 1981, 523.

［8］Менащаев X M, Liu H Q Ttans. The Application of Rare Earth in Catalysis. Beijing：Published by Academic Press, 1987, 251：289.

［9］Scientific Technology Information Institute of Chemical Industry Bureau of Shanghai. The Application of Rare Earth as Catalysts. Shanghai：Published by Scientific Technology Literature Press of Shanghai, 1982, 1.

Electrocatalysis of the Carbon Anode in Aluminium Electrolysis*

Abstract The anodic overvoltage of the carbon anode in aluminum electrolysis is of the order of 0.6V at normal current densities. However, it can be reduced somewhat by doping the anode carbon with various inorganic compounds. A new apparatus was designed to improve the precision of such measurements. Anodes were doped with $MgAl_2O_4$ and AlF_3 both by impregnation of the coke and by adding powder, and the measured overvoltages were compared with undoped samples. For prebake type anodes baked at around 1150℃, the anodic overvoltage was reduced by 40~60mV, and for Soderberg type anodes, baked at 950℃, by 60~80mV.

1 Introduction

The specific energy consumption, P, in aluminium electrolysis can be expressed as:

$$P = \frac{2.980 \cdot V}{CE} \quad (kW \cdot h/kg\text{-}Al) \qquad (1)$$

where V is the cell voltage; CE is the current efficiency. This expression demonstrates that the energy consumption can be reduced either by lowering the cell voltage or by increasing the current efficiency. According to equation (1), if we use $V = 4.2V$, $CE = 0.94$, and $P = 13.30 kW \cdot h/kg\text{-}Al$ as typical values for modern prebake cells, a 0.05V reduction in V or a 0.01(1%) increase in CE would reduce P by 1.2% and 1%, respectively.

The cell reaction in aluminum electrolysis is:

$$Al_2O_3 + 3/4C = Al + 3/4CO_2 \qquad (2)$$

where a cryolite-based melt (Na_3AlF_6-AlF_3-CaF_2) at ~960℃ serves as solvent for the alumina. The anodic overvoltage is defined by the following equation:

$$\eta_a = E - |E_{rev}| \qquad (3)$$

where E is the measured polarization potential; E_{rev} is the standard reversible potential for the cell reaction (2), which is 1.187V at 970℃. The anodic reaction is associated with considerable overvoltage (0.4~0.6V), which represents a significant contribution to the overall energy consumption. Hence it is of interest to find ways to reduce the overvoltage.

The performance of the carbon anode in aluminium electrolysis is an important factor with respect to carbon consumption, current efficiency, metal quality and energy consumption. Therefore, much attention has been paid to anode problems, and a number of studies have been carried out. According to Grjotheim et al.[1] and Muftuoglu et al.[2], impurities or additives in the anode contained in the anode raw materials or introduced during the anode manufacturing process, may substantially affect the anode performance.

---

* Copartner: Lai Yanqing, Yang Jianhong and Thonstad Jomar. Reprinted from Rare metals, 2002, 21(1):117-122.

Since 1980, several studies have been reported on the electrocatalytic effects of carbon anodes doped with additives. The research has mostly been focused on reducing the anodic overvoltage(η_a) and thereby possibly reducing the energy consumption. Electrocatalysis is an interesting and promising way of saving energy.

2 Literature Review

Since the concept "electrocatalysis" was first introduced in the 1960s[3], great progress has been made in this field, especially in the chlor-alkali industry, fuel cells, photoelectrocatalysis, organic synthesis etc. In the following, literature data on the effect of additives on the anodic overvoltage(η_a) in aluminum electrolysis is summarized.

Thonstad and Hove[4~6] studied the effect of some additives in carbon anodes on the anodic overvoltage(η_a). Amounts of 0.4%~1.3% were added to the coke. The results showed that η_a decreased slightly by the addition of Fe_2O_3 and Na_2CO_3, while H_3BO_3 led to a slightly higher overvoltage, the differences being of the order of 5mV.

Liu and Thonstad[7] showed that for oxygen evolution on SnO_2-based anodes in cryolite-alumina melts at 1000℃, η_a was lowered by adding dopants to the anode. The electrocatalytic effectiveness was ranked as follows:

$$Ru > Fe, Cr > Co, Mn > Ni$$

At a current density of $1A/cm^2$, the η_a value for a Ru-doped anode was only about 1/3 of that of an identical undoped electrode.

Liu[8] proposed to apply some additives to the carbon anode to improve its electrochemical activity so as to reduce the η_a. Since then several research works have been carried out in China, and a number of electrocatalysts have been proposed and tested.

Liu and Xiao[8-10] tested the dopants $CrCl_3$, $MnCl_2$, $CoCl_2$, $NiCl_2$, $RuCl_2$, Li_2CO_3 and NaCl. In a cryolite-alumina melt at 1000℃, it was found that the η_a of anodes doped with $CrCl_3$, Li_2CO_3 and $RuCl_2$ could be reduced by 275, 181 and 148mV, respectively at $0.85A/cm^2$, as compared to an identical undoped anode. Yao et al.[11] carried out similar experiments, adding 1.5 wt% Li_2CO_3 to Soderberg anodes, and the η_a was reportedly reduced by 380mV at $0.96A/cm^2$ (the original η_a value was not given). Conversely, Feng et al.[12,13] claimed that tests with 1.3 wt% Li_2CO_3 addition showed no measurable effect on the anodic overvoltage. Yu[14] found that Na_2CO_2, MgF_2 and NH_4VO_3 possessed some catalytic effect on the anode reaction, while HBO_3, $Na_2B_4O_7$ and Li_2CO_3 acted as inhibitors.

Duan et al.[15] presented results of tests in a 100A laboratory cell. Soderberg type anodes containing 0.3%~0.5%(wt) Li_2CO_3 showed the best electrocatalytic effect, i.e. at $0.88A/cm^2$, η_a was reduced from originally 444mV to 285mV. The experiments were conducted at 1000℃ in a melt containing 0.5 wt% Li_2CO_3 and CR = 2.9(molar NaF/AlF_3 ratio). Yan[16] studied the effect of adding rare earth elements. In the range of industrial current densities(0.5~$1.0A/cm^2$), η_a was reduced by 220~340mV, e.g. that of undoped graphite was 585mV at 1000℃ and $1.04A/cm^2$, compared to 245mV after doping with yttrium salt.

Li[17] investigated the effects of $LiCO_2$, Ba-Li, Li-Na-Fe and Mg-Fe complex salts as anode dopants. The results showed that, compared with the η_a value (454mV at 0.85A/cm^2) for undoped anodes, the η_a of Soderberg type anodes was reduced by 261mV for Ba-Li salt addition, 198mV for Li-Na-Fe, 125mV for Mg-Fe and 147mV for Li$_2$CO$_3$ addition. For prebaked type anodes, the following reduction of η_a was found: Li$_2$CO$_3$ 22mV, Mg-Fe salts 64mV, and Ba-Li salts 66mV. These results indicate that the electrocatalytic effect was diminished when the doped anode was baked at higher temperature (1200℃). Liu[18] studied the electrocatalytic effect of addition of complex salts such as Ca-Mg, Li-Ca-Mg etc. The η_a of an undoped Soderberg type anode was 440mV at 0.85A/cm^2, being reduced by 213-275mV after doping. The catalytic activity of the anode decreased after being baked at 1250℃, and η_a was reduced by only 88~130mV (the original η_a value of a prebaked type undoped anode was 446mV).

Hu[19] tested the electrocatalytic activity of carbon anodes doped with complex salts, such as Fe, Ca, K-Ca, K-Ca-Fe and K-Ca-Mg-Fe. The doped anodes contained 0.2 wt% complex salts and were baked at 1250℃. The η_a of an undoped prebaked type anode was 638mV at 0.85A/cm^2, being reduced by 99~159mV after doping.

The authors mentioned above[16-19] also tried to interpret the anodic reaction mechanism. Haarberg et al.[20] presented result related to industrial type carbon anodes with Fe$_2$O$_3$ as additive. The η_a of a carbon anode was reduced by more than 100mV from an initial value of 0.8V at 1A/cm^2. Liu et al.[21] used several additives separately as dopants. Samples baked at 1200℃ still exhibited electrocatalytic activity. Ba-Fe salts lowered η_a by 200mV, Mg-Al salts 170mV, and Ca-K salts 150mV (at 1000℃, saturated cryolite-alumina melt and 0.8A/cm^2, the η_a of an undoped prebaked type anode was 590mV). It was suggested that perovskite and spinel compounds formed from these dopants at high temperature could lead to the formation of active sites at the electrode surface, promoting the electron transfer so as to accelerate the anode reaction rate and reduce the η_a value. The experimental results obtained were summarized in[22,23].

In addition to the research on anodic overvoltage (η_a), there are many works dealing with other properties of carbon anodes containing additives. A few relevant papers will be mentioned here.

Liu et al.[24] and Qiu et al.[25,26] studied the effect of dopants on the anode effect and on the critical current density (ccd) for the initiation of the anode effect[26,27] on carbon anodes in aluminum electrolysis. Liu's results showed that doping with Pb, Cr and Ru salts increased the ccd and inhibited the ocurrence of the anode effect. Qiu's experiments showed that carbon paste containing lithium salt showed better wettability by the bath, higher ccd and lower η_a.

Kuang[28] and Solli[29] tested the effect of impurities of V, S, Al, Fe, Ni, Li and Na on the carbon consumption. The results showed that the compounds Na$_2$CO$_3$, NiO, Fe$_2$O$_3$, AlF$_3$, LiCl, V$_2$O$_5$ increased the electrolytic consumption of the anodes, while Al$_2$O$_3$ showed a weak catalytic effect and sulphur behaved as inhibitor. However, sulphur increased the dusting of the anode.

The effect of lithium salt additives, and anode paste containing lithium salt has been discussed in China since 1987[12-14,30-35]. The main arguments were as follows:

(1) The η_a values which appeared in some works[8,11] was not correct.

(2) Shen and Li[30-32] maintained that the reduction in η_a by 380mV reported in paper [11]

was unlikely.

(3) Feng and Yu[12-14,33] did not find any measurable effect on η_a by Li_2CO_3 addition.

(4) The good performance of industrial cells employing paste containing lithium salt was not due to reduction in η_a[34,35], but resulted from the lithium salt that was transferred into the bath, improving the properties of the bath, and thereby the electrolysis process.

As a result of these discussions, some conclusions were given[23,36]:

(1) The calculation of η_a was not correct at the very beginning of these studies, but it was corrected later on. However, the experimental results, i.e. the differences of η_a obtained by comparing doped and undoped anodes were correct.

(2) The choice of carbon materials, the preparation of electrodes and the measurement techniques were rather different among various authors. This could be the cause of the discrepancies.

(3) Industrial practice in cells using lithium bath showed that anodes made from paste containing lithium salt performed better than normal carbon anodes.

(4) Electrocatalysis in high temperature melts is quite a new field, making effective energy savings possible. More work is needed, in particular to explain discrepancies.

Since 1987, industrial tests of carbon anodes containing Li_2CO_3 have been carried out in China. Cooperation was established between the Central-South University of Technology and the Liancheng Aluminium Smelter[37,38], and between the North-East Institute of Technology and the Shandong Aluminium Smelter[39] to perform such tests for nine months on two pot-lines in each smelter. Good results were obtained, the energy saving in the Shandong Smelter being 460kW·h/t-Al and in the Liancheng Smelter 305kW·h/t-Al. Since then this technique of doping the anode paste with lithium salt has spread to many aluminium smelters in China.

In 1989, the former China National Nonferrous Metals Corporation(CNNC) organized an evaluation and appraisal of this technique by Chinese specialists in the field of aluminium electrolysis. It was concluded that the average cell voltage of the test pot-lines was reduced by 50~63mV, the current efficiency was increased by 0.54% and the overall energy savings were 246kW·h/t-Al(Shandong Smelter) and 152kW·h/t-Al(Liancheng Smelter), respectively. The energy savings were obvious, and the economical benefit was also good.

By the end of 1991, there were about 2000 horizontal stud Soderberg(HSS) cells belonging to 16 aluminium smelters in China using anode paste containing lithium salts. About 50 million kW·h electrical energy per year was saved(according to statistical data provided by CNNC). Owing to the outstanding work on energy saving, Liu and his co-workers and other contributors were awarded the first Technical Progress Prize of 1992.

On this background, a cooperation concerning a study of electrocatalysis in aluminium electrolysis was established in 1997 between the Central-South University of Technology, China and the Norwegian University of Science and Technology. It was supported financially by the Norwegian aluminium industry, the Norwegian Research Council and the China National Key Fundamental Research Development Project. In the present work, doped anodes were prepared with various doping methods, and an improved current interruption technique with good reproducibility was used for overvoltage measurements.

3 Experimental

3.1 Preparation of doped anodes by dipping

Carbon anode materials were prepared at the Carbon Laboratory of Elkem A/s Research in Kristiansand, Norway. The coke rather than the pitch was doped with additives for the purpose of increasing the electrochemical reactivity of the coke. Industrial grade petroleum coke was crushed and mixed to a specified recipe. It was then soaked in a solution with additives to impregnate the coke phase with catalyst(for example, to obtain the $MgAl_2O_4$ catalyst, an aqueous solution containing aluminum and magnesium salts was used). The solution was evaporated and the coke was dried and heat-treated at 450~500℃ in argon. Green paste was fabricated from the doped or undoped coke by mixing it with vacuum distilled coal tar pitch, and it was baked at a given temperature.

3.2 Preparation of doped anodes by mechanical mixing

Among the additives that have been tested (AlF_3, $MgAl_2O_4$, $LiAlO_2$, LiF, $LiCoO_2$ and $CaO+La_2O_3$ etc.), only two will be treated here, i.e. $MgAl_2O_4$ and AlF_3. Ultrafine powder of $MgAl_2O_4$ was prepared from an aqueous solution of $Mg(NO_3)_2$ and $Al(NO_3)_3$ by evaporating to dry the co-precipitate as hydroxides or as amorphous mixed salts, while technical grade AlF_3 was used without any pretreatment.

The carbon anode materials doped with AlF_3 were prepared at the Carbon Laboratory, Department of Inorganic Chemistry(abbreviated as CLT below), Norwegian University of Science and Technology. The anode material doped with $MgAl_2O_4$ powder was prepared at the Carbon Laboratory of Elkem A/S Research(abbreviated as CLE below) in Kristiansand, Norway, and at the Zhengzhou Research Institute of Light Metals(abbreviated as CLZ below), Zhengzhou, China. Industrial grade petroleum coke was crushed and mixed to a specified recipe, and mixed mechanically with the prepared ultrafine powder of $MgAl_2O_4$ or technical grade aluminum fluoride. Preheated pitch was added and mixed in a sigma-mixer. After molding, the green anodes were baked at a given temperature for a certain time. Cylindrical anode samples used for overvoltage studies were core-drilled from these carbon materials to a diameter of either 15mm or 10mm.

3.3 Measurements of overvoltage with a modified current interruption technique

As reviewed by Grjotheim et al.[1], the key problems to be solved for measuring anodic overvoltage in aluminium electrolysis are two-fold. Firstly, current and/or potential oscillations are quite pronounced on horizontal anodes facing downwards, imitating the industrial anode. These oscillations may seriously affect the precision of measurements at high current densities. Secondly, unless the reference electrode is located very close to the working electrode and the current density is low, the results must be corrected for the ohmic voltage drop.

Although the anodic process and measurement of anodic overvoltage in cryolite-alumina melts have been the subjects of numerous investigations, the difficulties involved in performing reliable measurements of anodic overvoltage have not been completely overcome. The reported data concerning overvoltage of carbon anodes are rather scattered and partly conflicting.

Four changes were made in this study to improve the measurement of anodic overvoltage by means of the current interruption method in cryolite-alumina melts.

(1) As shown in Fig. 1, vertical anodes whose upper sheath was cut to a conical shape were used to facilitate bubble detachment and thus decrease the potential oscillations at constant currents. The inner wall of the graphite crucible served as counter electrode (cathode), and the bottom of the graphite crucible was insulated with sintered alumina disks. The potential oscillations were less than 15mV at constant current. On horizontal anodes facing downwards the amplitude of oscillations are often more than 100mV.

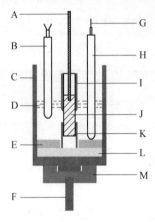

Fig. 1 The experimental cell and anode design used for overvoltage studies

A—3mm steel rod; B—Pt-Pt10% Rh thermocouple in a closed-end tube of sintered alumina;
C—Graphite crucible (Inner diameter: 61mm); D—Melt; E—Sintered alumina plate with a concentric hole;
F—Steel lead for cathode; G—Tungsten wire; H—Aluminium reference electrode in a
closed-end alumina tube with a small hole; I, K—Sintered alumina sheaths; J—Carbon anode;
L—Sintered alumina plate; M—Graphite support

(2) A high frequency digital oscilloscope (LS140, 200M) was used with a sampling time of only 5 ns per point. The oscilloscope made it possible to trace the whole potential decay curves just a few microseconds after the current was interrupted. Customary recording equipment used for this purpose, such as X-Y recorders, start responding to the potential change at least 50μs or even 1 ms after the current is interrupted. Decay curves and extrapolated ohmic drops obtained in this way will not be exact.

(3) Current interrupters with fast switching times were used. Two models of current interrupters were used (Model 800 with max. current 10A and Model 805 with max. current 50A, IR Measurement System, Scribner Associates INC., Virginia, USA). For the former, the switching time was less than 1μs and for the latter 1~3μs. The duration of current interruption is quite short (10~40μs), and hence the state of the anode polarization will not change much because of the current interruption. In comparison, mechanical commutators and even ordinary galvanostats with function generators, which have been used by many investigators, cannot attain such a

resolution, and the switching times tend to be more than 20μs, as tested in this study. Furthermore, mechanical commutators and ordinary galvnostats have much longer off periods.

(4) The last difficulty is the determination of the start point (the potential at time zero without ohmic drop) of potential decay curves, or the determination of the ohmic drop. In most cases, graphical extrapolation to time zero has been used. This procedure was applied in the previous stages of our studies. The precision of this procedure is questionable, especially in the presence of ringing (damping oscillations), due to inductance in the circuit at fast current interruption. A mathematical simulation was conducted to solve this problem.

Experiments were performed in Na_3AlF_6-Al_2O_3 (sat.) at 1100℃ and in Na_3AlF_6-Al_2O_3 (sat.)-10.9wt.% AlF_3-5wt.% CaF_2 melts at 970℃. The measurements were started 30~40 minutes after the anode was immersed into the melt, and two or three anodes were tested one by one for every experimental run. The cell current was decreased in steps from 12.0A (1.28A/cm^2) to 0.6A (0.064A/cm^2) under galvanostatic conditions. A steady state potential was normally attained in 120~180 seconds at 12.0A, and the duration of the rest of the steps was only 30 seconds. Due to the use of a vertical anode, the gas-induced potential oscillations at constant current were less than 15mV, when the potential had reached steady state after a certain time (30 to 180 seconds). After the measurements, the diameter of the carbon anodes had decreased from 15mm to about 14.6mm, and the geometrical surface area exposed to the melt was thus lowered by about 3%.

4 Results and Discussion

The anodic overvoltage (η_a) can be obtained by Tafel curves,

$$\eta_a = a + b\lg i \tag{4}$$

where a and b are Tefel coefficients; i is current density.

Table 1 lists the Tafel coefficients a and b for anodes with various additives doped by dipping. The undoped anodes N1 and N3 were prepared from two different batches, but were otherwise identical. There are slight differences in the Tafel coefficients, illustrating the difficulty of precise measurement of anodic overvoltage. Hence, to avoid any such influence, comparisons between undoped and doped anodes were made only for the same batch of anodes. The following information can be obtained from Table 1. A comparison between doped anodes (D1-D4 vs. N1) with additives of different molar ratios of MgO and Al_2O_3, indicates that the MgO/Al_2O_3 ratios have an effect on the anodic overvoltage. The closer the ratio was to the spinel composition, $MgAl_2O_4$, the more effective it was in decreasing the anodic overvoltage.

LiF has a positive effect in decreasing the anodic overvoltage (39mV at 1A/cm^2) compared to undoped anodes (L2 vs N3), while the effect of Li_2CO_3 is not so obvious (L1 vs N3). Liu et al.[37,38] found a positive effect of anodes doped with Li_2CO_3, being more than 100mV at 0.85A/cm in cryolite-alumina (sat.) melt at 1000℃. It should be mentioned that Qiu et al.[25,26] found that anodes doped with Li_2CO_3 showed better wettability and higher critical current density than undoped anodes, which may indicate that anodes doped with Li_2CO_3 decrease the voltage drop resulting from shielding the anode surface by gas bubbles.

Table 1 Tafel coefficients for carbon anodes with additives applied by dipping, in Na_3AlF_6-Al_2O_3(sat.) at 1100℃, $\eta_a = a + b\lg i$, Current density range: 0.06~1.30 A/cm²

Sample No.	Additives	Content of Additives(wt.)/%	Baking Temp /℃	a/V	b/V	i_0/A·cm⁻²
N1	Undoped	0	1000	0.481	0.175	1.78
D1	2MgO-Al_2O_3	0.1	1000	0.490	0.180	1.89
D2	2MgO-Al_2O_3	0.3	1000	0.478	0.197	3.74
D3	MgO-Al_2O_3	0.3	1000	0.439	0.200	6.38
D4	4MgO-Al_2O_3	0.3	1000	0.498	0.184	1.97
N3	Undoped	0	1000	0.497	0.165	0.97
L2	LiF	0.6	1000	0.458	0.144	0.66
L1	Li_2CO_3	0.6	1000	0.490	0.155	0.690
N5	Undoped	0	1170	0.500	0.139	0.25
D3*	MgO-Al_2O_3	0.3	1170	0.480	0.139	0.35

The difference in anodic overvoltage between N5 and D3* baked at 1170℃ is 22mV less than that between N1 and D3 of identical compositions, but baked at 1000℃.

Fig. 2 shows plots of y versus log i for a series of runs obtained with undoped anodes and an-

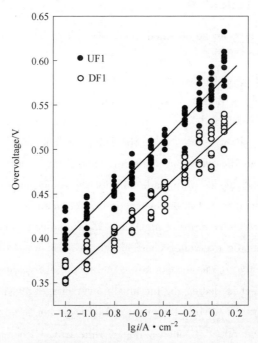

Fig. 2 Tafel plots for undoped and doped anodes with 1 wt.% AlF_3 (UF_1 and DF_2 in Table 1). The anode surface area exposed to the melt was 4.71cm². The anodic overvoltage data was obtained by using both models 800 and 805 current interruptors

odes doped with 1 wt. % AlF_3 (UF_1 and DF_2 in Table 2, baked at 1100℃). According to the linear regression lines shown in the figure, the overvoltage of the doped anodes at 1A/cm² was in average 61mV (±15mV) lower than that of the undoped anodes. This shows that AlF_3 has a positive effect in decreasing the anodic overvoltage. Table 2 lists the Tafel coefficients and the exchange current densities for anodes with various contents of AlF_3 added by mechanical mixing. Tafel curves for undoped and doped carbon anodes (NF_1, NF_2, AF_1, AF_2 and AF_3) made with the same recipe and identical compositions, but baked at 1250℃, are also presented in Table 2. It is noticed that the effect of AlF_3 in lowering the anodic overvoltage obviously was less than for the anodes baked at 1100℃.

Table 2 Tafel coefficients for carbon anodes doped with AlF_3 in Na_3AlF_6-Al_2O_3 (sat.)-10.9wt.%AlF_3-5wt.% CaF_2 melts at 970℃. $\eta_a = a + b\lg i$,
Current density range: 0.06~1.27A/cm²

Sample No.	Additives	Content of Additives(wt.)/%	Baking Temp. /℃	a/V	b/V	i_0 /A·cm⁻²	Made by
UF1	Undoped	0	1100	0.566	0.140	9.06	CLT
DF1	AlF_3	0.5	1100	0.557	0.132	6.03	CLT
DF2	AlF_3	1.0	1100	0.505	0.125	9.12	CLT
DF3	AlF_3	1.5	1100	0.551	0.130	5.77	CLT
NF1	Undoped	0	1250	0.565	0.135	6.53	CLT
NF2	Undoped	0	1250	0.561	0.134	6.51	CLT
AF1	AlF_3	0.5	1250	0.544	0.123	3.78	CLT
AF2	AlF_3	1.0	1250	0.558	0.125	3.44	CLT
AF3	AlF_3	1.5	1250	0.567	0.125	2.91	CLT

As expected, $MgAl_2O_4$ also has a positive effect in decreasing the anodic overvoltage, compared to undoped anodes. Fig. 3 shows plots of η versus $\lg i$ for a series of runs obtained with undoped anodes and anodes doped with 0.5 wt.% $MgAl_2O_4$ (B4 and M3 in Table 3, baked at 970℃). According to the linear regression lines shown in the figure, the overvoltage of the doped anodes at 1A/cm² was 79mV (±29mV) lower than that of the undoped anodes. Table 3 lists the Tafel coefficients and exchange current densities for undoped anodes and anodes doped with $MgAl_2O_4$. From the Tafel lines for undoped and doped carbon anodes with the same recipe and identical composition, but baked at 1200℃, the effect of $MgAl_2O_4$ in lowering the anodic overvoltage was far less than for anodes baked at a lower temperature. This is in agreement with the results mentioned above. The anodes U7, U8 and M7, M8 vs B4 and M3 were prepared under similar conditions but by two different makers (CLE and CLZ, respectively) from different raw materials. There is a difference in the Tafel coefficients and in the effect of $MgAl_2O_4$ in lowering the anodic overvoltage, which may indicate that the effect of additives is sensitive to the source of coke and pitch, or perhaps to slight differences in preparation.

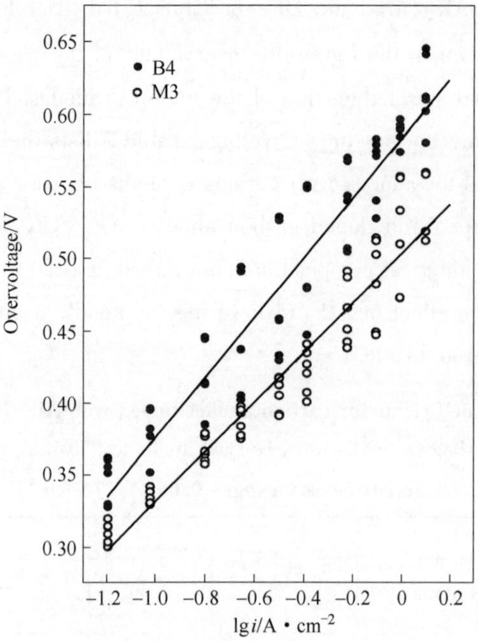

Fig. 3 Tafel plots for undoped anodes and anodes doped with 0.5wt.% MgAl$_2$O$_4$(B4 and M3 in Table 1). The anode surface area exposed to the melt was 9.42cm^2. The data was obtained by using the model 805 current interruptor

Table 3 Tafel coefficients for carbon anodes doped with MgAl$_2$O$_4$ in Na$_3$AlF$_6$-Al$_2$O$_3$(sat.)-10.9 wt.%AlF$_3$-5wt.% CaF$_2$ melts at 970℃ $\eta_a = a + b\lg i$, Current density range: 0.06~1.27A/cm^2

Sample No.	Additives	Content of Additives(wt.)/%	Baking Temp. /℃	a/V	b/V	i_0/A·cm^{-2}	Made by
U7	Undoped	0	1010	0.541	0.134	0.918	CLE
U8	Undoped	0	1010	0.541	0.125	0.470	CLE
M7	MgAl$_2$O$_4$	0.5	1010	0.519	0.146	2.79	CLE
M8	MgAl$_2$O$_4$	0.5	1010	0.511	0.113	0.301	CLE
B2	Undoped	0	1200	0.549	0.174	7.00	CLZ
M1	MgAl$_2$O$_4$	0.5	1200	0.534	0.165	5.80	CLZ
B4	Undoped	0	970	0.582	0.206	15.3	CLT
M3	MgAl$_2$O$_4$	0.5	970	0.503	0.171	11.4	CLT

5 Conclusion

(1) It has been clearly demonstrated that anodic overvoltage can be lowered by adding inorganic dopant to the anode paste in aluminium electrolysis.

(2) The present investigation indicates that the effect tends to be less than previously published by many authors. Typical values are 40~60mV for prebake type anodes baked at around 1150℃ and 60~80mV for Soderberg type anodes baked at 950℃.

(3) The mechanism of the doping effect is not well understood.

References

[1] K Grjotheim et al. Aluminium Electrolysis-Fundamentals of the Hall-Heroult Process. 2nd edition, Dusseldorf: Aluminium-Verlag, 1982: 194-256.

[2] T. Muftuoglu and H. A. Øye. Reactivity and Electrolytic Consumption of Anode Carbon with Various Additives. Light Metals, 1987, 471-476.

[3] J. O'M Bockris et al. Electrocatalysis: Past, Present and Future. Electrichim. Acta, 1994, 39.

[4] J. Thonstad et al. On the Anodic Overvoltage in Aluminum Electrolysis. Can. J. Chem., 1964, 42: 1542-1550.

[5] V. A. Braunworth et al. The Addition of Aluminum Fluoride to Soderberg Paste. Light Metals 1975: 325-337.

[6] M. Sorlie et al. Gas Reactivity and Electrolytic Consumption of Aluminum Cell Anodes with Aluminum Fluoride Additions. Proc. 21st Conference on carbon, Buffalo, NY, USA, 1993: 677-678.

[7] Y. Liu and J. Thonstad. Oxygen Overvoltage on SnO_2-Based Anodes in NaF-AlF_3-Al_2O_3 Melts: Electrocatalytic Effects of Doping Agents. Electrochim. Acta, 1983, 28(1): 113-116.

[8] H. Xiao and Y. Liu. Study on the Electrolytic Activity of Doped Carbon Anodes in Aluminum Electrolysis. Proc. 6th Meeting of Physico-Chemistry of Metallurgical Processes, Xian, China (in Chinese), 1986.

[9] Y. Liu and H. Xiao. A Study of the Electrocatalytic Activity of Doped Carbon Anodes in Cryolite-Alumina Melts. Proc. Joint International Symp. Molten Salt, The Electrochemical Soc. Inc. Honolulu, Hawaii, USA, 1987: 744-750.

[10] H. Xiao and Y. Liu. Studies on the Electrocatalytic Activity of Doped Carbon Anodes in Aluminum Electrolysis. J. Cent-South Inst. Min. & Metall., (in Chinese), 1988, 19(3): 241-248.

[11] G. Yao et al. The Electrocatalysis of Lithium Carbonate in Anodic Reaction in Hall-Heroult Process. Light Metals(in Chinese), 1988(12): 22-26.

[12] N. Feng et al. The Effect of Additives on the Anodic Overvoltage of Carbon Anodes in Aluminum Electrolysis. Light Metals, (in Chinese), 1989(7): 26-30.

[13] N. Feng et al. Influence of Lithium Carbonate Addition to Carbon Anodes in a Laboratory Aluminum Electrolysis Cell. Carbon, 1991, 29(1): 39-42.

[14] Z. Yu et al. The Effect of Some Additives on the Overvoltage of Carbon Anode in Aluminum Electrolysis. Non-ferrous Metals(in Chinese), 1987, 39(4): 59-63.

[15] C. Duan et al. Study on the Relationship between Lithium Addition and Overvoltage of Carbon Anode uin Aluminum Electrolysis. Light Metals(in Chinese), 1990(6): 27-30.

[16] D. Yan et al. Study on the Carbon Anode with Dopants for Saving Energy in Aluminum Electrolysis. J. Cent-South Inst. Min. Metall(in Chinese), 1989, 20(5): 505-511.

[17] Z. Li, Master Degree Thesis, Cent-South University of Technology(CSUT), Changsha, China, 1990.

[18] H. Liu, Master Degree Thesis, CSUT, 1991.

[19] W. Hu, Master Degree Thesis, CSUT, 1992.

[20] G. M. Haaberg et al. Electrochemical Studuies of the Anode Reaction on Carbon In NaF-AlF_3-Al_2O_3 Melts. Light Metals 1994(TMS, Warrendale, PA. USA): 227-231.

[21] Y. Liu et al. New Type Electrocatalysts for Energy Saving in Aluminum Electrolysis. Light. Metals 1995 (TMS, Warrendale, PA. USA): 247-251.

[22] Y. Liu et al. New Progress of Study on Lithium Salt Containing Carbon Anode Paste. Light Metals(in Chinese), 1994(5): 28-30.

[23] Y. Liu et al. Progress in Study of Electrocatalysis and Doped Carbon Anode in Aluminum Electrolysis Cells. Trans. Nonferrous Metals Soc. of China 1994,4(2):92-94.

[24] Y. Liu et al. The Inhibition of Anode Effect in Aluminum Electrolysis Process by Anode Dopants: a Laboratory Study. Light Metals 1991(TMS,Warrendale,PA. USA): 489-494.

[25] Z. Qiu et al. Carbon Anode with Lithium Salts Addition. Light Metals 1995(TMS,Warrendale,PA. USA): 749-751.

[26] Z. Qiu et al. Aluminum Electrolysis with Lithium Salt-Containing Carbon Anode. Aluminum 1995,71(3): 343-345.

[27] Z. Qiu et al. ,the study of critical current densities on lithium salt containing carbon anode, light metals (in chinese),1994(5):31-33.

[28] Z. Kuang. On the Consumption of Carbon Anode in Aluminum Electrolysis. Dr. Thesis, Norwegian Inst. Technology,Trondheim,1994:46-62.

[29] L. N. Solli. Carbon Anodes in Aluminum Electrolysis Cells: Factors Affecting Anode Potential and Carbon Consumption. Dr. Thesis,Norwegian Inst. Technology,Trondheim,1994:71-92.

[30] S. Shen. On the Measurement of Anodic Overvoltage in Aluminum Electrolysis by Using Reference Electrode. Light Metals(in Chinese),1989(10):24-29.

[31] D. Li. On Laboratory Measurement of Anode Overvoltage in Aluminum Electrolysis. Light. Metals(in Chinese),1990(6):31-35.

[32] S. Shen. On the Lithium Salt Paste in the Aluminum Electrolysis industry of China. Non-ferrous Metals(in Chinese),1991(3):41-43.

[33] D. Li and X. Wang. The Laboratory Measurement on Overvoltage of Anode Containing Lithium Salt in Aluminum Electrolysis. Proc. of a Special Conference on Lithium Salt Paste(in Chinese),Shenyang,China, November,1992.

[34] S. Shen. The Practical Effect of Lithium Salt Pasre on Depression of Anodic Overvoltage in the Aluminum Industrial Cells. Light Metals(in Chinese),1993(12):31-35.

[35] S. Sun. On the Practical Effectiveness of Overvoltage Reduced by Lithium Salt Paste Used in Aluminum Electrolysis Cells. Light Metals(in Chinese),1993(12):31-35.

[36] J. Yang et al. Discuss f Views and Divergences on Electrocatalysis of Lithium Salt Paste in Aluminum Eklectrolysis Cells. J. Cent-South Inst. Min. Metall(in Chinese),1994,25(3):326-332.

[37] Y. Liu et al. A New Approach to Reduce the Anodic Overvoltage in the Hall-Heroult Process. Light Metals 1989,(TMS,Warrendale,PA. USA):275-280.

[38] Y. Liu et al. A New Field to Reduce Energy in Hall-Heroult Process:the Research and Application of an Anode Paste Containing Lithium Salt. Light Metals,1993(TMS,Warrendale,PA. USA):599-601.

[39] Z. Qiu. The Application of Lithium Salt Carbon Anode Paste in Aluminum Electrolysis. Non-ferrous Metals (in Chinese),1991(1):22-25.

四、湿法冶金电解过程电催化

锌电解节能惰性阳极的研究
——惰性阳极上的析氧超电压[*]

1 引言

金属的电积过程在有色冶金工业中正日益显示其重要性,这不仅因为它实质上是一个无污染的过程,能顺应环境保护要求日高的形势,而且,考虑到一些金属的电积(例如 Cu 和 Zn)将逐渐取代火冶方法,这就使金属电积工业本身面临新的挑战——必须迈向新的技术高度,实现高产、优质、节能、无污染。达到此目的的关键之一是寻求功能优异的电极材料。

现代金属电积工业中,用于酸性电解液的 Pb 或 Pb 合金阳极存在一系列缺点,主要是:①投入使用时,要求镀膜及预调整的时间长,且工作不稳定,由此引起的浪费较多;②电解时,在膜上析氧超电压很高,在工业电流密度下接近 1V,这是电能无谓耗费的主要根源;③阴极产品受 Pb 的污染严重,为了保证产品质量,还要相当的花费等。其唯一优点是耐酸腐蚀性强。得失相较,付出的代价太大,实有加以革新的必要。

以铁为基材,具有电催化活性镀层的形状稳定的阳极(Dimensional Stable Anode,简称 DSA),在氯碱工业及其他电化学工业中已得到成功的应用,全世界每年因此节电数十亿度,经济效益显著[1]。这种阳极的特点是,阳极电位低(在氯化盐电解液中析氯超电压很低,在硫酸盐电解液中析氧超电压低),槽电压低,使用的电流密度高,耐腐蚀性强,因而电能消耗低,设备产能大,产品纯度高[2]。应利用这一技术进步,加快有色金属电积工业的技术改造与发展。

在有色金属电积工业中,近年来研制 Ti 基 DSA 电积 Cu,Zn,Ni,Co 已成为新发展动向之一[1,3]。试验用于氯化物,硫酸盐或二者混合电解液的 DSA 大体可分为四类,即:Ti/PbO_2[4,5],Ti/贵金属/PbO_2(例如,Ti/Au/PbO_2)[6],Ti/铂族金属或其合金(例如,Ti/Pt-Ir)[7,8]以及 Ti/MnO_2[9]等。这些电极的性能大多数还不理想,或尚未投入工业应用。在研制这类电极的同时,对硫酸溶液中 Ti 基 DSA 的电化学行为也进行了大量的基础研究[10]。

在 Zn 电积中,目前仍使用 Pb-Ag(约 1%Ag)阳极。美国矿务局[11]近年曾试图把所研究的 Ti/PbO_2 阳极用于 Zn 电积,但未获成功。主要问题是阳极本身的电阻及其上析氧超电压都较高(槽电压比 Pb-Ag 阳极电解槽高 100~300mV),而且寿命不长(仅使用 1~7 周)。

我们知道,Zn 电积过程中槽电压($V_槽$)的组成可表示如下:

$$V_槽 = E + \eta_活 + \eta_浓 + \eta_欧 + I\sum R_阻$$

---

[*] 本文合作者:吴良薰。原发表于《有色金属》,1984(4):12。

式中，E 为电解反应产物所形成的原电池的电动势；$\eta_{活}$、$\eta_{浓}$、$\eta_{欧}$ 分别为活化、浓差和欧姆超电压；$\sum R_{阻}$ 为电解槽上离子导体和电子导体的电阻总和。欲节电，就应在一定的工作电流密度下尽可能减小槽电压。在阳极上引进对析氧反应有利的电催化剂，就有可能减小活化超电压，达到节电的目的。本研究就是由此着眼进行探索的。

2 电催化活性阳极的研制

2.1 电极基材的选择

对铝，不锈钢，铁，SnO_2 基陶瓷和 Fe_2O_3 基陶瓷材料进行了试验筛选。在酸性 $ZnSO_4$ 电解液中，SnO_2 基阳极和 Fe_2O_3 基阳极在低于 100℃ 及室温下的电阻率较大，不适于作电极材料。金属铝阳极，在阳极极化时生成了不导电的 Al_2O_3 厚膜，后者不易与活性镀层牢固地结合，造成镀层过早脱落，使试验失败。根据材料本身的电阻率，耐腐蚀性，基材与活性镀层之间的相容性，以及阳极极化特性等表现，以铁最好，不锈钢次之。因此选定这两者作进一步试验。

2.2 镀覆中间层

钛基 DSA 在酸性电解液中使用时，存在的主要问题是寿命不长。当电极的活性丧失后，阳极的电位急剧升高，这时阳极的服役期即告终结。现已查明，丧失活性的原因，除因镀层组分被溶解损失之外，主要还在于阳极极化时，析出的氧渗入到活性镀层与基材之间，使基材表面氧化并生成钝化膜，增大了电极的电阻，严重时可使活性镀层崩落。E. R. Cole, Jr. 等[11] 在 Zn 电积中试验 Ti/PbO_2 失败，阳极寿命不长的原因，正是由于 Ti 与镀层 PbO_2 之间结合不牢。为了提高使用寿命，有必要在金属阳极表面镀覆一层有特殊性能要求的中间层。目前所研制的贵金属镀层[7,10] 费用高而效果不够理想。经我们多次试验，以采用掺 Sb-SnO_2 中间镀层为优。这是一层结构致密，具有电子导电性的薄膜，能达到阻止电极钝化，减小电极电阻和提高电极寿命的目的。

2.3 采用电催化活性高的 RuO_2-TiO_2 外部镀层

我们对 RuO_2/TiO_2 比，镀层溶剂及其浓度，镀覆方式，热处理温度制度等进行了研究。制成的电催化活性阳极为 $\phi = 1.5mm$ 的铁丝和不锈钢丝，中间镀层为掺 Sb-SnO_2 层，外部活性层为 RuO_2-TiO_2 层。两种镀层的总厚度为 $12\sim18\mu m$。可表示为 Ti/掺 Sb-SnO_2/RuO_2-TiO_2 阳极。有效工作面积控制在 $0.7\sim1.0cm^2$。

3 试验测定

用稳态恒电位法测定极化曲线，HDV-7 型恒电位仪由 KS-1 型快扫描讯号发生器控制，以 2.5mV/s 的速度向待测系统施加直流电势，所得结果分别显示在 x-y 函数记录仪和数字电压表上。电解槽容量为 300mL，铝阴极面积为 $4.0cm^2$。参比电极用饱和甘汞电极，其 Luggin 毛细管管嘴紧靠待测阳极，以使阳极-参比电极之间电解液的 IR 降最小。电解液含 Zn 50g/L，H_2SO_4 150g/L。电解槽置于恒温水浴内，试验温度为 (35±0.1)℃。图 1 为 Ti/掺 Sb-SnO_2/RuO_2-TiO_2 阳极（以下简称 T_2 阳极）的典型极化曲线。

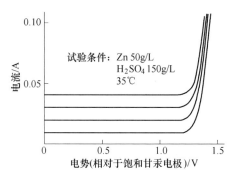

图 1　Ti/掺 Sb-SnO$_2$/RuO$_2$-TiO$_2$ 阳极极化曲线（电流至 0.20A）

4　结果及讨论

对上述制备的钛基和不锈钢基阳极进行了测定。所得极化曲线数据经处理，得到了超电压与电流密度的对数的关系，见图 2 和表 1，有关的电极过程动力学数据列于表 2。

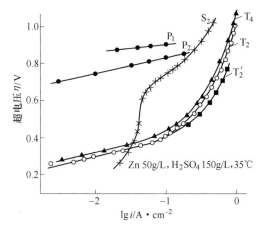

图 2　阳极析氧超电压（η）与电流密度对数（$\lg i$）的关系

P_1—Pb-Ag 阳极（即 PbO$_2$ 阳极），工业生产条件下的实测数据[12]；
P_2—PbO$_2$ 阳极（石墨基材），实验室测定数据[13]； T_2—Ti/掺 Sb-SnO$_2$/RuO$_2$-TiO$_2$ 阳极；
T_2'—T_2 阳极，使用后放置 100 小时再测定之结果；T_4—Ti/RuO$_2$-TiO$_2$ 阳极（无中间层）；
S_2—不锈钢/掺 Sb-SnO$_2$/RuO$_2$/TiO$_2$ 阳极

表 1　锌电解不同阳极上的析氧超电压

阳极	电流密度/A·m^{-2}				
	析氧超电压/V				
	400	500	600	800	1000
PbO$_2$，工业实测[12]	0.88	0.887	0.89	0.897	0.90
PbO$_2$，实验室测定[13]	0.80	0.805	0.81	0.82	0.83
T_2（有中间层）	0.346	0.36	0.37	0.39	0.41
T_4（无中间层）	0.37	0.38	0.39	0.415	0.43
S_2（有中间层）	0.42	0.63	0.67	0.715	0.745

电积条件：Zn 50g/L，H$_2$SO$_4$ 150g/L，35℃。

"电催化"一词，就实际应用来说是通过相互比较来鉴别的。在一定的超电压下，某电极上进行的给定反应（此处是析氧反应）要比另一电极的快，那么，前一电极的电催化性能就好[14]。电催化的标志,可借电极过程动力学的有关参数表示,即塔菲尔直线的斜率 b 值较低,交换电流密度 i_0 值较大,以及在相同表观电流密度下的超电压 η 较小。

我们所得的动力学数据表明(见表2),T_2 电极的塔菲尔斜率 b 值比 P_2 电极的小,而交换电流密度 i_0 值比 P_2 电极高出4个数量级,且恒定电流密度下的超电压 η 远比 P_2 电极为小(见表1),这是电催化活性很高的表现。

表2 Ti 基 DSA(T_2) 与 PbO_2 阳极(P_2)的析氧反应动力学参数

电极名称	a/V	b/V	i_0/A·cm^{-2}	电流密度范围/A·cm^{-2}
P_2[15]	1.18	0.134	1.7×10^{-9}	0.003~0.178
T_2	0.455	0.10	2.85×10^{-5}	0.003~0.042

注:电积条件同表1。

从表1也可以看出,在工业电流密度范围内,我们所研究的阳极都显著优于 Pb-Ag 阳极,特别是 T_2 电极表现优异。以 500A/m^2 为例,T_2 阳极的析氧超电压比 PbO_2 阳极小 0.527V,T_4 比 PbO_2 阳极小 0.507V,S_2 则小 0.257V。如此大幅度地降低析氧超电压,正是此种电极具有高的电催化活性的结果。可以预期,采用这种阳极取代 Pb-Ag 阳极,能够大幅度地节约电能。例如,用 Pb-Ag 阳极在 500A/m^2 下电解时,在平均槽电压为 3.5V,电流效率为 90% 时,每吨锌的电能消耗为 3190 度。如用 T_2 阳极代替,在相同条件下,每吨锌的电能消耗降为 2710 度,即每吨锌可节电 480 度,节电达 15%。

使用 T_2 阳极还能在很高的电流密度下工作。此外,还将完全消除产品受 Pb 阳极的污染。根据这种特性,可望使用在其他金属的电积生产(如 Ni,Cu,Co 等),多种电化学工业以及环境保护方面。

不锈钢基材的 DSA(S_2 电极)也有相当好的表现,特别是价格远比金属钛低廉。但在电流密度达到 7400A/m^2 时,活性镀层脱落,电极遭到破坏。

上述电极的使用寿命,电解液中杂质对电极的催化活性的影响,贵金属钌化合物的最佳用量及代用品,以及 DSA 上析氧过程机理等问题,均待今后继续研究。

参 考 文 献

[1] Vittorio de Nora. ソーダと塩素,1983,No.3,1-7.
[2] Vittorio de Nora. Diaphragm Cell Chlorine prod. Proc. Symp. London,1976(Pub.1977),15-21.
[3] G. Bewer et al. J. of Metals,1982,34(1),37,ibid,1983,35(4),49.
[4] L. W. Higley,et al. USBMI RI,Ri-8111,1976.
[5] D. Gilroy. J. Appl. Electrochem. 1982,12:171-183.
[6] A. J. Scarpellino,et al. J. Electrochem. Soc.,1982,129:515-521,522-525.
[7] C. L. Comninellis,et al. J. Appl. Electrochem,1982,12:399-404.
[8] D. Wensley,et al. Met. Trans. B.,1979,10B:503-511.
[9] Н. ИФурман и др. Ц. М.,1974,5,22-24.
[10] 高桥正雄. エネルギ——材料研究. No.4 别刷,横浜国立大学工学部,1982.
[11] E. R. Cole,Jr.,T. J. O'Keefe. USBMI RI,RI-8531,1981.
[12] 赵天从. 重金属冶金学(下). 北京:冶金工业出版社,1981,70.
[13] M. S. V. Pathy,et al. Electrochim. Acta. 1965,10:1185-1189.
[14] J. O'M. Bockris,et al. Electro-chemical Science,1972.

用于 Ni 和 Zn 电积的节能形稳阳极（DSA）
——实验室研究*

摘　要　在金属电积生产中，降低阳极上的析氧（或析氯）超电位是节能的重要途径之一。采用形稳阳极（Dimensionally Stable Anodes，简称 DSA）可以实现。以 Ti 为基材的 DSA 具有抗腐蚀性强，析氧超电压低，不污染阴极产品等优点，可望取代现行 Pb 阳极，达到节能目的。研制了四种新型阳极，即 Ti 基 RuO_2-TiO_2（Ⅰ），Ti 基 MnO_2（Ⅱ），石墨基 RuO_2-TiO_2（Ⅲ）及陶瓷基 RuO_2-TiO_2（Ⅳ）。用稳态恒电位法测定了极化曲线，获得了不同电流密度下析氧超电压数据和电极过程动力学数据。表明阳极（Ⅲ）的电催化活性最好，析氧超电压最低。在电解含 H_2SO_4 的 $NiSO_4$ 时，60℃及 250A/m^2 下，阳极析氧超电压的增大次序为Ⅲ<Ⅳ<Ⅰ<Ⅱ。在电解含电解含 H_2SO_4 的 $ZnSO_4$ 时，在 35℃ 及 500A/m^2 下，此次序为Ⅳ<Ⅲ<Ⅰ<Ⅱ。这些阳极均优于 Pb-Ag 阳极。若以阳极（Ⅲ）代替 Pb-Ag 阳极电积 Ni 和 Zn，在其他条件不变时，由于超电压减小而获得的节电效果均大于 15%。若以阳极（Ⅰ）代替 Pb-Ag 阳极，节电效果分别为 11% 和 14%。

金属电积生产中节能的重要途径之一，就是降低阳极上的析氧（或析氯）超电压。过去对这一途径的重视和研究是很不够的。目前，在酸性硫酸盐电解液中电积金属所用的 Pb 合金阳极，在工业电流密度下，其上析氧超电压普遍很高，有的接近 1V（例如 Zn 电积）。这样高的超电压造成了很大的电能浪费，应采取措施予以降低，此中的节能潜力很大，例如以电解含硫酸的 $ZnSO_4$ 溶液而言，在其他条件不变时，仅降低 Pb 合金阳极上的析氧超电压 100mV，即可获得每吨产品节电 2.86% 的效益，对电解含硫酸的 $NiSO_4$ 溶液，每减小阳极超电压 100mV，可节电 3.05%[①]，节能的效果是很显著的。采用新型的形状稳定阳极（Dimensionally Stable Anodes，简称 DSA）就可实现这一点。以 Ti 为基体材料的 DSA 具有抗腐蚀性强，析氧（或氯）超电压低，不污染阴极产品，机械强度好及重量轻等优点。近年来，金属电积工业中研制 Ti 基 DSA 用于电解铜、锌、镍、钴等成为发展新动向之一[1-3]。E. R. Coler, Jr 等研究了 Ti/PbO_2 阳极，用于 Zn 电积时由于析氧超电压高和电极寿命短而未获成功[4]。我们对 Zn 和 Ni 电积用的 DSA 也作了研究[5,6]。采用镀中间层的措施，制成了 Ti/SbO_x-SnO_2/RuO_2-TiO_2 阳极，在降低析氧超电压和提高使用寿命方面效果较好。A. J. Scarpellino 等[7] 研制了用于 Ni 电积的 Ti 基中间镀钯的钌-铱不溶阳极（Ti/Pd/Ru-Ir），在实验室条件下获得了析氧超电压低，使用寿命长的效果。考虑到在少用或不用贵金属，以及选用性能较好的基体材料方面还有开发的必要，我们在前面工作的基础上，研究了可用于 Zn 和 Ni 电积的 4 种新型 DSA 阳极，它们是：Ti 基 RuO_2-TiO_2 阳极（Ⅰ型），Ti 基 MnO_2 阳极（Ⅱ型），石墨基 RuO_2-TiO_2 阳极（Ⅲ型）及陶瓷基 RuO_2-TiO_2 阳极（Ⅳ型）。其目的是，获得具有较高电催化活性的新型阳极，能显著降低析氧超电压，可以取代传统的 Pb 合金阳极，研究它们的电化学行为并对其节能的潜力作出评价。

---

*　本文合作者：吴良蕙、袁炳年。原发表于《有色金属》，1985, 37 (3)：53-57。

①　按 Zn 电解槽槽电压为 3.5V，Ni 电解槽槽电压为 2.5V 计算。

1 实验部分

1.1 电极的制备

根据我们早先的工作[5~7],我们选定了钛、石墨和化学陶瓷作为电极的基体材料。

工业级 Ti 丝（$\phi1.0~1.5$mm）经过洁净处理,机械的和化学的粗化处理,而后镀覆一含 SbO_x-SnO_2 的中间层,经适当的热处理后,在其外表分别涂镀含钌和钛的溶液和酸性锰盐溶液,在经热烧结后获得 Ti/SbO_x-SnO_2/RuO_2-TiO_2（Ⅰ）和 Ti/SbO_x-SnO_2/MnO_2（Ⅱ）阳极。

将光谱纯石墨棒（$\phi5.5~6.0$mm）经去脂洁净及中温处理,多次涂镀含钌、钛的有机溶液,再经过适当热处理,即制得石墨基 RuO_2-TiO_2（Ⅲ）阳极。

将化学陶瓷棒（$\phi5.2~5.3$mm）去污脱脂洁净处理后,进行适当的导电处理,选择在室温下导电良好的坯体为基材,在经多次涂镀含钌、钛的有机溶液,而后热烧结,制得陶瓷基 RuO_2-TiO_2（Ⅳ）。

为了便于比较,我们也制备了 Pb-Ag（含 Ag 0.78%）阳极,采用了文献[8,9]的镀膜方法,在该阳极上形成 PbO_2 膜。

上述各类阳极的有效工作面积控制在 $0.3~0.6$cm^2。

1.2 实验测定

用稳态恒电位法测定极化曲线,采用了 HDV-7 型电位仪和 KS-1 型快扫描讯号发生器,扫描速度用 2.5mV/s 或 10mV/s。所得结果分别显示在 X-Y 函数仪和数字电压表上。玻璃电解池容量为 300mL。纯铝板辅助电极的面积为 4.0cm^2,参比电极为饱和甘汞电极,其 Luggin 毛细管管嘴紧靠待测阳极,以使阳极与参比电极之间电解液的 IR 降最小。每种电极测试 6 次以上。

选择接近工业电解液的组成：其中 Ni 电解液含 Ni 70g/L, H_2SO_4 40g/L,溶液 pH 值为 0.3~0.5。Zn 电解液含 Zn 50g/L, H_2SO_4 150g/L。电解池置于恒温水浴内,试验温度分别为（60±0.1）℃（Ni 电积）和（35±0.1）℃（Zn 电积）。

2 结果及讨论

4 种新型阳极及 Pb-Ag 阳极的典型极化曲线示于图 1。由极化曲线数据得到的超电压与电流密度对数的关系见图 2。有关电极过程动力学参数列于表 1。工业电流密度下各电极析氧超电压的对比列于表 2 和表 3。

2.1 阳极的电催化活性问题

从应用的观点看,最好的电催化剂表现出高的交换电流密度（i_0）和低的 Tafel 斜率 b 值[10,11],由表 1（a）可见,在 Ni 电解液中,新型阳极以石墨基 RuO_2-TiO_2 阳极的 i_0 值最大, b 值最小,其上析氧反应的速度最大,因而催化活性最高,（Ⅰ）型阳极次之。所有这些阳极都优于 Pb-Ag 阳极。在 Zn 电解液中[表 1（b）],以陶瓷和石墨基 RuO_2-TiO_2 阳极表现最优越。它们的 i_0 比 Pb-Ag 阳极要大 6~7 个数量级。

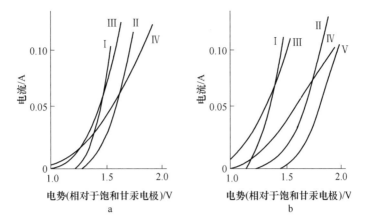

图 1 新型阳极和 Pb-Ag 阳极的典型极化曲线

a—在 Ni 电解液中（70g/L Ni，40g/L H_2SO_4，40g/L Na_2SO_4，pH=0.3~0.5，60℃）；

b—在 Zn 电解液中（50g/L Zn，150g/L H_2SO_4，35℃）

Ⅰ—Ti/SbO_x-SnO_2/RuO_2-TiO_2 阳极；Ⅱ—Ti/SbO_x-SnO_2/MnO_2 阳极；Ⅲ—石墨/RuO_2-TiO_2 阳极；

Ⅳ—陶瓷/RuO_2-TiO_2 阳极；Ⅴ—Pb-Ag（~1% Ag）阳极

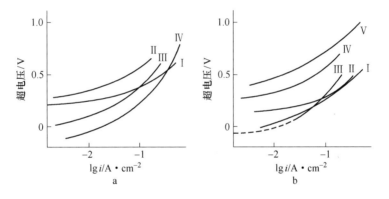

图 2 新型阳极和 Pb-Ag 阳极的阳极超电压—半对数图

a—在 Ni 电解液中；b—在 Zn 电解液中（符号同图 1）

表 1 新型阳极和 Pb-Ag 阳极的析氧反应动力学参数

(a) Ni 电积

阳 极	a/V	b/V	i_0/A·cm^{-2}	电流密度范围/A·cm^{-2}
Ti 基 RuO_2-TiO_2（Ⅰ）	0.33	0.087	0.16×10^{-3}	5.0×10^{-3}~1.26×10^{-2}
Ti 基 MnO_2（Ⅱ）	0.56	0.117	1.64×10^{-5}	5.0×10^{-3}~1.26×10^{-2}
石墨基 RuO_2-TiO_2（Ⅲ）	0.14	0.087	2.46×10^{-2}	5.0×10^{-3}~1.26×10^{-2}
	0.485	0.269	1.57×10^{-2}	1.58×10^{-2}~2.5×10^{-2}
陶瓷基 RuO_2-TiO_2（Ⅳ）	0.44	0.209	7.85×10^{-3}	5.0×10^{-3}~1.26×10^{-2}
Pb-Ag（~1%Ag）（Ⅴ）	0.95	0.234	8.7×10^{-5}	5.0×10^{-3}~2.0×10^{-2}

试验条件：电解液含 Ni 70g/L，H_2SO_4 40g/L，Na_2SO_4 40g/L，pH=0.3~0.5，60℃。

(b) Zn 电积

阳极	a/V	b/V	i_0/A·cm^{-2}	电流密度范围/A·cm^{-2}
Ti 基 RuO$_2$-TiO$_2$(Ⅰ)	0.45	0.095	1.8×10^{-5}	6.3×10^{-3}~6.3×10^{-2}
Ti 基 MnO$_2$(Ⅱ)	0.63	0.138	2.7×10^{-5}	3.0×10^{-3}~1.0×10^{-2}
石墨基 RuO$_2$-TiO$_2$(Ⅲ)	0.39	0.132	1.1×10^{-3}	3.0×10^{-3}~1.0×10^{-2}
陶瓷基 RuO$_2$-TiO$_2$(Ⅳ)	0.20	0.120	2.2×10^{-2}	3.0×10^{-3}~1.0×10^{-2}
Pb-Ag(~1% Ag)(Ⅴ)	1.18	0.134	1.6×10^{-9}	3.0×10^{-3}~1.78×10^{-1}

试验条件：电解液含 Zn 50g/L，H$_2$SO$_4$ 150g/L，35℃。

表 2 Ni 电解液中阳极上的析氧超电压

阳极 \ 超电压/V	电流密度/A·m^{-2}				
	250	300	400	500	1000
Ⅰ	0.196	0.210	0.222	0.240	0.290
Ⅱ	0.390	0.407	0.435	0.465	0.565
Ⅲ	0.055	0.080	0.120	0.170	0.310
Ⅳ	0.105	0.120	0.050	0.170	0.250
Ⅴ	0.570	0.590	0.620	0.640	0.735

电积条件：电解液含 Ni 70g/L，H$_2$SO$_4$ 40g/L，Na$_2$SO$_4$ 40g/L，pH=0.3~0.5，60℃。

表 3 Zn 电解液中阳极上的析氧超电压

阳极 \ 超电压/V	电流密度/A·m^{-2}				
	400	500	600	800	1000
Ⅰ	0.305	0.320	0.330	0.350	0.375
Ⅱ	0.450	0.470	0.495	0.530	0.565
Ⅲ	0.235	0.265	0.290	0.330	0.375
Ⅳ	0.075	0.110	0.140	0.185	0.230
Ⅴ	0.800	0.805	0.810	0.820	0.830

电积条件：电解液含 Zn 50g/L，H$_2$SO$_4$ 150g/L，35℃。

对比恒定表现电流密度下不同阳极上的超电压大小，是区别电催化优劣的有力判据[11]。以超电压小者催化活性高。由表2、表3列出的数据可见，在 Ni 电积的工业电流密度范围内（250~400A/m^2），析氧超电压以石墨 RuO$_2$-TiO$_2$ 阳极最低，仅 55~130mV。这4种电极按电压增大的排列次序是：Ⅲ<Ⅳ<Ⅰ<Ⅱ。在 Zn 电积的工业电流密度范围内（400~1000A/m^2），以陶瓷和石墨基 RuO$_2$-TiO$_2$ 阳极的析氧超电压最低，该排列次序是：Ⅳ<Ⅲ<Ⅰ<Ⅱ。这些阳极的析氧超电压都远比 Pb-Ag 阳极为小。

综上所述，可以认为这4种新型阳极的电催化活性，无论在 Ni 电积或在 Zn 电积中，都优于 Pb-Ag 阳极。

2.2 陶瓷和石墨 RuO$_2$-TiO$_2$ 阳极的优异性能

我们在试验中观察到，陶瓷和石墨 RuO$_2$-TiO$_2$ 阳极进行极化时，氧的析出，均在低于高酸度酸性溶液中氧的平衡电极电位下，即低于 1V 的情况下发生（见图1）。对这种现象的初步解释可能是由于阳极反应发生了改变，因而使电极反应在比原反应的平衡电位更低的电位下进行。由于石墨和化学陶瓷都是多孔材料，经一系列电极制备处理后，其真实表面积将由很大增加，易吸附反应物参加电极反应。这一现象是饶有兴趣的，留待下一步深入研究。需要指出的是，尽管陶瓷基 RuO_2-TiO_2 阳极的电催化活性甚高，无论在 Ni 和 Zn 电积时，其上析氧超电压均很小，但这种材料的电阻率很高（$0.2\Omega \cdot cm$，比石墨基 RuO_2-TiO_2 阳极高出 5 个数量级），影响到它的节电效果。

2.3 降低超电压节能的效益估算

金属的电积生产，电解槽的平均电压可用下式表示：

$$V_c = E_r + \eta_c + \eta_a + I\sum R$$

式中，E_r 为电解反应产物所形成的原电池的电动势；η_c、η_a 分别为阴极超电压和阳极超电压；I 为电流；$\sum R$ 为电解槽上电子导体与离子导体的电阻总和。

当其他条件不变时，降低 η_a（此处是降低阳极析氧超电压）即可节约电能。若采用我们所研制的 4 种阳极代替 Pb-Ag 阳极，在 $250A/m^2$ 下电解含 H_2SO_4 的 $NiSO_4$ 溶液和 $500A/m^2$ 下电解含 H_2SO_4 的 $ZnSO_4$ 溶液，每吨金属经计算可节约电能列于表 4 和表 5。

表 4 Ni 电积使用新型阳极的节电效益

阳极	$250A/m^2$ 下的超电压/V	与 Pb-Ag 阳极比较，超电压减少/V	直流电耗 /$kW \cdot h \cdot t^{-1}$	节约电能 $kW \cdot h/t$	节约电能 %
石墨（Ⅲ）	0.055	-0.515	2603	485	15.7
陶瓷（Ⅳ）	0.105	-0.465	2650	438	14.2
Ti-Ru（Ⅰ）	0.196	-0.374	2736	352	11.4
Ti-Mn（Ⅱ）	0.390	-0.180	2919	169	5.5
Pb-Ag	0.570	0	3088	0	0

表 5 Zn 电积使用新型阳极的节电效益

阳极	$500A/m^2$ 下的超电压/V	与 Pb-Ag 阳极比较，超电压减少/V	直流电耗 /$kW \cdot h \cdot t^{-1}$	节约电能 $kW \cdot h/t$	节约电能 %
石墨（Ⅲ）	0.265	-0.54	2698	492	15.4
陶瓷（Ⅳ）	0.11	-0.695	2157	633	19.9
Ti/Ru-Ti（Ⅰ）	0.32	-0.485	2748	442	13.9
Ti/MnO_2（Ⅱ）	0.47	-0.335	2885	305	9.6
Pb-Ag(PbO_2)[8]	0.805	0	3190	0	0

计算结果表明，用上述 4 种阳极代替现行的 Pb-Ag 阳极，在 Ni 电积时节电最少为 5.5%，最高可达 15.7%；在 Zn 电积时，节电最低为 9.6%，最高接近 20%。由此可见，降低阳极析氧超电压节电的潜力很大，特别是采用阳极（Ⅲ）和（Ⅰ）两种，将获得显著的节电效果。

实验室研究还表明，这 4 种新型阳极都能使用在较高电流密度下而超电压较小。例如，可用在 $500\sim1000A/m^2$，如果从工程上配合，使提高电流密度后电路上的电压降增加不多，那么，提高电流密度可以增加金属产量，在不增加设备的情况下多产金属，实现既节电又增产。

今后需要分别对上述新型阳极作深入研究，特别是石墨与陶瓷阳极的基本特征与析氧机理，研究更高催化活性的镀层组成与制备工艺，研究这些电极的使用寿命，以及降低制造成本等，使之早日投入工业应用。

参 考 文 献

[1] G. Bewer, et al. J. of Metal. 1982, 34 (1)：37；1983, 35 (4)：49.

[2] Vittorio de Nora. ソ一ダ と 塩素, 1983, 3：1-7.

[3] 113th AIME Annual Meeting. 1984, Feb.

[4] E. R. Cole, Jr., T. J. O' Keefe. USBM Report of Investigations. RI-8531, 1981.

[5] 刘业翔，等. 有色金属（冶炼部分）. 1984, 4：28-31.

[6] 刘业翔，等. 全国第五周冶金物理化学年会论文集（下）. 1984：313-318.

[7] A. J. Scarpellino, et al. J. Electrochem. Soc, 1982, 129：515-525.

[8] M. S. V. Pathy, et al. Electrochim. Acta. 1965, 10：1185-1189.

[9] L. W. Higley, et al. USBM. RI. RI-8111, 1976.

[10] S. Trasatti. Electrodes of Conductive Metallic Oxides. part B. Elsevier Scientific pub. Co., 1981：523-527.

[11] J' O'M Bockris, et al. Electro-chemical science. Taylor & Francis Ltd., 1972：167.

A Novel Porous Pb-Ag Anode for Energy-saving in Zinc Electrowinning Part Ⅰ: Preparation and Property Tests in Laboratory*

Abstract A novel porous Pb-Ag(0.8wt.%) anode was introduced into zinc electrowinning, and six kinds of these anodes with different controllable pore size and homogeneous pore structure were prepared by negative pressure infiltration. Compared with the traditional flat plate anode, their electrochemical properties were studied by chronopotentiometry in the electrolyte of $ZnSO_4$-H_2SO_4 and $ZnSO_4$-$MnSO_4$-H_2SO_4, respectively. The results show that, with the decrease of pore size the anodic potential decreases first and then increases in $ZnSO_4$-H_2SO_4 electrolyte. And the lowest value is 1.729V when the pore size is 1.25-1.60mm, which is 106mV lower than that of flat plate anode, while the anodic corrosion rate reaches the lowest value at the pore size of 1.60-2.00mm. When in the electrolyte of $ZnSO_4$-$MnSO_4$-H_2SO_4, the formation of dense PbO_2/MnO_2 protective layer leads to further decrease of the anodic potential and corrosion rate while has little effect on cathode current efficiency. What's more, the porous anode is beneficial to reduce Mn^{2+} precipitation. Compared with flat plate anode, the behavior of porous anode is more similar to flat plate anode under a lower current density.

Key words zinc electrowinning, negative pressure infiltration, porous anode, anodic potential, corrosion rate, Mn^{2+} precipitation

1 Introduction

In zinc electrowinning industry, Pb-Ag(0.5wt.%-1.0wt.%) anode is widely used (Petrova M. et al., 1999). It can meet the basic requirements but still has several problems such as high oxygen evolution over-voltage (about 860mV), unsatisfactory corrosion resistance, lead contamination of cathode product and high consumption of noble metal Ag and so on.

In order to reduce the energy consumption, enhance the corrosion resistance, improve the product quality and reduce the consumption of Ag or even dispense with it, much research work has been done mainly focused on Pb-based alloy anode and Ti-based coating anode (named DSA®). Detailed investigations have already been carried out on binary alloys (Pb-Ca, Pb-Co), ternary alloys (Pb-Ag-Ca, Pb-Ag-Ti, Pb-Ag-Sn, Pb-Sr-Sn, Pb-Ca-Sn) and quaternary alloys (Pb-Ag-Ca-Sr, Pb-Ag-Ca-Ce, Pb-Ag-Sn-Co) etc (Stefanov Y. and Dobrev T., 2005; Ivanov I. et al., 2000; Rashkov S. et al., 1999). Although these alloy anodes effectively reduce the consumption of Ag, only Pb-Co and Pb-Ag-Sn-Co anodes performed well in the perspective of reducing anodic potential. Regretfully, their complicated manufacturing conditions restricted their further commercial use. Inspired by the successful use of Ti-based electrode in the chlor-

---

* Copartner: Lai Yanqing, Jiang Liangxing, Li Jie, Zhong Shuiping, Lü Xiaojun, Peng Hongjian. Reprinted from Hydrometallurgy, 2010(102): 73-80.

alkali industry, different kinds of Ti-based electrode (Ti/IrO$_2$, Ti/RuO$_2$, Ti/PbO$_2$, Ti/TiO$_2$/PbO$_2$, Ti/SnO$_2$+Sb$_2$O$_3$+MnO$_2$/PbO$_2$ etc.) were investigated (Li B. S. et al. ,2006; Hu J. M. et al. ,2004; Stefanov Y. and Dobrev T. ,2005) for zinc electrowinning. However, since these electrodes are costly and can't avoid the passivation of base metal when anodized in H$_2$SO$_4$ solution, their further application is confined. Meanwhile, new techniques such as combination electrolysis (Verbaan B. , Mullinder B. , 1981), methanol oxidation electrode (Wesselmark M. et al. ,2005) and gas diffusion electrode (Nikolova V. , Nikolov I. , Vitanov T. ,1987; Jin B. J. and Yang X. W. ,2007) were also studied extensively, but there was no report of industrial. Therefore, Pb-Ag(0.5~1.0%) is still the only anode that can be used in industry (Felder A. and Prengaman R. D. ,2006).

Aiming to provide large reaction rates per unit volume, high surface area porous electrodes have been used in many fields, such as flow through electrode (Richard Alkire and Brian Gracon,1975), lead-acid battery grid (Dai Ch. , Yi T. , Wang D. , Hu X. ,2006) and fuel cell (Yuh C. Y. and Selman J. R. , 1992). However, there was no report of porous lead and lead alloy electrode for zinc electrowinning. If the anode can be made porous, it will effectively reduce the anodic current density without changing the cathodic current density.

In this paper, a porous Pb-Ag anode was introduced and prepared by negative pressure infiltration, its electrochemical properties was studied by comparing with flat plate anode in the electrolyte of ZnSO$_4$-H$_2$SO$_4$ (Zn^{2+} 60g/L, H$_2$SO$_4$ 160g/L) and ZnSO$_4$-MnSO$_4$-H$_2$SO$_4$ (Zn^{2+} 60g/L, H$_2$SO$_4$ 160g/L, Mn^{2+} 4g/L), respectively.

2 Experimental

2.1 Preparation of porous Pb-Ag anode

The main raw material includes Pb-Ag (0.8wt.%) alloy, filler particles (a water-soluble metallic sulfate bonding with a portion of CaSO$_4$) and release agent. The setup sketch for negative pressure infiltration is shown in Fig. 1.

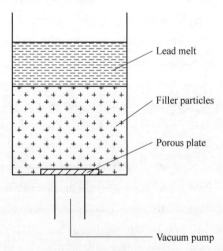

Fig. 1 Setup sketch for negative pressure infiltration

The whole process of negative pressure infiltration can be divided into the following steps:

(1) Pretreatment and classification of filler particles; (2) Preheat the mold filled with filler particles to a certain temperature; (3) Press Pb-Ag alloy melt into the infiltration mold under a certain negative pressure, the time needed is usually 3~9s; (4) Strip the samples after solidification, and then remove the filler particles in water. In this paper, all the porous materials were prepared under the following conditions: the casting temperature of lead alloy melt is 450℃, the preheating temperature of filler particles is 280℃ and the negative pressure is 0.03MPa.

2.2 Measurements

Porosity, an important parameter, directly influences the specific surface area and the actual anodic current density of porous anode. All samples with different pore size were wire-cut to an appropriate size and shape. The porosity(P) can be calculated by the following equation:

$$P = 1 - \frac{m}{V\rho_0} \tag{1}$$

where m is the weight of sample; V is the volume of sample; ρ_0 is the density of Pb-Ag (0.8wt.%) alloy.

Before electrochemical test, the Pb-Ag porous materials with different pore size were cut into cubic samples (10mm×10mm×5mm), welded with a copper conductor and sealed by epoxy resin left 1cm^2 surface to be tested. And then the electrodes were degreased by alkali and ethanol and washed with double-distilled water.

For porous anode, the anodic potential was tested by chronopotentiometry(CP) under the apparent current density of 500A/m^2. For flat plate anode, the chronopotentiometry was carried out at different current densities (50A/m^2, 100A/m^2 200A/m^2, 300A/m^2, 400A/m^2, 500A/m^2, 600A/m^2, 700A/m^2). All the tests were carried out in a glass three-electrode system. The anodic potential was measured against Hg/Hg$_2$Cl$_2$(SCE) reference electrode, and the counter electrode is Al or Pt electrode depending on the electrolyte system. Pt electrode was used in ZnSO$_4$-H$_2$SO$_4$ electrolyte, while Al electrode was used in ZnSO$_4$-MnSO$_4$-H$_2$SO$_4$ electrolyte. The ZnSO$_4$-H$_2$SO$_4$ system is simple and easy to analyze the experimental phenomenon, while ZnSO$_4$-MnSO$_4$-H$_2$SO$_4$ electrolyte is the simulation of zinc electrowinning industry. All electrolytes were prepared with analytically pure grade chemicals and double-distilled water, and the volume of electrolytes is 1000mL in every test to keep the ions variation comparable. The temperature was kept constant (37±0.5℃) by means of an HH-1 thermostat.

As we know, the corrosion of anode is caused by the dissolution of lead into electrolyte. Some of the dissolved lead co-deposits with zinc in the cathode, some intermixes with anode slime, and others remain in electrolyte. The corrosion rate of anode was usually measured with weight loss method by comparing the anode weight before and after polarization. However, this method may be unsuitable for porous anode owing to the difference of real surface area when boiling in the solution of NaOH and glucose. So another method called Pb balance method is utilized, which uses the variation of Pb content in the electrolyte and zinc product to determine the corrosion rate. Atomic absorption spectrophotometer (Hitachi, Z-5000) was used to test the concentration of Pb^{2+} in electrolyte after anodic polarization, and quartz spectrograph (ИСП-30) was used to test Pb content in Zn. The corrosion rate measured by Pb balance method was calculated by the following equation:

$$Corr = \frac{C \cdot V + w \cdot m}{S \cdot t} \quad (2)$$

where C is the concentration of Pb in electrolyte; V is the electrolyte volume; w is the mass percentage of Pb in zinc product; m is the weight of zinc product; S is the apparent area of anode; t is the polarization time.

As the anode slime on the surface of anode and in the electrolyte is of small quantity and difficult to collect on laboratory-scale experiment, this part of Pb is neglect in the Pb balance method. It is easy to conclude that the corrosion rate measured by Pb balance method is lower than weight loss method (see Fig. 2). However, the variation tendency is similar, that is, the corrosion rate increases with the increase of current density. So it is feasible to compare the corrosion rate by Pb balance method.

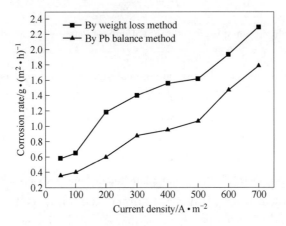

Fig. 2　Comparison of the corrosion rate measured by different method in $ZnSO_4$-H_2SO_4 electrolyte

After polarization for 72h, the anode was removed from electrolyte, washed with double-distilled water and dried immediately. Then, the microstructure of the passive layer was observed by electron microscopy (JEOL Japan, JSM-6360LV).

3　Results and Discussion

3.1　Morphology and structure

The pore size distribution of porous material prepared by negative pressure infiltration was obtained from optical microscopy (Langlois S., Coeuret F., 1989), and the histogram of pore size distribution for the porous material prepared using filler particles of 1.43-1.60mm is shown in Fig. 3a. Before the measurement, the pores were filled with epoxy resin to eliminate the influence of pores in the cell walls. We can see from Fig. 3a that the pore size mainly distributes in the range of 1.50-1.80mm. What's more, the fraction of pore size larger than 1.80mm is little. Therefore, we can conclude that the pores are evenly arranged and its size is determined by the size of filler particles, and the negative pressure infiltration techniques can be used to prepare lead-based porous materials with uniform pore structure and controllable pore size.

Fig. 3b shows the microstructure of porous anode prepared by using the filler particle size of

0.80-1.00mm. After observing all the porous materials of different pore size, we find that the smaller the pore size, the greater the possibility of generating structural defects. It is because that the smaller the filler particles are, the greater the lead flow resistance will be. Additionally, the wettability between the filler particles and lead melt is poor. All of these make more structural defects in porous materials with smaller filler particles.

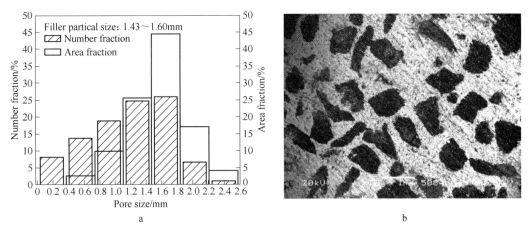

Fig. 3 Pore size distribution and (b) Microstructure of the porous Pb-Ag(0.8wt.%) material

3.2 Anodic potential

According to Tafel equation ($\eta_{O_2} = a + b\lg i$, where η_{O_2} is overpotential, a and b are constants related to the reaction mechanism, i is current density) (Bagotsky V. S., 2006), it is well understandable that porous anode can reduce the anodic potential. But as a novel anode in zinc electrowinning, it should be proved in practical application. Firstly, the anodic potential of anode $0^{\#}$-$6^{\#}$ (see Table 1) was tested in $ZnSO_4$-H_2SO_4 electrolyte by a 120h's galvanostatic polarization. While in the electrolyte of $ZnSO_4$-$MnSO_4$-H_2SO_4, the polarization time was 24h. The results are shown in Fig. 4 and Fig. 5.

Table 1 Types and porosity of porous anode

Anode	$0^{\#}$	$1^{\#}$	$2^{\#}$	$3^{\#}$	$4^{\#}$	$5^{\#}$	$6^{\#}$
Pore size/mm	—	2.00-2.50	1.60-2.00	1.25-1.60	1.00-1.25	0.80-1.00	0.60-0.80
Porosity/%	0	60.36	56.89	56.11	56.02	56.86	57.16

When a Pb-Ag anode was polarized under a certain current densities in a H_2SO_4 containing electrolyte, a nonconductive $PbSO_4$ layer is firstly generated on the fresh anode surface and then the current density and potential of uncovered surface increased. Subsequently, the generated $PbSO_4$ and uncovered Pb will transform into PbO_2 and oxygen evolution takes place. When the reaction reaches balance, the anodic potential becomes stable at the macroscopic level. This makes the anodic potential decreases rapidly at the beginning and then stabilizes gradually (see Fig. 4). When Mn^{2+} exists in the electrolyte, the oxidization of Mn^{2+} and Pb will happen simultaneously and the anodic potential will be stable only when the steady layer of PbO_2/MnO_2 has formed. Therefore, the anodic potential first increases and then stabilizes (see Fig. 5).

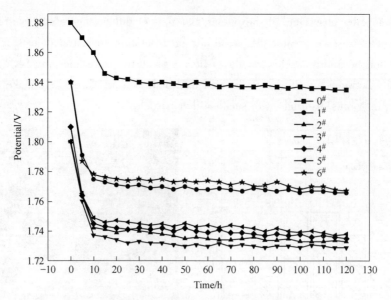

Fig. 4　Anodic potential in $ZnSO_4$-H_2SO_4 electrolyte

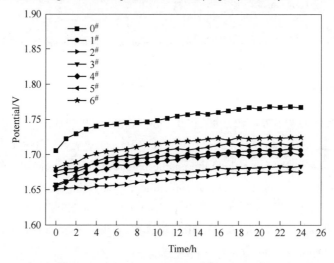

Fig. 5　Anodic potential in $ZnSO_4$-$MnSO_4$-H_2SO_4 electrolyte

From Fig. 4 and Fig. 5, we can also conclude that the stable anodic potential of porous anode is lower than flat plate anode and it decreases firstly and then increases with the decrease of pore size irrespective of Mn^{2+}. And the anode $2^{\#}$ and $3^{\#}$ have the lowest value. The reason is that the anodic potential is determined by current density and escape resistance of bubbles. With the decrease of pore size, the specific surface area of anode increases and then the actual current density reduces, thus the anodic potential decreases. However, since the gas bubbles produced inside the pores are not very easy to escape, not all the electrode surface can be expected to participate in oxygen evolution reaction. Some of the bubbles would stay in and push the electrolyte out of the pores, rendering some inner part of electrodes inactive. As a result, the electrochemical reaction will be concentrated at the rest electrode(Tseung A. C. C. ,1985). Therefore, there is an optimum thickness and pore size of porous anode with which the most effective utilization of electrode area is obtained. From this point of view, the optimum pore size is 1.25-1.60mm and 1.60-2.00mm in electrolyte of $ZnSO_4$-H_2SO_4 and $ZnSO_4$-$MnSO_4$-H_2SO_4, respectively.

The stable anodic potential of flat plate anode as a function of current density is shown in Fig. 6. According to these data, the Tafel curves are obtained. We can see that the Tafel curves are parallel, which indicates that the oxygen evolution mechanism is the same both in the electrolytes of $ZnSO_4$-H_2SO_4 and $ZnSO_4$-$MnSO_4$-H_2SO_4, the existence of Mn^{2+} in electrolyte has no effect on it. Meanwhile, the difference between the parameter a is 52mV, the depolarization of Mn^{2+} is obvious. In zinc electrowinning industry, there is always a pre-treating process when new anodes are to be used. The purpose of pre-treatment is to use low-temperature, low-current density to form a dense passive layer in a short time to protect anode from corrosion. For porous anode, due to its large specific surface area, the actual current density will be reduced when it works. Thus, porous anode is in favor of the reactions on the surface to reach balance, and then the anodic potential can stabilize faster than flat plate anode under the same apparent current density. Comparing the stable anodic potentials of porous anode and flat plate anode, and presuming that the anodic potential is only determined by current density, we may conclude that the actual current density on porous anode is between $80A/m^2$ and $170A/m^2$ when the apparent current density is $500A/m^2$.

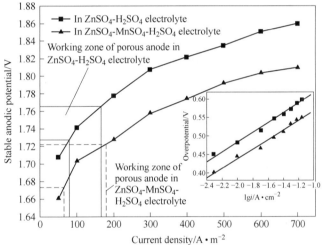

Fig. 6　Stable anodic potential of flat plate anode under different current densities

3.3　Corrosion rate

In the polarization state, the major factors that influence the corrosion rate of anode are the actual current density, the microstructure of anode and the electrolyte composition. As the same electrolyte was used in this work, the major factors are the former two. Also, the corrosion process can be classified into three types as chemical corrosion, electrochemical corrosion and structural defects corrosion. Chemical corrosion is caused by the reaction between Pb and sulphuric acid, which is determined by the area of anode. Electrochemical corrosion is caused by the anodic current applied by external power source, which is mainly determined by the current density. Structural defects corrosion is caused by the structural defects formed in the process of casting. The electrolyte can infiltrate into the defects and then corrosion happens (both chemical and electrochemical corrosion).

For flat plate anode, its corrosion rate is only determined by current density as they were cut from the same ingot and had the same surface area (1cm^2). Therefore, we can conclude that the corrosion rate increases with the increase of current density on flat plate anode (see Fig. 7).

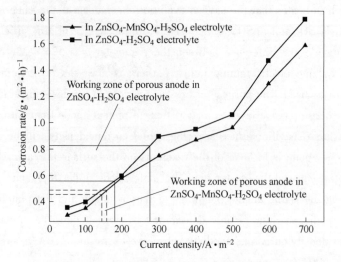

Fig. 7 Corrosion rate of flat plate anode measured by Pb balance method

For porous anode, its corrosion rate may co-determined by the actual current density, specific surface area and structural defects. Thanks to the large specific surface area of porous anode, the current density decreased and then the electrochemical corrosion rate decreased too. However, large specific surface area will increase the chemical corrosion rate, which is harmful to the total corrosion rate. Besides, the structural defects will increase the corrosion rate of porous anode. Fortunately, it seems that current density is the most important parameter by comparing the corrosion rate of porous anode (see Table 2) and flat plate anode. We can conclude that the actual current density of porous anode in $ZnSO_4$-$MnSO_4$-H_2SO_4 electrolyte is less than 270A/m^2 in the aspect of corrosion rate. According to the analysis above, it is understandable that the evaluated actual current density is greater than that evaluated by stable anodic potential irrespective of Mn^{2+}.

Table 2 Corrosion rate of different anode in different electrolyte

Anode		0$^\#$	1$^\#$	2$^\#$	3$^\#$	4$^\#$	5$^\#$	6$^\#$
In $ZnSO_4$-H_2SO_4 electrolyte	Corrosion rate /g · (m^2 · h)$^{-1}$	1.065	0.695	0.555	0.715	0.775	0.795	0.825
	Relative corrosion rate (vs. flat plate anode)/%	100	65.26	52.11	67.14	72.77	74.65	77.46
In $ZnSO_4$-$MnSO_4$-H_2SO_4 electrolyte	Corrosion rate /g · (m^2 · h)$^{-1}$	0.962	0.460	0.465	0.460	0.480	0.485	0.488
	Relative corrosion rate (vs. flat plate anode)/%	100	47.82	48.34	47.82	49.90	50.42	50.73

It can be seen that Mn^{2+} in electrolyte does not change the variation tendency of corrosion rate. Moreover, the corrosion rate in $ZnSO_4$-$MnSO_4$-H_2SO_4 electrolyte is lower than that in $ZnSO_4$-H_2SO_4 electrolyte, which is coincident with the former researches (Yu P. and O'Keefe

T. J. , 2002) and industrial practices. The reason is that Mn^{2+} can be oxidized to permanganic acid on the anode:

$$4MnSO_4+6H_2O+5O_2 = 4HMnO_4+4H_2SO_4 \quad (3)$$

Then, the permanganic acid in the solution reacts with Mn^{2+} to form MnO_2 according to the following equation:

$$2HMnO_4+3MnSO_4+2H_2O = 5MnO_2+3H_2SO_4 \quad (4)$$

A part of MnO_2 precipitates to the bottom of electrolytic tank, which is usually recycled and reused in the leaching process. The rest adheres to the anode and forms a protective layer with PbO_2 on the anode surface and then reduce the corrosion rate.

3.4 Passive layer and Mn^{2+} precipitation

According to above discussion, the current density may have great effects on the anodic passive layer which influences the anodic potential and corrosion rate directly. In order to observe the effects of current density on the morphology of passive layer, the surface morphology of flat plate anode was observed by SEM after a 72h's polarization in electrolyte of $ZnSO_4$-H_2SO_4 and $ZnSO_4$-$MnSO_4$-H_2SO_4 under current densities of $50A/m^2$, $100A/m^2$ and $500A/m^2$, respectively. For porous anode, the same observation was carried out only under the apparent current density of $500A/m^2$.

As we can see in Fig. 8, the surface morphology changes a lot with the change of current density. When the current density is $50A/m^2$, dense surface morphology is observed and the passive layer is well combined with the base body after 72h's polarization. When the current density increases to $500A/m^2$, the anode surface morphology becomes loose as shown in Fig. 8c. Therefore, we can predict that low current density can produce dense PbO_2 coating on the anode surface, which is beneficial for protecting the anode from corrosion. While large current density will produce loose oxidized layer and the electrolyte can infiltrate into the base metal easily, and then increase the corrosion rate.

When in $ZnSO_4$-$MnSO_4$-H_2SO_4 electrolyte, the surface morphology also changes a lot with the change of current density(see Fig. 9). The particles on the anode surface are less, coarse and well combined with the base body under $50A/m^2$. When the current density is $500A/m^2$, the particles are more, fine, loose and not well combined with the base body. Comparing Fig. 8 and Fig. 9, we can find that Mn^{2+} has significant influence on the microstructure of passive layer. It is propitious to form a dense passive layer and then reduce the corrosion rate as mentioned above.

According to the above results, it is predictable that current density will have some influence on the quantity of anode slime. Because of the small anode surface, limited polarization time and uncollectible of anode slime, the variation of Mn^{2+} concentration in electrolyte after electrolysis was used to determine the formation of anode slime, and the results are shown in Fig. 10.

It can be seen that when the current density is low(such as $50A/m^2$), the Mn^{2+} content changes little after 24h's electrolysis(from 4.000g/L to 3.969g/L). With the increase of current density, the Mn^{2+} precipitation phenomenon becomes more severe, which indicates that the quantity of anode slime increases(the main component of the anode slime is manganese

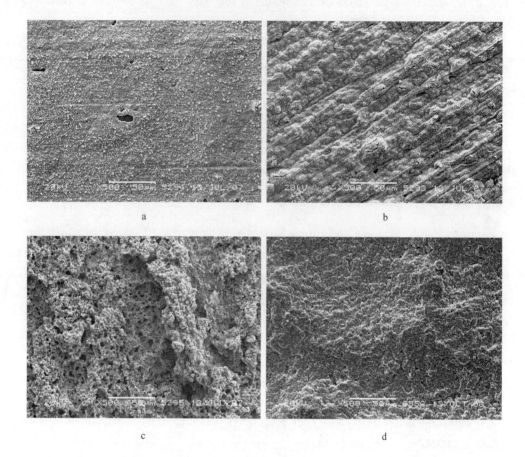

Fig. 8 SEM observation of flat plate Pb-Ag(0.8wt.%) anode under the current density of (a) 50A/m^2, (b) 100A/m^2 and (c) 500A/m^2 and of (d) porous anode with the filler particle size of 1.60-2.00mm under apparent current density of 500A/m^2 after 72h's polarization in ZnSO$_4$-H$_2$SO$_4$ electrolyte

dioxide).

Comparing the results on porous anode and flat plate anode, we find that the morphology of passive layer and the precipitation of Mn^{2+} of porous anode are similar to that of flat plate anode under low current density. However, there are some differences. In ZnSO$_4$-H$_2$SO$_4$ electrolyte, the passive layer on porous anode seems denser than on flat plate anode under low current density (see Fig. 8d). And in ZnSO$_4$-MnSO$_4$-H$_2$SO$_4$ electrolyte, the passive layer on porous anode has more cracks and less particles (see Fig. 9d). According to the Mn^{2+} variation data (see Table 3), the working zone of porous anode can be evaluated which is greater than that evaluated by corrosion rate and stable anodic potential. It seems that not only the current density, but also the porous structure has effects on the passive layer. Low current density can decrease the precipitation of Mn^{2+}. However, the current density on porous anode is not uniform. It is higher on the edge of pores than on the inner surface. In addition, maybe porous surface is propitious for the crystallization of oxide particles, which leads to a denser passive layer and then reduces the corrosion rate as mentioned above.

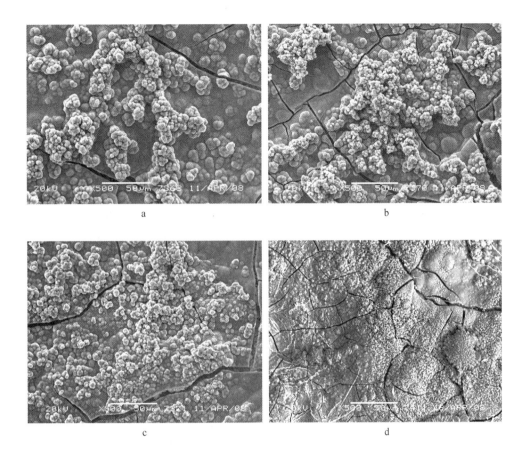

Fig. 9 SEM observation of flat plate Pb-Ag(0.8wt.%) anode under the current density of (a)50A/m^2, (b)100A/m^2 and (c)500A/m^2 and of (d) porous anode with the filler particle size of 1.60-2.00mm under apparent current density of 500A/m^2 after 72h's polarization in $ZnSO_4$-$MnSO_4$-H_2SO_4 electrolyte

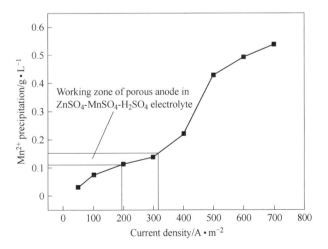

Fig. 10 Mn^{2+} precipitation of flat plate anode as a function of current density

Table 3 The variation of Mn^{2+} concentration in $ZnSO_4$-$MnSO_4$-H_2SO_4 electrolyte after 24h's electrolysis

Anode	0#	1#	2#	3#	4#	5#	6#
Before electrolysis/g · L^{-1}	4.000	4.000	4.000	4.000	4.000	4.000	4.000
After electrolysis/g · L^{-1}	3.568	3.881	3.862	3.843	3.849	3.843	3.848
Variation/g · L^{-1}	0.432	0.119	0.138	0.157	0.151	0.157	0.152

3.5 Cathode current efficiency and quality of zinc product

Current efficiency on cathode is mainly influenced by the electrolyte composition, temperature, cathodic current density, cathode surface state, electrolyte circulation and zinc stripping cycle etc. In this paper, the current efficiencies of porous anode and flat plate anode should be the same because all the mentioned factors are kept unchanged. Table 4 confirms this postulate, the current efficiency are both between 91% and 92% and there is no obvious difference found among porous.

Table 4 Current efficiency and quality of zinc product of anode 0#-6# in $ZnSO_4$-$MnSO_4$-H_2SO_4 electrolyte

Anode	1#	2#	3#	4#	5#	6#
Pb content in cathode zinc	30.9×10^{-6}	32.3×10^{-6}	30.9×10^{-6}	32.6×10^{-6}	33.6×10^{-6}	31.4×10^{-6}
Cathode current efficiency/%	91.32	91.35	91.42	91.45	91.32	91.35

It can be seen from Table 5 that the cathode current efficiency increases with the increase of current density, while Pb content in zinc product represents the inverse trend. Compared with Fig. 7, we can find that the variation tendency of corrosion rate and Pb content of zinc product according to current density is inverse. The reason may be that the precipitation speed of Pb is not only relative to the precipitation potential but also to its concentration. When its concentration reaches a certain low degree, the diffusion step is the rate-determining step, and the limiting current density (J_{lim}) can be calculated by the following equation:

$$J_{lim} = \frac{nFDC}{d} \quad (5)$$

where n is charge transfer number; F is Faraday constant; D is the diffusion coefficient of Pb^{2+}; C is the concentration of Pb^{2+}; d is the thickness of diffusion layer.

Table 5 Cathode current efficiency and quality of Zn under different anodic current densities in $ZnSO_4$-$MnSO_4$-H_2SO_4 electrolyte

Anodic current density/A · m^{-2}	50	100	200	300	400	500	600	700
Pb content in cathode zinc	350×10^{-6}	244×10^{-6}	156×10^{-6}	118×10^{-6}	73×10^{-6}	67×10^{-6}	61×10^{-6}	70×10^{-6}
Cathode current efficiency/%	67.10	75.06	84.18	88.45	90.36	91.60	93.69	93.80

The precipitation speed of Pb will be determined by its limiting current density, and its content in cathode zinc can be calculated by the following equation (Peng R. Q. et al., 2003):

$$w_{Pb} = \frac{J_{lim} M_{Pb}}{M_{Zn} J \eta} \quad (6)$$

where w_{Pb} is the Pb content in zinc product, M_{Pb} is atomic weight of lead; M_{Zn} is atomic weight of zinc; η is the cathode current efficiency; J is the current density.

Combining equation(5) and(6), the following equation is obtained:

$$w_{Pb} = \frac{nFDM_{Pb}}{M_{Zn}d} \cdot \frac{C}{J\eta} \tag{7}$$

In equation(7), the thickness of diffusion layer(d) is considered to be constant because of stirring effects of the hydrogen bubbles generated on the surface of cathode. Thus, the Pb content in zinc product can be determined simply by cathode current efficiency, current density and its concentration in electrolyte, and we can calculate the value of $C/J\eta$ to simulate the variation tendency of w_{Pb}.

It can be seen from Fig. 11 that the tendency of simulated curve is accordant with experimental curve. Therefore, we can conclude that the increase of current density is beneficial to improve the quality of zinc product.

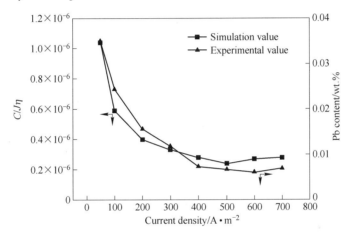

Fig. 11 Simulation of the variation tendency of Pb content in zinc product

According to the above results, we find that if we want a higher cathode current efficiency and lower Pb content in zinc product, the higher cathode current density is required. From Table 2, we can find that porous anode can reduce the corrosion rate by about 50%, which leads to a considerable decrease of the Pb^{2+} concentration in electrolyte. Therefore, it is reasonable that the Pb content in zinc product of porous anode is a half of that of flat plate anode under $500A/m^2$.

4 Conclusions

The lead-based porous anodes with homogeneous pore structure and controllable pore size can be prepared by negative pressure infiltration.

No matter Mn^{2+} exists or not in electrolyte, the anodic potential of porous anode is obviously lower than that of flat plate anode. What's more, porous anode has no effect on cathode current efficiency. So, porous anode can reduce energy consumption of zinc electrowinning.

Under the influence of actual current density and structural defects, the corrosion rate of porous anode is lower than flat plate anode. When Mn^{2+} exists in the electrolyte, the corrosion rate of porous anode will be reduced further more. Meanwhile, porous anode can reduce the Mn^{2+}

precipitation, and then prevent the blockage of anode pores. All of these can guarantee a long lifetime for porous anode and an improvement of the quality of zinc product. However, as the accumulation of anode slime in long-time electrowinning, the blockage is inevitable. It is necessary to remove them and a possible way is to dip the anode in a reductive electrolyte (such as $FeSO_4$ containing electrolyte).

In a word, porous anode has great potential in energy saving and improving the quality of zinc product. It is worthy of further research.

Acknowledgements

This work was supported by the Chinese National Natural Science Foundation under grant No. 50954006.

References

[1] Bagotsky V. S., 2006. Fundamentals of electrochemistry (second edition). A. N. Frumkin Institute of Physical Chemistry and Electrochemistry, Russian Academy of Sciences, Moscow, Russia.

[2] Dai Ch., Yi T., Wang D., Hu X., 2006. Effects of lead-foam grids on performance of VRLA battery. Journal of Power Sources 158, 885-890.

[3] Felder A., Prengaman R. D., 2006. Lead alloys for permanent anode in the nonferrous metals industry. J. O. M. 58(10), 28-31.

[4] Hu J. M., Zhang J. Q., Cao Ch. N., 2004. Oxygen evolution reaction on IrO_2-based DSA type electrodes: kinetics analysis of Tafel lines and EIS. International Journal of Hydrogen Energy 29(8), 791-797.

[5] Ivanov I., Stefanov Y., Noncheva Z., Petrova M., Dobrev T., Mirkova L., Vermeersch R., Demaerel J. P., 2000. Insoluble anodes used in hydrometallurgy. Part I. Corrosion resistance of lead and lead alloy anodes. Hydrometallurgy 57(2), 109-124.

[6] Langlois S., Coeuret F., 1989. Flow-through and flow-by porous electrodes of nickel foam. I. Material characterization. Journal of Applied Electrochemistry 19, 3-50.

[7] Li B. S., Lin An, Gan F. X., 2006. Preparation and electrocatalytic properties of Ti/IrO_2-Ta_2O_5 anodes for oxygen evolution. Trans. Nonferrous Met. Soc. China 16(5), 1193-1199.

[8] Nikolova V., Nikolov I., Vitanov T., 1987. Utilization of gas-diffusion electrodes catalyzed with tungsten carbide as anodes for zinc electrowinning. Journal of Applied Electrochemistry 17, 322-328.

[9] Petrova M., Stefanov Y., Noncheva Z., Dobrev T., Rashkow S., 1999. Electrochemical behavior of lead alloys as anodes in zinc electrowinning. British Corrosion Journal 34(3), 198-200.

[10] Peng R. Q., Ren H. J., Zhang X. P., 2003. Metallurgy of Lead and Zinc. Chinese Science Press, Beijing, China (in Chinese).

[11] Rashkov S., Dobrev T., Noncheva Z., Stefano Y., Rashkova B., Petrova M., 1999. Lead-cobalt anodes for electrowinning of zinc from sulphate electrolytes. Hydrometallurgy 52, 223-230.

[12] Richard Alkire and Brian Gracon, 1975. Flow-through porous electrodes. Journal of Electrochemical Society 122(12), 1594-1601.

[13] Stefanov Y., Dobrev T., 2005. Developing and studying the properties of Pb-TiO_2 alloy coated lead composite anodes for zinc electrowinning. Transactions of the Institute of Metal Finishing 83(6), 291-295.

[14] Stefanov Y., Dobrev T., 2005. Potentiodynamic and electronmicroscopy investigations of lead-cobalt alloy coated lead composite anodes for zinc electrowinning. Transactions of the Institute of Metal Finishing 83(6), 296-299.

[15] Tseung A. C. C., 1985. Gas evolution on porous electrodes. Journal of applied electrochemistry 15,

575-580.

[16] Verbaan B., Mullinder B., 1981. The simultaneous electrowinning of manganese dioxide and zinc from purified nutral zinc sulphate at high current efficiencies. Hydrometallurgy 7(4):339-352.

[17] Wesselmark M., Lagergren C., Lindergh G., 2005. Methanol oxidation as anode reaction in zinc electrowinning. Journal of Electrochemical Society 152(11), 201-207.

[18] Yuh C. Y. and Selman J. R., 1992. Porous-electrode modeling of the molten-carbonate fuel-cell Electrodes. Journal of Electrochemical Society 139(5), 1373-1379.

[19] Yu P., O'Keefe T. J., 2002. Evaluation of lead anode reactions in acid sulfate electrolytes. II. Manganese reactions. Journal of Electrochemical Society 149(5), 558-569.

A Novel Porous Pb-Ag Anode for Energy-saving in Zinc Electrowinning Part Ⅱ: Preparation and Pilot Plant Tests of Large Size Anode*

Abstract Aiming to develop a novel porous Pb-Ag anode for energy-saving in zinc electrowinning, previous laboratory tests in part Ⅰ showed that porous Pb-Ag anode has no negative effects on current efficiency, but lower the anodic potential, corrosion rate, Mn^{2+} precipitation significantly and enhance the quality of the cathodic product. In this Part, a new method called counter-gravity infiltration was developed, and four kinds of large size porous Pb-Ag anodes (200mm×100mm×6mm) with different pore size were prepared with this method. These anodes were tested at a pilot plant for 24 days with the electrolyte drawn from the zinc electrowinning plant. The results show that with little effect on the electrowinning system, the weight of porous anode is 60% less than flat plate anode, the corrosion rates of porous anodes are only one fifth of traditional flat plate anode, and the pass rate of $0^{\#}$ Zn almost reach to 100%; The cell voltages matching each porous anode are lower than flat plate anode with no difference in the current efficiencies, which lead to the reducing of energy consumption. Among the porous anodes, the one with the pore size of 1.6-2.0mm gets the most effective energy conservation, reducing energy consumption 78kW·h/t-Zn.

Key words: zinc electrowinning, porous anode, counter-gravity infiltration, corrosion rate, anode slime, energy consumption

1 Introduction

Pb-Ag alloy was used as anodic materials in zinc electrowinning industry for a long time for its excellent anti-corrosion performance. But, at the same time, it has many shortcomings, such as the high oxygen evolution over-potential (about 860mV), the contamination of cathodic products for the dissolution of lead and the consumption of abundant noble metal Ag. So, it's necessary to improve the Pb-Ag anode or seek novel anodes with better performances.

Porous electrode, which can increase the specific surface area and reduce the usage of materials notably, has attracted great interests in electrochemical researchers. The part Ⅰ of this work introduced this kind of electrode into zinc electrowinning, and fabricated porous Pb-Ag anode by negative pressure infiltration in laboratory successfully. The laboratory study shows that, this novel anode has great potential for saving energy and improving the quality of zinc product in zinc electrowinning.

In zinc electrowinning industry, the real electrolyte contains many impurity ions, especially the Cl^-、F^- and Mn^{2+}, which have great effects on the anodic corrosion rate and the formation of anode slime. It's necessary to evaluate this kind of anode in zinc electrowinning plant. Therefore,

---

* Copartner: Lai Yanqing, Jiang Liangxing, Li Jie, Zhong Shuiping, Lü Xiaojun. Reprinted from Hydrometallurgy, 2010 (102):81-86.

large size porous anode which can satisfy the operation in industry is required.

In order to improve the performance of lead-acid battery, porous lead and lead alloy was widely studied as supports for the active material in replacement of conventional lead grid(Dai Ch. S. et al.,2006;Gyenge E. et al.,2002). However,there was no report on the study of porous lead and lead alloy for zinc electrowinning. Porous lead and lead alloy was always prepared by electro-deposition(Dai C. S. et al.,2006) and infiltration casting. The materials with high porosity and specific surface area can be obtained by electro-deposition method, but the substrates used are always copper(Dai Ch. S.,Yi T. F. and Wang D. L.,2006),RVC(Reticulated Vitreous Carbon)(Gyenge E. et al.,2003) and so on,which are not suitable for zinc elctrowinning. The negative pressure infiltration used in part I of this paper, which foams the metal by infiltrating the melt into the bulk of filler particles, has many advantages, such as easy control of pore size and porosity, isotropic foam properties and so on. Although it was widely studied, it can only obtain small size porous metal for its lack in infiltration length and easy generation of structural defect.

Combining the negative pressure infiltration method with the counter-gravity casting technology used in the preparation of large and complex thin-walled components, we developed a new method called counter-gravity infiltration to prepare large size porous Pb based anodes for Zn electrowinning. The main advantage of this method is that the melt was pushed by compressed air into the bulk of filler particles from the bottom of the mold, which can eject the air conveniently and keep the melt arising levelly for its own gravity. The lack in infiltration length, which restricts the use of negative pressure infiltration, would not be a problem any more. And the large size porous metal can be fabricated by designing a mould of desired geometry.

In this part, the porous Pb-Ag alloy anodes, which have the size of 200mm×100mm×6mm, were prepared by counter-gravity infiltration under the optimum conditions, and a 24 days continuous pilot plant test was carried out in the pilot plant of Hunan Zhuzhou Smelter Group Co. Ltd. In the test, the anodic potential, anodic corrosion rate, anode slime and zinc product were studied.

2 Experimental

2.1 Preparation of electrodes

The setup sketch for counter-gravity infiltration is shown in Fig. 1. The whole process is similar to negative pressure infiltration. And the only difference is that the melt is pushed into the infiltration chamber(A) along the lift tube(D) by compressed air. The air is introduced into the pressure crucible(J) from the inlet valve(C) by an air compressor.

Four kinds of porous Pb-Ag materials(see Table 1) with the size of 250mm×160mm×30mm were prepared by counter-gravity infiltration under the optimum conditions, i. e. the melt temperature, crystallization pressure and preheating temperature of fillers were 490℃, 0.12MPa and 320-330℃, respectively.

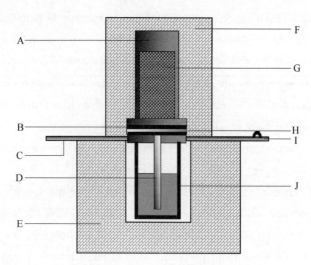

Fig. 1 Setup sketch for counter-gravity infiltration

A—infiltration chamber; B—sealring; C—inlet value; D—lift tube; E—melting furnace;
F—pre-heating furnace; G—filler particles; H—diaphragm; I—vent value; J—pressure crucible

Table 1 Anodes studied in pilot plant tests

Anode	1#	2#	3#	4#	5#
Pore size/mm	Flat plate	2.0-2.5	1.6-2.0	1.25-1.6	1.0-1.25
Porosity/%	0	64.36	63.48	60.00	63.57

The porous materials were cut into pieces of 200mm×100mm×6mm by wire-cutting method, and joint to a lead sealed copper hanger bar of 220mm×30mm×18mm, then the test anode was obtained. There were 5 test anodes, and the anode 1# was a flat plate anode used as a comparison. Electrolytic cells were "wired" in series while the electrodes in a cell were "wired" in parallel. In every single cell, there was a piece of anode sandwiched by two pieces of rolled Al plates used as cathodes. The rear side and the edges of the cathode were sealed with plastic sheets to prevent the precipitation of zinc, so only the side face to the anode worked.

2.2 Pilot plant tests

2.2.1 Electrowinning cells and conditions

The tests were performed continuously during 24 days in the electrowinning pilot plant of Hunan Zhuzhou Smelter Group Co. Ltd, China. And the main operating conditions were the same as industrial zinc electrowinning, which were listed as follows:

(1) The concentration of Zn^{2+} and H_2SO_4 of electrolyte before electrowinning was about 55g/L and 165g/L, respectively.

(2) Keeping the circulation of electrolyte 50mL/min · cell (or 360L/day · cell).

(3) Applying current 18.36A (i.e. 500A/m² of anodic surface) by a rectifier.

(4) Maintaining the temperature of electrolyte at 38-42℃ and the anode and cathode 30mm apart.

(5) Bone glue and strontium carbonate were added by 1.1g/day · cell.

(6) Electrowinning period was 24 hours, and the electric circuit was kept through when stripping zinc.

2.2.2 Measurements

The principal aim of the tests was to evaluate the effects of porous anode through anodic potential, cell voltage, anodic corrosion rate, anode slime, current efficiency and energy consumption in the industrial electrolyte.

(1) Anodic potential and cell voltage.

In order to reflect the change of anodic potential quickly and exactly, the anodic potential was collected online in a three-electrode system by computer and UNI-T70D multimeter every 1 minute, and the data were relative to saturated calomel electrode (SCE). At the same time, the cell voltage was recorded every 1 hour manually.

(2) Anodic corrosion rate.

The weight loss method is not reliable to evaluate the anodic corrosion rate of porous anode, so a lead-equilibrium method described in part I of this paper was adopted. Because the lead dissolved from anode can exist in zinc product, anode slime and electrolyte, the Pb contents of them were analyzed, and the anodic corrosion rate ($Corr$) of each anode was calculated using the following equation:

$$Corr = \frac{w_1 \cdot m_1 + w_2 \cdot m_2 + C \cdot V}{S \cdot t} \quad (1)$$

where w_1 and w_2 are the mass percentage of Pb in zinc product and anode slime; m_1 and m_2 are the weight of them; C is the concentration of Pb in electrolyte; V is the volume of electrolyte; S is the apparent area of anode; t is electrowinning time. As for the Pb suspended in the electroyte in the form of PbO_2, they were filtered and collected into the anode slime.

(3) Current efficiency and energy consumption.

According to the data collected from the tests, the current efficiency (η) and energy consumption (W) were calculated using the following equations:

$$\eta = \frac{m_1}{I \cdot t \cdot q} \quad (2)$$

$$W = \frac{U}{q \cdot \eta} \times 1000 \quad (3)$$

where I is the applied current; q is the electrochemical equivalent of zinc; U is the average cell voltage of electrowinning cell.

3 Results and Discussion

3.1 Anodic potential and cell voltage

After the 24 days continuous electrowinning, the data of anodic potential and cell voltage were obtained in detail. Fig. 2 gives the anodic potential-time curves of anode $1^\#$-$5^\#$ in the 16th day, the average cell voltage matching each anode is show in Fig. 3.

Fig. 2 shows that there is little fluctuation in the anodic potential, which indicates that the whole electrowinning system is stable, and the influence of the environment is negligible. Also,

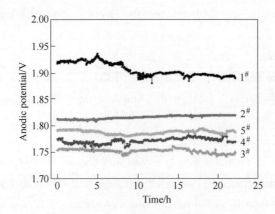

Fig. 2 Anodic potential of each anode as a function of time on the 16th day
(1# is flat plate anode used as a comparison, 2#-4# are porous anodes with the pore size of
2.0-2.5mm, 1.6-2.0mm 1.25-1.6mm and 1.0-1.25mm, respectively)

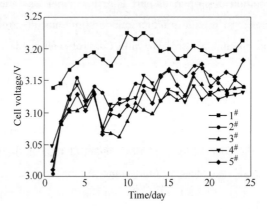

Fig. 3 Average cell voltage matching each anode as a function of time
(1# is flat plate anode used as a comparison, 2#-4# are porous anodes with the pore size of 2.0-2.5mm,
1.6-2.0mm, 1.25-1.6mm and 1.0-1.25mm, respectively)

we can see that the anodic potentials of porous anodes are lower than that of flat plate anode, which is consistent with the results in laboratory tests.

As a function of time, the fluctuation of average cell voltage happened (see Fig. 3). This may due to the following reasons: firstly, the formation of anodic oxide film is slow, and under the influence of stripping zinc everyday, the surface of the anode is difficult to fully achieve a stable status. Secondly, because of the complexity of ore source, although Zn^{2+}, H_2SO_4 concentration are constant, however, changes in ion concentration of impurities make the solution resistance and the anodic oxygen evolution status unstable. Additionally, the anode slime, formed through the oxidation of Mn^{2+}, partially attached on the anode surface, resulting in increased resistance of the electrode surface and electrolyte, and blocking part of the holes of porous anode gradually. Subsequently, the cell voltage increased slowly after a sharp increase in the first few days, which is more obvious on porous anodes.

The average anodic potentials and cell voltages in 24 days are calculated and compared to the flat plate anode (anode 1#), and the results are show in Fig. 4 and Fig. 5. In which we can see that the aim of reducing anodic potential and cell voltage by the use of porous anode has achieved. And the change of cell voltage is in line with the anodic potential, which indicates that anodic potential(or oxygen evolution potential)is the key factor that affects the cell voltage.

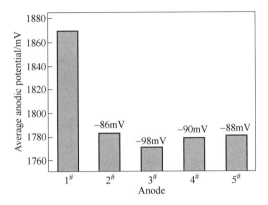

Fig. 4 Average anodic potential of each anode

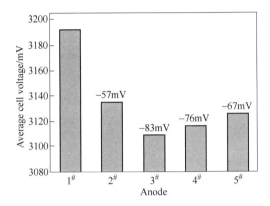

Fig. 5 Average cell voltage matching each anode

Commonly, the specific surface area increases with the decrease of pore size of porous anode. From the point of actual current density, the anode potential and cell voltage should decreased with the increase of specific surface area, but the data did not show such regularity actually, which is similar to the laboratory tests. The only difference is the pore size where the minimum values reached. In laboratory tests, the minimum values are got at the pore size of 1.25-1.6mm, while in the pilot plant test, the minimum values, which are 98mV and 83mV lower than that of flat plate anode, are got at the pore size of 1.60-2.00mm. The reason to make this difference should be the generation of anode slime on the inner wall of holes, which reduces the size of them. On one hand, too small aperture size is not conducive to oxygen evolution, on the other hand, the specific surface area decreases. Fortunately, the blocking of holes is limited, this will be discussed below.

3.2 Current efficiency and energy consumption

The current efficiency and energy consumption corresponding to each anode are calculated using equation(2) and(3) and listed in Table 2. The results indicate that porous anode has no negative effect on the current efficiency. However, the values are not perfect in this test, which is just in the range of 84%-85% and lower than the value in the laboratory and industry. This is because the parameters are more difficult to control in this scale cells. What's worse, in the testing period, the electrolyte of the industrial electrowinning system in this plant was unstable, and "plate-burning" took place frequently. However, this does not affect the comparison between the anodes. On the basis of similar current efficiency, low cell voltage means low energy consumption. According to Table 2, we can see that the anode $3^{\#}$ gets the most effective energy saving, reducing energy consumption 78kW · h/t-Zn.

Table 2 Elemental analysis of the electrolyte before electrowinning

Elements	Mn^{2+}	F^-	Cl^-	Cu^{2+}	Cd^{2+}	Ca^{2+}	As^{3+}	Sb^{3+}	Ge^{4+}	Co^{2+}	Ni^{2+}	Al^{3+}	Mg^{2+}	Fe^{3+}
Conc./g·L^{-1}	3.50	0.054	0.55	0.0004	<0.0002	0.91	0.0006	<0.0003	<0.0002	0.00058	<0.0005	0.028	15.87	0.010

3.3 Anode slime

During this test, the generation of anode slime is inevitable because there is a certain concentration of Mn^{2+} (3-5g/L) in the industrial electrolyte. Some attach to the surface of anode and some suspend in the electrolyte and flow out through the outlet, others which are the most, precipitate at the bottom of the cell. All the anode slime is collected, dried and weighed in this test, so as to examine the effects of porous anode on the formation of it.

We found an interesting phenomenon that the anode slime generated on flat plate anode and porous anodes are quite different in terms of quantity and morphology. The one generated on the flat plate anode is abundant and scale-like, while the one formed on the porous anodes is little and fine powder-like. Fine powder-like anode slime is propitious for it to be brushed off the surface of anode by the evolution of O_2, avoiding incrustation and blockage of pores. The quantity of anode slime generated on porous anode is only 10% of that on flat plate anode (see Table 3), which indicates that porous anode can reduce the precipitation of Mn^{2+}, and this is consistent with the conclusions of laboratory tests in part Ⅰ. The reduction of anode slime can reduce the frequency of cell cleaning and labor intensity.

Table 3 Current efficiency and energy consumption corresponding to each anode

Anode	$1^{\#}$	$2^{\#}$	$3^{\#}$	$4^{\#}$	$5^{\#}$
Current efficiency/%	84.6	84.7	84.5	84.5	84.4
Energy consumption/kWh·$(t\text{-}Zn)^{-1}$	3095	3035	3017	3024	3038
Energy conservation①/kWh·$(t\text{-}Zn)^{-1}$	0	60	78	71	57

① aversus flat plate anode.

The content analysis of certain elements of anode slime indicates that the composition of anode slime on porous anodes and flat plate anode is very different, especially in terms of Ag and Mn, which get the higher contents on flat plate anode. In these elements, Ag might originate from the electrolyte or the corrosion product of Pb-Ag anode. Because the electrolytes before electrowinning are the same, we can conclude that the difference of Ag content in anode slime between each anode is caused by the difference of anode corrosion rate. So we can assume that the corrosion rate of porous anode is smaller than that of flat plate anode, the total content of Pb in anode slime confirms this prediction. According to the Mn content, we can conclude that the mass percentage of MnO_2 in anode slime formed on flat plate anode is more than that on porous anode, which lead to a deeper color of the former.

The X-ray diffractions of anode slime indicate that the main composition of anode slime is MnO_2 and $SrSO_4$, in which $SrSO_4$ is the reaction product of $SrCO_3$ added into the electrolyte during electrowinning to reduce the concentration of Pb. Also, we can see that the diffraction peek of $SrSO_4$ in anode slime formed on porous anode is much higher than that on flat plate anode, which further indicate that the former contains less MnO_2.

3.4 Anodic corrosion rate and quality of zinc product

Anodic corrosion rate is a very important indicator for evaluation of anode in zinc electrowinning, which is directly related to the lifetime of the anode and the quality of zinc product. The lower the corrosion rate is, the longer the anodic lifetime and the lower the Pb content in zinc product will be.

In zinc electrowinning industry, a pretreatment under low temperature and low current density on the anodes are always performed before electrowinning. After that, a dense PbO_2 film is formed, which can protect the anode from corrosion by H_2SO_4. Because of the large specific surface area of porous anode, the working current density is much lower than flat plate anode all the time. In this case, the porous anode is always under a near-pretreatment condition, and the anodic protection film will be denser than on the flat plate anode, which is favor to reduce the anodic corrosion rate. Newnham(Newnham R. H., 1992) has investigated the corrosion rates of anodes made from various Pb-Ag alloys in various conditions. In the presence of manganese in electrolyte, the corrosion rate decreased with decreasing of current density. Also, the Pb content in electrolyte and anodic corrosion rate(see Table 4, Table 5) in this work confirm the analysis above. From Table 5, we can see that corrosion rate of porous anode is more or less one fifth of that of flat plate anode.

Table 4 Anode slime mass balance

Anode	Wt. anode slime $/m_2$, g	% mass elements in anode slime$/w_i$, wt. %			Total Pb in anode slime $/m, g(m=m_2 \cdot w_2)$
		Pb	Ag	Mn	
1#	336.9	0.68	0.036	49.0	2.29
2#	32.2	0.70	0.014	11.1	0.23
3#	35.5	0.49	0.017	12.4	0.17
4#	33.8	0.49	0.014	10.6	0.17
5#	35.5	0.56	0.025	7.2	0.20

Table 5 Concentration of lead in electrolyte after electrowinning

Anode	1#	2#	3#	4#	5#
Pb content$\times 10^{-4}$(g/L)	3.25	1.52	1.77	1.97	1.89

So we can conclude that porous anode can reduce the anodic corrosion rate and Pb content in electrolyte, which is favor to enhance the quality of zinc product. Table 6 show that the average Pb contents in zinc product corresponding to each anode all meet the $0^{\#}$ Zn standard (Pb content is no more than 25×10^{-6}), especially the Pb content matching porous anode is almost an order of magnitude less than that of flat plate anode. As a function of time, the Pb content has a notable fluctuation in the first few days and then gradually stabilized. And the $0^{\#}$ Zn pass rates of zinc product corresponding to anode $2^{\#}$-$4^{\#}$ are 100%, significantly higher than the flat plate anode.

Table 6 Corrosion rates of anodes

Anode	$1^{\#}$	$2^{\#}$	$3^{\#}$	$4^{\#}$	$5^{\#}$
Corrosion rate($g/m^2 h$)	1.75	0.30	0.29	0.31	0.35
Relative corrosion rate[①]/%	100	17.1	16.6	17.7	20.0

① Versus flat plate anode.

Table 7 Average lead content in zinc cathode product

Anode	$1^{\#}$	$2^{\#}$	$3^{\#}$	$4^{\#}$	$5^{\#}$
Pb content/ppm	15.2	6.3	6.6	6.1	7.9

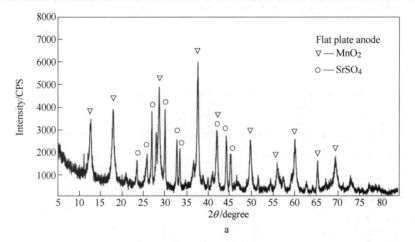

a

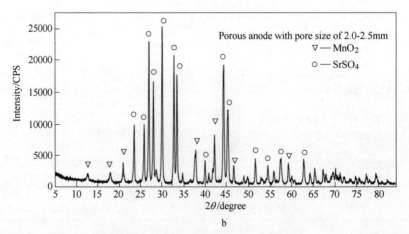

b

Fig. 6 X-ray diffraction spectra of anode slime
(a) flat plate anode (anode $1^{\#}$); (b) porous anode (anode $2^{\#}$)

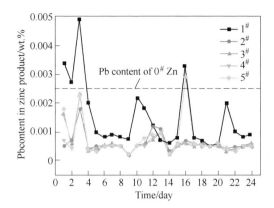

Fig. 7 Lead contents in zinc cathode product as a function of time
(1[#] is flat plate anode used as a comparison, 2[#]-4[#] are porous anodes with the pore size of
2.0-2.5mm, 1.6-2.0mm, 1.25-1.6mm and 1.0-1.25mm, respectively)

4 Conclusions

Porous anode can reduce the anodic potential and cell voltage with no negative effect on cathode current efficiency. It has good potential for energy-saving. Especially, the one with the pore size of 1.6-2.0mm gets the most effective energy conservation, reducing energy consumption 78kW · h/t-Zn. Besides, the weight of porous anode is 40% of flat plate anode, which means the utility of porous anode can reduce the capital investment by 60%.

Porous anode can reduce the precipitation of Mn^{2+} by about 90% and is in favor of forming powder-like anode slime which avoids the incrustation and blockage of pores, and then the anode can remain its porous structure and has a long lifetime. However, as the accumulation of anode slime in long-time working, the blockage is inevitable. It is necessary to remove them and a possible way is to dip the anode in a reducing electrolyte.

Compared with flat plate anode, the corrosion rate of porous anode studied decline significantly, the relative corrosion rates (vs. flat plate anode) is less than 20%. The quality of zinc product are enhanced, the Pb content is 60% less than that of flat plate anode. 0[#] Zn pass rate of flat plate anode is 80%, in contrast, those of porous anodes, excepting 5[#] anode, are all reach to 100%. So the porous anode has the potential to produce zinc of low Pb content.

The results of pilot plant tests indicate that porous anode is a good candidate for further testing as a possible replacement for the conventional Pb-Ag anode. A key consideration is the mechanical strength and electric conductivity. One of the ways to solve this problem is reinforcing it by embedding a metal sheet of higher mechanical strength and electric conductivity than lead alloy in the porous anode. It is anticipated that after testing of the additional prototypes, a whole cell commercial test will be carried out to completely evaluate the porous anode.

References

[1] Dai Ch. S., Yi T. F., Wang D. L., 2006. Effects of lead-foam grids on performance of VRLA battery. Journal

of Power Sources 158,885-890.

[2] Dai Ch. S. ,Zhang B. ,Wang D. L. ,Yi T. F. ,2006. Preparation and performance of lead foam grid for negative electrode of VRLA battery. Materials Chemistry and Physics 99,431-436.

[3] Dai Ch. S. ,Zhang B. ,Wang D. L. ,Yi T. F. ,Hu X. G. ,2006. Study of influence of lead foam as negative electrode current collector material on VRLA battery charge performance. Journal of alloys and components 422,332-337.

[4] Gyenge E. ,Jung J. ,Mahato B. ,2003. Electroplated reticulated vitreous carbon current collectors for lead-acid batteries: opportunities and challenge. Journal of Power Sources 113,388-395.

[5] Gyenge E. ,Jung J. ,Splinter S. ,Snaper A. ,2002. High specific surface area, reticulated current collectors for lead-acid batteries. Journal of Applied Electrochemistry 32,287-295.

[6] Newnham R. H. ,1992. Corrosion rates of lead-based anodes for zinc electrowinning at high current densities. Journal of Applied Electrochemistry 22,116-124.

Oxygen Evolution and Corrosion Behaviors of Co-deposited Pb/Pb-MnO₂ Composite Anode for Electrowinning of Nonferrous Metals*

Abstract The oxygen evolution and corrosion behaviors of Pb/Pb-MnO$_2$ composite anode in H$_2$SO$_4$ solution were investigated comparing with Pb and industrial Pb-Ag(1.0%) alloy. Its electrocatalytic activity towards oxygen evolution reaction(OER) and corrosion resistance during the galvanostatic polarization were evaluated in detail. The microcosmic morphology of the anodic layer during the polarization was observed using scanning electron micrographs(SEMs). The results indicate that this kind of composite anode displays a particular polarization behavior during the polarization due to the changes of surface state: A "potential valley" with the minimal potential of about 300mV lower than that of Pb anode can be observed at the very beginning of polarization, then the anodic potential gradually decrease. The stable potential after 72h is 100~150mV lower than that of Pb anode, and is comparable to industrial Pb-Ag (1.0%) anode. Furthermore, the corrosion resistance of the composite anode largely depends on MnO$_2$ content in deposit. The dissolution of Pb is very little at the early stage, then, after standing a period of heavy intergranular corrosion, a stationary PbO$_2$-MnO$_2$ composite layer will formed on anode surface, which is predicted to present satisfactory corrosion resistance in the following long-time electrolysis.

Key words Pb/Pb-MnO$_2$ composite anode, oxygen evolution reaction(OER), electrocatalytic activity, corrosion resistance

1 Introduction

Oxygen evolution reaction (OER) is one of the most important electrochemical processes during the electrowinning of nonferrous metals(Aromaa and Evans, 2007). Pb alloys have been widely used as the anode materials due to its high corrosion resistance in acidic sulfate solution. However, Pb alloys have also presented several serious problems such as high OER over-potential, Pb contamination to cathodic products, short-circuit resulting from lead distortion and great consumption of noble metal(e.g. Ag). Pointed at these shortcomings, several metal elements such as Ca, Sn, Co, Sb, In, Mn were added into Pb to improve its corrosion resistance, enhance its mechanical strength, and decrease its OER over-potential (Ivanov et al., 2000; Cifuentes et al., 2005; Rerolle and Wiart, 1996a; Felder and Prengaman, 2006; Nguyen and Atrens, 2008). However, these researches made little effective progress when their complications to the anode preparation process were taken into consideration.

Aimed at decreasing the high over-potential of OER, some researchers paid their attention to

---

* Copartner: Y. Li, L. X. Jiang, X. J. Lv, Y. Q. Lai, H. L. Zhang, J. Li. Reprinted from Hydrometallurgy, 2011(109): 252-257.

the development of electrocatalytic oxides, which were typically coated on a titanium plate or mesh by thermal-decomposition or electrodeposition, called Dimensional Stable Anode (DSA) (Chniola et al., 2003; Moats, 2008). Besides, these materials were also coated on Pt, Ni, steel and other substrates (Cattarin et al., 2000; Singh et al., 2009; Palmas et al., 2008). The main approach to improve these OER catalysts involved the synthesis of simple or composite metal oxides. RuO_2, IrO_2 and several other noble metal oxides have been firstly studied because of their tempting electrocatalytic activity (Oliveira-Sousa et al., 2000; Hu et al., 2004; Ma et al., 2006; Song et al., 2008). In the subsequent investigations, they were combined with some base metals to prepare composite oxide catalysts such as PbO_2-RuO_2, $Ru_{0.8}Co_{0.2}O_{2-x}$, $Ru_{0.9}Ni_{0.1}O_{2-\delta}$, RuO_2-PdO_x, $Ir_xSn_{1-x}O_2$, IrO_2-MnO_2 and so on, which were found to present much better OER activity than the simple ones (Muaiani et al., 1999; Jirkovsky et al., 2006; Macounová et al., 2009; Shrivastava and Moats, 2008; Marshall et al., 2006; Ye et al., 2008). Considering the high cost of these noble metals, electrocatalysts only containing simple or composite oxides of base metals were explored. For instance, MnO_2, PbO_2, Co_3O_4, SnO_2, $M_xCo_{3-x}O_4$ (M = Ni, Cu, Zn), $MMoO_4$ (M = Fe, Co, Ni), $MFe_{2-x}Cr_xO_4$ (M = Ni, Cu, Mn) (Cattrin et al., 2001; Dalchiele et al., 2000; Palmas et al., 2009; Mohd and Pletcher, 2006; Singh et al., 2000) have been proved to exhibit favorable electrocatalytic properties toward OER, as well. However, when all of these catalysts were combined with the substrate to prepare DSA, their practical applications in a metallurgical plant were seriously limited by its high fabricating cost (both Ti substrate and precious metal oxides) and short effective life due to the unavoidable passivation of Ti in H_2SO_4 system.

In order to avoid the passivation problem of DSA, recently, the association of a conducting matrix and an active dispersed phase has been explored as a new type of anode for OER. This design couples the high electrocatalytic activity of dispersed phase to the high electrical conductivity, good mechanical and chemical stability of the matrix (Cattarin and Musiani, 2007). In these researches, lead was one of the most frequently adopted matrix materials due to its high corrosion resistance in H_2SO_4 solution. Musiani et al. (1999) showed that Pb-RuO_2 anode has a better electrocatalytic activity than Pb anode for both anodic oxygen evolution and cathodic hydrogen evolution processes. Hrussanova et al. (2001) studied the anodic behaviors of Pb/Pb-Co_3O_4 composite coatings in copper electrowinning, and found these anodes show a depolarizing effect on OER and its corrosion rate during prolonged polarization is approximately 6.7 times lower than that of Pb-Sb anode. Meanwhile, Stefanov and Dobrev (2005) and Dobrev et al. (2009) investigated the Pb-TiO_2, Pb-$CoTiO_3$ composite as new anode materials in zinc electrowinning and indicated that these anodes exhibit good corrosion resistance despite its anode potential is negligibly higher than that of Pb-Ag(1.0%) anode. However, the dispersed particles in these studies were either precious metal oxides (RuO_2), or those hardly stable in H_2SO_4 solution (Co_3O_4), or those with little catalytic activity (TiO_2), which largely confined their further popularization. Moreover, the electrocatalytic properties and formation mechanism of anodic layer for this kind of composite anode have not been evaluated in detail. Since they are of great importance in comprehensively understanding the new anode material, it is necessary to perform deeper theoretical studies.

As one of the most frequently used OER catalysts, MnO$_2$ has been widely brought in the researches of anodic process towards OER. Yu and O'Keefe (2002) and Zhang and Cheng (2007) studied the influence of manganese in zinc electrowinning, and showed that the insoluble MnO$_2$ formed on anode surface during electrowinning could both depolarize anodic reaction and retard anode corrosion to a certain extent. (Carbon-polymer)-MnO$_2$ composite was also developed using hot press method. Its over-potential for OER was 0.2V lower than that of lead anode(Brungs et al., 1996). Recently, Schmachtel et al. (2009) cold-sprayed the Pb-MnO$_2$ composite as a new oxygen evolution anode, and indicated that when the mass fraction of CMD particles was 5%, the obtained composite anode was 0.35V more active than Pb-Ag (0.6%) anode in a short time galvanostatic test.

According to the excellent electrocatalytic activity of MnO$_2$, our former study employed the simple, cost-effective composite electrodeposition(CED) method to explore a new Pb/Pb-MnO$_2$ composite anode in which lead act as the conducting matrix and active MnO$_2$ particles as the electrocatalytic dispersed phase(Li et al., 2010). Compared with aforesaid hot press and cold spray methods, composite electrodeposition is more effective in enhancing the associativity of matrix and particles and can obtain a more uniform deposit. Besides, lead electrodeposition on lead substrate is simple and well-developed, which can easily obtain a good combination between deposit and substrate.

This paper presents a further electrochemical study of this kind of matrix-dispersed phase composite anode in the typical H$_2$SO$_4$ solution. The electrocatalytic activity, corrosion resistance and anodic layer properties during 72h galvanostatic polarization were investigated by different techniques. The same experiments have also been performed on Pb(casted) and industrial Pb-Ag(1.0%) anode as comparison.

2 Experimental Details

2.1 Preparation of Pb/Pb-MnO$_2$ composite anode

The Pb-MnO$_2$ composite coatings were co-deposited on 1cm^2 casted pure lead substrate. The basic composition of the electrodeposition bath was as follows: 120~150g/L Pb(BF$_4$)$_2$; 30~40g/L HBF$_4$; 12~15g/L H$_3$BO$_3$; 0.5g/L gelatin and 0.2g/L diethanolamine(DEA). These experiments were carried out in a 250mL beaker, where the horizontal electrodes were used to guarantee adequate particles embedded in the deposit. The operating temperature and agitation rate were controlled at (35 ± 0.5)℃ and (400 ± 20) r/min by the DF-101S constant temperature magnetic agitator. Unless other specified, the constant current density of cathode and the deposition time were 40mA/cm^2 and 60min, respectively. The deposition solution was prepared with analytical reagents and double-distilled water, and the dispersed particles used were 2~7μm γ-MnO$_2$(CMD) of high purity(>95%, HUITONG Co. Ltd., China).

2.2 Morphology and composition measurement

The microscopic morphologies of Pb, industrial Pb-Ag(1.0%) alloy (YGGL Co. Ltd., China) and Pb/Pb-MnO$_2$ composite anodes were separately observed on the fresh surface (only for Pb/Pb-MnO$_2$ anode) and on the surfaces at the 0.5th, 24th, and 72th hour of galvanostatic polarization using JSM-6360 type Scanning Electron Microscope (SEM). The element composition of the Pb/Pb-MnO$_2$ composite anodes was detected by GENESIS60S type Energy Dispersive Spectrometry (EDS). The mass fraction of MnO$_2$ (w(MnO$_2$)) in deposit was approximately calculated with formula (1):

$$w(\text{MnO}_2) = \left(1 + \frac{M_{\text{Pb}}}{M_{\text{Mn}} + 2M_{\text{O}}} \times \frac{N_{\text{Pb}}}{N_{\text{Mn}}}\right)^{-1} \times 100\% \qquad (1)$$

where M_{Pb}, M_{Mn} and M_{O} represents the Relative Atomic Mass of Pb, Mn and O, respectively, and $N_{\text{Pb}}/N_{\text{Mn}}$ is the atomic ratio of Pb and Mn detected by EDS.

2.3 Electrochemical measurements

2.3.1 Galvanostatic method

Electrolysis experiments of aforesaid Pb, Pb-Ag and Pb/Pb-MnO$_2$ anodes (The particles content is within 6% ~ 8% unless other specified.) were performed in 160g/L H$_2$SO$_4$ solution at an anodic current density of 50mA/cm^2 and constant temperature of 35℃. The standard three-electrode system was adopted. A Pt plate of 4cm^2 and the KCl saturated calomel electrode (SCE) were used as the counter electrode and the reference electrode, respectively. Then a multimeter with a high ohmic imput ($5 \times 10^{12}\Omega$) indicated the potential values detailedly during the galvanostatic polarization.

2.3.2 Corrosion rate determination

The corrosion of lead-based anode in H$_2$SO$_4$ solution is mainly caused by the dissolution of lead. Other parts also include the formation of sedimentary anode slime, which can be neglected in such laboratory experiments. Therefore, the corrosion rate can be measured with the ionic equilibrium method (Lai et al., 2010). When it comes to the case of Pb/Pb-MnO$_2$ anode, the dissolution of manganese also has to be considered. As there is little Pb and Mn deposited on Pt cathode, the average corrosion rate of these anodes in a certain time interval can be calculated by formula (2):

$$Corr = \frac{(\Delta C_{\text{Pb}} + 1.58\Delta C_{\text{Mn}}) \cdot V}{A \cdot \Delta t} \qquad (2)$$

where ΔC_{Pb} and ΔC_{Mn} respectively represents the concentration variation of lead and manganese in electrolyte between a certain time interval; V is the electrolyte volume; A is the anode area; Δt is the time distance.

In this paper, the concentrations of lead and manganese in electrolyte were measured by an atomic absorption spectrophotometer (Hitachi, Z-5000) at several time points during the galvanostatic polarization. The average corrosion rates between every two adjacent points were separately calculated to analyze the electrochemical corrosion disciplines of the three kinds of anodes.

3 Results and Discussion

3.1 Surface morphology and composition of Pb/Pb-MnO₂ composite anode

Fig. 1 shows the secondary electron image (SEI) and back-scattered electron image (BEI) of electrodeposited Pb/Pb-MnO$_2$ composite anode before electrolysis, of which the mass fraction of MnO$_2$ particles is 6.13%. Analogous to that reported by Dobrev et al. (2009), pyramidal crystal grains with clearly outlined boundaries was observed, and the surface is relatively heterogeneous, which makes the specific surface area of the composite anode considerably large. Moreover, individual particles can also be detected partially embedding in lead matrix, and they are proved to be MnO$_2$ phase (black parts in Fig. 1b) by EDS analysis.

Fig. 1 Surface morphology (×500) of Pb/Pb-MnO$_2$(6.13%) composite anode before electrolysis
a—SEI; b—BEI

3.2 Galvanostatic polarization

Since industrial electrowinning processes are operated at constant current, the anodic potential is one of the most significant criteria to estimate the energy consumption. Fig. 2 presents the alteration of anodic potential as a function of time for Pb, Pb-Ag and several Pb/Pb-MnO$_2$ anodes with different MnO$_2$ content during the galvanostatic polarization for 72h. We can find in the inset that anodic potential of these composite anodes drastically decreased at the very beginning of the polarization (about 300mV lower than that of Pb anode). However, during the subsequent electrolysis, the potential reversely increased to the degree of Pb anode, and the anodic activity was obviously diminished. Then, their anodic potential slowly dropped, and the stable potential after 72h was 100~150mV lower than that of Pb anode, and is comparable to the industrial Pb-Ag(1.0%) anode.

Magnifying the very beginning of the polarization curves, a visible "potential valley" can be found in the curves of Pb/Pb-MnO$_2$ anodes. We defined the minimal potential at the valley bottom as "Dented Potential (DP)", since it is an important symbol of electrocatalytic activity at the fresh surface. The formation of this valley is probably due to the surface state changes in H$_2$SO$_4$ solution. As a fresh anode, the specific surface area is considerably enlarged and there

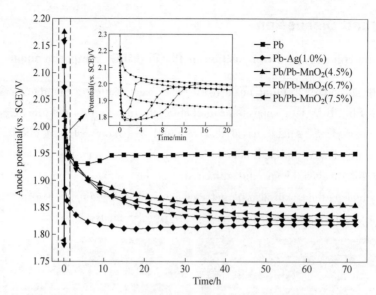

Fig. 2 Dependence of the anode potential of Pb, Pb-Ag and Pb/Pb-MnO$_2$ anodes on time during 72h galvanostatic polarization

are plenty of partially exposed MnO$_2$ particles, contributing the anode a highly active surface. When such lead-based anodes are polarized galvanostaticly, an anodic layer mainly composed of PbO$_2$ and PbSO$_4$ will be irreversibly formed(Czerwinski et al. ,2000). This process, probably, will destroy the original surface state to some extent: on the one hand, the anodic layer covers some partially exposed particles; on the other, some particles may become slack and even drop into the bulk solution. Meanwhile, the brushing effect of evolved O$_2$ further exacerbates the dropping of MnO$_2$. Therefore, the amounts of effective MnO$_2$ particles were markedly diminished, rendering the electrocatalytic activity of these composite anodes visibly decreased after several minutes' efficient electrolysis. Notwithstanding, the anodic potential after polarization for 72h(we define as "Stable Potential(SP)") of such composite anodes still had a considerable decrease, which can probably attribute to the formation of a PbO$_2$-MnO$_2$ composite layer.

The relationships of anodic potentials, particles content in deposit and CED suspension are shown in Fig. 3. The particles embedded in deposit regularly increase with the increase of particle content in the suspension. The corresponding DP and SP values present the trend of firstly decreasing and then increasing, and the change scope of DP values is much larger than that of SP. When the particles content is in the range 6% ~ 8%, the composite anodes show the best electrocatalytic activity.

The relationships of anodic potentials, particles content in deposit and deposition time(represent the deposit thickness) are shown in Fig. 4. The superficial particles content in deposit generally decreases with the prolonging of deposition time. This phenomenon is usually ascribed to the partial agglomeration of suspended particles and the diminution of deposit coarseness(Li et al. ,2010). Meanwhile, the corresponding DP and SP values also express the trend of first slightly decreasing and then increasing, and the change scope of DP values are analogously lar-

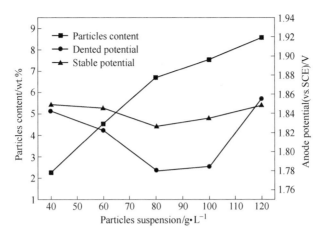

Fig. 3 Dependence of the anodic potential and particles content in deposit on the particles content in CED suspension

ger than that of SP. When the deposition time is controlled within 60~90min, the composite anodes obtain the best electrocatalytic activity.

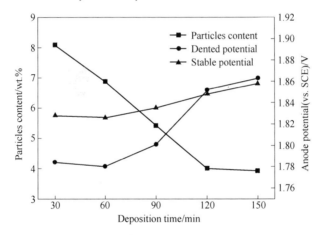

Fig. 4 Dependence of the anodic potential and particles content in deposit on the deposition time

These aforesaid dependences of anodic potentials on particles content in deposit and deposit thickness (indicated by the deposition time) are probably ascribed to the integrative influence of high electrocatalytic activity and poor conductivity of embedded MnO_2 particles. When the particles are either overfull or deficient in deposit, it is unfavorable for the composite anodes to exert the best electrocatalytic activity. Thus, only when the particles content in suspension and the deposition time are controlled within appropriate values, the obtained composite anode can present the lowest anodic potential.

3.3 Corrosion rate

Lead and manganese concentration in the electrolyte of Pb, Pb-Ag and Pb/Pb-MnO_2 anodes were detected at several time points during the galvanostatic polarization. The results are shown in Table 1. We can find that Pb anode exerts the highest dissolution speed and the lead concentration reaches the saturation level (8.5mg/L, practically measured in a $PbSO_4$ saturated solution at 35℃ using aforesaid atomic absorption spectrophotometer) in a relatively short time (before 48h). After that, the corrosion went on in the form of black PbO_2 slime, which was observed beginning to settle in the beaker bottom. As compared with Pb anode, Pb-Ag anode presented much lower Pb concentration values with a small variation during the electrolysis. The lead concentration in the Pb/Pb-MnO_2 electrolyte gradually increased during the electrolysis, indicating the sustaining corrosion of lead matrix. The manganese concentration slightly increased at the first stage of polarization, and then the formation of anodic PbO_2 layer prevented the inner MnO_2 particles to contact the electrolyte, rendering the concentration values came to a constant. Furthermore, we can observe that the concentration of Pb and Mn largely depends on the MnO_2 content in deposits. The composite anode with high MnO_2 content shows much lower Pb concentration values, revealing its corrosion resistance was effectively enhanced.

Table 1 Concentration (mg/L) of Pb and Mn in the electrolyte during the galvanostatic polarization

Anodes	Measured elements	Polarization time/h								
		6	12	18	24	30	36	48	60	72
Pb	Pb	1.7	3.4	4.8	5.8	6.7	7.5	8.3	8.5	8.5
Pb-Ag(1.0%)	Pb	0.6	0.9	1	1	1	1	1	1.1	1.5
Pb/Pb-MnO_2 (1.4%)	Pb	0.4	0.7	1.3	2.1	3	3.8	5.2	6.4	7.4
	Mn	0.03	0.04	0.05	0.05	0.05	0.05	0.06	0.05	0.05
Pb/Pb-MnO_2 (6.8%)	Pb	0.3	0.5	0.8	1.0	1.3	1.5	1.6	1.7	1.8
	Mn	0.1	0.2	0.2	0.2	0.2	0.2	0.2	0.2	0.2

The average values of anodic corrosion rate between every two adjacent times points, as shown in Fig. 5, are separately calculated with formula (2). Obviously, Pb anode presented the highest corrosion rate at the first polarization stage, and it reduced a lot later due to the formation of PbO_2 layer. The corrosion rate of Pb-Ag anode had a relatively higher value at first, and then it nearly decreases to zero. But it began to increase again after polarization for about 48h, which is probably due to the change of surface state. For Pb/Pb-MnO_2(1.4%) anode, the corrosion rate gradually increased from a relative low value in the early stage. after reaching the maximal value, it began to decrease slowly. The Pb/Pb-MnO_2 (6.8%) anode generally presented a decrease trend. The relatively higher corrosion rate at the beginning was mainly contributed by the dissolution of Mn^{2+}. In addition, both of two composite anodes reached the highest corrosion rate approximately at the 24th hour, indicating this is a heavy corrosion period. However, their corrosion rates after 72h were relatively small and one even lower than that of Pb-Ag anode. Therefore, it is predicable that Pb/Pb-MnO_2 composite anode will exhibit the best corrosion resistance in the subsequent long-time electrolysis.

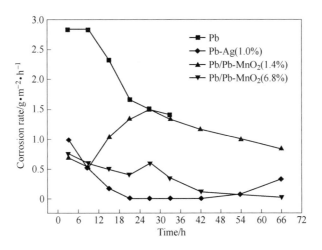

Fig. 5 Corrosion rate variation of Pb, Pb-Ag and Pb/Pb-MnO$_2$ anodes during 72h galvanostatic polarization

3.4 Scanning electron micrographs

The micrographs (SEI) obtained on the surface of Pb, Pb-Ag and Pb/Pb-MnO$_2$ anodes at the 0.5th, 24th and 72th hour of galvanostatic polarization are shown in Fig. 6, respectively. After 0.5h polarization, the anodic layers of Pb and Pb-Ag(especially) anode are relatively smooth, with prismatic crystals of PbSO$_4$ dispersing on the fine-grained PbO$_2$ layer (Rerolle and Wiart, 1996b). However, compared with the fresh surface (Fig. 1), although there is a thin PbO$_2$ membrane without any observable PbSO$_4$ crystals, the morphology of Pb/Pb-MnO$_2$ anode has not changed too much. It is probably due to the quite small real current density. This compact membrane may be very effective to prevent the inner metallic lead to contact the electrolyte, leading to the small corrosion rate at beginning of galvanostatic polarization. Furthermore, we can find that part of the embedded MnO$_2$ particles have been brushed off or covered as conjectured above, which result in an obvious decrease of electrocatalytic activity.

When the polarization reaches the 24th hour, the surface morphologies of three anodes have changed to a large extent. The anodic layer of Pb-Ag anode represents the most compact and smoothest surface with the absence of prismatic PbSO$_4$. Such a structure may be very propitious to protect the inner substrate from corrosion, corresponding to the very low corrosion rate in Fig. 5. Meanwhile, the surfaces of Pb and Pb/Pb-MnO$_2$ anode are rather rough. Pb/Pb-MnO$_2$ anode, especially, presents a chaotic coral structure with numerous inverted pyramidal type holes in them, which is probably a remnant of the fresh surface due to the intergranular corrosion at the boundaries between the typical pyramidal grains of Pb deposit and between MnO$_2$ and Pb matrix. This structure is certainly prone to heavy corrosion, corresponding to the maximal corrosion rate in Fig. 5.

After polarization for 72h, the stationary anodic layers are considered to have formed. Pb anode presents the loosest surface with a typical coral structure. The film of Pb-Ag anode is denser than Pb, and presents a relatively fine-grained structure. However, this layer is much looser and more porous than that obtained at the 24th hour, rendering its corrosion rate secondly begins to

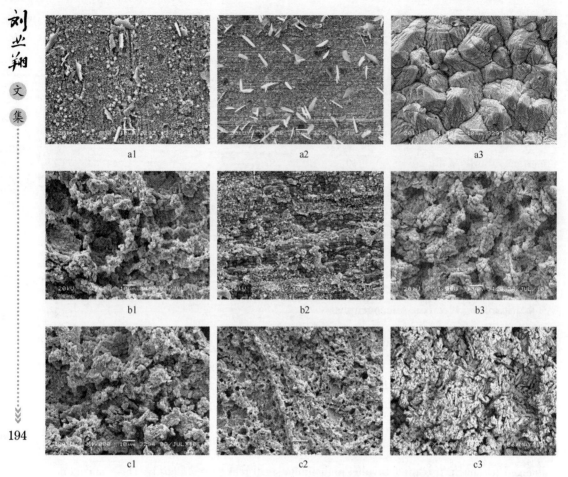

Fig. 6 Morphology(×1000) of Pb(1), Pb-Ag(2) and Pb/Pb-MnO$_2$ anodes
(3) obtained at the 0.5th(a), 24th(b) and 72th(c) hour of galvanostatic polarization

increase, as showed in Fig. 5. However, the surface of Pb/Pb-MnO$_2$ anode becomes the densest after 72 hours' polarization. This result may be consistent with that reported by Dai et al. (1995), which showed that the corrosion layer formed on electrodeposited Pb alloy is relatively homogeneous and compact comparing with casted ones. In our paper, the uniform and fine-grained anodic layer, containing typical needle-like β-PbO$_2$(Dobrev et al., 2009), is thought to be able to present satisfied corrosion resistance (as showed in Fig. 6). Since there are still some amount of MnO$_2$ can be detected in the surface, we can deduce that the anodic layer finally turns out to be a composite of PbO$_2$-MnO$_2$.

4 Conclusion

The electrocatalytic activity and corrosion behaviors of the matrix-dispersed phase kind Pb/Pb-MnO$_2$ composite anode were investigated during 72h galvanostatic polarization in H$_2$SO$_4$ solution. The results indicate that Pb/Pb-MnO$_2$ composite anodes present preferable oxygen evolution activity at the beginning of galvanostatic polarization. The "Dented Potential" is about 300mV lower than that of Pb anode. Then, with the formation of anodic layer, the effective MnO$_2$ amount is partially diminished, rendering the anode potential increase to the level of Pb anode. In the prolonged polarization, the anodic potential gradually decreases, and the "Stable Potential" after 72h is 100~150mV lower than that of Pb anode, and is comparable to Pb-Ag (1.0%) anode. Further, the DP and SP values largely depend on the particle content in deposit and the deposition time. When the particle contents is within 6%~8% and the deposition time is in the range of 60~90min, the composite anode displays the best electrocatalytic activity.

The corrosion rate of $Pb/Pb-MnO_2$ composite anodes largely depends on the MnO_2 content in deposit. The anode with higher MnO_2 content shows an effectively enhanced corrosion resistance. Further, the corrosion of $Pb/Pb-MnO_2$ anodes will stand a period of heavy intergranular corrosion, but then it seem to be preferable to form a stationary, dense and fine-grained PbO_2-MnO_2 composite layer as compared with Pb and Pb-Ag anodes. Accordingly, it is predicable that this kind of $Pb/Pb-MnO_2$ composite anode is promising to present satisfactory corrosion resistance in long-time electrowinning.

References

[1] Aromaa J, Evans J M, 2007. Electrowinning of metals. Encyclopedia of Electrochemistry 5, 159-265.

[2] Brungs A, Haddadi-asl V, Skylias-kazacos M, 1996. Preparation and evaluation of electrocatalytic oxide coatings on conductive carbon-polymer composite substrates for use as dimensionally stable anodes. J. Appl. Electrochem. 26, 1117-1123.

[3] Cattarin S, Musiani M, 2007. Electrosynthesis of nanocomposite materials for electrocatalysis. Electrochim. Acta 52, 2796-2805.

[4] Cattrin S, Guerriero P, Musiani M, 2001. Preparation of anodes for oxygen evolution by electrodeposition of composite Pb and Co oxides. Electrochim. Acta 46, 4229-4234.

[5] Cattarin S, Frateur I, Guerriero P, Musiani M, 2000. Electrodeposition of PbO_2+CoO_x composites by aimultaneous oxidation of Pb^{2+} and Co^{2+} and their use as anodes for O_2 evolution. Electrochim. Acta 45, 2273-2288.

[6] Chniola J, Gogotsi Y, Ferdman A, 2003. Mechanically stable insoluble titanium-lead anodes for sulfate electrolytes. Sci. Sinter. 35, 75-83.

[7] Cifuentes L, Astete E, Crisostomo G, 2005. Corrosion and protection of lead anodes in acidic copper sulphate solutions. Corros. Eng. , Sci. Techn. 40, 321-327.

[8] Czerwinski A, Zelazowska M, Grden M, Kuc K, Milewski J D, Nowacki A, Wojcik G, Kopczyk M, 2000. Electrochemical behavior of lead in sulfuric acid solutions. J. Power Sources 85, 49-55.

[9] Dai C S, Wang Z H, Wang D L, Ding F, Hu X G, 2005. Study on CV behavior of electrodeposited lead alloy in sulfuric acid. Journal of harbin institute of technology 37, 530-532.

[10] Dalchiele E A, Cattarin S, Musiani M, 2000. Electrodeposition studies in the $MnO_2 + PbO_2$ system: formation of $Pb_3Mn_7O_{15}$. J. Appl. Electrochem. 30, 117-120.

[11] Dobrev Ts, Valchanova I, Stefanov Y, Magaeva S, 2009. Investigations of new anodic materials for zinc electrowinning. Trans. Inst. Met Finish. 87, 136-140.

[12] Felder A, Prengaman R D, 2006. Lead alloys for permanent anodes in the nonfereous metals industry. JOM 58, 28-31.

[13] Hu J M, Zhang J Q, Cao C N, 2004. Oxygen evolution reaction on IrO_2-based DSA type electrodes: kinetics analysis of Tafel lines and EIS. Int. J. Hydrogen Energy 29, 791-797.

[14] Hrussanova A, Mirkova L, Dobrev Ts, 2001. Anodic behavior of the $Pb-Co_3O_4$ composite coating in copper electrowinning. Hydrometallurgy 60, 199-213.

[15] Ivanov I, Stefanov Y, Noncheva Z, Petrova M, Dobrev Ts, Mirkova L, Vermeersch R, Demaerel J-P, 2000. Insoluble anodes used in hydrometallurgy Part I. Corrosion resistance of lead and lead alloy anodes. Hydrometallurgy 57, 109-124.

[16] Jiang L X, Zhong S P, Lai Y Q, Lv X J, Li J, Liu Y X, 2010. Effect of current density on the electrochemical behavior of a flat plate Pb-Ag anode for zinc electrowinning. Acta Phys. -Chim. Sin. 26, 2369-2374.

[17] Jirkovsky J, Makarova M, Krtil P, 2006. Particle size dependence of oxygen evolution reaction on nanocrystalline RuO_2 and $Ru_{0.8}Co_{0.2}O_{2-x}$. Electrochem. Commun. 8, 1417-1422.

[18] Lai Y Q, Jiang L X, Li J, Zhong S P, Lv X J, Peng H J, Liu Y X, 2010. A novel porous Pb-Ag anode for energy-saving in zinc electro-winning. Part I Laboratory preparation and properties. Hydrometallurgy 102, 73-80.

[19] Li Y, Jiang L X, Ni H F, Lv X J, Lai Y Q, Li J, Liu Y X, 2010. Preparation and properties of Pb/Pb-MnO_2 composite anode for zinc electrowinning. The Chinese Journal of Nonferrous Metals 20, 2357-2365.

[20] Ma H C, Liu C P, Liao J H, 2006. Study of ruthenium oxide catalyst for electrocatalytic performance in oxygen evolution. J. Mol. Catal. A:Chem. 247, 7-13.

[21] Macounová K, Jirkovský J, Marina V, 2009. Oxygen evolution on $Ru_{1-x}Ni_xO_{2-y}$ nanocrystalline electrodes. J. Solid State Electrochem. 13, 959-965.

[22] Marshall A, Borresen B, Hagen G, Sunde S, Tsypkin M, Tunold R, 2006. Iridium oxide-based nanocrystalline particles as oxygen evolution electrocatalysts. Russ. J. Electrochem. 42, 1134-1140.

[23] Moats M S, 2008. Will lead-based anodes ever be replaced in aqueous electrowinning? . JOM 60, 46-49.

[24] Mohd Y, Pletcher D, 2006. The fabrication of lead dioxide layers on a titanium substrate. Electrochim. Acta 54, 786-793.

[25] Muaiani M, Furlanetto F, Bertoncello R, 1999. Electrodeposited $PbO_2 + RuO_2$: a composite anode for oxygen evolution from sulphuric acid solution. J. Electroanal. Chem. 465, 160-167.

[26] Nguyen T, Atrens A, 2008. Influence of lead dioxide surface films on anodic oxidation of a lead alloy under conditions typical of copper electrowinning. J. Appl. Electrochem. 38, 569-577.

[27] Oliveira-Sousa A de, Silva M A S da, Machado S A S, Avaca L A, Lima-Neto P de, 2000. Influence of the preparation method on the morphological and electrochemical properties of Ti/IrO_2-coated electrodes. Electrochim. Acta 45, 4467-4473.

[28] Palmas S, Ferrara F, Mascia M, Polcaro A M, Rodriguez Ruiz J, Vacca A, Piccaluga G, 2009. Modeling of oxygen evolution at Teflon-bonded Ti/Co_3O_4 electrodes. Int. J. Hydrogen Energy 34, 1647-1654.

[29] Palmas S, Polcaro A M, Ferrara F, Rodriguez Ruiz J, Delogu F, Bonatto-Minella C, Mulas G, 2008. Electrochemical performance of mechanically treated SnO_2 powders for OER in acid solution. J. Appl. Electrochem. 38, 907-913.

[30] Rashkov St, Stefanov Y, Noncheva Z, 1996. Investigation of the processes of obtaining plastic treatment and electrochemical behavior of lead alloys in their capacity as anodes during the electroextraction of zinc II. Electrochemical formation of phase layers on binary Pb-Ag and Pb-Ca, and ternary Pb-Ag-Ca alloys in a sulphuric-acid electrolyte for zinc electro-extraction. Hydrometallurgy 40, 319-334.

[31] Rerolle C, Wiart R, 1996a. Kinetics of oxygen evolution on Pb and Pb-Ag anodes during zinc electrowinning. Electrochim. Acta 41, 1063-1069.

[32] Rerolle C, Wiart R, 1996b. Kinetics of oxygen evolution on Pb and Pb-Ag anodes during zinc electrowinning II. Oxygen evolution at high polarization. Electrochim. Acta 41, 83-90.

[33] Schmachtel S, Toiminen M, Kontturi K, Forsén O, Barker M H, 2009. New oxygen evolution anodes for metal electrowinning: MnO_2 composite electrodes. J. Appl. Electrochem. 39, 1835-1848.

[34] Shrivastava P, Moats M S, 2008. Ruthenium palladium oxide-coated titanium anodes for low-current-density oxygen evolution. J. Electrochem. Soc. 155, E101-E107.

[35] Singh R N, Awasthi M R, Sinha A S K, 2009. Preparation and electrochemical characterization of a new NiMoO$_4$ catalyst for electrochemical O$_2$ evolution. J. Solid State Electrochem. 13, 1613-1619.

[36] Singh R N, Pandey J P, Singh N K, Lal B, Chartier P, Koenig J F, 2000. Sol-gel derived spinal M$_x$Co$_{3-x}$O$_4$ (M=Ni, Cu; 0≤x≤1) films and oxygen evolution. Electrochim. Acta 45, 1911-1919.

[37] Song S D, Zhang H M, Ma X P, Shao Z G, Baker R T, Yi B L, 2008. Electrochemical investigation of electrocatalysts for the oxygen evolution reaction in PEM water electrolyzers. Int. J. Hydrogen Energy 33, 4955-4961.

[38] Stefanov Y, Dobrev Ts, 2005. Developing and studying the properties of Pb-TiO$_2$ alloy coated lead composite anodes for zinc electrowinning. Trans. Inst. Met. Finish. 83, 291-295.

[39] Yamamoto Y, Fumino K, Ueda T, Nambu M, 1992. Apotentiodynamic study of the lead electrode in sulphuric acid solution. Electrochim. Acta 37, 199-203.

[40] Ye Z G, Meng H M, Sun D B, 2008. New degradation mechanism of Ti/IrO$_2$+MnO$_2$ anode for oxygen evolution in 0.5M H$_2$SO$_4$ solution. Electrochim. Acta 53, 5639-5643.

[41] Yu P, O'Keefe Y J, 2002. Evaluation of lead anode reactions in acid sulfate electrolytes II. Manganese reactions. J. Electrochem. Soc. 149, 558-569.

[42] Zhang W S, Cheng C Y, 2007. Manganese metallurgy review. Part III: Manganese control in zinc and copper electrolytes. Hydrometallurgy 89, 178-188.

Electrochemical Performance of a Pb/Pb-MnO$_2$ Composite Anode in Sulfuric Acid Solution Containing Mn^{2+} *

Abstract The influence of Mn^{2+} on oxygen evolution kinetics and corrosion behaviour of Pb/Pb-MnO$_2$ composite anode in sulfuric acid electrolyte was investigated using SEM, XRD, and several electrochemical methods. The results indicate that a high concentration of Mn^{2+} (e.g. 3.0g/L) in the electrolyte resulted in the formation of a MnO$_2$ layer on the surface of the anode. This layer decreased the oxygen evolution activity of the anode, but at the same time made the underlying PbO$_2$ layer more compact and flat, effectively improving the anodic corrosion resistance. When the Mn^{2+} concentration was low (e.g. 0.1g/L), no MnO$_2$ layer was formed but the structure of the PbO$_2$ anodic layer was modified. As a result, the oxygen evolution activity and corrosion resistance were both significantly improved. In addition, Mn^{2+} in the electrolyte did not change the kinetic mechanism of oxygen evolution reaction. The reaction was exclusively controlled by the formation and adsorption of first intermediate, and the adsorption resistance played a dominant part in the whole reaction resistance.

Key words Pb/Pb-MnO$_2$ composite anode, oxygen evolution reaction, corrosion resistance, Mn^{2+} ion, zinc electrowinning

1 Introduction

In the zinc electrowinning industry Pb-Ag alloys are widely used as the anode material (Aromaa and Evans, 2007; Felder and Prengaman, 2006). However, this alloy has several disadvantages, such as high anodic over-potential, Pb contamination of the cathode zinc, short-circuits resulting from lead distortion, and high consumption of silver. Therefore, recently composite anodes of lead and an active dispersed phase has been widely studied (Muaiani et al., 1999; Schmachtel et al., 2009; Chang et al., 2007; Dobrev et al., 2009; Stefanov et al., 2005; Hrussanova et al., 2002; Hrussanova et al., 2004a, 2004b). Such anodes couple the high electrocatalytic activity of the dispersed phase and good chemical stability of lead matrix. As a result, the oxygen evolution activity (Muaiani et al., 1999; Schmachtel et al., 2009; Chang et al., 2007) and corrosion resistance (Dobrev et al., 2009; Stefanov et al., 2005; Hrussanova et al., 2002) of the lead matrix were significantly improved. Based on its fairly low material and preparation cost compared with conventional Pb-Ag alloys, such composite anodes are promising alternatives to the present cast lead anodes.

When these composite anodes are applied in industry, the first key problem which needs to be recognized is the influence of Mn^{2+}. This additive is a double-edge sword in the traditional zinc electrowinning electrolyte with the casted or rolled Pb-Ag alloy as anode materials (Krup-

---

* Copartner: Y. Q. Lai, Y. Li, L. X. Jiang, X. J. Lv, J. Li. Reprinted from Journal of Electroanalytical Chemistry, 2012, 671:12-23.

kowa et al., 1997; Verbaan and Mullinder, 1981; Zhang and Cheng, 2007). The Mn^{2+} can be oxidized to form a compact and strongly adhesive MnO_2 anodic layer during the electrolysis, which effectively reduces the corrosion of anodes and minimizes the contamination of cathodic zinc by lead(MacKinnon and Brannen, 1991; Schierle and Hein, 1993; Newnham, 1992; Saba and Elsherief, 2000). The MnO_2 slimes also adsorb detrimental ionic impurities, such as Cu^{2+}, Co^{2+}, Ni^{2+}, Sb^{3+} (Ivanov, 2004; Ivanov and Stefanov, 2002), and decrease the effect of Cl^- (Kelsall, 2000).

On the other hand, the complex oxidation and reduction reactions among manganese ions of different valences will considerably decrease the current efficiency(MacKinnon and Brannen, 1991; Cathro, 1991). Also, the MnO_2 layer can passivate the anodic surface and retard the oxygen evolution(Rerolle and Wiart, 1996). Furthermore, Yu et al. (2002) studied the electrochemical impact of Mn^{2+} to Pb-Ag and Pb-Ca-Sn anodes and found that the presence of Mn^{2+} depolarizes anodic reactions and decreases PbO_2 formation. Cachet et al. (1999) studied the influence of Mn^{2+} on oxygen evolution on lead anodes, and stated that the kinetics for lead and lead-silver anodes were affected in the presence of Mn^{2+}.

Previous studies investigated the electrochemical performance of the co-deposited Pb/Pb-MnO_2 composite anode in sulfuric acid solution and found that its oxygen evolution activity and corrosion rate were largely improved by the association of lead and MnO_2 particles(Li et al., 2010; Li et al., 2011). The current paper presents further electrochemical studies to investigate the influence of Mn^{2+} in the electrolyte on the oxygen evolution behaviour and corrosion resistance of this kind of Pb/Pb-MnO_2 composite anode.

2 Experimental Details

Pb-MnO_2 composite coatings were electrochemically co-deposited on $1cm^2$ casted pure lead substrate. The basic composition of the deposition bath was as follows: 120~150g/L $Pb(BF_4)_2$; 30~40g/L HBF_4; 12~15g/L H_3BO_3; 0.5g/L gelatin; 0.2g/L diethanolamine(DEA) and 80~120g/L MnO_2(in suspension). The experiments were conducted in a 250mL beaker, where up-faced horizontal cathode was used to guarantee MnO_2 particles were incorporated in the deposit. The plating temperature and agitation rate were controlled at (35 ± 0.5)°C and (400 ± 20)r/min using a DF-101S constant temperature magnetic stirrer. The cathodic current density was maintained at $40mA/cm^2$ for 60min producing a deposit 200~300μm thick containing 6wt%~8wt% MnO_2. All solutions were prepared with analytical reagents and double-distilled water, and the MnO_2 particles used were 2~7μm γ-MnO_2 of high purity(>95%, HUITONG Co,. Ltd.).

What was the approximate surface area coated?

The electrochemical measurements for the Pb/Pb-MnO_2 composite anodes were performed in 160g/L H_2SO_4 solution with different contents of Mn^{2+} at (35 ± 0.5)°C using a standard three-electrode system. A platinum plate of $4cm^2$ and a saturated calomel electrode(SCE) were used as counter electrode and reference electrode, respectively. All potentials shown in the figures are against SCE. The constant current density used during galvanostatic electrolysis was $50mA/cm^2$ which is equivalent to the $500A/m^2$ used in industry. The anode potential was measured using a high impedance input (5×10^{12} Ω) multimeter during the 72h electrolysis. After electrolysis, cyclic voltammograms, Tafel parameters and electrochemical impedance spectroscopy (EIS) were immediately carried out on the anodic layers using an EG&G Princeton applied research model 2273 potentiostat/galvanostat controlled using PowerSuite software. The scanning range of the voltammograms was $-1.0 \sim 2.1$V at a sweep rate of 10mV/s (did you start at the anodes rest potential or another value?). The quasi-stationary polarization curve for Tafel measurement was recorded from 1.7 to 2.0V with a scanning rate of 0.2mV/s. The frequency interval of EIS measurement was from 10^5 to 10^{-2}Hz and the AC Amplitude is 5mV rms.

The corrosion resistance of the Pb/Pb-MnO$_2$ composite anode was measured by the ionic equilibrium method. The corrosion of lead-based anode in H$_2$SO$_4$ solution is mainly caused by the dissolution of lead. Other parts also include the formation of sedimentary anode slime, which can be neglected in such laboratory experiments due to the short duration (Lai et al., 2010). As there is little lead deposited on the platinum cathode, the corrosion resistance can be evaluated by the concentration of Pb^{2+} in the electrolyte during the electrolysis. The Pb^{2+} concentration in the electrolyte was measured by atomic absorption spectroscopy (Hitachi, Z-5000).

In addition, JSM-6360F Scanning Electron Microscope (SEM) and TTR-III X-ray Diffraction (XRD) were used to detect the microscopic morphology and phase composition of the Pb/Pb-MnO$_2$ anode before and after electrolysis, respectively.

3 Results and Discussion

3.1 Galvanostatic polarization

Fig. 1 shows the influence of Mn^{2+} in electrolyte on the galvanostatic polarization curves of the composite anodes. As described in earlier work (Li et al., 2010; Li et al., 2011), there is a "potential valley" on the polarization curve at the very beginning of electrolysis due to the formation of a stable surface. Over the 72h polarization time the layers came to an equilibrium state and the behaviour of the anode became consistent.

As shown in the inset of Fig. 1, in the electrolytes with high Mn^{2+} content (e.g. 3g/L and 5g/L), the time to stability was extended. This may be due to the more ready formation of a MnO$_2$ layer by oxidation of Mn^{2+} in solution, which to some extent aided retention of the MnO$_2$ particles in the deposit. During the following electrolysis, this MnO$_2$ layer gradually became thick, rigid and cracked, and then covered most of the PbO$_2$ surface (visually observed), and as a result the anodic potential markedly increased. When the MnO$_2$ layer became thick enough, it began to break off bit by bit and drop into the electrolyte in the form of anodic slime. Low-concentration Mn^{2+} (1.0g/L or 0.1g/L) had very little influence on the polarization curve at the beginning. In the prolonged electrolysis, there was no apparent MnO$_2$ layer formed on the anodic surface, and powdery MnO$_2$ slime was found to gradually deposit in the cell bottom. When the MnO$_2$ Mn^{2+} concentration decreased to 0.1g/L, the stable anodic potential to some extent reduced. This is probably due to the influence of Mn^{2+} on the structure and composition of the anodic layer.

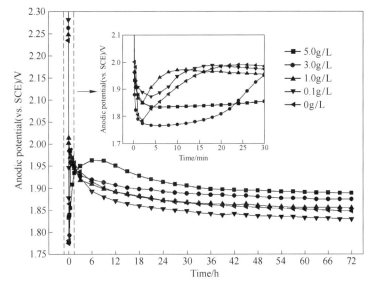

Fig. 1　Influence of Mn^{2+} concentration in the electrolyte on the anodic potential of Pb/Pb-MnO$_2$ composite anode during the 72hours' galvanostatic polarization

3.2　Surface morphology and phase composition

Fig. 2 shows the microscopic morphology of a fresh composite anode and the surface after 72 hours galvanostatic polarization in an electrolyte without Mn^{2+}. On the fresh surface, pyramidal crystal grains of lead with clearly outlined boundaries can be observed, and the surface is relatively heterogeneous, which makes the specific surface area of the composite anode considerably larger than the geometric area. Moreover, individual particles partially embedded in lead matrix can also be observed, and were found to be Mn-rich by EDS. When the anode was polarized for 72h, although the macro surface was still smooth and flat, the microscopic morphology was relatively loose and porous. Such a structure will not protect the inner lead base from further corrosion with increasing time of electrolysis.

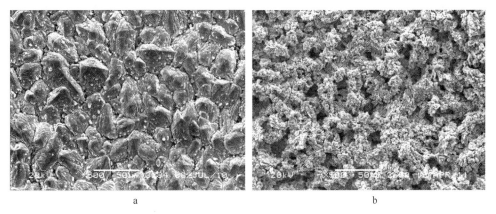

Fig. 2　Surface morphology(×500) of the Pb/Pb-MnO$_2$ anode before(a) and after(b) 72hours' electrolysis in the electrolyte without Mn^{2+}

The MnO_2 layer formed at high Mn^{2+} content after 72hours' electrolysis was rigid, cracked and weakly adherent to the PbO_2 surface. It should be noted that this MnO_2 layer is quite different from MnO_2 layers formed on traditional cast or rolled Pb-Ag anode, where the MnO_2 layers are smooth, compact and firmly combined with the PbO_2 substrate(MacKinnon and Brannen, 1991; Yu and O'Keefe, 2002). Fig. 3a and b show the surface morphology of Pb/Pb-MnO_2 anode after the MnO_2 surface layer was washed away(how did you do this without disturbing the underlayers?). Compared with that formed in the absence of Mn^{2+}, the PbO_2 layer was a hard, compact and flat surface, which can be expected to protect the lead substrate. When the Mn^{2+} content was 1.0g/L(Fig. 3c) or 0.1g/L(Fig. 3d), even though there was no apparent MnO_2 layer formed on anode surface, the anodic layer also became more compact and flat compared with that without Mn^{2+}. In the electrolyte with 0.1g/L Mn^{2+}, the anodic surface was relatively ordered and the structure of the fresh surface was partially maintained. Such structure was found to present favorable oxygen evolution activity and high corrosion resistance(did EDS show any incorporation of Mn into this layer?).

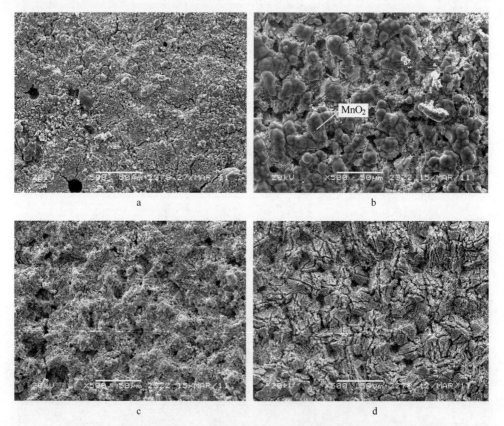

Fig. 3 Influence of Mn^{2+} in electrolyte on the surface morphology(×500) of Pb/Pb-MnO_2 anode after 72h galanostatic polarization
a—5.0g/L; b—3.0g/L; c—1.0g/L; d—0.1g/L

Corresponding to Fig. 3, XRD patterns were collected to analyze the phase composition of these anodic layers. Fig. 4 shows that the layers were mainly composed of PbO_2, PbO_2 normally exists in two forms: α-PbO_2 and β-PbO_2. β-PbO_2 is thought to be a favorable electrode material for oxygen evolution reaction due to its better conductivity, chemical stability, and electrocatalytic activity in H_2SO_4 system (Ruetschi, 1992). According to Fig. 4, in the electrolyte with the highest Mn^{2+}, the major phase of anodic layer was α-PbO_2. At the Mn^{2+} decreased, the proportion of β-PbO_2 gradually increased. When Mn^{2+} was 0.1g/L, the highest fraction of β-PbO_2 was present, this is considered to be the reason that the anodes in this cell presented the best oxygen evolution activity.

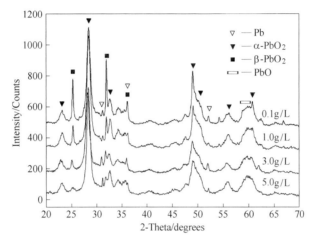

Fig. 4 XRD patterns of the anodic layers corresponding to Fig. 3

The influence of Mn^{2+} on the phase composition of anodic layer can be ascribed to the formation environment of α-PbO_2 and β-PbO_2. It has been reported that α-PbO_2 is prone to form in basic environment while β-PbO_2 in acid environment (Devilliers et al., 2004). When anode surface was covered by a layer of MnO_2, the diffusion of electrolyte was largely impeded and so the inner layer existed in a higher pH environment, which is favorable for the formation of α-PbO_2. At low Mn^{2+} concentrations, there was no MnO_2 layer formed, and the Pb directly contacted H_2SO_4 solution, which prompted the formation of β-PbO_2 (does this make sense? the anode produces O_2 and protons, so if the O_2 is evolved beneath the MnO_2 layer then the pH will decrease making it more acidic not more basic).

3.3 Corrosion resistance

According to above discussion, the experiments can be divided into two cases: MnO_2 layer formed on anode surface at high Mn^{2+} content and no MnO_2 layer formed at low Mn^{2+} concentration. Thus, in the following electrochemical investigations, detailed studies were carried out when the electrolyte contained 0, 0.1 and 3.0g/L Mn^{2+}.

Lead concentration in electrolyte was measured during the galvanostatic electrolysis, Fig. 5, to evaluate the corrosion resistance of Pb/Pb-MnO_2 anode in the electrolyte with different concentrations of Mn^{2+}. It can be seen that the addition of Mn^{2+} to the electrolyte significantly decreases the corrosion of the composite anode. In the electrolyte with 3.0g/L Mn^{2+}, the MnO_2 layer formed on the anode surface would effectively decrease the diffusion of H_2SO_4 solution, and the enhanced PbO_2 layer (Fig. 2b) was also capable of preventing acid reaching the lead sub-

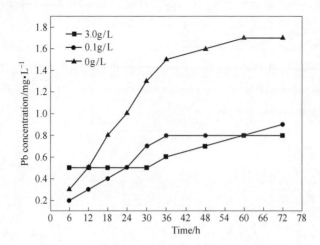

Fig. 5 Influence of Mn^{2+} content in the electrolyte on the corrosion resistance of Pb/Pb-MnO_2 anodes

strate. As a result, the oxidation and dissolution of the lead was largely diminished. When Mn^{2+} was 0.1g/L, the surface PbO_2 layer (Fig. 3d) was also enhanced compared to that formed without Mn^{2+} (Fig. 2b), so the anode presented satisfactory corrosion resistance, as well. Such corrosion behaviour is quite different from that of traditional Pb-Ag anode, for which only high-concentration Mn^{2+} can improve the corrosion resistance through formation of an anodic MnO_2 layer (Krupkowa et al., 1977; MacKinnon and Brannen, 1991).

3.4 Electrochemical behaviour of the stable anodic layer

3.4.1 Cyclic voltammetry

Fig. 6 shows the cyclic voltammograms obtained from the stable layers on the composite anodes. The voltammograms are overall characterized by one anodic peak (a) and anodic branch (b) during scanning in the positive direction and by one cathodic peak (d) and one cathodic branch (f) at the negative scanning. According to Czerwinski et al., (2000), Yamamoto et al., (1992) and Rashkov et al., (1996), the anodic peak (a) represents the oxidation of Pb to $PbSO_4$; the anodic branch (b) is due to the overlapping of the oxidation reaction of $PbSO_4$ to PbO_2 and oxygen evolution reaction; the cathodic peak (d) represents the reduction of PbO_2 to $PbSO_4$; and the cathodic branch (f) is both due to the reduction of $PbSO_4$ to Pb and hydrogen evolution reaction. In addition, on the analogous voltammograms, former literatures (Czerwinski et al., 2000; Rashkov et al., 1996) have reported the existence of a special anodic peak at 1.4~1.6V (vs. SCE) during the negative scan. This peak is normally ascribed to the oxidation reaction of metallic Pb and SO_4^{2-} that diffused through the anodic layer. However, in Fig. 6 this peak is absent for all the voltammograms, implying that the anodic layers were sufficiently compact to prevent the diffusion of electrolyte to the lead substrate. In addition, the voltammogram on the anodic layer formed in the electrolyte containing 3.0g/L Mn^{2+} was somewhat noisy, this was probably due to the porous and cracked MnO_2 layer. The anodic branches (b) show that the three anodes had different oxygen evolution characteristics. However, this is only a general and approximate trend because the anodic layers have been largely reduced in the low potential region.

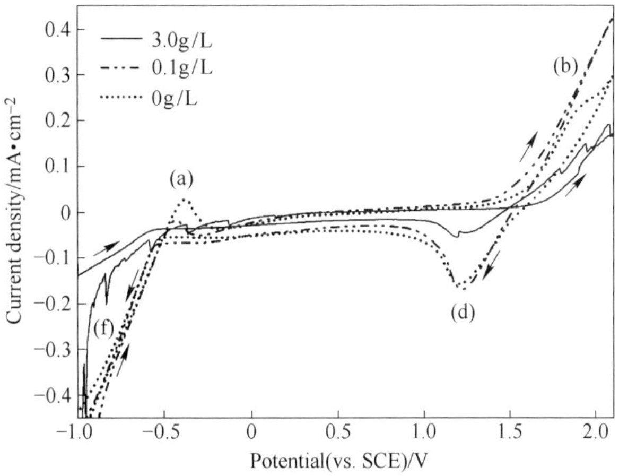

Fig. 6 Cyclic voltammograms of Pb/Pb-MnO$_2$ anodes on the stable anodic layers formed in the electrolyte containing different amount of Mn^{2+}

3.4.2 Tafel characterization

The reaction of oxygen evolution by decomposition of water in acid medium can be described with formula(1):

$$2H_2O \longrightarrow 4H^+ + O_2 + 4e^- \tag{1}$$

The following reaction steps have been proposed as the mechanism for oxygen evolution on active metal oxide electrodes(Shrivastava and Moats, 2009; Aromaa and Foesen, 2006; Da Silva et al., 2000):

$$S + H_2O \longrightarrow S\text{-}OH_{ads} + H^+ + e^- \tag{2}$$

$$S\text{-}OH_{ads} \longrightarrow S\text{-}O_{ads} + H^+ + e^- \tag{3a}$$

$$2S\text{-}OH_{ads} \longrightarrow S\text{-}O_{ads} + S + H_2O \tag{3b}$$

$$S\text{-}O_{ads} \longrightarrow S + 0.5O_2 \tag{4}$$

where S represent active sites on the oxide surface and S-OH$_{ads}$, S-O$_{ads}$ are two adsorbed intermediates. Step(3a) was considered to be obeyed by compact morphology and step(3b) by cracked morphology(reference this, it's important). Among three steps, the rate determining step for a specific electrode normally corresponds to its Tafel slope for oxygen evolution, since the Tafel slope to a large extent depends on the type, composition, and physical properties of the oxide electrodes(Aromaa and Foesen, 2006; Morimitsu et al., 2000). The same factors also govern the adsorption strength of the intermediates, which effectively determine the rate determining step(Martelli et al., 1994). For lead dioxide electrode, it can be generally summarized that when the Tafel slope, b, is ≥120mV/dec, step(2) is rate determining; when b is ~40mV/dec, step(3) was rate determining; if the Tafel slope was ~15mV/dec the rate determining step was step(4) (Franco et al., 2006; Da Silva et al., 2003; Da Silva et al., 2001).

In this paper, the corrected potential(E_c) is used to represent the real potential for oxygen evolution. The anodic polarization curve for Tafel analysis was corrected by formula(5):

$$E_c = E_{appl} - iR_s \tag{5}$$

where E_{appl} is the applied potential, i is the faradaic current and R_s is the electrolyte resistance. Fig. 7 shows the iR_s-corrected Tafel lines obtained for the stable anodic layers of Pb/Pb-MnO_2 anodes. Consistent with that reported by several authors in the studies of metal-oxide anodes (Shrivastava and Moats, 2009; Da Silva et al., 2000), these Tafel lines all presented two distinct linear segments in the low potential region and high potential region (double-slope behaviour). Some authors have attributed the increase of the slope value to the influence of partially evolved O_3 (Franco et al., 2006). As in the voltammetry, the Tafel line of the stable anodic layer formed in 3.0g/L Mn^{2+} electrolyte was to some extent noisy due to the porous and cracked MnO_2 layer.

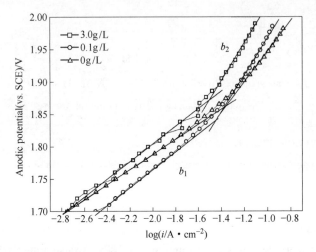

Fig. 7 Tafel lines of Pb/Pb-MnO_2 anodes on the stable anodic layers formed in the electrolyte containing different amount of Mn^{2+}

The slopes and intercepts of the three lines were separately analyzed and the values shown in Table 1. The over-potential (η_a) for specific current densities was calculated with Tafel formula (6):

$$\eta_a = a + b\log i \qquad (6)$$

where $a = a_0 - \varphi$ (a_0 represents the intercept value and φ represents the equilibrium electrode potential of the oxygen evolution (1)). Comparing formula (6) with the Butler-Volmer formula in high anodic polarization region ($\eta_a \geq 0.116V$), the exchange current density J_0 can be described as formula (7):

$$\log J_0 = -\frac{a}{b} \qquad (7)$$

In Table 1, the values of J_0 were extremely small and are often considered to be meaningless in evaluating the electrocatalytic activity of anode materials (reference them being meaningless). As a result, the over-potentials η were identified as the only important criterion (surely the slope is more important than overpotential as you know what the different slopes may mean). The oxygen evolution over-potential on the stable anodic layer formed in 3.0g/L Mn^{2+} electrolyte was visibly increased due to the low conductivity MnO_2 layer, while that in the electrolyte with 0.1g/L Mn^{2+} and absence of Mn^{2+} present similar values. Furthermore, with the addition of

Mn^{2+}, either 3.0g/L or 0.1g/L, the Tafel slopes were increased to some extent, probably due to the change in layer composition and structure. Besides, the slopes of the three anodes were all ≥ 120mV/dec in the whole over-potential region, revealing that the reaction was controlled by step(2). Under this case, step(3) and (4) are regarded as the fast-rate steps. Thus, it may be true that the addition of Mn^{2+} or the formation of MnO_2 layer can only influence the oxygen evolution activity of the Pb/Pb-MnO_2 anodes, but would not be expected to change the reaction mechanism.

Table 1　OER dynamic parameters of Pb/Pb-MnO_2 anode corresponding to the Tafel lines in Fig. 7.

Mn^{2+}/g·L^{-1}	b_1/mV·dec^{-1}	b_2/mV·dec^{-1}	η_a(i=0.05A/cm^2, vs. SCE)/V	J_0/A·cm^{-2}
0	116	265	0.625	1.02×10^{-5}
0.1	143	300	0.623	6.84×10^{-5}
3.0	144	294	0.667	26.8×10^{-5}

3.4.3　EIS characterizations

Since oxygen evolution on the stable anodic layers was kinetically controlled by reaction(2), and reactions(3) and (4) were fast-rate steps, we can simply assume that only the first intermediate, S-OH_{ads}, makes a significant contribution to the faradaic impedance (Hu et al., 2004). On the basis of Hu and Cao's expression (Hu et al., 2004; Cao, 1990), the faradaic impedance, Z_F, of such irreversible electrode reactions with one state variable in addition to electrode potential can be written as

$$Z_F = R_t + \frac{R_a}{1 + j\omega R_a C_a} \quad (8)$$

where R_t is the charge transfer resistance of the electrode reaction and always has a positive value; R_a and C_a are equivalent resistance and capacitance associated with the adsorption of intermediate, respectively. Based on Eq. (8), equivalent electrical circuit (EEC) shown in Fig. 8 can be predicted for the oxygen evolution reaction, in which C_{dl} represents the double-layer capacitance.

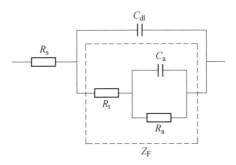

Fig. 8　Equivalent electrical circuit(EEC) based on reaction mechanism for OER on oxide anodes

EIS spectra of Pb/Pb-MnO_2 anodes on the stable anodic layers were measured and the results were showed in Fig. 8. Before recording each EIS spectrum, the electrode was maintained at its rest potential for 300s(check this is correct). It can be seen that the complex planes are mainly characterized by two generally separated semicircles. The large semicircle in the low frequency domain is indicative of adsorption processes and the small semicircle in the high frequency domain is due to the resistive/capacitive behaviour associated with the charge transfer processes (Amadelli et al., 2002; Ho et al., 1994). For the anodic layer formed in 3.0g/L Mn^{2+}, the EIS points were relatively noisy and the spectrum was very unsmooth, which were probably due to the MnO_2 layer, as well.

These experimental EIS data were mathematically simulated using above-proposed equivalent circuit. In the simulation, Constant Phase Elements (CPEs) were used instead of capacitors (C) to fit the experimental data. According to Da Silva et al. (2000) and Franco et al. (2006), the introduction of CPEs can give a good approximation of the surface roughness, physical non-uniformity or the non-uniform distribution of surface reaction sites. The impedance of a CPE can be written as:

$$Z_{CPE} = \frac{1}{Q(j\omega)^n} \tag{9}$$

where $j = (-1)^{1/2}$ and n represents the deviation from the ideal behaviour, n being 1 for the perfect capacitors. The simulation results, shown in Fig. 9, present a satisfactory agreement. The obtained parameters are presented in Table 2. As conjectured above, the adsorption resistance of the intermediate, R_a, plays a major part in the whole reaction resistance. The other two resistances, R_t and R_s, are relatively small. Moreover, the resistances for the stable anodic layer formed in 0.1g/L Mn^{2+} electrolyte are all slightly lower than those in the absence of Mn^{2+}, revealing the former layer is more favorable to the oxygen evolution reaction. However, the values for the layer obtained in 3.0g/L Mn^{2+} electrolyte were all largely increased due to the porous and cracked nature of the MnO_2 layer. The layer inhibited the mass transfer of electrolyte, leading to an increased solution resistance, and meanwhile the charge transfer and intermediates adsorption were also impeded.

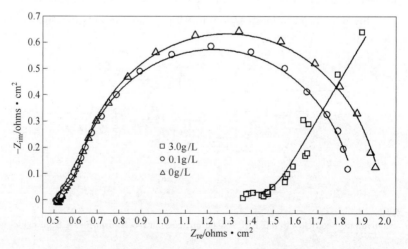

Fig. 9 EIS spectra of Pb/Pb-MnO_2 anodes on the stable anodic layers formed in the electrolyte containing different amount of Mn^{2+}

Table 2 EIS parameters of Pb/Pb-MnO$_2$ anode corresponding to the spectra in Fig. 9

Mn^{2+}/g·L^{-1}	R_s/Ω·cm^2	Q_{dl}/S-s^n·cm^{-2}	n	R_t/Ω·cm^2	Q_a/S-s^n·cm^{-2}	n	R_a/Ω·cm^2
0	0.527	0.184	0.89	0.183	0.274	0.97	1.281
0.1	0.522	0.208	0.87	0.175	0.347	0.95	1.175
3.0	1.415	0.148	0.59	0.231	0.278	0.70	—

4 Conclusions

The oxygen evolution activity and corrosion behaviour of the Pb/Pb-MnO$_2$ anode are strongly influenced by Mn^{2+} in the sulfuric acid electrolyte. At high Mn^{2+} concentrations a porous and cracked MnO$_2$ layer formed, despite being weakly adhered to the anode surface the layer improved the structure of the underlying PbO$_2$ layer, which was compact and flat. As a result, the corrosion resistance of the composite anode was improved. When the Mn^{2+} concentration was low, there was no apparent MnO$_2$ layer formed, but the PbO$_2$ layer was more ordered and less porous than in the absence of Mn^{2+}. Even at 0.1g/L Mn^{2+}, the corrosion of the composite anode was significantly diminished, despite there being no Mn in the surface layer. These influences are largely different from that on the traditional Pb-Ag anode.

Although Mn^{2+} in the electrolyte either improves (at low concentration) or impedes (at high concentration) the oxygen evolution from the composite anode, the reaction mechanism was unchanged. Regardless of the formation of MnO$_2$ layer or not, oxygen evolution on the composite anode was exclusively controlled by the formation and adsorption of the first intermediate, S-OH$_{ads}$, of which the adsorption resistance takes a dominant part in the whole reaction resistance.

References

[1] Amadelli R, Maldotti A, Moluiari A, Danilov F I, Velichenko A B, 2002. Influence of the electrode history and effects of the electrolyte composition and temperature on O$_2$ evolution at β-PbO$_2$ anodes in acid media. J. Electroanal. Chem. 534,1.

[2] Aromaa J, Evans J M, 2007. Electrowinning of metals. Encyclopedia of Electrochemistry 5,159-265.

[3] Aromaa J, Foesen O, 2006. Evaluation of the electrochemical activity of a Ti-RuO$_2$-TiO$_2$ permanent anode. Electrochem. Acta 51,6104-6110.

[4] Cachet C, Le Pape-rerolle C, Wiart R, 1999. Influence of Co^{2+} and Mn^{2+} ions on the kinetics of lead anodes for zinc electrowinning. J. Appl. Electrochem. 29,813-820.

[5] Cao C N, 1990. On the impedance plane displays for irreversible electrode reactions based on the stability conditions of the steady-state-I. One state variable besides electrode potential. Electrochim. Acta 35, 831-836.

[6] Cathro K J, 1991. Electrowinning of zinc at high current density. Chemistry in Australia 58,490.

[7] Chang Z W, Cuo Z C., Pan J Y, Xu R D, 2007. Studies on electrochemical properties of Al/Pb-WC-ZrO$_2$ composite electrode. Journal of Yunnan University 29,272-277. (In Chinese)

[8] Czerwinski A, Zelazowska M, Grden M, 2000. Electrochemical behaviour of lead in sulfuric acid solutions. J. Power Sources,85,49-55.

[9] Da Silva L M, Boodts J F C, De Faria L A, 2000. Oxygen evolution at RuO$_2$(x) +Co$_3$O$_4$(1−x) electrodes

from acid solution. Electrochim. Acta. 46,1369-1375.

[10] Da Silva L M,de Faria L A. Boodts J F C,2003. Electrochemical ozone production:influence of the supporting electrolyte on kinetics and current efficiency. Electrochim. Acta 48,699.

[11] Da Silva L M,de Faria L A. Boodts J F C. 2001. Green processes for environmental application. Electrochemical ozone production. Pure Appl. Chem. 73,1871.

[12] Devilliers D,Dinh Thi M T,Mahe E,Daruiac V,Lequeux N,2004. Electroanalytical investigations on electrodeposited lead dioxide. J. Electroanal. Chem. 573,227-239.

[13] Dobrev T,Valchanova I,Stefanov Y,Magaeva S,2009. Investigations of new anodic materials for electrowinning. Trans. Inst. Met Finish. 87,136-140.

[14] Felder A,Prengaman R D,2006. Lead alloys for permanent anodes in the nonfereous metals industry. JOM 58,28-31.

[15] Franco D V,Da Silva L M,Jardim W F,2006. Influence of the electrolyte composition on the kinetics of the oxygen evolution reaction and ozone production processes. J. Braz. Chem. Soc. 17,746-757.

[16] Ho J C K,Tremiliosi Filho G,Simpraga R,Conway B E,1994. Structure influence on electrocatalysis and adsorption of intermediates in the anodic O_2 evolution at dimorphic α-and β-PbO_2. J. Electroanal. Chem. 366,147.

[17] Hrussanova A,Mirkova L,Dobrev Ts,2002. Electrochemical properties of Pb-Sb,Pb-Ca-Sn and Pb-Co_3O_4 anodes in copper electrowinning. J. Appl. Electrochem. 32,505-512.

[18] Hrussanova A,Mirkova L,Dobrev Ts,Vasilev S,2004a. Influence of temperature and current density on oxygen overpotential and corrosion rate of Pb-Co_3O_4,Pb-Ca-Sn,and Pb-Sb anodes for copper electerwinning:Part Ⅰ. Hydrometallurgy 72,205-213.

[19] Hrussanova A,Mirkova L,Dobrev Ts,2004b. Influence of additives on the corrosion rate and oxygen overpotential of Pb-Co_3O_4,Pb-Ca-Sn,and Pb-Sb anodes for copper electerwinning:Part Ⅱ. Hydrometallurgy 72,215-224.

[20] Hu J M,Zhang J Q,Cao C N,2004. Oxygen evolution reaction on IrO_2-based DSA type electrodes: kinetics analysis of Tafel lines and EIS. Int. J. Hydrogen Energy 29,791-797.

[21] Ivanov I,2004. Increased current efficiency of zinc electrowinning in the presence of metal impurities by addition of organic inhibitors. Hydrometallurgy 72,73-78.

[22] Ivanov I,Stefanov Y,2002. Electroextraction of zinc from sulfate electrolytes containing antimony ions and hydroxyethylated-butyne-2-diol-1, 4: part 3. The influence of manganese ions and a divided cell. Hydrometallurgy 64,181-186.

[23] Kelsall G H,Guerra E,Li G,Bestetti M,2000. Effects of manganese(Ⅱ)and chloride ions in zinc electrowinning reactors. Proceedings-Electrochemical Society,2000-14(Electrochemistry in Mineral and Metal Processing V),pp. 350-361.

[24] Krupkowa D,Jurczyk H,Krezel K,Nosel J,Gluszczyszyn A. and Mendyka M,1977. Zinc electrowinning. Pl Patent No. 89821.

[25] Lai Y Q,Jiang L X,Li J,Zhong S P,Lv X J,Peng H J,Liu Y X,2010. A novel porous Pb-Ag anode for energy-saving in zinc electrowinning. Part Ⅰ Laboratory preparation and properties. Hydrometallurgy 102,73-80.

[26] Li Y,Jiang L X,Ni H F,Lv X J,Lai Y Q,Li,J,Liu,Y X,2010. Preparation and properties of Pb/Pb-MnO_2 composite anode for zinc electrowinning. The Chinese Journal of Nonferrous Metals 20,2357-2365. (In Chinese)

[27] Li Y,Jiang L X,Lv X J,Lai Y Q,Zhang H L,Li J,Liu Y X,2011. Oxygen evolution and corrosion behaviour of co-deposited Pb/Pb-MnO_2 composite anode for electrowinning of nonferrous metals. Hydrometallurgy 109,252-257.

[28] MacKinnon D J, Brannen J M, 1991. Effect of manganese, magnesium, sodium and potassium sulfates on zinc electrowinning from synthetic acid sulfate electrolytes. Hydrometallurgy 27, 99-111.

[29] Martelli G N, Ornelas R, Faita G, 1994. Deactivation mechanism of oxygen evolving anodes at high current density. Electrochim. Acta. 39, 1551.

[30] Morimitsu M, Otogawa R, Matsunaga M, 2000. Effects of cathodizing on the morphology and composition of IrO_2-Ta_2O_5/Ti anodes. Electrochim. Acta. 46:401.

[31] Muaiani M, Furlanetto F, Bertoncello R, 1999. Electrodeposited PbO_2 + RuO_2: a composite anode for oxygen evolution from sulphuric acid solution. J. Electroanal. Chem. 465, 160-167.

[32] Newnham R H, 1992. Corrosion rates of lead-based anodes for zinc electrowinning at high current densities. J. Appl. Electrochem. 22, 116-124.

[33] Rashkov St, Stefanov Y, Noncheva Z, 1996. Investigation of the processes of obtaining plastic treatment and electrochemical behaviour of lead alloys in their capacity as anodes during the electroextraction of zinc II. Electrochemical formation of phaes layers on binary Pb-Ag and Pb-Ca, and ternary Pb-Ag-Ca alloys in a sulphuric-acid electrolyte for zinc electro-extraction. Hydrometallurgy. 40, 319-334.

[34] Rerolle C, Wiart R, 1996. Kinetics of oxygen evolution on Pb and Pb-Ag anodes during zinc electrowinning. Electrochim. Acta. 41, 1063-1069.

[35] Ruetschi P, 1992. Influence of Crystal Structure and Interparticle Contact on the Capacity of PbO_2 Electrodes. J. Electrochem. Soc. 139, 1347.

[36] Saba A E, Elsherief A E, 2000. Continuous electrowinning of zinc. Hydrometallurgy. 54, 91-106.

[37] Schierle, T., Hein, K., 1993. Anode depolarization in the metal winning electrolysis. Erzmetall World of Metallurgy 46, 164-169.

[38] Schmachtel S, Toiminen M, Kontturi K, Forsén O, Barker M H, 2009. New oxygen evolution anodes for matal electrowinning: MnO_2 composite electrodes. J. Appl. Electrochem. 39, 1835-1848.

[39] Shrivastava P, Moats M S, 2009. Wet film application techniques and their effects on the stability of RuO_2-TiO_2 coated titanium anodes. J. Appl. Electrochem. 39, 107-116.

[40] Stefanov Y, Dobrev Ts, 2005. Developing and studying the properties of Pb-TiO_2 alloy coated lead composite anodes for zinc electrowinning. Trans. Inst. Met Finish. 83, 291-295.

[41] Verbaan B, Mullinder B, 1981. The simultaneous electrowinning of manganese dioxide and zinc from purified neutral zinc sulfate at high current efficiencies. Hydrometallurgy. 7, 339-352.

[42] Yamamoto Y, Fumino K, Ueda T, Nambu M, 1992. A potentiodynamic study of the lead electrode in sulphuric acid solution. Electrochim. Acta 37, 199-203.

[43] Yu P, O'Keefe T J, 2002. Evaluation of lead anode reactions in acid sulfate electrolytes II. Manganese reactions. J. Electrochem. Soc. 149, A558-A569.

[44] Zhang W, Cheng C Y, 2007. Manganese metallurgy review. Part III: Manganese control in zinc and copper electrolytes. Hydrometallurgy 89, 178-188.

Electrochemical Behaviors of Co-deposited Pb/Pb-MnO$_2$ Composite Anode in Sulfuric Acid Solution-Tafel and EIS Investigations[*]

Abstract The oxygen evolution kinetics and anodic layer properties of Pb/Pb-MnO$_2$ composite anode during the 72h galvanostatic electrolysis in H$_2$SO$_4$ solution were investigated with Quasi-stationary polarization(Tafel) and Electrochemical Impedance Spectroscopy(EIS) techniques. The results revealed that the anode activity and reaction kinetics changed a lot during the 72h of electrolysis that represents the period for anodic layer forming and stabilizing. At the very beginning of electrolysis, the composite anode exhibited very high oxygen evolution activity for the reaction was controlled by the transformation step of intermediates. Then, its oxygen evolution activity was largely diminished and the rate determination step(rds) became the formation and adsorption of first intermediate, S-OH$_{ads}$. In the prolonged electrolysis, the anodic potential gradually decreased and the final stable value was comparable to industrial Pb-Ag(1.0%) anode. On the stable anodic layer after 72h, the oxygen evolution reaction(OER) was still controlled by the formation and adsorption of intermediate, S-OH$_{ads}$, and the adsorption resistance took a dominant part in the whole impedance. Besides, compared with Pb-Ag and Pb anode, although their OERs at this stable state were all controlled by the intermediate adsorption process, the adsorption resistance of Pb/Pb-MnO$_2$ anode was much smaller than the other two due to, probably, the existence of MnO$_2$ particles and large amount of β-PbO$_2$ in the stable anodic layer.

Key words Pb/Pb-MnO$_2$ composite anode, electrochemical behaviors, Tafel, EIS

1 Introduction

Recently, the association of the conducting lead matrix and an active dispersed phase has been widely studied as a new type of composite anode for the oxygen evolution reaction(OER) in the electrowinning of nonferrous metals[1-9]. It has been found that this design effectively couples the high electrocatalytic activity of the dispersed phase to the high electrical conductivity and chemical stability of the matrix. As a result, the oxygen evolution activity[1-3] and corrosion resistance[4-6] of the lead matrix are significantly improved. In the field of zinc electrowinning, this kind of composite anode has been regarded as one of the most promising substitutes to the traditional Pb-Ag anode due to its favorable corrosion resistance and fairly low material and preparation cost comparing with the expensive silver and high-temperature casting cost of Pb-Ag anode, even though its anodic potential for OER is equivalent to or negligibly higher than that of Pb-Ag(1.0 wt.%) anode[5,9,10]. Currently, the studies on such composite anode are mainly focused on the direct evaluation of anodic performances, but involved little on the oxygen

---

[*] Copartner: Yanqing Lai, Yuan Li, Liangxing Jiang, Wang Xu, Xiaojun Lv, Jie Li. Reprinted from Journal of Electroanalytical Chemistry, 2012, 671: 16-23.

evolution kinetics and anodic layer properties.

Tafel and EIS techniques have been widely applied in the studies of oxygen evolution process on metal-oxide anodes. In general, Tafel analysis presents the information of exchange current density and over-potentials[11-13]. High exchange current density and low Tafel slope are considered to be beneficial to OER. Meanwhile, EIS investigations are of great use in characterizing the surface properties and oxygen evolution kinetics of such anode material[14-22]. Some literature pointed out that the response in the frequency domain of the OER can be simulated based on an intuitive but rationally chosen equivalent circuit[14,15]. Ji-ming et al[16] showed that the model deduced from the mathematical simulation of current-potential curve and EIS data can successfully interpret the double Tafel lines, and the mathematical expression of faradic impedance based on one state variable besides electrode potential can fit well the experimental EIS data. Towards the lead dioxide anode, several literatures have reported the relevant Tafel and EIS investigations dealing with its oxygen evolution kinetics in sulfuric acid solution [23-31]. Franco et al[23] showed that, on electrodeposited PbO_2 anode, the Tafel slope obtained from the impedance data and quasi-stationary polarization revealed a very good agreement, and thus supporting that the experimental findings are reliable from the kinetic point of view. Ho et al[24] conducted detailed Tafel and EIS experiments to discuss the importance of the electrochemical adsorption behaviors of chemisorbed intermediates involved in the electrocatalytic multistep mechanism of OER at electrodeposited α- and β-PbO_2 electrodes. However, when it comes to the studies of oxygen evolution kinetics at the Pb/PbO_2 electrode formed by steady-state polarization in the sulfuric acid solution (especially that formed on the electrodeposited lead-based anode), there was little literature concerning Tafel or EIS investigations [32].

In our former studies[10,33], we investigated the electrochemical performances of the co-deposited Pb/Pb-MnO_2 composite anode in sulfuric acid solution. The results indicate that, during the 72h galvanostatic electrolysis, the composite anode was prone to form a stationary and dense PbO_2 anodic layer, and thus presented favorable OER activity and corrosion resistance compared with Pb and industrial Pb-Ag(1.0 wt.%) anode. Accordingly, this paper conducts further Tafel and EIS investigations on the composite anode to study its oxygen evolution kinetics and anodic layer properties during the 72h galvanostatic electrolysis in H_2SO_4 solution. The same experiments are also carried out on cast Pb and industrial Pb-Ag(1.0 wt.%) alloy anode as comparison.

2 Experimental Details

Pb-MnO_2 composite coatings were electrochemically co-deposited on 1cm^2 cast pure lead substrate. The basic composition of the deposition bath was as follows: 0.30~0.40mol/dm^3 Pb(BF_4)$_2$; 0.35~0.45mol/dm^3 HBF_4; 0.20~0.25mol/dm^3 H_3BO_3; 0.5g/dm^3 gelatin; 0.2g/dm^3 diethanolamine(DEA) and 80~120g/dm^3 MnO_2(suspension). The experiments are conducted in a 250mL beaker, where a 4cm^2 pure lead electrode was used as the anode. Both of the electrodes were horizontally arranged to guarantee adequate particles to incorporate in the deposit. A sketch of the electrolytic cell and co-deposition mechanism are shown in Fig. 1. The plating temperature and agitation rate were respectively controlled at (35±0.5)℃ and (400±20)r/min using DF-101S constant temperature magnetic agitator. The cathodic constant current density was 40mA/cm² and the deposition time was 60min. All solutions were prepared with analytical reagents and double-distilled water, and the dispersed particles adopted are 2~7μm γ-MnO₂(CMD) of high purity(>95%, HUITONG Co., Ltd.).

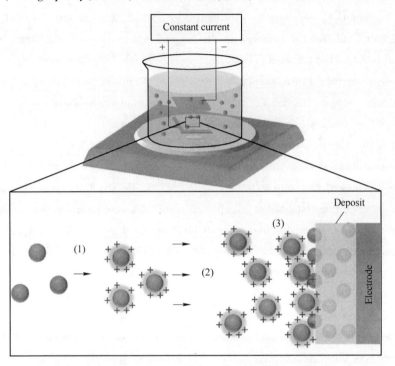

Fig. 1 Sketch of the electrodeposition cell and co-deposition mechanism: (1) dispersing and charging of particles; (2) charged particles transfer to the electrode surface; (3) particles co-deposit with Pb matrix after weak adsorption and strong adsorption

The electrochemical measurements for Pb/Pb-MnO₂ composite anodes, of which the mass fraction of MnO₂ is within 6%~8%, were performed in 1.63mol/dm³ H₂SO₄ solution at a constant temperature of 35℃. The standard three-electrode system was used. A platinum plate of 4cm² and the 217[#] double salt bridge saturated calomel electrode(SCE) were used as counter electrode and reference electrode, respectively. All potentials shown in the figures are against SCE. The constant current density of the galvanostatic electrolysis was 50mA/cm². The anode potentials were detected by a high ohmic imput($5×10^{12}Ω$) multimeter during the 72h electrolysis. When the pre-designed electrolysis time (0h, 0.5h, 24h and 72h) reached, Tafel or EIS measurements were immediately carried out on the EG&G Princeton applied research model 2273 potentiostat/galvanostat. Each Tafel or EIS measurement corresponded to an individual new anode. The quasi-stationary polarization curve for Tafel measurement was recorded from 1.7 to 2.0V with a scanning rate of 0.2mV/s. The frequency interval of EIS measurement was from 10^5 to 10^{-2}Hz, and the AC Amplitude is 5mV root mean squared (rms). In addition, the TTR-Ⅲ X-Ray Diffratometer (Rigaku Co., Japan) using CuKα (k = 1.54056Å) as radiation source was used to analyze the phase composition of anodic layers after 72h electrolysis. The scanning range was from 10° to 80°(2-theta) in steps of 0.02° and 0.6s/step.

3 Results and Discussion

3.1 Galvanostatic electrolysis

Fig. 2 shows the typical potential-time curves of Pb/Pb-MnO$_2$, Pb-Ag and Pb anodes during the 72h galvanostatic electrolysis. In the inset diagram, Pb/Pb-MnO$_2$ composite anode exhibited excellent electrocatalytic activity for OER at the very beginning of the electrolysis. Then the activity diminished largely due to the change of surface state[10]. In the prolonged electrolysis, the anodic potential gradually decreased and the stable potential after 72h was comparable to industrial Pb-Ag(1.0%) anode. The whole electrolysis process can be expressed by the following four critical time points: the very beginning when the composite anode exhibited high OER activity; the 0.5th hour when its activity largely diminished; the 24th hour in the transition stage and the 72th hour when the electrolysis reached a steady state. Therefore, the Tafel and EIS experiments were separately carried out at these time points. The anodic performances at the 72th hour were especially investigated in detail for a stable PbO$_2$ layer was thought to have been formed at this moment.

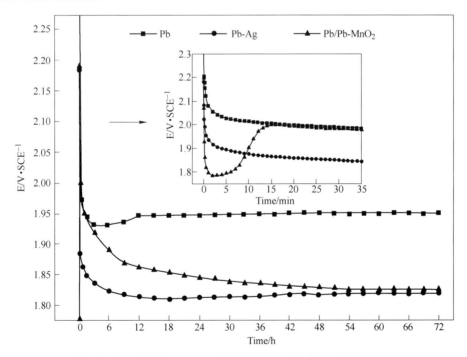

Fig. 2 Potential-time curves of Pb/Pb-MnO$_2$, Pb-Ag and Pb anodes during the galvanostatic electrolysis for 72h

3.2 Tafel characterization

The reaction of oxygen evolution by decomposition of water in acid medium can be described with formula(1):

$$2H_2O - 4e^- \longrightarrow 4H^+ + O_2 \quad (1)$$

The following reaction steps were generally proposed as the mechanism of OER on active metal oxides electrodes[11,34,35]:

$$S + H_2O \longrightarrow S\text{-}OH_{ads} + H^+ + e^- \quad (2)$$

$$S\text{-}OH_{ads} \longrightarrow S\text{-}O_{ads} + H^+ + e^- \quad (3a)$$

$$2S\text{-}OH_{ads} \longrightarrow S\text{-}O_{ads} + S + H_2O \quad (3b)$$

$$S\text{-}O_{ads} \longrightarrow S + 1/2 O_2 \quad (4)$$

where S stands for active sites on oxide surface and $S\text{-}OH_{ads}$, $S\text{-}O_{ads}$ are adsorption intermediates. Step(3a) was normally considered to be obeyed by compact morphology and step(3b) by cracked morphology [34,36]. Among the three steps, the reaction determination step (*rds*) for a specific electrode is normally thought to correspond to its Tafel slope of OER[23,27], since the Tafel slope to a large extent depends on the type, composition and physical properties of the oxide electrodes [37,38], and meanwhile, these factors also govern the adsorption strength of the intermediates, which in essence determines the *rds*[13]. For the lead dioxide electrode, one can generally summarized that when the Tafel coefficient *b* is equivalent to or higher than 120mV/dec, step(2)-the formation and adsorption of the first intermediate, $S\text{-}OH_{ads}$, was suggested to be the *rds*; when *b* is approximate to 40mV/dec, the *rds* was considered to be step(3)-the transformation of intermediates; and when 15mV/dec, the *rds* was regarded as step(4)-the generation of O_2[27-29].

In present paper, we employed the corrected potential (E_c) to present the real potential value of OER. The anodic polarization curve for Tafel analysis was corrected by formula(5):

$$E_c = E_{appl} - iR_s \quad (5)$$

where E_{appl} is the applied potential, i is faradaic current and R_s is the uncompensated electrolyte resistance [23,39]. Fig. 3 shows the iR_s-corrected Tafel lines of the Pb/Pb-MnO$_2$ composite anode at the very beginning, the 0.5th, 24th and 72th hour of the galvanostatic electrolysis. At the very beginning, the Tafel slope firstly presented a very small value (33mV/dec). According to above proposed mechanism, the OER was controlled by the intermediates transformation step (3) at this moment. This is probably because that on the fresh anode there was no any anodic layer formed, and the OER happened on these active MnO_2 particles, so the formation and adsorption of the first intermediate, $S\text{-}OH_{ads}$, was fairly easy, making the composite anode present favorable OER activity[40]. Then, with the electrolysis going on, a cracked and porous anodic layer mainly composed of $PbSO_4$ and PbO_2 was irreversibly formed, rendering the surface MnO_2 particles become slack or even drop into the electrolyte. Meanwhile, some other partially embedded MnO_2 would be covered by this layer[10,33]. As a result, the activity of the composite anode began to largely diminish and the Tafel line presented a negative slope. This turning rightly corresponds to the upturn of anodic potential at the very beginning of the potential-time curve in Fig. 2. Afterwards, the anodic layer basically formed, and OER occurred on both PbO_2 layer and residual MnO_2 particles, so the Tafel line presented a second positive slope (156mV/dec), indicating the OER had been controlled by the step(2).

At the 0.5th hour, the Tafel line presented relatively high potential intercept and large slope (150mV/dec), both of which are thought to be unfavorable to OER. The reaction was limited by

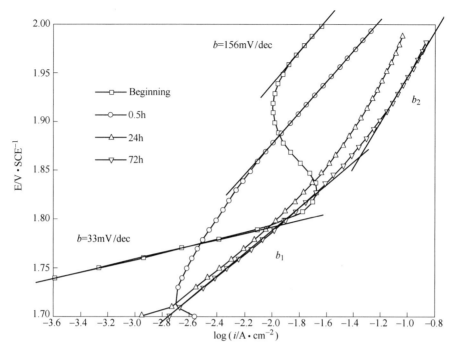

Fig. 3 Tafel lines of Pb/Pb-MnO$_2$ composite anodes at the very beginning, the 0.5th, 24th and 72th hour of the galvanostatic electrolysis

the formation and adsorption step(2) of intermediate. With the electrolysis prolonging, more and more PbO$_2$ gradually formed on the anode surface and the Tafel slope and intercept both decreased. After 72hours' electrolysis, it can be presumed that a permanent stable anodic layer has been formed[5,9]. The layer was mainly composed of PbO$_2$ and MnO$_2$, of which the mass fraction of MnO$_2$ was 1%~2%[10,33]. Consistent with that reported by several authors in the studies of metal-oxide anodes[11,12,23], the Tafel line on this stable anodic layer presented two distinct linear segments in the low and high potential regions(double-slope behavior). Some authors have attributed the increase of slope value in the high over-potential region to the influence of partially evolved O$_3$[23,27].

Fig. 4 shows the iR_s-corrected Tafel lines of Pb/Pb-MnO$_2$, Pb-Ag and Pb anodes on the stable anodic layers after 72h galvanostatic electrolysis. It is observable that all lines presented the double-slope behavior. Based on the aforesaid OER mechanism, these double slope values and the potential intercepts of the three lines were separately analyzed with the Origin software. Part of the results was displayed in Table 1. The over-potential(η_a) under specific current density was calculated with Tafel formula(6):

$$\eta_a = a + b\log|i| \tag{6}$$

where $a = a_0 - \varphi$ (a_0 represents the intercept value obtained with Origin and φ represents the equilibrium electrode potential of OER(2)). Comparing formula(6) with the Butler-Volmer formula in the high anodic polarization region($\eta_a \geqslant 0.116V$), the exchange current density J_0 can be described as formula(7):

$$\log J_0 = -\frac{a}{b} \tag{7}$$

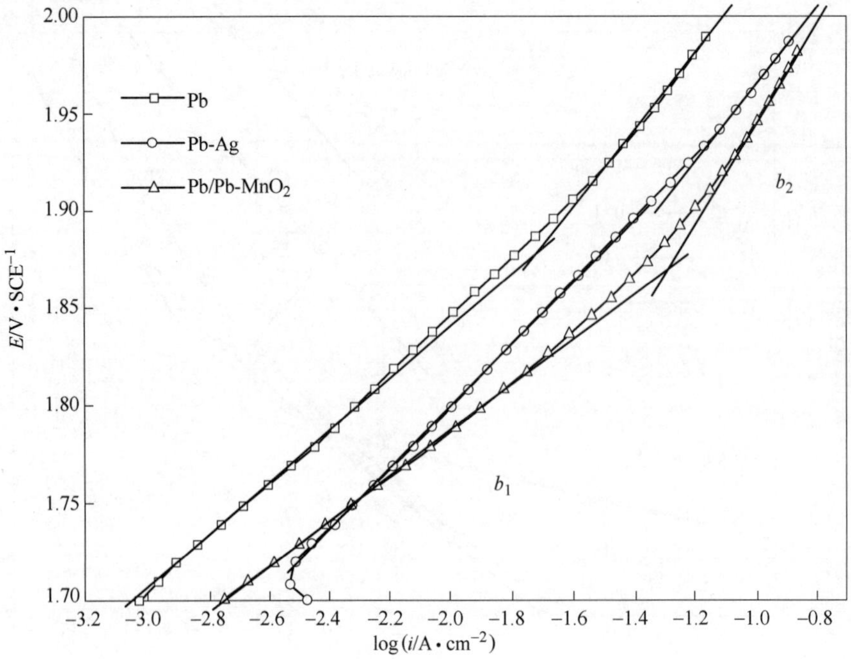

Fig. 4 Tafel lines of Pb/Pb-MnO$_2$, Pb-Ag and Pb anodes on the stable anodic layers formed after 72h galvanostatic electrolysis

Table 1 OER dynamic parameters of Pb/Pb-MnO$_2$, Pb-Ag and Pb anodes obtained from the Tafel lines in Fig. 3

	b_1/mV·dec^{-1}	b_2/mV·dec^{-1}	$\eta_a(i=0.05\text{A/cm}^2)$/V	J_0/A·cm^{-2}
Pb/Pb-MnO$_2$	116	265	0.655	2.93×10^{-7}
Pb-Ag	165	182	0.652	4.21×10^{-6}
Pb	147	198	0.700	5.93×10^{-7}

Generally, we can observe in Table 1 that the values of J_0 for OER were negligibly small and often considered to be meaningless in evaluating the electrocatalytic activity of anode materials. As a result, the over-potentials η were identified as one of the most important criterions. Among the three anodes, the Pb/Pb-MnO$_2$ anode showed a low over-potential comparable to Pb-Ag anode, indicating that its anodic layer was to some extent more preferable to OER. Furthermore, Pb/Pb-MnO$_2$ anode presents relatively lower Tafel slope b_1 comparing with Pb-Ag and Pb anodes. However, its b_2 value is obviously larger, so we conjecture that this composite anode is also more prone to the evolution of O_3 in the high over-potential region. Further, the slopes of the three anodes were all approximate or larger than 120mV/dec in the whole o-ver-potential regions, revealing that OER on the stable anodic layers was controlled by the step (2). Under this case, step (3) and (4) are regarded as fast-rate steps [22,23]. Thus, we can deduce that although the stable anodic layers of the three kinds of anodes presented quite different structure and electrocatalytic activity, the kinetic mechanism of OER would not be changed in essence.

3.3 EIS characterization

According to the mechanism proposed above, the OER was kinetically controlled by the step (2) during the 72h electrolysis except the very beginning, and step (3) and (4) were fast-rate processes. Therefore, we can simply assume that apart from the potential only the first intermediate, S-OH$_{ads}$, makes a significant contribution to the faradaic impedance of OER[16]. On the basis of the expressions deduced by some authors[14,16,41], the faradaic impedance, Z_F, of such irreversible electrode reactions with one state variable besides electrode potential can be written as:

$$Z_F = R_t + \frac{R_a}{1 + j\omega R_a C_a} \quad (8)$$

where R_t is the charge transfer resistance of the electrode reaction and always has a positive value; R_a and C_a are equivalent resistance and pseudo-capacitance associated with the adsorption of intermediate (S-OH$_{ads}$), respectively. Eq. (8) predicts an equivalent electrical circuit (EEC) for the OER, as shown in Fig. 5a[16,24], in which C_{dl} represents the double-layer capacitance, and R_s represents the uncompensated solution resistance. Generally, based on this circuit two capacitance loops will be expected to display in the impedance diagram.

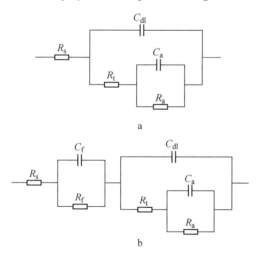

Fig. 5 Electrical equivalent circuit (EEC) used to simulate the impedance data for OER on active metal oxide anodes

In this paper, EIS patterns of the Pb/Pb-MnO$_2$ anode were respectively measured at the very beginning, the 0.5th, 24th and 72th hour of the galvanostatic electrolysis. The applied anodic potential is 1.85V. Before recording each EIS pattern, the electrode was conditioned for 300s under this potential. The same experiments have been also carried out on Pb-Ag and Pb anodes as comparison. The results were shown in Fig. 6 (the scatters). Inconsistent with above proposed EEC, most of the spectra showed only one capacitance arc in the whole frequency domain except that obtained on the stable anodic layer of Pb/Pb-MnO$_2$ anode. Such one-arc spectra have also been found in the study of IrO$_2$-based DSA[16]. By theoretical calculation, the authors pointed out that the traditional EEC, at this case, misunderstood the OER mechanism and led to overlapping the ($R_a C_a$) combination at the OER frequency domain. Accordingly, a modified EEC as shown in Fig. 5b was applied, where the additional combination ($R_f C_f$) takes into account the capacitive response of the oxide layer[14,16] and R_f, C_f represents the resistance and capacitance of the oxide layer, respectively.

In the simulation of those spectra with one capacitive arc, we also found that only when the layer impedance ($R_f C_f$) was added into the circuit (Fig. 5b), the simulated and experimental data can reach a good agreement. Such phenomenon may be explicable: It has been widely proposed that the formation of anodic layer on lead-based anode is mainly according to the following two processes: (a) Pb → PbSO$_4$ → β-PbO$_2$; (b) Pb → PbO → α-PbO$_2$ → β-PbO$_2$[47,48]. Therefore, during the formation process of anodic layers (beginning, the 0.5th and 24th hour, even the 72th hour for Pb anode), the considerable layer resistance can be found, which probably due to the non-conducting PbSO$_4$ and incompletely oxidized products.

Based on above consideration, EEC b and a in Fig. 5 were respectively used to simulate the impedance data with one arc and two arcs (actually only that on the stable anodic layers of Pb/Pb-MnO$_2$ anodes). In the simulation, Constant Phase Elements (CPEs) were used instead of capacitors (C_{dl} and C_a) to fit the experimental data. According to the literatures[11,14,23], the introduction of CPE can obtain a good approach to the surface roughness, physical nonuniformity or the nonuniform distribution of surface reaction site. The impedance of CPE can be written as:

$$Z_{CPE} = \frac{1}{Q(j\omega)^n} \quad (9)$$

Q representing the capacity parameter expressed in S/(cm$^2 \cdot$ s^n) and n accounting for the deviation from the ideal behaviour, n being 1 for the perfect capacitors. Moreover, the double layer capacitance (C_{dl}) is coupled with the uncompensated solution resistance (R_s) and the charge transfer resistance (R_t) according to the following equation [42]:

$$Q_{dl} = (C_{dl})^n [(R_s)^{-1} + (R_t)^{-1}]^{1-n} \quad (10)$$

Thus, using the model described by the Eq. (10), C_{dl} were calculated with Q_{dl} values obtained from the CNLS fit[13,23,43]. Since Eq. (10) was specially proposed for the double layer capacitance, we used Q_a parameters to describe the variation of adsorption pseudo-capacitance (C_a) during the electrolysis, which actually reflects the variation of intermediate coverage on anode surface[24,30,44]. Such consideration is based on that when n is close to 1, the Q parameters adequately describe the pseudo-capacitance[45,46].

The simulated EIS patterns and parameters were respectively displayed in Fig. 6 (the lines) and Table 2. As shown in Table 2, the electric transfer resistance R_t, at all cases, only takes little part in the whole resistance, while the film resistance R_f and/or adsorption resistance R_a dominantly determine the final resistance value. In general, The R_f and R_a values of the three anodes largely decreased during the electrolysis, indicating that the adsorption of intermediate become easier with the forming of PbO$_2$ layer. The film resistance of Pb-Ag anode in the latter electrolysis became very small and could be neglected due to the existence of highly conducting Ag. Impedance data for Pb/Pb-MnO$_2$ anode at the 72th hour of electrolysis was obtained using

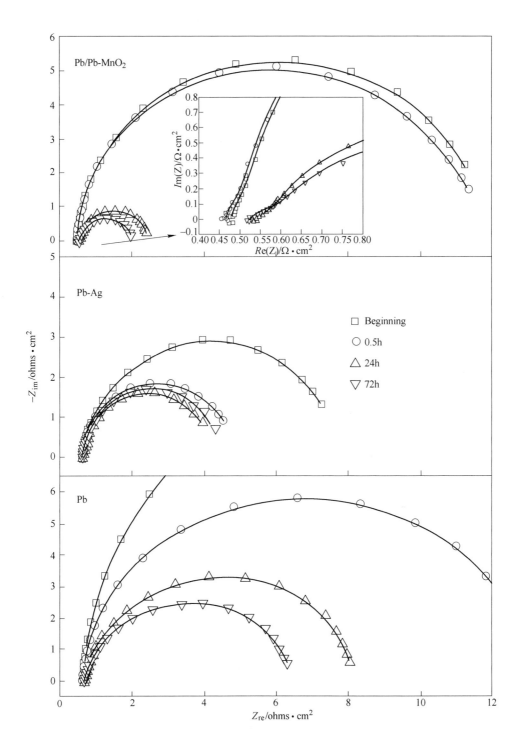

Fig. 6 EIS patterns of Pb/Pb-MnO$_2$, Pb-Ag and Pb anode obtained at the very beginning, the 0.5th, 24th and 72th hour of the galvanostatic electrolysis

EEC a in Fig. 5, the adsorption resistance at this moment presented the smallest value among the three anodes, which is thought to be beneficial to OER.

Table 2 Equivalent circuit parameters for Pb/Pb-MnO$_2$, Pb-Ag and Pb anodes according to the EIS spectra shown in Fig. 5

Anodes	Time/h	R_s /$\Omega \cdot cm^2$	$10^2 C_f$ /F·cm^{-2}	R_f /$\Omega \cdot cm^2$	$10^2 C_{dl}$ /F·cm^{-2}	R_t /$\Omega \cdot cm^2$	$10^2 Q_a$ /S·(s^n·cm^2)$^{-1}$	n	R_a /$\Omega \cdot cm^2$
Pb/Pb-MnO$_2$	0	0.478	0.652	10.15	0.575	0.163	0.780	0.81	1.319
	0.5	0.463	2.703	5.903	1.003	0.039	0.500	0.86	5.262
	24	0.538	0.125	0.005	7.247	0.122	16.44	0.98	1.777
	72	0.497	—	—	13.03	0.184	27.90	0.97	1.281
Pb-Ag	0	0.605	1.367	2.340	0.226	0.092	1.200	0.72	4.862
	0.5	0.611	—	—	0.987	0.285	0.824	0.80	4.175
	24	0.626	—	—	1.689	0.098	2.977	0.88	3.495
	72	0.621	—	—	3.366	0.572	1.145	1	3.199
Pb	0	0.614	0.618	12.87	0.58	0.257	0.750	0.88	11.98
	0.5	0.616	4.868	7.355	1.321	0.174	0.391	0.98	4.960
	24	0.665	9.288	3.953	3.825	0.043	2.126	0.90	3.500
	72	0.658	16.97	2.972	4.075	0.080	8.448	1	2.679

As proposed above, the double layer capacitance, C_{dl}, was calculated with Eq. (10). The generally increasing C_{dl} values of the three anodes in Table 2 reveal that with the prolonging of electrolysis the anodic surface became more prone to the adsorption of anions such as HSO_4^-/SO_4^{2-}[49]. The behavior of adsorption pseudo-capacitance during the electrolysis was evaluated by the Q_a values obtained in the CNLS fit. Since the pseudo-capacitance represents the coverage of intermediate on the anodic surface [24,44], its generally increasing trend during the electrolysis indicates that the intermediate, S-OH$_{ads}$, can be more easily adsorbed on the stable PbO$_2$ layer, as well. Besides, among the three anodes, Pb/Pb-MnO$_2$ anode presented obviously larger Q_a values in the latter electrolysis, this is probably due to that the thick, micro-porous and β-PbO$_2$-riched anodic layer [33]. Thus, summarizing above Tafel and EIS investigations, we can find that even though the OER on the stable anodic layers of the three anodes was all controlled by the intermediate adsorption step, the adsorption activity on the Pb/Pb-MnO$_2$ anode is much higher than the other two. This may be the essence that the three anodes present different electrocatalytic activity.

The experimental(scatters) and simulated(lines) Bode patterns for the three kinds of anodes on the stable anodic layer formed after 72h electrolysis(corresponding to Fig. 6) were shown in Fig. 7. The experimental and simulated data reach a very good agreement. The impedance values generally showed in the Bode diagram is consistent with that obtained by Nyquist patterns.

EIS patterns of Pb/Pb-MnO$_2$ anode at different anodic potentials were measured on the stable anodic layer after 72hours galvanostatic electrolysis. The results were shown in Fig. 8. Before recording each EIS pattern, the electrode was kept at the corresponding potential to condition for 300s. Under these cases, the complex planes are mainly characterized by two generally separated semicircles, distributed in the low and high frequency domains. The high-frequency

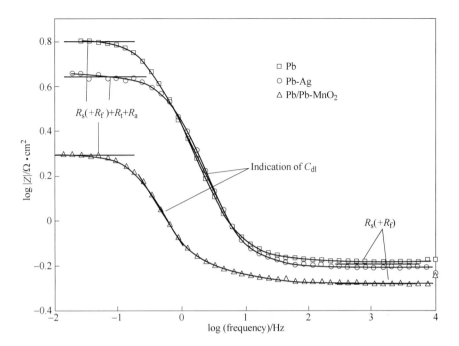

Fig. 7 Bode diagram of Pb/Pb-MnO$_2$, Pb-Ag and Pb anode on the stable anodic layers corresponding to Fig. 6 (Applied potential: 1.85 V)

semicircles are relatively small and show a weak dependence on potential, while the diameter of low-frequency semicircles shrinks obviously with the increase of potential.

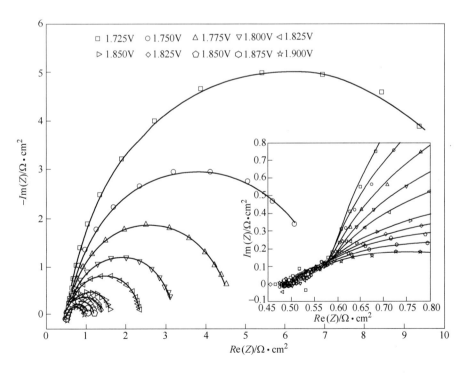

Fig. 8 EIS patterns at different anodic potentials obtained on the stable anodic layer of Pb/Pb-MnO$_2$ anode after 72h galvanostatic electrolysis

The EEC shown in Fig. 5a was used to simulate the experimental data. The results in Fig. 8 indicate the simulated data (the solid lines) present a good agreement with the experimental ones (the scatters). The EEC parameters at different anodic potentials are shown in Table 3. It is observable that R_s presented constant values independent of potential. The values of R_t are overall very small and negligibly affected by the potential, revealing that charge transfer process only plays a minor influence on the reaction impedance. Consistent with the Tafel analysis, the adsorption resistance, R_a, is relatively larger and presents a well-defined exponential decrease trend with the increase of potential, which also indicates that the OER is mainly governed by the formation and adsorption of intermediate [23,26]. The parameters C_{dl} and Q_a were also obtained by the above-proposed method. The results in Table 3 show that C_{dl}, in general, only presented a very weak decreasing trend with the increase of potential, while Q_a values were almost independent of potential [23].

Table 3 Equivalent circuit parameters for Pb/Pb-MnO$_2$ anode according to the EIS patterns shown in Fig. 6

Potential/V	$R_s/\Omega \cdot cm^2$	$10^2 C_{dl}/F \cdot cm^{-2}$	$R_t/\Omega \cdot cm^2$	$10^2 Q_a/S \cdot (s^n \cdot cm^2)^{-1}$	n	$R_a/\Omega \cdot cm^2$
1.725	0.490	14.01	0.187	39.39	0.97	10.77
1.750	0.499	13.70	0.160	39.62	0.96	6.368
1.775	0.499	19.47	0.268	29.41	1	3.879
1.800	0.494	13.38	0.187	35.47	0.97	2.528
1.825	0.499	16.42	0.223	32.02	0.98	1.653
1.850	0.497	16.04	0.188	34.82	0.96	0.945
1.875	0.495	12.70	0.162	36.93	0.96	0.750
1.900	0.489	13.03	0.150	38.50	0.94	0.576
1.925	0.491	10.40	0.111	39.03	0.88	0.475
1.950	0.486	9.21	0.088	38.77	0.94	0.386

Further, according to literatures [24,44,50], the Tafel behavior of metal-oxide anode can be approximately described by the experimental profile, E vs. $\log R_a^{-1}$, and the Tafel coefficient (b) was normally calculated from the linear segment according to the definition, $b \equiv (\partial E/\partial \log R_a^{-1})_T$. Fig. 9 shows the Tafel plots calculated basing on the R_a values in Table 3. It is clearly observed that two linear segments separately distribute in the low and high potential domains, which present a well correspondence to the Tafel curve obtained in quasi-stationary polarization (as shown in Fig. 4).

3.4 XRD analysis

As shown in Fig. 10, XRD detection was separately carried out on the stable anodic layers of Pb/Pb-MnO$_2$, Pb-Ag and Pb anodes formed after the 72h galvanostatic electrolysis. The results indicate that these anodic layers are mainly composed of α-PbO$_2$, β-PbO$_2$ and PbO. The anodic layer of Pb anode presents a high content of α-PbO$_2$ phase, and the existence of Pb peak reveals that the layer is relatively thin. The major phase of Pb-Ag anode is also α-PbO$_2$. However, compared with Pb anode, its anodic layer might be more compact and thicker for the peaks

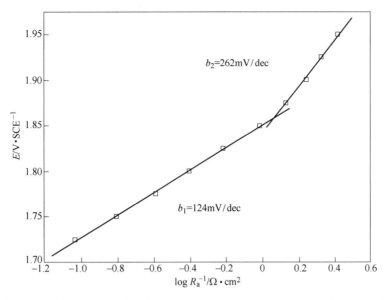

Fig. 9 Calculated Tafel plots for OER according to EIS data displayed in Table 3

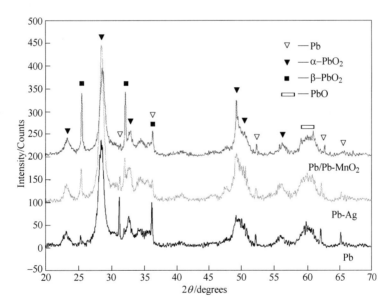

Fig. 10 XRD patterns of Pb/Pb-MnO$_2$, Pb-Ag and Pb anode obtained on the stable anodic layers after 72h galvanostatic electrolysis (The intensity of Pb-Ag and Pb/Pb-MnO$_2$ anode has been respectively added by 100 and 200 counts to make a comparable figure)

of Pb largely diminished. For Pb/Pb-MnO$_2$ anode, the peaks of α-PbO$_2$ were relatively lower and the peaks of Pb nearly disappeared, instead the peaks of β-PbO$_2$ obviously heightened. In the anodic layer formation processes proposed above, (a) is thought to happen on the acid surface layer while (b) is in the relatively basic inner layer[47,48]. Accordingly, the formation of β-PbO$_2$ on Pb/Pb-MnO$_2$ anode is probably ascribed to the following two reasons: (1) the MnO$_2$ particles on the deposit surface supplied large amount of active centers for the transformation of Pb to PbSO$_4$ and PbSO$_4$ to β-PbO$_2$ due to the free-state Mn^{3+} existing in the crystal lattice of γ-MnO$_2$, which prompted the reaction by accepting and transferring electrons[51,52]; (2) numerous dislocations and the largely refined crystal grains in the quasi-crystal lead deposit effectively improved the corrosion resistance of the composite anode by forming a homogenous and stable anodic layer[53], in which PbO$_2$ existed in the more stable β- form. Meanwhile, these crystal defects in the deposit also provide numerous active centers for the oxidations from Pb to PbSO$_4$ and further to PbO$_2$. Since β-PbO$_2$ is a preferable anode material due to its high conductivity and electrocatalytic activity, we consider that the stable anodic layer finally formed on Pb/Pb-MnO$_2$ anode is more beneficial to OER[24]. Moreover, such results can be also used to explain the fairly small R_f and R_a values on its stable anodic layer.

4 Conclusions

During the galvanostatic electrolysis, the OER kinetics and anodic layer properties of Pb/Pb-MnO$_2$ composite anode were largely changed due to the transformation of surface state. At the very beginning, the composite anode firstly presented fairly high oxygen evolution activity and the reaction was controlled by the transformation of intermediates due to the existence of superficial MnO$_2$. Then with the formation of PbSO$_4$-PbO$_2$ anodic layer, the amount of effective MnO$_2$ particles decreased and the electrode activity were largely diminished, rendering the formation and adsorption step of intermediate become the *rds*. With the prolonging of electrolysis, PbSO$_4$ in the anodic layer transformed to PbO$_2$ and the values of R_f and R_a gradually decreased. On the stable anodic layer formed after 72h, the film impedance nearly disappeared and the reaction resistance was dominantly composed by the adsorption resistance, indicating OER was still controlled by the intermediate formation and adsorption process.

Compared with Pb-Ag and Pb anodes, despite the *rds* of OER on the stable anodic layers was similarly controlled by the intermediate formation and adsorption step, the adsorption resistance and pseudo-capacitance of the Pb/Pb-MnO$_2$ anode was respectively much smaller and larger than the other two. Such phenomenon can be well interpreted by the XRD analysis on these stable anodic layers, which shows that Pb/Pb-MnO$_2$ anode presented obviously higher content of more conducting and active β-PbO$_2$ phase in the anodic layer. These results may reveal the kinetic essence of the different oxygen evolution activity presented by the three kinds of anodes.

References

[1] M. Muaiani, F. Furlanetto, R. Bertoncello. J. Electroanal. Chem. 465(1999)160-167.

[2] S. Schmachtel, M. Toiminen, K. Kontturi. J. Appl. Electrochem. 39(2009)1835-1848.

[3] Z. W. Chang, Z. C. Cuo, J. Y. Pan, R. D. Xu. Journal of Yunnan University 29(2007) 272-277. (In Chinese)

[4] S. Schmachtel, S. E. Pust, K. Kontturi. J. Appl. Electrochem. 40(2010)581-592.

[5] Y. Stefanov, Ts. Dobrev, Trans. Inst. Met. Finish. 83(2005)291-295.

[6] A. Hrussanova, L. Mirkova, Ts. Dobrev. J. Appl. Electrochem. 32(2002)505-512.

[7] A. Hrussanova, L. Mirkova, Ts. Dobrev, S. Vasilev. Hydrometallurgy. 72(2004)205-213.

[8] A. Hrussanova, L. Mirkova, Ts. Dobrev. Hydrometallurgy. 72(2004)215-224.

[9] T. Dobrev, I. Valchanova, Y. Stefanov, S. Magaeva. Trans. Inst. Met. Finish. 87(2009)136-140.

[10] Y. Li, L. X. Jiang, H. F. Ni et al. T Nonferr Metal Soc. 20(2010)2357-2365. (In Chinese)

[11] L. M. Da Silva, J. F. C. Boodts, L. A. De Faria. Electrochim. Acta 46(2001)1369-1375.
[12] P. Shrivastava, M. S. Moats. J. Appl. Electrochem. 39(2009)107-116.
[13] G. N. Martelli, R. Ornelas, G. Faita. Electrochim. Acta 39(1994)1551-1558.
[14] S. Palmas, A. M. Polcaro, F. Ferrara et al. J. Appl. Electrochem. 37(2008)907-913.
[15] L. M. Da Silva, K. C. Fernandes, L. A. de Faria. Electrochim. Acta 49(2004)4893-4906.
[16] J. M. Hu, J. Q. Zhang, C. N. Cao. Int. J. Hydrogen Energy 29(2004)791-797.
[17] V. A. Alves, L. A. Da Silva, J. F. C. Boodts. Electrochim. Acta 44(1998)1525-1534.
[18] Z. G. Ye, H. M. Meng, D. B. Sun. Electrochim. Acta 53(2008)5639-5643.
[19] R. N. Singh, M. R. Awasthi, A. S. K. Sinha. J. Solid State Electrochem. 13(2009)1613-1619.
[20] T. A. F. Lassali, J. F. C. Boodts, L. O. S. Bulhoes. Eletrochim. Acta 44(1999)4203-4216.
[21] S. Palmas, F. Ferrara, M. Mascia. Int. J. Hydrogen Energy 34(2009)1647-1654.
[22] J. M. Hu, H. M. Meng, J. Q. Zhang. Corros. Sci. 44(2002)1655-1668.
[23] D. V. Franco, L. M. Da Silva, W. F. Jardim et al. J. Braz. Chem. Soc. 17(2006)746-757.
[24] J. C. K. Ho, T. Filho, R. Simpraga, B. E. Conway. J. Electroanal. Chem. 366(1994)147-162.
[25] R. Amadelli, A. B. Velichenko. J. Serb. Chem. Soc. 66(2001)835-845.
[26] C. N. Ho, B. J. Hwang. J. Electroanal. Chem. 377(1994)177-190.
[27] L. M. Da Silva, L. A. de Faria, J. F. C. Boodts. Electrochim. Acta 48(2003)699-709.
[28] L. M. Da Silva, L. A. de Faria, J. F. C. Boodts. Pure Appl. Chem. 73(2001)1871-1884.
[29] E. R. Kötz, S. Stucki. J. Electroanal. Chem. 228(1987)407-415.
[30] L. Bai, D. A. Harrington, B. E. Conway. Electrochim. Acta 32(1987)1713-1731.
[31] L. Bai, B. E. Conway. J. Electrochem. Soc. 137(1990)3737-3747.
[32] D. Pavlov, B. Monahov. J. Electrochem. Soc. 143(1996)3616-3629.
[33] Y. Li, L. X. Jiang, X. J. Lv, Y. Q. Lai, H. L. Zhang, J. Li, Y. X. Liu. Hydrometallurgy 109(2011)252-257.
[34] S. Trasatti, G. Lodi. Amsterdam: Elsevier(1981)521-626.
[35] J. Aromaa, O. Foesen. Electrochim. Acta 51(2006)6104-6110.
[36] B. E. Conway, G. Ping, A. de Battisti, J. Mater. Chem. 1(1991)725-734.
[37] C. P. De Pauli, S. Trasatti. J. Electroanal. Chem. 538-539(2002)145-151.
[38] M. Morimitsu, R. Otogawa, M. Matsunaga. Electrochim. Acta 46(2000)401-406.
[39] D. M. Shub, M. F. Reznik. Elektrokhimiys 21(1985)855-861.
[40] S. Ardizzone, S. Trasatti. Adv. Colloid Interface Sci. 64(1996)173-251.
[41] C. N. Cao. Electrochim. Acta 35(1990)831-836.
[42] G. J. Brug, A. L. G. van den Eeden, M. Sluyters-Rehbach. J. Electroanal. Chem. 176(1984)275-295.
[43] B. Piela, P. K. Wrona, J. Electroanal. Chem. 388(1995)69-79.
[44] D. A. Harrington, B. E. Conway. Electrochim. Acta 32(1987)1703-1712.
[45] R. K. Shervedani, A. Lasia. J. Electrochem. Soc. 144(1997)2652-2657.
[46] C. Hitz, A. Lasia. J. Electroanal. Chem. 500(2001)213-222.
[47] A. Czerwinski, M. Zelaowska, M. Grden, et al. J. Power Sources. 85(2000)49-55.
[48] St. Rashkov, Y. Stefanov, Z. Noncheva, et al. Hydrometallurgy, 40(1996)319-334.
[49] R. Amadelli, A. Maldotti, A. Moluiari, F. I. Danilov. J. Electroanal. Chem. 534(2002)1-12.
[50] R. Amadelli, L. Armelao, A. B. Velichenko, N. Nikolenko. Electrochim. Acta 45(1999)713-720.
[51] S. B. Kanungo. J. Catalysis 58(1979)419-435.
[52] S. B. Kanungo, K. M. Parida, B. R. Sant. Electrochim. Acta 26(1981)1157-1167.
[53] C. S. Dai, Z. H. Wang, D. L. Wang, F. Ding, X. G. Hu. Journal of harbin institute of technology 37(2005) 530-532. (In Chinese)

五、电化学储能材料与器件

天然石墨中嵌/脱锂离子过程的研究[*]

摘 要 本文主要研究了锂离子在天然石墨中的嵌入/脱出过程，以及电流密度对锂离子嵌入石墨过程的影响。非现场 XRD 结果显示，石墨的层间距随锂的嵌入与脱出发生先膨胀后收缩复原的变化；SEM 结果显示，小电流密度下石墨表面的 SEI 膜较致密、均匀，而大电流密度下的则疏松、不均一；恒电流充放电测试结果显示，在相同的嵌锂电位下锂与微晶石墨和鳞片石墨所形成的层间化合物（Li-GICs）不完全相同；合理的充放电制度有利于小幅度改善材料的大电流充放电性能。

早期的二次锂电池由于采用金属锂作为负极材料，存在着"枝晶锂"、"死锂"等现象，严重制约了其应用[1]，后来开发出了人造石墨、热解炭等负极材料[2-5]，这些材料要么制备工艺复杂、制造成本高，要么存在着电位高（vs. Li/Li$^+$）、电压滞后及不可逆容量大等缺点。因此，人们把天然石墨作为负极材料的研究重点[6-10]。天然石墨作为负极材料时，具有很多的优点，但是它的大电流充放电性能差，为了能将天然石墨用作锂动力电池的负极材料，研究锂离子在其中的嵌入/脱嵌过程及影响该过程的因素具有非常重要的意义。本文作者主要采用了非现场 XRD 技术、恒电流充放电及 SEM 等测试方法研究了锂的嵌入/脱出对材料结构的影响，以及影响锂嵌入/脱出过程的因素。

1 实验

1.1 实验样品

所采用的样品为经过纯化后的 <38μm 的微晶石墨样 CZ 及鳞片石墨样 JD，样品的含碳量大于 99%，主要成分及结构参数见表 1。

表 1 石墨样品的主要组分

Sample \ Constituent	Fe	Si	Ca	Mg	Al	Structural parameters			
						d_{002}/nm	L_a/nm	L_c/nm	Graphitic degree/%
CZ	0.001	0.25	0.002	0.002	0.021	0.33641	27.042	17.166	88.4
JD	trace	0.01	trace	trace	0.001	0.33484	50.310	28.222	98.7

1.2 模拟电池的组装及恒电流充放电性能的测试

1.2.1 实验电池的组装

首先制备电极片：将所制备的粉末样品（<38μm）与 60% 的 PTFE 乳化液以 90:8（质量比）的比例在丙酮（AR，>99.5%）中充分混合，然后在对辊压片机（自制）

---

[*] 本文合作者：周向阳、李劼、胡国荣。原发表于《中国有色金属学报》，2002，12（6）：1257-1262。

上碾压成厚度约为 0.2mm 的炭膜，将该炭膜置于 170℃ 的真空干燥箱中干燥 24h 后取出，用冲孔器冲出所需直径大小的电极片。

然后组装模拟电池：将所得电极片作为三电极体系中的工作电极，锂片作为参比电极和电极。在充满干燥氩气的手套箱（自制）中以 Celgard 2400 微孔聚丙烯膜为隔膜，以 1mol/L LiPF$_6$/EC+DMC（1∶1）（德国默克公司配制）为电解液，组装成模拟实验电池。

1.2.2 恒电流充放电性能测试

将装配好的模拟电池接在 Land 电池测试系统中进行恒电流充放电。充、放电制度为：充电终止电压为 0V，放电终止电压为 2.8V，充、放电电流强度均为 15mA/g。

1.3 X-射线衍射（XRD）测试

XRD 测试在 D/max-rA 自动 X 射线衍射仪（日本理学电机）上进行，阳极靶材为 Cu，管电压 50V，电流 100mA。

用作非现场 XRD 测试的试样制备：

（1）测 XRD 图谱样品的制备：先将充放电到不同状态的模拟电池静置 10h 以上，然后在充满干燥氩气的手套箱中将电极片拆出后在 PC 溶剂中清洗，并用电吹风吹干，密封后去作 XRD 测试；

（2）测晶体尺寸 L_c 样品的制备：将烘干的在不同状态下的电极片捣碎，加入适量（20%质量分数）的 Si 粉作为内标，密封后去测试，整个样品的制备过程均在充满干燥氩气的手套箱中进行。

1.4 扫描电镜（SEM）测试

采用日本电子公司（JEOL）JSM5600 型扫描仪来观察所测试样的形貌。将以不同电流密度首次恒电流放电至 0V（vs. Li/Li$^+$）的模拟电池静置 10h 以上，然后在充满干燥氩气的手套箱中将待测电极片拆出，密封后立即作 SEM 测试。

2 结果与讨论

2.1 锂离子的嵌入对石墨结构参数的影响

图 1 为鳞片石墨样品 JD 充放电至不同状态下的 XRD 结果。图 1a~e 分别为石墨电极充放电反应进行前、经过第一次嵌锂反应、经过第一个循环、经过第三次嵌锂反应以及经过第三个循环后的 XRD 图谱。从图中可以看出，经过第一、第三个完整的循环后的曲线 c、e 与未经过充放电循环反应的曲线 a 基本相同，在 $2\theta \approx 26°$ 处有一个明显的石墨的衍射峰（002 峰），这说明石墨电极在电解液体系 LiPF$_6$/EC+DMC（1∶1）中的嵌脱锂循环性能较好；经过第一、第三次嵌锂后的曲线 b、d 非常相似，它们与其他三条曲线相比，嵌锂前以及每一个完整的循环完成后的在 $2\theta \approx 26°$ 处的衍射峰消失，在 2θ 为 24° 附近出现了 LiC$_6$ 与 LiC$_{12}$ 的新峰，这一方面说明锂的嵌入影响了石墨的结构参数，同时也说明了嵌锂后的石墨电极中有多阶 Li-GIC(Li-石墨层间化合物)共存的现象。

表 2 中列出了放电（嵌锂）前及放电到 0V 时 CZ 和 JD 的 XRD 测试结果，从表中可知，完全嵌锂后，石墨层间距 d_{002} 值增大，层片堆积高度也同样随锂的嵌入而增大，嵌锂后的（002）峰处的 2θ 角发生了偏移，这说明锂的嵌入导致了石墨结构的变化。

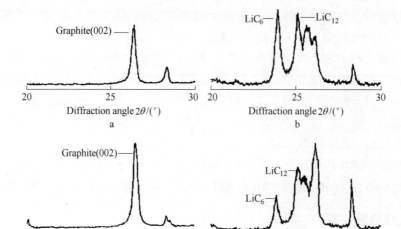

图 1 鳞片石墨样品 JD 在不同充、放电状态下的 XRD 测试结果

a—循环前；b—第一次嵌锂反应后；c—第一个循环后；d—第三次嵌锂反应后；e—第三个循环后

表 2 石墨样品嵌锂前后的结构参数

Structural Parameters	Before intercalation Lithium ion		After intercalation Lithium ion to 0V	
	CZ	JD	CZ	JD
d_{002}/nm	0.3364	0.3348	0.3645	0.3692
L_c/nm	17.166	28.222	19.326	32.179
$2\theta/(°)$	26.53	26.48	24.42	24.65

表 3 中示出了鳞片石墨样品 JD 充、放电到不同电位下的 d_{002} 值，不难发现随嵌锂电位的降低，层间距逐渐增大，层间距的增大会导致石墨 c 轴方向的膨胀，这种膨胀将会导致石墨层离，在低电位下的层离将会出现新的未成膜面，在新的未成膜面上的再次成膜将进一步扩大首次不可逆容量；在锂的脱嵌过程中，石墨的层间距又逐渐减小；从表中还可看出，JD 试样在第一个充、放循环中，当电位相同时，脱嵌时的 d_{002} 值就已大于嵌锂时的 d_{002} 值；但当一个充、放循环完成后，石墨的层间距又基本回到了充、放电前的值；这些说明，在本实验条件下，锂离子在 JD 试样中的嵌入/脱嵌过程尽管不完全可逆，但还是有相当高的可逆程度，这也就是选择石墨作为锂离子电池负极材料的基本依据。

图 2 为微晶样品 CZ 首次嵌锂后的 XRD 结果，同 JD 样品一样，锂的嵌入导致了石墨电极结构参数的变化，同时锂嵌入后的电极中不仅有 LiC_6 与 LiC_{12} 两种新物质共存，而且从 XRD 图上也可看出，另外还有些小的衍射峰，Song 等[11]认为这些小峰是由于嵌锂后的材料中还含有 LiC_{16} 和 LiC_{40} 等物相，那就是说嵌锂后的石墨中含有多种多阶

锂-碳层间化合物（Li-GICs）。由此也可知，锂嵌入石墨层间是一个复杂的过程。

表3　JD样在第一次充、放循环中、不同嵌/脱锂电位下的d_{002}值变化

Items	Voltage V(vs. Li/Li$^+$)			
	2.8	0.15	0.07	0
d_{002}(nm) in 1st intercalation Li$^+$ process		0.3446	0.3507	0.3692
d_{002}(nm) in 1st deintercalation Li$^+$ process		0.3350	0.3452	0.3511

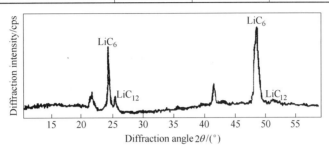

图2　微晶样品CZ首次嵌锂后的XRD结果

2.2　影响嵌锂过程因素的分析

由前面可知，石墨中锂的嵌入过程是复杂的，影响嵌锂过程的因素也非常多。电解质的组分、锂离子电池的使用温度等都是影响锂嵌入过程的重要因素[12-14]，因为Li嵌入石墨层间形成能可逆释放Li的Li$_x$C$_6$化合物的电位在0.2V(vs. Li/L$^+$) 左右[15]，为了排除不可逆反应等因素的干扰，取JD与CZ在第四个周期0~0.3V(vs. Li/Li$^+$) 电位范围内的嵌锂曲线（见图3）进行研究。从图中可知，在0.2V及0.1V(vs. Li/Li$^+$) 左右，微晶石墨中形成的新相是LiC$_{48}$与LiC$_{24}$，而鳞片石墨中的为LiC$_{84}$与LiC$_{30}$，并且当放电到0V时，两种材料中的嵌锂量也不相同，x分别为0.688，0.874，这些都说明微晶石墨与鳞片石墨的嵌锂过程是有差别的，这些差别与两种石墨在结构方面的差异紧密相关。

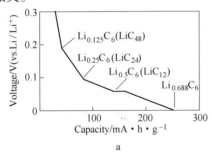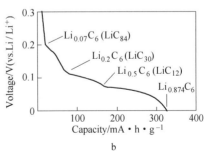

图3　样品CZ与JD的第四次嵌锂曲线
a—Sample：CZ；b—Sample：JD

电流密度对嵌锂过程也是一个重大的影响因素。图4为鳞片石墨JD在不同电流密度下的第四次放电曲线。图中曲线1的电流密度为15mA/g，从它可知，在此电流密度下锂的嵌入具有明显的分级行为，但随电流密度的加大，这种分级行为越来越模糊，表现为平台电压的降低、倾斜和缩短，相应造成嵌锂容量的下降，这种变化主要体现在图中的B到C与C到D的相变上。

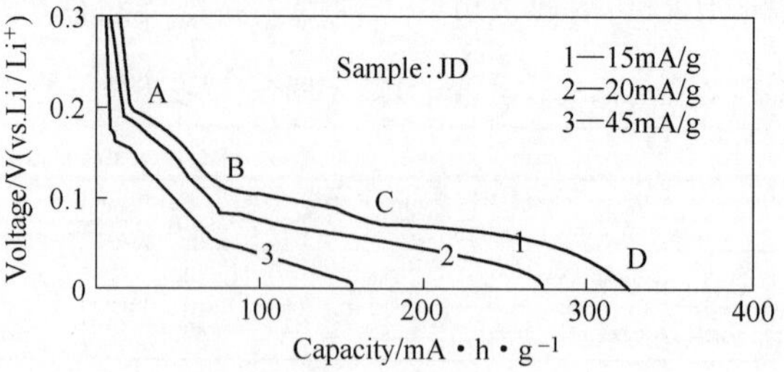

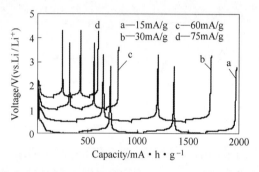

图 4　不同电流密度下的第四次嵌锂曲线

图 5 中示出了以 15mA/g、30mA/g、60mA/g 与 75mA/g 的电流密度恒电流充、放电时的前 3 次循环曲线，为了清楚区别 4 条曲线，大电流下的曲线分别向上平移了 0.5V、1.0V、1.5V；表 4 中列出了其相应的充、放电性能。从图 5 和表 2 可知，随电流密度的提高，石墨的前 3 次循环的嵌/脱锂容量都减小；并且第一个循环的效率，也随电流密度的增大而降低。

图 5　石墨阳极样品在不同充、放电流密度下的前 3 个循环的恒电流充、放电曲线
（b，c，d 曲线分别向上平移了 0.5V、1.0V、1.5V，以示区别）

图 6 中示出了充、放制度充放电 40 个循环的脱嵌容量变化。先以 15mA/g 的电流密度充、放 10 个循环，再以 150mA/g 充、放 10 个循环，最后将电流密度回复到 15mA/g 充、放 10 个循环。从图中可以清楚发现，随电流密度的升高，脱嵌容量依次减小，当电流密度为 150mA/g 时，脱嵌容量降到了 50mA·h/g 以下。随电流密度增大石墨的脱嵌容量减小的原因是由于碳电极是多孔电极，且它具有复杂的几何形貌，这就意味着在大电流时电极上的不同部位的压降各

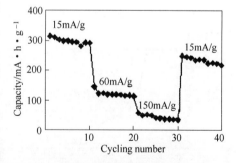

图 6　不同充、放电流密度对样品石墨阳极样品 JD 脱嵌容量的影响

不相同，致使不同部位形成的 SEI 膜不均匀，这可从不同电流密度下的电极表面的形成的 SEI 膜的扫描电镜图中看出（见图 7），不均匀致密的 SEI 膜不能有效阻挡溶剂化锂离子的嵌入及溶剂在电极表面的分解，因此首次充放电效率、脱嵌容量随电流密度增大而减小。在较大电流密度时，由于电极中各部位的压降的不同，电极中的有些部位可能到了 Li 沉积的电位，另外的部位中可能还没有 Li 的嵌入，这样材料中的总的嵌锂量便处于非常不饱和状态，材料的表观比容量（嵌锂量）便减小了。当电流密度降到 15mA/g 时，脱嵌容量有较大幅度的回升，但没有回升到前 10 次循环的水平，第 40 次循环的脱

嵌容量为 217mA·h/g，只有第 10 次循环容量的 75%；出现这种现象的主要原因是，当充、放电流密度又减小时，锂离子又开始慢慢嵌入在大电流下未嵌入锂的部位，自然，脱嵌容量又会回升，但由于大电流下所形成的 SEI 膜不能有效阻挡溶剂化锂离子的嵌入，使得石墨中崩溃的部位较多、能有效嵌/脱锂位置减少，这样最终便导致了嵌脱容量的减少。

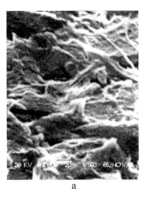 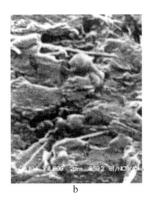

图 7　以不同电流密度第一次放电至 0V 时，
在石墨电极表面形成的钝化膜（SEI 膜）的 SEM 图
a—i_0 = 15mA/g；b—i_0 = 60mA/g

比较图 6 与表 4 中在 60mA/g 电流密度下的脱嵌容量可知，脱嵌容量在先小电流后大电流充、放电制度下的，大于直接大电流下的，这就是说合理的充放电制度有利于改善石墨的大电流性能，这主要是小电流密度下形成了较致密、均匀的 SEI 膜的缘故。但从图 6 知道，当电流密度升至 150mA/g 时，脱嵌容量急剧下降，也就是说，充放电制度对天然石墨的大电流充、放电性能不可能起到根本的改善作用。因此，为了能将天然石墨用作锂动力电池的负极，对它经过改性来较大地改善其大电流性能是必要的。

表 4　石墨阳极样品在不同电流密度下的前 3 次恒电流充放电性能

Item		Ⅰ			Ⅱ			Ⅲ		
		D_1	C_1	η_1	D_2	C_2	η_2	D_3	C_3	η_3
Current density /mA·g^{-1}	15	413	314	76	325	310	95	324	304	94
	30	386	270	70	268	252	94	264	248	94
	60	192	125	65	111	103	93	112	105	94
	75	160	96	60	87	82	94	82	78	95

D_i—Discharge (intercalation Li$^+$) capacity (mA·h/g)；Ⅰ、Ⅱ、Ⅲ—Cycle number；C_i—Charge (deintercalation Li$^+$) capacity (mA·h/g)；η_i—Cycle efficiency (%)。

3　结论

（1）锂离子在石墨中的嵌入/脱出会引起层间距先膨胀变大后又基本恢复原状的变化，这也是把石墨作为锂离子电池负极材料研究的基本依据之一；但是，这种膨胀会引起石墨的层离，从而导致材料的首次不可逆容量较大，且循环性能变差。

（2）非现场 XRD 结果显示，锂嵌入石墨的过程非常复杂，微晶石墨和鳞片石墨的嵌锂过程亦有区别，嵌锂后的石墨中存在 LiC_{12} 和 LiC_6 以及其他多阶 Li-GICs 共存的现象。

（3）电流密度是影响锂在石墨中嵌入/脱嵌的重要因素，随电流密度的增大，材料的比容量、循环效率降低；合理的充放电制度对材料的大电流性能有较小幅度的改善，但要取得很大程度的改善，还必须对天然石墨进行进一步的改性研究。

参 考 文 献

［1］马树华. 作为锂二次电池的炭材料［J］. Carbon Techniques（炭素技术），1995，3：19.

［2］D. Guerard, A. Herold. Intercalation of Lithium into Graphite and Other Carbons［J］. Carbon.，1975，13：337-342.

［3］T. Zheng, Y. Liu, Fuller E W, et al. Lithium Insertion in High Capacity Carbonaceous Materials［J］. J. Electrochem. Soc.，1995，142（8）：2581-2590.

［4］M. J. Mattews, M. S. Dresselhaus, et al. Characterization of Polyparaphenylene（PPP）-Based Carbons［J］. J. Mater. Res.，1996，11：3099-3106.

［5］K. Sato, M. Noguchi, A. Demachi, et al. A Mechanism of Lithium Storage in Disordered Carbons［J］. Science，1994，264：556-558.

［6］董建，周伟，刘旋，等. 微晶石墨作为阳极材料对二次锂离子电池电化学性能的影响［J］. 炭素技术，1999，105：1-6.

［7］Ohzuku T, Iuakoshi Y, Sawai K. Formation of lithium-graphite intercalation compounds in nonaqueous electrolytes and their application as a negative electrode for a lithium ion.（shuttlecock）cell［J］. J. Electrochem. Soc.，1993，140（9）：2490-2498.

［8］Nakamura H, Komastsu H, Yoshio M, Suppression of electrochemical decomposition of propylene carbonate at graphite anode in lithium-ion cells［J］. J. Power Sources，1996，62：219-227.

［9］Besenhard J O, Wagner M W, Winter M. Inorganic film-forming electrolyte additives improving the cycling behavior of metallic lithium electrodes and the self-discharge of carbon lithium electrodes［J］ J. Power sources，1993，43-44：413-420.

［10］Shu Z X, McMillan R S, Murrary J. Electrochemical intercalation of lithium into graphite［J］. J Electrochem. Soc.，1993，140（4）：922-927.

［11］X. Y. Song, K. Kinoshita. Microstructural Characterization of Lithiated Graphite［J］. J. Electrochem. Soc.，1996，143（6）：L120-L122.

［12］Barsonkov, E. Jong H K, Chul oh Yooh et al. Effect of Low-Temperature Conditions on Passive Layer Growth on Li Intercalation Materials［J］. J. Electrochem. Soc.，1998，145（8）：2711-2717.

［13］M. N. Richard, J. R. Dahn. Accelerating Rate Calorimetry Study on the Thermal Stability of Lithium Intercalated Graphite in Electrolyte［J］. J. Electrochem. Soc.，1996，149（6）：2068-2077.

［14］M. B. Smart, Ratnakumar R, Surampudis, et al. Irreversible Capacities of Graphite in Low-Temperature Electrolytes for Lithium-Ion Batteries［J］. J. Electrochem. Soc.，1999，146（11）：3963-3969.

［15］M. Winter, Novak P, Monnier A. Graphites for Lithium-Ion Cells：The Correlation of the First-Cycle Charge Loss with the Brunauer-Emmett-Teller Surface Area［J］. J. Electrochem. Soc.，1998，145（2）：428-437.

Effect of Cooling Modes on Microstructure and Electrochemical Performance of LiFePO$_4$*

Abstract LiFePO$_4$ was prepared by heating the pre-decomposed precursor mixtures sealed in vacuum quartz-tube. Three kinds of cooling modes including nature cooling, air quenching, and water quenching were applied to comparing the effects of cooling modes on the microstructure and electrochemical characteristics of the material. The results indicate that the water quenching mode can control overgrowth of the grain size of final product and improve its electrochemical performance compared with nature cooling mode and air quenching mode. The sample synthesized by using water quenching mode is of the highest reversible discharge specific capacity and the best cyclic electrochemical performance, demonstrating the first discharge capacity of 138.1mA · h/g at 0.1C rate and the total loss of capacity of 3.11% after 20 cycles.

1 Introduction

Lithium-ion batteries with the characteristics of high voltage, long cycling life, high energy density, non-memory effect and pollution-free, introduced in the early 1990s, are widely applied in the field of personal electronic devices, such as cell phones, laptops and digital cameras[1]. Furthermore, the latent crisis of energy and environment requires the batteries with the basic requirements of good thermal stability, low raw materials cost and environmental friendliness for making large-size cells used in power tools, electric bikes (E-bikes) and even hybrid electric vehicle (HEVs) or electric vehicle (Evs). Among the cathode materials, the key parts in the Li-ion batteries, LiFePO$_4$ cathode material with a structure of olivine-type might meet the requirements above[2-6]. For LiFePO^4, the key barriers limiting its application are extremely low electronic conductivity and low ion diffusive rate[2]. Although the problems have been solved to some extent by coating carbon[7-9] or suitable preparation procedures to minimize the particle of the sample[10-12] or doping supervalent metals ions to Li$^+$ sites[13,14] or Fe^{2+} sites[15,16], the controversies of which is the nature in improving the electrochemical performance still exist[17,18]. Chung et al[13] believed that electron conductivity can be improved by crystal lattice doping to decrease the forbidden band width[13]. Ravet et al[17] and Subramanya et al[18] found that the electronic conductivity can be improved by carbon or residual carbon in final products or some compounds such as Fe$_2$P etc with high electronic conductivity. Although the method of coating carbon is the most effective, it will not only lower the content of active material in samples but also decrease their tap density greatly. Meanwhile, the routes by doping supervalent metals ions cannot effectively avoid the overgrowth of single crystal in the calcining process, which makes

---

* Copartner: Hu Guorong, Gao Xuguang, Peng Zhongdong, Tan Xianyan, Du Ke, Deng Xinrong. Reprinted from J. Cent. South Univ. Technol. ,2007,14(5):647-650.

the material with poor electrochemical performance. According to the mechanism of diffusion limit in this material, only the sample with uniform and fine particle size can obtain a higher availability ratio of active material and decrease electrode polarization. It is believed that choosing of cooling mode is also very important in controlling the crystallization process and avoiding the overgrowth of the final product. In this study, the cooling processes, including nature cooling(NC), air quenching(AQ), and water quenching(WQ) and their effects on the microstructure and electrochemical behaviors of pure $LiFePO_4$ were investigated.

2 Experimental

$LiFePO_4$ was prepared by the solid-state reaction using stoichiometric Li_2CO_3 (99.95%), $FeC_2O_4 \cdot 2H_2O$ (99%) and $NH_4H_2PO_4$ (98%) as raw materials. The synthesis was performed in four consecutive steps. The starting materials were first fired at 400℃ for 6h in flowing ultra-pure Ar_2 atmosphere to pre-decompose the oxalate and phosphate. After cooling down, the pre-decomposed precursor mixtures were pressed into pellets (d20mm×10mm) at a load of 10kN and transferred to a quartz-tube and sealed in vacuum. Then the vacuum-tight samples were finally calcined at 650℃ for 8h in muffle furnace. Finally, the samples treated with different cooling modes were available by taking the quartz-tube out of the furnace and putting them in air or in water or in the furnace with cutting off the power at the calcining temperature. The final products were obtained by breaking the quartz-tube into pieces and separating the sample with the fragile quartz piece after the temperature cooled down.

The process of sealing samples was as follows. The pellets of predecomposed mixture were transferred into a piece of quartz tube (d20mm×150mm) with one edge sealed by oxyacetylene welding flame. The air in the quartz-tube was expelled by flowing ultra-pure argon for 30min, then the other edge was sealed by oxyacetylene welding flame while the vacuum pump was used to pump the ultra-pure argon. The schematic diagram of vacuum-tight sample is shown in Fig. 1.

The phases and cell parameters of all samples were determined by X-ray diffraction (XRD, D/max-r A type Cu $K_α$, 40kV, 300mA, 10°~70°, Japan).

Scanning electron microscope (SEM) was used to examine the microstructures of the cathode material (SEM, KYKY 2800, Japan).

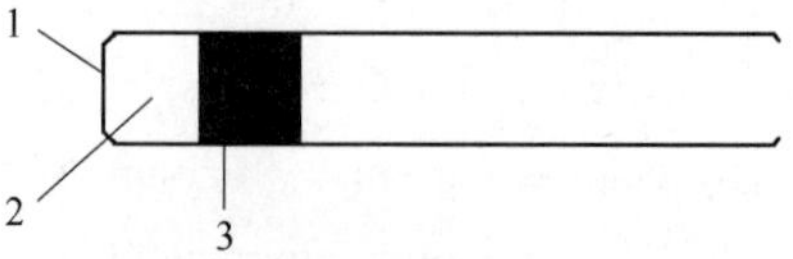

Fig. 1 Schematic diagram of reactive system
1—Quartz-tube; 2—Vacuum state;
3—Pre-decomposed mixture

The electrochemical characterization of the cathode powders was evaluated at 23℃. $LiFePO_4$ electrodes for electrochemical testing were prepared by slurrying $LiFePO_4$ powder (80%) with 10% acetylene black(AB) and 10% polyvinylidene difluoride(PVDF, in N-methyl pyrolidinone(NMP)), and then casting the mixture onto an aluminum foil. After being vacuum dried at 120℃ for 8h, the electrode disks with diameter of 12mm were punched and weighed. The cathodes were incorporated into laboratory-scale 2025 coin-type cells. A microporous polypropylene film (Celgard2400) was used as a separator and 1mol/L $LiPF_6$ solution with the volume ratio of EC to DEC of 1:1 was used as electrolyte solution. Lithium metal foil was used as counter electrode. All cells were assembled inside a glove box filled with ultra-pure argon. Charge/discharge characteristics of the active material were recorded at 0.1C(15mA/g) rate over a voltage range of 2.5~4.1V using a battery test system (LAND CT2001A).

An EG&G PAR potentiostat/galvanostat model 273A, operated by model 270 software EG&G was used to scan the potential at 0.5mV/s by using the powder microelectrode assembled in argon in dry box as the working electrode, lithium metal foils as the counter electrode and 1mol/L $LiPF_6$/EC+DMC (1:1, in volume ratio) as the electrolyte solution. The potential window was from 2.5 to 4.2V.

3 Results and Discussion

Fig.2 displays the powder X-ray diffraction patterns of $LiFePO_4$ sintered at 650℃ for 8h, followed by three different cooling processes at 650℃. All the patterns fit the olivine structure (PDF #83-2092) well with the relatively high intensity ratio of (I_{311}/I_{211} > 1.2) (Table 1). This result indicates that the olivine phase $LiFePO_4$ can be obtained by heating the pre-decomposed precursor mixtures sealed in vacuum quartz-tube under various cooling conditions. Besides, compared with the diffraction peak position of powder under NC, those of other samples shift to the left observed from the part-zooming patterns, which shows the slightly increase of unit cell with increasing cooling speed.

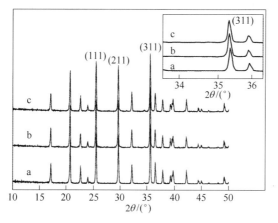

Fig.2 XRD patterns for samples prepared under various cooling conditions
a—NC; b—AQ; c—WQ

Table 1 XRD data for various $LiFePO_4$ phases

Sample	Lattice parameter				$X_{s,311}$/nm	d_{311}/nm	I_{311}/I_{211}
	a/nm	b/nm	c/nm	V/nm³			
PDF#83-2092	1.0334	0.6010	0.4693	0.2915	—	0.2521	1.2005
NC	1.0328	0.6001	0.4697	0.2914	71.7	0.2521	1.2232
AQ	1.0337	0.6008	0.4699	0.2918	71.1	0.2524	1.2248
WQ	1.0338	0.6010	0.4702	0.2922	66.0	0.2525	1.2356

$X_{s,311}$ and d_{311} stand for the grain size of the sample on <311> crystal plane and the inter-planar distance of <311> crystal plane, respectively.

The XRD data in Table 1 can also reflect the same information. The grain size of sample under WQ at (311) crystal plane is 66.0nm, which is smaller than those of the sample under NC (71.7nm) and sample under AQ(71.1nm). The sample WQ has larger peak intensity ratio(I_{311}/I_{211}) of 1.2356. The lattice parameters are $a = 1.0334$nm, $b = 0.6012$nm, $c = 0.4702$nm and $V = 0.2922$nm^3, which are larger than the corresponding parameters of samples under NC and AQ.

The SEM images of LiFePO$_4$ prepared in three different cooling ways are shown in Fig. 3. In Fig. 3a, there is many sub-micrometer grains aggregated to regular particles within a diameter range of 1~4μm, in which the grains have no obvious interface. More fine grains are observed in Fig. 3b and the particle size range is 0.5~2.0μm with irregular and rough shape. The particle distribution of sample under WQ is the most uniform. Meanwhile the particle becomes round and smoother, with a diameter down to 0.2~1.0μm. In addition, with the increase in particle size and uneven particle distribution, the electrochemical activity and the polarization of the larger particles will decrease greatly. The SEM micrograph of the sample synthesized by a water quenching method shows that this method is beneficial to improving the electrochemical performance and diminish the polarization when the material prepared under WQ is used as an electrode for lithium ion battery. So, choosing more rapid cooling speed and shortening cooling time can effectively control the overgrowth of pure LiFePO$_4$.

Fig. 3 SEM images of LiFePO$_4$ prepared under different conditions
a—NC; b—AQ; c—WQ

Fig. 4a shows the first charge-discharge cyclic curves for LiFePO$_4$ with three different falling temperature method. The charge-discharge current density is 0.1C rate (15mA/g) with cut-off voltage of 2.5~4.1V at room temperature.

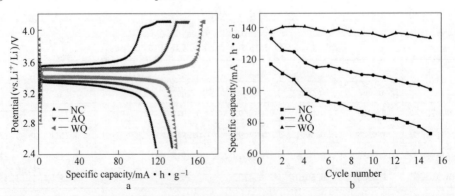

Fig. 4 First-cycle voltammograms (a) and electrochemical cycling stability (b) for LiFePO$_4$ cycled at 0.1C rate

The sample prepared by water quenching(sample WQ) has the highest discharge specific capacity of 138.1mA·h/g, while the samples NC and AQ display the reversible specific capacity of 117.3 and 133.8mA·h/g, respectively, and the separate distance between charge plateau and discharge plateau of WQ is also the shortest, which means the lowest electrode polarization. The cycling stability of samples NC, AQ and WQ is shown in Fig. 4b. The discharge capacity losses of samples NC, AQ and WQ are 38.2%, 24.4% and only 3.11%, respectively after 15 times cycles. The sample WQ processes the highest discharge capacity and the most excellent cycling performance.

Fig. 5 shows the cyclic voltammogram curves of samples NC, AQ and WQ in the first cycle at the scan rate of 0.5mV/s between 2.5 and 4.2V. All curves have a pair of peaks, consisting of an anodic and a cathodic peak, observed around 3.45V(vs Li$^+$/Li), corresponding to the two-phase charge-discharge reaction of the Fe^{2+}/Fe^{3+} redox couple. The oxidation and reduction peak positions for samples NC, AQ and WQ occurred at 4.09 and 2.97, 4.01 and 2.92, 3.78 and 3.18V(vs Li$^+$/Li), respectively. The sample WQ has the narrowest peak separation, indicating excellent kinetic reversibility of the two-phase transformation.

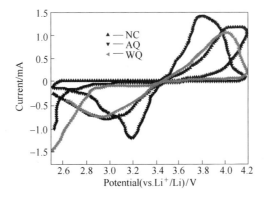

Fig. 5 CVs of LiFePO$_4$ cathode material in electrolyte of
1mol/L LiPF$_6$ in 1∶1 EC∶DEC solvent at sweep rate of 0.5mV/s

4 Conclusions

(1) The grain size of the sample has great effect on its electrochemical capability. The cooling processes including nature cooling, air quenching and water quenching were applied to preparing pure LiFePO$_4$ and studying their effects on microstructure and electrochemical capability.

(2) LiFePO$_4$ with perfect olivine phase can be obtained by heating the predecomposed precursor mixtures sealed in vacuum quartz-tube under various cooling conditions.

(3) Rapid cooling speed and short cooling time can effectively control the overgrowth of pure LiFePO$_4$ and improve its electrochemical performance. Sample WQ(prepared by water quenching) has uniform and fine grain size and good electrochemical performance. The grain size of sample WQ on (311) crystal plane is 66nm, and its grain size is smaller than those of sample NC (71.7nm) and sample AQ (71.1nm). Meanwhile, sample WQ has higher crystalline degree.

(4) Cooling mode is also a key factor in controlling the overgrowth of pure LiFePO$_4$ and improving its electrochemical performance in synthesizing process. The method can avoid the loss of bulk density in contrast to others.

References

[1] Nishi Y. Lithium ion batteries: Past 10 years and the future[J]. J Power Sources, 2001, 100(1/2): 101-106.

[2] Padhi A K, Najundaswamy K S, Goodenough J B. Phospho-olivines as positive-electrode materials for rechargeable lithium batteries[J]. J Electrochem Soc, 1997, 144(4): 1188-1194.

[3] Bauer E M, Bellitto C, Pasquali M, et al. Versatile synthesis of carbon-rich LiFePO$_4$ enhancing its electrochemical properties [J]. Electrochem Solid-State Lett, 2004, 7(4): A85-A90.

[4] Zaghib K, Striebel K, Guerfi A, et al. LiFePO$_4$/polymer/natural graphite: Low cost Li-ion batteries[J]. Electrochimica Acta, 2004, 50(2/3): 263-270.

[5] Huang H, Yin S C, Nazar L F. Approaching theoretical capacity of LiFePO$_4$ at room temperature at high rates[J]. Electrochem Solid-State Lett, 2001, 4(10): A170-A172.

[6] Guerfi I A, Kaneko M, Petitclerc M, et al. LiFePO$_4$ water-soluble binder electrode for Li-ion batteries[J]. J Power Sources, 2007, 163(2): 1047-1052.

[7] Hu Guorong, Gao Xuguang, Peng Zhongdong, et al. Synthesis of LiFePO$_4$/C composite electrode with enhanced electrochemical performance[J]. Trans Nonferrous Met Soc China, 2005, 15(4): 795-798.

[8] Myung S T, Komaba S, Hirosaki N, et al. Emulsion drying synthesis of olivine LiFePO$_4$/C composite and its electrochemical properties as lithium intercalation material [J]. Electrochimica Acta, 2004, 49(24): 4213-4222.

[9] Ho C S, Won C, Ho J. Electrochemical properties of carbon-coated LiFePO$_4$ cathode using graphite, carbon black, and acetylene black [J]. Electrochimica Acta, 2006, 52(4): 1472-1476.

[10] Yang Shoufeng, Zavalij P Y, Whittinggham M S. Hydrothermal synthesis of lithium iron phosphate cathodes[J]. Electrochemistry Commun, 2001, 3(9): 505-508.

[11] Park K S, Kang K T, Lee S B, et al. Synthesis of LiFePO$_4$ with fine particle by coprecipitation method[J]. Materials Research Bulletin, 2004, 39(12): 1803-1810.

[12] Yamada A, Chung S C, Hinokuma K. Optimized LiFePO$_4$ for lithium battery cathodes[J]. J Electrochem Soc, 2001, 148(3): A224-A229.

[13] Chung S Y, Blocking J T, Chiang Y M. Electronically conductive phospho-olivines as lithium storage electrodes[J]. Nature Mater, 2002, 2: 123-128.

[14] Hu Guorong, Gao Xuguang, Peng Zhongdong, et al. Influence of Ti ion doping on electrochemical properties of LiFePO$_4$/C cathode material for lithium-ion batteries [J]. Trans Nonferrous Met Soc China, 2007, 17(2): 296-300.

[15] Hong Jian, Wang Chunsheng, Kasavajjula U. Kinetic behavior of LiFeMgPO$_4$ cathode material for Li-ion batteries[J]. J Power Sources, 2006, 162(2): 1289-1296.

[16] Wang Deyu, Li Hong, Shi Siqi, et al. Improving the rate performance of LiFePO$_4$ by Fe site doping[J]. Electrochimica Acta, 2005, 50(14): 2955-2958.

[17] Ravet N, Abouimrane A, Armand M, et al. On the electronic conductivity of phospho olivines as lithium storage electrodes[J]. Nature Mater, 2003, 2: 702-720.

[18] Subramanya H P, Ellis B, Coombs N, et al. Nano-network electronic conduction in iron and nickel olivine phosphates[J]. Nature Mater, 2004, 3: 147-152.

Coating of LiNi$_{1/3}$Co$_{1/3}$Mn$_{1/3}$O$_2$ Cathode Materials with Alumina by Solid State Reaction at Room Temperature[*]

Abstract Alumina coated LiNi$_{1/3}$Co$_{1/3}$Mn$_{1/3}$O$_2$ particles were obtained by a simple method of solid state reaction at room temperature. The reaction mechanism of solid state reaction at room temperature was investigated. The structure and morphology of the coating materials were investigated by XRD, SEM and TEM. The results show that. The electrochemical performances of uncoated and Al$_2$O$_3$-coated LiNi$_{1/3}$Co$_{1/3}$Mn$_{1/3}$O$_2$ cathode materials were studied within a voltage window of 3.00~4.35V at current density of 30mA/g. SEM, TEM and EDS analytical results indicate that the surface of LiNi$_{1/3}$Co$_{1/3}$Mn$_{1/3}$O$_2$ particles is coated with very fine Al$_2$O$_3$ composite, which leads to the improved cycle ability though a slight decrease in the first discharge capacity is observed. It is proposed that surface treatment by solid state reaction at room temperature is a simple and effective method to improve the cycle performance of LiNi$_{1/3}$Co$_{1/3}$Mn$_{1/3}$O$_2$ particles.

1 Introduction

The synthesis of composite particles, consisting of core particle covered by a coating of different materials, has opened new promising directions for materials research. It can be expected that the coating materials may stabilize the core materials to prevent possible undesirable interactions with the environment or may improve electrochemical, optical, magnetic, conductive, adsorptive, and surface reactive properties of the dispersed matter to meet certain requirements[1,2].

The cathode material LiNi$_{1/3}$Co$_{1/3}$Mn$_{1/3}$O$_2$ is one of the most promising cathode for rechargeable lithium ion batteries owing to its high reversible capacity at high potential. In order to develop high cyclability of layered cathode material for advanced lithium-ion batteries, one of the approaches is to coat it with a thin layer of electrochemically inactive metal oxide materials[3-5]. The improved cycling performance and capacity retention of the coated cathode materials are thought to be successful in minimizing the side reactions within the batteries by placing a protective barrier layer between cathode material and the liquid electrolyte during (de)intercalation process[6].

The coating can be obtained by a variety of methods including impregnation, chemical vapor deposition method[7], precipitation and sol-gel techniques[8,9]. These methods all require the core material to be suspended in solvent (e.g. water). Coating by solid state reaction at room temperature is a new method that is simple and environmentally friendly. The advantages such as high yield, strong selectivity, low cost, no need of solvent and eliminating reunite problem[10-13] make it easy to be industrialized, and it can also reduce by-products to the greastest

---

  * Copartner: Peng Zhongdong, Deng Xinrong, Du Ke, Hu Guorong, Gao Xuguang. Reprinted from J. Cent. South Univ. Technol., 2008, 15:34-38.

extent. Solid state reaction at room temperature is an efficient method for synthesizing nano-particles[14], atom cluster compounds[15], coordination compounds[16]. However, no reports on coating lithium-ion battery cathode material by this method have been found until now.

In this work, the method of solid state reaction at room temperature was introduced to synthesize Al_2O_3-coated $LiNi_{1/3}Co_{1/3}Mn_{1/3}O_2$ lithium ion cathode materials. The possible reaction mechanism of preparing Al_2O_3-coated $LiNi_{1/3}Co_{1/3}Mn_{1/3}O_2$ lithium ion cathode materials through solid state reaction at room temperature was studied by thermodynamics and kinetics.

2 Experimental

$LiNi_{1/3}Co_{1/3}Mn_{1/3}O_2$ particles in micrometer were prepared in our laboratory[17]. $Al_2(SO_4)_3 \cdot 18H_2O$ and $LiOH \cdot H_2O$ were reagent grade. $LiNi_{1/3}Co_{1/3}Mn_{1/3}O_2$ powder was mixed and ground softly with $Al_2(SO_4)_3 \cdot 18H_2O$ (total metal ion molar ratio, $n(Ni+Mn+Co)/n(Al) = 10:1$). Stoichiometric ground $LiOH \cdot H_2O$ was then added to the mixture and ground softly for 40min. During the process of grinding, the reactants became wet. Then the mixture was aged for 2h and washed with distilled water. At last, the product ($Al(OH)_3$-coated $LiNi_{1/3}Co_{1/3}Mn_{1/3}O_2$) was dried at 60℃ and then heated at 700℃ for 30min to obtain the Al_2O_3 coated $LiNi_{1/3}Co_{1/3}Mn_{1/3}O_2$ particles.

Composite electrodes were prepared by mixing 80% $LiNi_{1/3}Co_{1/3}Mn_{1/3}O_2/Al_2O_3$- coated $LiNi_{1/3}Co_{1/3}Mn_{1/3}O_2$ particles, 10% acetylene black (AB) and 10% polyvinylidene fluoride (PVDF). The mixture was spread on an aluminum foil and dried at 120℃ for 24h. The charge-discharge cycling tests were performed using the CR2025 coin-type cell. The separator was a Celgard 2400 microporous polylene membrane. Lithium metal was used as anode in this study. The electrolyte was 1.0mol/L $LiPF_6$/EC+DEC(1:1, volume ratio). The cells were assembled in a glove box filled with ultra-pure argon gas. The charge-discharge capacity and cycling performance of the cells were galvanostatically performed at constant current density of 30mA/g with a voltage window of 3.00 ~ 4.35V (vs Li/Li^+) using a LAND CT2001A computer-controlled battery testing system at room temperature. The crystal structure of cathode materials was identified from an X-ray diffraction (XRD) pattern obtained by diffractometry (Rigaku D/max 2550 VB^+). TEM analysis was performed on Tecnai $G^2$12 operated at 200kV. The morphology and EDS were observed by scanning electron microscopy (SEM, JEOL JSM-6360LV).

3 Results and Discussion

3.1 Thermodynamic and kinetic analysis of alumina coating

The solid state reactions at room temperature are always ignored. Similar to the chemical reaction in solution, the chemical reaction in solid must satisfy the thermodynamics law that the change of the whole reaction Gibbs energy should be less than zero. When the reaction thermodynamics condition is satisfied, reactant structure will be the key condition that determines the solid state reactions at room temperature will happen or not. The effect of the solid structure on chemical reactions can also be known from the reaction temperature. If the solid state reactions happen, the reactant molecule must long-range move and collide each other. However, the force of the long-range motion and colliding each other can be known from the melting point of the solid reactant.

Usually, the temperature under which the solid state reaction can happen is decided by the lower TAMMANN temperature (T_m) of reactant[18]. Here, the TAMMANN temperature is the temperature when the solid interior diffusion becomes prominent. TAMMANN and MANSURI[18] pointed out that the temperature has something to do with solid melting point (t_A) of the denotation of absolute temperature standard: $T_m = k(t_A + 273)$, k was a coefficient, 0.3 to metal, 0.5 to inorganic compound and 0.9 to organic compound. Actually, in order to get a high reaction speed, usually a higher reaction temperature will be used. For example, to inorganic compound, it is 2/3, as shown in Fig. 1. Therefore, the reaction of molecule solid state will happen at room temperature when the TAMMANN temperature is less than 300K.

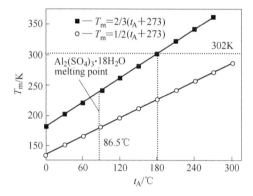

Fig. 1 Correlation of melting point and TAMMANN temperature

To sum up, the thermodynamic and kinetic conditions that the solid state reaction can happen are as follows:

$$\Delta_r G_m = \sum_{B=1}^{N} \nu_B \mu_B < 0 \quad (1)$$

$$T_m \leqslant 300K \quad (2)$$

Eq. (1) denotes the thermodynamic condition, and Eq. (2) denotes the kinetic condition. Only meeting the two conditions together can the solid state reaction at room temperature happen.

$Al_2(SO_4)_3 \cdot 18H_2O$ reacts with $LiOH \cdot H_2O$ at room temperature as follows:

$$Al_2(SO_4)_3 \cdot 18H_2O + 6LiOH \cdot H_2O \longrightarrow 2Al(OH)_3 + 3Li_2SO_4 + 24H_2O \quad (3)$$

The $\Delta_r G_m$ of Eq. (3) is less than zero and the melting point of $Al_2(SO_4)_3 \cdot 18H_2O$ is 86.5℃, as shown in Fig. 1. Therefore, viewing from both the thermodynamic and kinetic analysis point, the reaction can happen. The above theoretical analysis will be proved in the following EDS and TEM experiment.

3.2 Analysis of experimental results

Fig. 2 shows the X-ray diffraction pattern of the uncoated and coated $LiNi_{1/3}Co_{1/3}Mn_{1/3}O_2$ powders. The spectrum of uncoated $LiNi_{1/3}Co_{1/3}Mn_{1/3}O_2$ powders indicates that the material is similar to that of $LiCoO_2$ (α-NaFeO$_2$ type, space group R3m) and no impure peaks appear. It can be observed that the spectrum of the coated $LiNi_{1/3}Co_{1/3}Mn_{1/3}O_2$ powders is almost same as that of the uncoated one. The absence of any other signals in the spectrum indicates the coated Al_2O_3 is probably not crystal, but amorphous in nature.

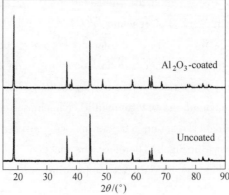

Fig. 2 XRD patterns of uncoated and Al_2O_3-coated $LiNi_{1/3}Co_{1/3}Mn_{1/3}O_2$ powders

Fig. 3 shows the surface morphologies of the uncoated and Al_2O_3-coated $LiNi_{1/3}Co_{1/3}Mn_{1/3}O_2$ composites. The surface morphologies of the uncoated $LiNi_{1/3}Co_{1/3}Mn_{1/3}O_2$ composites are very clean and smooth. By comparison, the surface morphology of Al_2O_3-coated $LiNi_{1/3}Co_{1/3}Mn_{1/3}O_2$ materials is blurry.

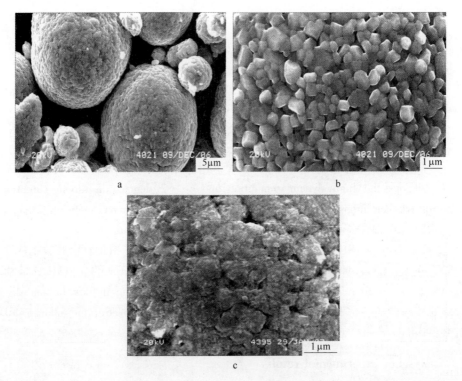

Fig. 3 SEM images of uncoated (a, b) and Al_2O_3-coated (c) $LiNi_{1/3}Co_{1/3}Mn_{1/3}O_2$ composite

Fig. 4 shows the TEM image of Al_2O_3-coated $LiNi_{1/3}Co_{1/3}Mn_{1/3}O_2$ materials. The crystalline grains are obviously coated with small particles and the thickness of the kernel is around 20nm. The above layer in morphology should be the result of presence of the Al_2O_3 coating.

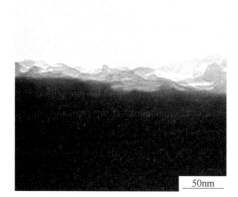

Fig. 4 TEM image of Al_2O_3-coated $LiNi_{1/3}Co_{1/3}Mn_{1/3}O_2$ particles

The surface composition of the coated $LiNi_{1/3}Co_{1/3}Mn_{1/3}O_2$ powders was analyzed by scanning electronic microscope energy spectrum. The relative surface compositions of the coated particles expressed quantitatively as molar fraction, derived from the EDS intensities, are shown in Fig. 5. Al and O components appear on the surface of coated particles. It can be concluded that alumina is coated on the surface of the $LiNi_{1/3}Co_{1/3}Mn_{1/3}O_2$ particles. It can also prove that the reaction of $Al_2(SO_4)_3 \cdot 18H_2O$ with $LiOH \cdot H_2O$ in solid state reaction route at room temperature is feasible.

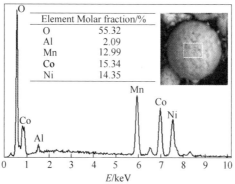

Fig. 5 Scanning electronic microscope energy spectrum analysis of coated $LiNi_{1/3}Co_{1/3}Mn_{1/3}O_2$ particles

The electrochemical performances of uncoated and Al_2O_3-coated $LiNi_{1/3}Co_{1/3}Mn_{1/3}O_2$ materials were tested by Li half-cell. The typical discharge curves of different samples are presented in Fig. 6. The initial discharge specific capacity of the Al_2O_3-coated $LiNi_{1/3}Co_{1/3}Mn_{1/3}O_2$ cathode material (147.6mA · h/g) is lower than that of the uncoated $LiNi_{1/3}Co_{1/3}Mn_{1/3}O_2$ cathode material (156.5mA · h/g). However, the capacity retention of the $LiNi_{1/3}Co_{1/3}Mn_{1/3}O_2$ cathode material is significantly improved by coating a small amount of Al_2O_3. As shown in Figs. 6 and 7, the uncoated $LiNi_{1/3}Co_{1/3}Mn_{1/3}O_2$ material loses about 10.29% of the initial discharge specific capacity (140.4mA·h/g) after 20 cycles, while the Al_2O_3-coated $LiNi_{1/3}Co_{1/3}Mn_{1/3}O_2$ material behaves much better, losing about 2.44% of its initial discharge specific capacity (144.0mA·h/g) during the same number of cycles.

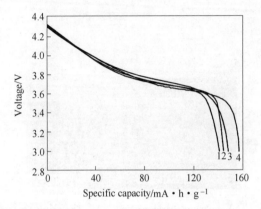

Fig. 6 Discharge curves of uncoated and Al_2O_3-coated $LiNi_{1/3}Co_{1/3}Mn_{1/3}O_2$ electrode

1— The 1st, uncoated $LiNi_{1/3}Co_{1/3}Mn_{1/3}O_2$; 2— The 20th, uncoated $LiNi_{1/3}Co_{1/3}Mn_{1/3}O_2$;
3— The 1st, coated $LiNi_{1/3}Co_{1/3}Mn_{1/3}O_2$; 4— The 20th, coated $LiNi_{1/3}Co_{1/3}Mn_{1/3}O_2$

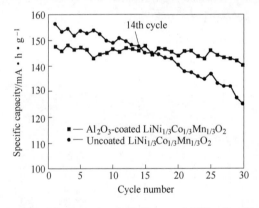

Fig. 7 Cycling stability of uncoated and Al_2O_3-coated $LiNi_{1/3}Co_{1/3}Mn_{1/3}O_2$ electrode

The 3rd cyclic voltammetry records for the uncoated as well as the Al_2O_3- coated $LiNi_{1/3}Co_{1/3}Mn_{1/3}O_2$ electrodes are shown in Fig. 8. Fig. 8 shows that the similar specific cyclic voltammograms from the uncoated and Al_2O_3-coated $LiNi_{1/3}Co_{1/3}Mn_{1/3}O_2$ electrodes indicate that the improvement of capacity retention is not due to suppression of the phase transition. Another coating effect, which was reported in Ref. [19], is that the prevention of electrode reactions with electrolyte since the oxide-coated layer isolates them. Generally speaking, the reaction at the interface induces impedance increase and causes the capacity fading of the cathode[20].

Fig. 9 shows the comparison of the electrochemical impedance speceroscopy(EIS) profiles of uncoated and Al_2O_3-coated $LiNi_{1/3}Co_{1/3}Mn_{1/3}O_2$ samples at a charge potential of 4.35V, respectively, as a function of cycle number. A high-frequency semicircle represents the impedance due to a solid-state interface layer formed on the surface of the electrodes, and a low-frequency semicircle is related to a slow charge transfer process at the interface and its relative double-layer capacitance at the film bulk oxide. The cell impedance is mainly determined by the cathode-side impedance[21], especially by the charge-transfer impedance. Thus, we focused on comparison of the a low-frequency semicircle that represents the charge-transfer impedance. The cell impedance of the uncoated cathode drastically increases during cycling, whereas the cell impedance of the coated cathode does not increase much. From these results, the smaller capacity loss of the coated $LiNi_{1/3}Co_{1/3}Mn_{1/3}O_2$ cathode to the uncoated $LiNi_{1/3}Co_{1/3}Mn_{1/3}O_2$ cathode appears due to the inactive Al_2O_3 coating layer on the cathode surface, which considerably decreases the impedance growth.

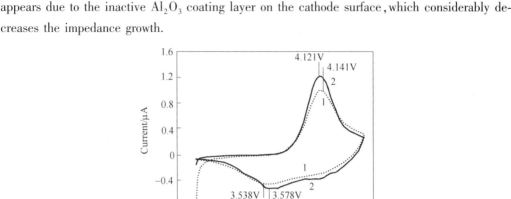

Fig. 8 The 3rd cyclic voltammetry curve of uncoated and Al_2O_3-coated $LiNi_{1/3}Co_{1/3}Mn_{1/3}O_2$ electrodes at scan rate of 0.1mV/s

1—Uncoated $LiNi_{1/3}Co_{1/3}Mn_{1/3}O_2$; 2— Al_2O_3-coated $LiNi_{1/3}Co_{1/3}Mn_{1/3}O_2$

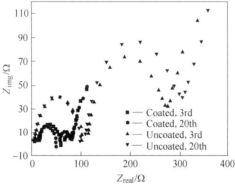

Fig. 9 Nyquist plots of uncoated and Al_2O_3-coated $LiNi_{1/3}Co_{1/3}Mn_{1/3}O_2$ samples

4 Conclusions

(1) Alumina coating on $LiNi_{1/3}Co_{1/3}Mn_{1/3}O_2$ particles is obtained by solid state reaction at room temperature. The possibility of preparing Al_2O_3-coated $LiNi_{1/3}Co_{1/3}Mn_{1/3}O_2$ lithium ion battery cathode materials through solid state reaction at room temperature is studied by thermodynamics and kinetics.

(2) The TEM, SEM, XRD and EDS experimental results show that Al_2O_3 is coated on the surface of $LiNi_{1/3}Co_{1/3}Mn_{1/3}O_2$ particles. The surface of $LiNi_{1/3}Co_{1/3}Mn_{1/3}O_2$ particles is coated with fine Al_2O_3 particles as a coating material for improving cyclic property at room tempera-

ture, though its first discharge specific capacity is slightly lower than that of the uncoated one. The uncoated $LiNi_{1/3}Co_{1/3}Mn_{1/3}O_2$ particles lose about 10.29% of its initial discharge capacity after 20 cycles, whereas the Al_2O_3-coated $LiNi_{1/3}Co_{1/3}Mn_{1/3}O_2$ behaves much better, only losing about 2.44% of its initial discharge capacity during the same cycles.

(3) The surface treatment of the solid state reaction at room temperature is a simple and effective way to improve the electrochemical performance of $LiNi_{1/3}Co_{1/3}Mn_{1/3}O_2$ material for lithium ion batteries.

References

[1] Liu GuiXia, Hong Guangyan. Synthesis of SiO_2/Y_2O_3: Eu core-shell materials and hollow spheres [J]. Journal of Solid State Chemistry, 2005, 178(5): 1647-1651.

[2] Guo Huajun, Li Xinhai, Zhang Xinming, Zeng Suming, Wang Zhixing. Characteristics of $LiCoO_2$, $LiMn_2O_4$ and $LiNi_{0.45}Co_{0.1}Mn_{0.45}O_2$ as cathodes of lithium ion batteries [J]. Journal of Central South University of Technology, 2005, 12(1): 44-49.

[3] Kim Y, Kim H S, Martin S W. Synthesis and electrochemical characteristics of Al_2O_3- coated $LiNi_{1/3}Co_{1/3}Mn_{1/3}O_2$ cathode materials for lithium ion batteries [J]. Electrochimica Acta, 2006, 52(3): 1316-1322.

[4] Tong Dongge, Lai Qiongyu, Wei Nini, Tang Aidong, Tang Lianxing, Huang Kelong, Ji Xiaoyang. Synthesis of $LiNi_{1/3}Co_{1/3}Mn_{1/3}O_2$ as a cathode material for lithium ion battery by water-in-oil emulsion method [J]. Materials Chemistry and Physics, 2005, 94(2/3): 423-428.

[5] Peng Zhongdong, Hu Guorong, Liu Yexiang. Influence on performance and structure of spinel $LiMn_2O_4$ for lithium-ion batteries by doping rare-earch Sm [J]. Journal of Central South University of Technology, 2005, 12(1): 28-32.

[6] Li C, Zhang H P, Fu L J, Liu H, Wu Y P, Rahm E, Holze R, Wu H Q. Cathode materials modified by surface coating for lithium ion batteries [J]. Electrochimica Acta, 2006, 51(19): 3872-3883.

[7] Elkasabi Y, Chen H Y, Lahann J. Multipotent polymer coatings based on chemical vapor deposition copolymerization [J]. Advanced Materials, 2006, 18(12): 1521-1526.

[8] Wang F H, Guo R S, Wei Q T, Zhou Y, Li H L, Li S L. Preparation and properties of Ni/YSZ anode by coating precipitation method [J]. Materials Letters, 2004, 58(24): 3079-3083.

[9] Phani A R, Gammel F J, Hack T. Structural, mechanical and corrosion resistance properties of Al_2O_3-CeO_2 nanocomposites in silica matrix on Mg alloys by a sol-gel dip coating technique [J]. Surface and Coatings Technology, 2006, 201(6): 3299-3306.

[10] Zhou Taoyu, Yuan Xin, Hong Jianming, Xin Xinquan. Room-temperature solid-state reaction to nanowires of zinc sulfide [J]. Materials Letters, 2006, 60(2): 168-172.

[11] Li Fa-shen, Wang Hai-bo, Wang Li, Wang Jianbo. Magnetic properties of $ZnFe_2O_4$ nanoparticles produced by a low-temperature solid-state reaction method [J]. Journal of Magnetism and Magnetic Materials, 2007, 309(2): 295-299.

[12] Liu Guixia, Hong Guangyan, Sun DuoXian. Coating Gd_2O_3: Eu phosphors with silica by solid-state reaction at room temperature [J]. Powder Technology, 2004, 145(2): 149-153.

[13] Cui Hongtao, Hong Guangyan, Wu Xueyan, Hong Yuanjia. Silicon dioxide coating of CeO_2 nanoparticles by solid state reaction at room temperature [J]. Materials Research Bulletin, 2002, 37(13): 2155-2163.

[14] Cui Hongtao, Hong Guangyan, You Hongpeng, Wu Xueyan. Coating of Y_2O_3:Eu^{3+} particles with alumina by a humid solid state reaction at room temperature [J]. Journal of Colloid and Interface Science, 2002, 252(1): 184-187.

[15] Shah J G, Patki V A, Shivakamy K, Wani B N, Patwe S J, Rao U R. On the material transport during solid state reactions at room temperature [J]. Applied Physics Communications, 1993, 12(1): 141-152.

[16] Yuan A Q, Liao S, Tong Zh F, Wu J, Huang Z Y. Synthesis of nanoparticle zinc phosphate dihydrate by solid state reaction at room temperature and its thermochemical study [J]. Materials Letters, 2006, 60 (17/18): 2110-2114.

[17] Yu Xiaoyuan. Study on layered $LiNi_{1/3}Co_{1/3}Mn_{1/3}O_2$ and modified spinel $LiMn_2O_4$ [D]. Changsha: Central South University, 2006 (in Chinese).

[18] Tammann G, Mansuri Q A. Hardness of amalgams Ag-Sn [J]. Anorg Chem, 1923, 132: 66-67 (in German).

[19] Zhang Z R, Liu H S, Gong Z L, Yang Y. Electrochemical performance and spectroscopic characterization of TiO_2-coated $LiNi_{0.8}Co_{0.2}O_2$ cathode materials [J]. Journal of Power Sources, 2004, 129(1): 101-106.

[20] Chen Z H, Dahn J R. Methods to obtain excellent capacity retention in $LiCoO_2$ cycled to 4.5V [J]. Electrochimica Acta, 2004, 49(7): 1079-1090.

[21] Chen C H, Liu J, Amine K. Symmetric cell approach and impedance spectroscopy of high power lithium-ion batteries [J]. Journal of Power Sources, 2001, 96(2): 321-328.

Synthesis of Nitrogen-containing Hollow Carbon Microspheres by A Modified Template Method as Anodes for Advanced Sodium-ion Batteries[*]

Abstract Nitrogen-containing hollow carbon microspheres (NHCSs) are prepared by a modified template method in the presence of resorcinol/formaldehyde as carbon precursors and ethylenediamine (EDA) as both a base catalyst and nitrogen precursor. The NHCSs are used as anode materials for sodium-ion batteries, showing a superior reversible discharge capacity of 334mA · h/g after 100 cycles at 50mA/g. A high reversible discharge capacity of 114mA · h/g can also be obtained even at an extremely high current density of 10A/g. Moreover, excellent long-term cycling stability (> 1200 cycles) is also observed even at 500mA/g. The results show that the NHCS electrode exhibits excellent electrochemical performance (superior reversible capacity, high rate capability, and long-term cycling stability). The excellent performance of NHCS electrode is most likely attributed to the unique nitrogen-containing hollow carbon structure. Furthermore, this study provides a novel route to produce NHCSs, which may find application in other fields.

1 Introduction

Among various energy storage devices, Li-ion batteries (LIBs) are important as power sources for consumer portable electronic devices because of their very high energy density and cyclic stability[1,2]. However, for large-scale energy storage (such as that for electric vehicles and grid-related applications), LIBs face substantial challenges associated with the high cost and limited terrestrial reserves of lithium[3,4]. There is an urgent need to explore alternative energy storage technologies for large-scale energy storage. Because of the abundant supply and widespread terrestrial reserves of Na mineral salts, sodium-ion batteries (SIBs) have attracted increasing attention as a low-cost alternative to LIBs. However, the progress of SIBs has been slow because of the lack of appropriate active materials for both cathodes and anodes. As a typical Na ion is approximately 55% larger than the Li ion, most materials do not have a sufficiently large interstitial space within their crystallographic structure to host Na ions[5,6]. Therefore, investigating suitable electrode materials is crucial for the development of SIBs.

Although much effort has been made in the field of cathode materials for SIBs, studies on the anode material are still at a very early stage[7,8]. In recent years, Na storage anode materials, such as Sn[9,10], Ge[11,12], Sb[13-15], and P[16-19], have been explored as anode materials for SIBs, but the application of these materials is hampered because of the large volumetric expansion during sodiation and the intrinsic low conductivity. Carbonaceous materials have attracted much attention since the first report of hard carbon as anode for SIBs[20], because of the versa-

---

[*] Copartner: Yaohui Qu, Zhian Zhang, Ke Du, Wei Chen, Yanqing Lai, Jie Li. Reprinted from Carbon, 2016, 105: 103-112.

tile preparation methods and large interlayer space for Na⁺ insertion[21]. In recent years, various kinds of carbon materials have been investigated, such as hollow carbon wires[21], carbon sheets[22], carbon fibers[23], hollow carbon spheres[24,25], and expanded graphite[26]. Among them, Wang et al.[24] synthesized hollow carbon spheres from ionic liquids, and found that a reversible capacity of approximately 120mA·h/g is achieved after 30 cycles at a current density of 100mA/g. Maier et al.[25] have prepared hollow carbon spheres that showed a high capacity of approximately 160mA·h/over 100 cycles under 100mA/g, suggesting that the hollow carbon spheres show extraordinary electrochemical properties. Therefore, the hollow carbon spheres are one of the most promising anode materials for the SIBs, and have attracted much attention.

In addition, recent studies have also shown that surface functional groups or doped heteroatoms play important roles in improving the performances of the carbon-based electrode materials[27-31]. In particular, nitrogen is one of the most attractive doped heteroatoms, which could enhance the carbon electrical conductivity, improve carbon wettability[32], enhance the reactivity and capacity of carbon[33,34], generate a pseudocapacitance[33,35], and improve the electrochemistry performance of the carbon anodes in the SIBs[36,37]. Wang et al.[36] synthesized porous nitrogen-doped nanosheets from graphene oxide (GO)-polypyrrole composites, and found that a reversible capacity of approximately 200mA·h/g is achieved after 250 cycles at a current density of 50mA/g. Dou et al.[37] reported nitrogen-doped graphene foams, which can produce a reversible capacity of 594mA·h/g after 150 cycles at the current density of 500mA/g, and could still maintain 137mA·h/g at the current of 5A/g.

Therefore, we are motivated to synthesize novel nitrogen-containing hollow carbon microspheres (NHCSs) for high-performance SIB anodes. As we know, the hollow carbon spheres can be prepared by the silica template method[38-40]. However, there are only few reports on the preparation of nitrogen-doped hollow carbon spheres, and their preparation is mostly focused on the surface chemical postmodification[41]. In this study, we successfully synthesized N-doped hollow carbon spheres by a modified template method in the presence of resorcinol/formaldehyde as carbon precursors and ethylenediamine (EDA) as both a base catalyst and nitrogen precursor[42]. The obtained NHCSs possess a large specific surface area, appropriate nitrogen doping and unique hollow structure. Because of their superior reversible capacity, high rate capability, and long-term cycling stability, they can be used as the anode material in SIBs. This novel NHCS electrode shows a superior reversible discharge capacity of 334mA·h/g after 100 cycles at a current density of 50mA/g. Even at an extremely high current density of 10A/g, a high discharge capacity of 114mA·h/g is obtained. Moreover, the asprepared NHCS electrode exhibits a long-term cycling stability up to 1200 cycles even at 500mA/g.

2 Experimental Section

2.1 Materials synthesis

NHCSs were synthesized by a modified template method. In a typical synthesis process, 3.2mL of ammonia aqueous solution (28 wt%) was added to a mixture of deionized water (11.0mL) and ethanol (75.6mL), and then stirred for 30min at room temperature. Subsequently, 3.0mL of tetraethoxysilane(TEOS) was added to this mixture, and further stirred for 10min at room temperature. Then, 0.21g of resorcinol, 0.31mL of formaldehyde solution (37 wt%), and 0.10mL of EDA were added to the stirred solution at intervals of 10min, respectively[42]. Then, the mixture was vigorously stirred for 24h at 30℃ and transferred to 60mL of Teflon-lined autoclave under static conditions (100℃) for another 24h. The solid products were obtained by centrifugation and dried at 100℃ for 18h. Finally, the solid samples were carbonized by heating at 750℃ for 2h under a N_2 environment in a tube furnace, the SiO_2 templates were removed by treating the material in dilute HF solution overnight, and the final NHCSs were obtained after drying.

For comparison, hollow carbon microspheres (HCSs, nitrogen-free) were also prepared using the same method, but without EDA precursors. Commercial activated carbons (ACs, YP17, Kuraray Co., Ltd.) are bulk-type carbon materials that are widely used in supercapacitors, and ACs were also prepared for comparison.

2.2 Materials characterization

Scanning electron microscopy (SEM, Nova SEM 230) and transmission electron microscopy (TEM, Tecnai G2 20ST) were used to investigate the material morphology. The structure of the NHCSs was characterized by X-ray diffraction (XRD, Rigaku3014). Nitrogen adsorption/desorption measurements were carried out on Quabrasorb SI-3MP (Quantachrome Instruments). Surface functional groups and bonding characterization were conducted using X-ray photoelectron spectroscopy (XPS, ThermoFisher ESCALAB250xi). The fitting errors of XPS test results are within ±1%.

2.3 Electrochemical measurements

In order to conduct electrochemical measurements of the NHCS electrode, a CR-2032-type coin cell was fabricated. Sodium pellet and Whatman glass fiber membrane served as counterelectrode and separator, respectively. A mixed slurry of NHCSs, Super P, and carboxymethyl cellulose (CMC) (8 : 1 : 1 wt%) in deionized water was spread onto a copper foil. The coating copper foil was first air-dried, then transferred into a oven at 80℃ for 12h, and finally the electrode was cut into pellets with a diameter of 1.0cm and dried for 12h in a vacuum oven at 60℃. The typical mass loading of active NHCSs was 0.50-0.60mg/cm^2. The thickness of the electrode was 0.045mm, including the thickness of the copper foil (~0.020mm). The assembly of the tested cells was carried out in a glove box under argon atmosphere. The electrolyte used in this study was 1 M $NaClO_4$ (Sigma-Aldrich) in a solvent mixture of ethylene carbonate and propylene carbonate (with a volume ratio of 1 : 1). Cyclic voltammetry (CV) and measurements were conducted using PARSTAT 2273 electrochemical measurement system. CV tests were performed at a scan rate of 0.1mV/s in the voltage range of 0.0 – 3.0V. Galvanostatic charge-discharge tests were performed in the potential range of 0.005 – 3.0V by using a LAND CT2001A battery-testing instrument. Electrochemical impedance spectroscopy (EIS) was also performed on PARSTAT 2273 electrochemical measurement system in the frequency range of 100kHz to 0.01Hz, and the voltage perturbation was controlled at 10mV. These electrochemical tests were carried out under a constant temperature of 25℃.

3 Results and Discussion

NHCSs were prepared by a modified template method in the presence of resorcinol/formaldehyde as carbon precursors and EDA as both a base catalyst and nitrogen precursor. The schematic illustration of the synthesis procedure of NHCSs is shown in Fig. 1. First, an aqueous-alcoholic solution was prepared by mixing ethanol, ammonia aqueous solution, and distilled water and stirring for 30min. TEOS was added to this mixture, and further stirred for 10min. Then, uniform colloidal SiO_2 spheres were obtained from the solution. Second, resorcinol, formaldehyde solution, and EDA were added to this mixture at intervals of 10min, respectively. Then, the mixture was vigorously stirred and transferred to Teflonlined autoclave under hydrothermal treatment, and the solid products (SiO_2@Nitrogen-containing polymer layers) were obtained by centrifugation. Third, the solid samples were carbonized, and the SiO_2@Nitrogen-containing C layers were obtained. Finally, the SiO_2 colloidal cores were removed by etching the products in HF solution, and pure NHCSs were obtained.

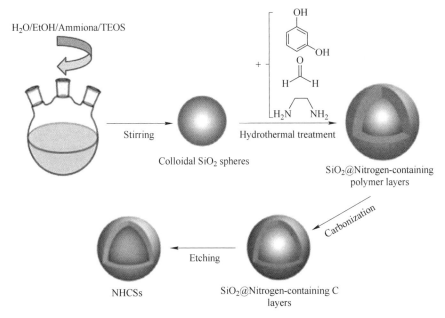

Fig. 1 Schematic illustration of the synthesis procedure of the NHCSs
(A colour version of this figure can be viewed online)

Field emission SEM and TEM were used to examine the morphology and microstructure of the as-prepared NHCSs. As shown in Fig. 2a and b, the majority of the products exhibit spherical morphology with diameters ranging from 200 to 400nm. A typical TEM image shown in Fig. 2c and d reveals that the NHCSs have a hollow structure with large internal void space and thin carbon shell (~18nm), and cross-linked with each other at the sphere surface, which is beneficial for both electron and sodiumion transfers[25]. The high-resolution transmission electron microscope (HRTEM) image (Fig. 2d, inset) of the NHCSs reveals a turbostratic carbon structure, indicating their amorphous structure due to the relatively low pyrolysis temperatures (750℃). Moreover, it has been reported that amorphous carbon is a suitable anode material for SIBs[22]. Fig

2e, e_1, and e_2 shows the annular dark-field TEM images of NHCSs and the corresponding elemental mappings of carbon and nitrogen, respectively. They are found to have a similar intensity across the NHCSs, indicating that nitrogen is evenly distributed throughout the carbon framework. Compared with the previous reports[25], the novel NHCSs contain a certain amount of nitrogen, which will favor the improvement of the electrochemistry performance of NHCSs.

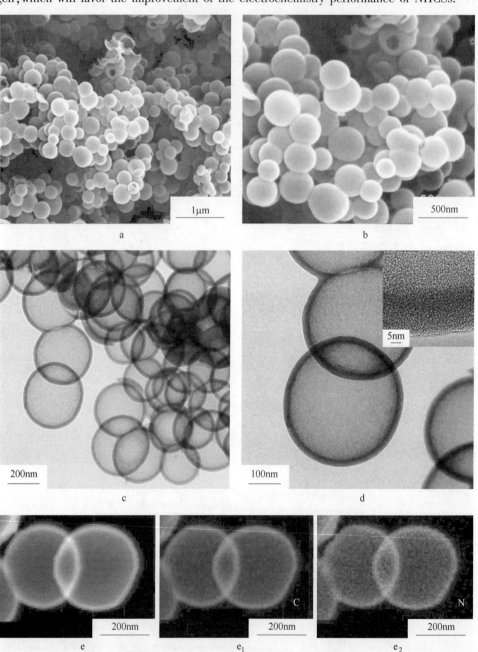

Fig. 2　SEM images of the NHCSs (a and b), TEM images of the NHCSs (c and d), and HRTEM (inset) image of the NHCSs (d). Annular dark-field TEM image of the NHCSs (e) and the corresponding elemental mappings of carbon (e_1) and nitrogen (e_2)

(A colour version of this figure can be viewed online)

XRD pattern of the NHCSs is given in Fig. 3a, which shows two broad peaks of 2θ at approximately 23.0° and 43.8°, corresponding to the (002) and (100)/(101) diffraction modes of the disordered carbon structure, which agrees well with the result of the HRTEM. On the basis of the XRD result, the interlayer spacing (d_{002}) of NHCSs is calculated to be 0.387nm. It is well known that the application of graphite for SIBs is hindered because of the limited interlayer spacing (0.336nm), and a large interlayer spacing makes NHCSs a suitable host material for SIBs to store sodium ions[21].

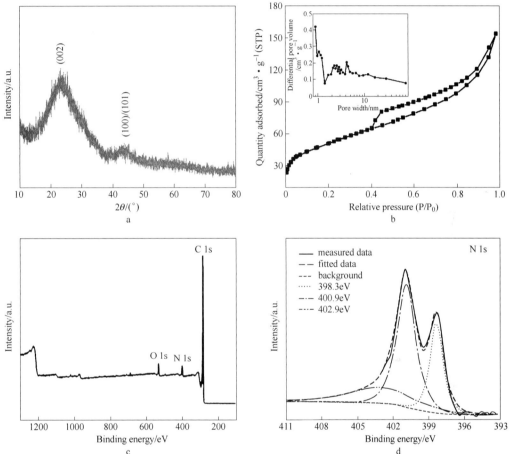

Fig. 3 XRD pattern of the NHCSs(a), the nitrogen adsorption/desorption isotherm and the pore size distribution (inset) of the NHCSs(b). XPS spectra of the NHCSs(c) and high-resolution spectra of N 1s for the NHCSs(d) (A colour version of this figure can be viewed online)

The specific surface area analysis of the as-prepared NHCSs was performed by nitrogen Brunauer-Emmett-Teller (BET) adsorption measurements. The pore size distribution of carbon was calculated by the Barrett-Joyner-Halenda (BJH) method. As shown in Fig. 3b, a type Ⅳ isotherm was observed, indicating the assembly of micro and mesoporous structures of the NHCSs. The specific BET surface area is approximately 182.6m²/g. Moreover, the pore size distribution (Fig. 3b, inset) of the NHCSs is mainly focused on the micro-and mesoporous, and it further verifies the assembly of micro-and mesoporous structures of the NHCSs. The average pore diameter of the NHCSs is approximately 5.2nm. These pores are probably attributed to the calcination process and the removal of SiO_2 colloidal cores embedded in the carbon matrix[43]. In addition, these pores can serve as paths for the electrolyte and facile transport channels for sodium ions during electrochemical cycling.

A TCH-600 Nitrogen/Oxygen/Hydrogen Determinator(USA) was used to determine the nitrogen content in the NHCSs, which was found to be approximately 12.40 wt%. Moreover, in order to confirm the chemical composition and surface properties of NHCSs, XPS measurements were performed and the result is shown in Fig. 3c. The XPS survey spectra of NHCSs (Fig. 3c) showed three peaks centering at 284.8, 400.9, and 532.3eV, corresponding to C 1s, N 1s, and O 1s, respectively. As estimated by the XPS results, the atom percent of nitrogen is 5.3 for NHCSs. The high-resolution N 1s peaks(Fig. 3d) in NHCSs can be divided into three fitted peaks with different binding energies. The peak centered at approximately (398.3 ±0.3)eV is due to pyridinic nitrogen(N-6)[44], the one at approximately (400.9±0.3)eV to pyrrolic nitrogen and/or pyridinic nitrogen(N-5)[32,45]. The contribution of the peak at approximately (402.9±0.3)eV derives from graphitic nitrogen[46]. According to the XPS result, it can be concluded that the nitrogen atoms of the NHCSs are structurally well bonded in the carbon matrix. Compared with the conventional method for the synthesis of nitrogen-doped hollow carbon spheres[47-49], this study provides a novel route to produce NHCSs.

In addition, recent studies have also shown that nitrogen doping in the carbon anodes plays an important role in improving the electrochemical performance of SIBs[36,37]. The role of nitrogen in enhancing the performance can be presented as follows. First, nitrogen doping in the carbon matrix can improve the electrical conductivity of carbon[32,50]. Second, nitrogen doping can generate extrinsic defects, and hence enhance the reactivity and capacity[51]. Third, after nitrogen doping, pseudo-capacitance can be generated due to the interaction between the electrolyte and the N species on the surface[35,52]. Therefore, the introduction of nitrogen atoms enhances the performance of NHCS electrodes, because of the enhancement of not only the electric conductivity but also pseudo-capacitance[37].

In order to study the electrochemical properties of the NHCS electrode, the CR-2032 coin cells with metallic sodium counter-electrode were assembled and evaluated. Fig. 4a shows the typical CV curves of the NHCS electrode in the potential range of 0.0-3.0V with a constant scan rate of 0.1mV/s. For the cathodic processes, three peaks at 2.08, 0.54, and 0V are observed during the first cycle. The reduction peak at 2.08V may be attributed to the reaction between the sodium ion and the surface functional group(s) of the carbon surface[25,53~55]. The peak at 0.54V can be attributed to the decomposition of the electrolyte and the formation of solid electrolyte interface(SEI) film[21,51,56]. These two peaks disappear in the following cycles, indicating the irreversibility of the reaction, while the peak located around 0V is attributed to Na^+ insertion into the carbonaceous materials. In the subsequent cycles, this peak still exists, suggesting that the insertion of Na^+ is reversible. For the anodic processes, Na^+ extraction occurs over a wide potential range (0-0.6V). Notably, the oxidation broad peak intensities of the subsequent cycles are larger than that of the first cycle, which should be attributed to the adsorption with charge transfer on both sides of single graphene layers[57] and the reaction of Na^+ with nitro group or carbonyl group[32,58]. Coinciding with the previous reports, the Na^+ insertion/extraction of the NHCS electrode occurs below 1.2V. Furthermore, according to the CV results, after the first cycle, the subsequent CV curves overlap each other, indicating a remarkable electrochemical reversibility of the NHCS electrode.

In order to investigate the cycling performance and rate capability of the NHCS electrode, galvanostatic charge-discharge cycling was performed. Fig. 4b depicts the charge-discharge curves of the NHCS electrode at different cycles at 100mA/g. During the first discharge-charge cycle, the NHCS electrode delivers specific discharge and charge capacities of 1249 and 303mA·h/g, respectively, showing a 24.3% initial Coulombic efficiency. The large irreversible capacity loss is commonly caused by the electrolyte decomposition and SEI layer formation. In the subsequent cycles, the reversible capacities are stabilized and the discharge-charge curves become more linear, which agrees well with the CV curves.

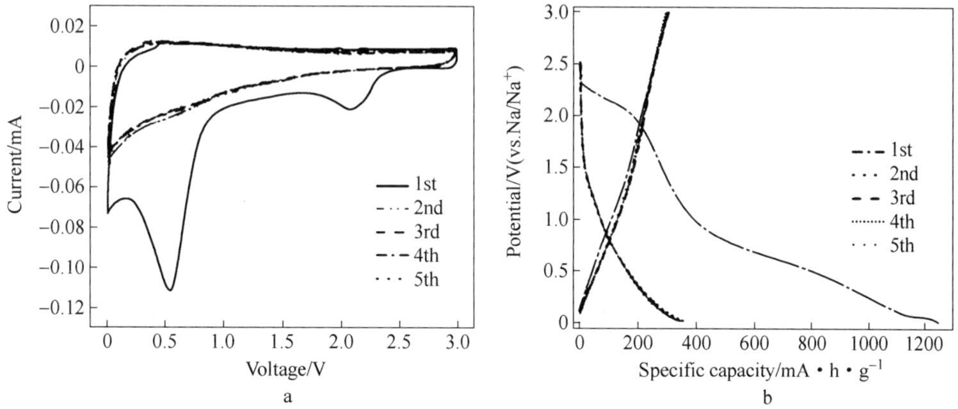

Fig. 4 Cyclic voltammograms of the NHCS electrode at a scanning rate of 0.1mV/s (a), charge/discharge curves of the NHCS electrode at a current density of 100mA/g(b)
(A colour version of this figure can be viewed online)

In addition, for the Na^+ insertion into the disordered carbon reaction mechanisms, three different views have been proposed: (i) It is widely believed that Na^+ storage in disordered carbon follows the sequential intercalation into carbon layers and pore filling into the voids between carbon layers. Namely, the sloping voltage profile can be attributed to the insertion of Na^+ between carbon layers, while insertion of Na^+ into nanopores takes place at the lowest potentials[5,23,25,36]. (ii) However, some experimental results showed discrepancies. For example, Cao et al. proposed that Na^+ intercalation into the carbon layers of hollow carbon nanowires corresponds to the potential plateau at low potentials[21]. Moreover, ex situ XRD by reported studies showed a reversible dilation and contraction of the carbon layers in the low-voltage plateau[22,59-61]. (iii) Recently, Bommier et al. have suggested that the Na^+ storage mechanism may be three-tiered, where the sloping capacity was attributed to defect sites, supported by ex situ total neutron scattering/associated pair distribution function (PDF) studies[62]. By using galvanostatic intermittent titration technique (GITT), the authors observed a substantial increase in diffusivity at voltages close to Na metal plating, which breaks the plateau region down to two possible Na^+ storage mechanisms. Therefore, on the basis of the above conclusion, it is difficult to conclude the Na^+ storage mechanism of the NHCS electrode, and further studies are demanded to fully understand the mechanisms of Na^+ storage in disordered carbon.

The cycling stability during Na⁺ insertion/extraction in the NHCS electrode was also investigated at the current density of 50mA/g. As illustrated in Fig. 5a, the NHCS electrode presents a reversible discharge capacity of 334mA · h/g after 100 cycles at the current density of 50mA/g. The reversible capacity of the NHCS electrode is superior, and only few studies have shown higher capacity at the same current density[63]. Although the initial Coulombic efficiency is low, it increases to approximately 95% after several cycles. Moreover, the small irreversible capacity during each cycle can be attributed to the incomplete stabilization of the SEI film during the charge-discharge processes[25]. In order to verify the improvement of electrochemical performances of N doping, HCSs(nitrogen-free) were also prepared using the same method, but without EDA precursors, and tested as anode materials for SIBs. As showed in Fig. 5a, at 50 mA/g,

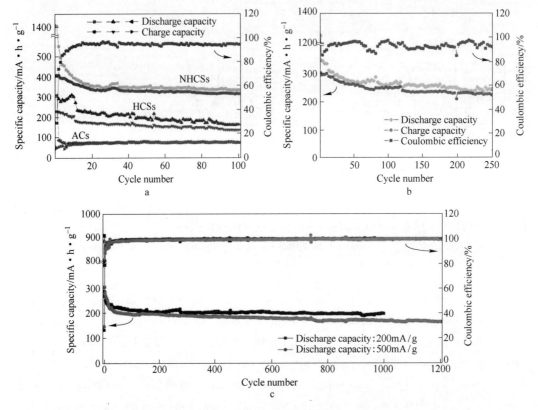

Fig. 5 Cycling performance of the NHCS, HCS, and AC electrodes at the current density of 50mA/g
(a), cycling performance of the NHCS electrode at the current density of 100mA/g
(b), and long term cycling performance of the NHCS electrode at different current
density(c) (A colour version of this figure can be viewed online)

the as-prepared HCSs exhibit a reversible discharge capacity of 163mA · h/g after 100 cycles, which is much lower than that of NHCSs. Thus, it is concluded that nitrogen doping can improve the capacity effectively.

Moreover, in order to verify the improvement of electrochemical performances of hollow structure, the bulk-type carbon is used for a comparison. The commercial ACs(YP17, Kuraray Co., Ltd.) are bulk-type carbon materials, and are widely used in supercapacitors. Therefore, we prepared AC electrode and tested as anode materials for SIBs. As shown in Fig. 5a, at 50mA/g, the as-prepared ACs show a reversible discharge capacity of 74mA · h/g after 100 cycles, which is much lower than that of HCSs. Thus, it can be concluded that the hollow structure improves the capacity effectively. Importantly, the NHCS electrode still exhibits a good cyclic performance at a high current density of 100mA/g, and the capacity still reaches 241mA · h/g after 250 cycles (Fig. 5b). Obviously, the NHCS electrode shows a higher capacity than the hollow carbon nanospheres[25], nitrogen-doped carbon nanosheets[36], and hollow carbon wires[21].

The long cycling performance of the NHCS electrode at different high current densities (200 and 500mA/g) was also investigated. As shown in Fig. 5c, at the current density of 200mA/g, after 100 cycles, the specific capacity stays around 211mA · h/g and a high Coulombic efficiency (~99%) is obtained. Even after 1000 cycles, the NHCS electrode still has a remarkable capacity of 196mA · h/g, corresponding to a capacity retention ratio of 92.9% (from 100 to 1000 cycles). The capacity decay mainly happens in the initial 10 cycles, which possibly results from the volume "adjustment" during the Na^+ insertion/extraction in the carbon structure. Moreover, even at current density as high as 500mA/g, the NHCS electrode still exhibits an excellent cycling stability, with its discharge capacity always stabilizing around 163mA · h/g after 1200 cycles. The long cycling performance of NHCS electrode can be comparable to the previous reports of expanded graphite and porous carbon nanofiber electrodes[23,26], suggesting excellent high rate capability and long-term cycling stability. Such a good long cycling performance of >1200 cycles has seldom been achieved in previous reports on carbon-based anode materials for SIBs[22]. The excellent long cycling performance of the NHCS electrode is most likely attributed to the unique hollow structure, which provides a buffering zone for effective release of mechanical stress caused by Na^+ insertion/extraction[21].

The rate capability of the NHCS electrode is shown in Fig. 6. When tested at different current densities, the NHCS electrode shows high reversible discharge capacities of 296, 235, 181, 155, and 131mA · h/g at 0.2, 0.5, 1, 2, and 5A/g, respectively. Even at an extremely high current density of 10A/g, a high discharge capacity of 114mA · h/g is obtained. When the current density returns to 0.1A/g after 36 cycles, the specific discharge capacity can be recovered up

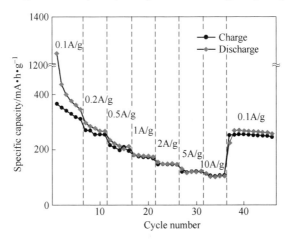

Fig. 6 Rate capability of the NHCS electrode (A colour version of this figure can be viewed online)

to 223mA·h/g, showing the excellent rate performance of the NHCS-based electrode. It is notable that the rate performance of the NHCSs is comparable or even better than those anode materials[22,25,36,64]. In addition, we compared the electrochemical performances of the SIBs based on the NHCS electrode with other typical carbon electrode materials (Table 1). The NHCS electrode is superior to those reported carbon electrode materials. Therefore, the NHCSs are promising for the practical large-scale application as anode materials.

Table 1 Comparison of the performances of the SIBs based on the NHCS electrode with those of other typically carbon electrode materials

Sample	Rate capacity	Cyclic stability	Ref.
Hollow carbon spheres	168mA·h/g at 0.2 A/g	160mA·h/g at 100mA/g(100 cycles)	[25]
	142mA·h/g at 0.5 A/g		
	120mA·h/g at 1 A/g		
	100mA·h/g at 2 A/g		
	75mA·h/g at 5 A/g		
	50mA·h/g at 10 A/g		
Hard carbon microspherules	201mA·h/g at 0.5℃	290mA·h/g at 25mA/g(100 cycles)	[65]
Nitrogen-doped carbon nanosheets	195mA·h/g at 0.2A/g	155mA·h/g at 50mA/g(260 cycles)	[36]
	140mA·h/g at 0.5A/g		
	89mA·h/g at 1A/g		
Carbon nanofibers	200mA·h/g at 1A/g	266mA·h/g at 50mA/g(100 cycles)	[23]
	164mA·h/g at 2A/g	140mA·h/g at 500mA/g(1000 cycles)	
	90mA·h/g at 5A/g		
	60mA·h/g at 10A/g		
Ordered porous carbon	Not reported	128mA·h/g at 25mA/g(100 cycles)	[64]
Nitrogen-doped carbon fibers	153mA·h/g at 1A/g	243mA·h/g at 50mA/g(100 cycles)	[37]
Carbon nanosheet frameworks	134mA·h/g at 2A/g		
	101mA·h/g at 5A/g		
	72mA·h/g at 10A/g		
	250mA·h/g at 0.2A/g	255mA·h/g at 100mA/g(210 cycles)	[22]
	203mA·h/g at 0.5A/g		
	150mA·h/g at 1A/g		
	106mA·h/g at 2A/g		
	66mA·h/g at 5A/g		
Nitrogen-containing hollow carbon microspheres(NHCSs)	296mA·h/g at 0.2A/g	334mA·h/g at 50mA/g(100 cycles)	This study
	235mA·h/g at 0.5A/g	241mA·h/g at 100mA/g(250 cycles)	
	181mA·h/g at 1A/g	196mA·h/g at 20mA/g(1000 cycles)	
	155mA·h/g at 2A/g	163mA·h/g at 500mA/g(1200 cycles)	
	131mA·h/g at 5A/g		
	114mA·h/g at 10A/g		

EIS was used to show the stable performance of the NHCS electrode. Fig. 7a shows the EIS spectra of the NHCS electrode after various discharge-charge cycles (fresh, 10th, 25th, 50th, 75th, 175th, and 250th) at a current density of 100mA/g. All the impedance spectra plots consist of a depressed semicircle in the high-and middle-frequency regions and a straight line in the low-frequency region. The depressed semicircle can be attributed to the SEI film and contact resistance at high frequencies and a charge transfer process in the middle frequency, while the linear increase in the low-frequency range may reflect Warburg impedance due to the diffusion of Na^+ within the NHCS electrode[66]. Thus, the equivalent circuit can be expressed as shown in Fig. 7b, where Rs represents the sum of all Ohmic resistances from the electrode and the electrolyte, Rct is the charge transfer resistance, CPE1 is used instead of double-layer capacitance (C_{dl}), and Wo represents the diffusion impedance (Warburg impedance). The charge transfer resistances (Rct) simulated from the equivalent circuits are 177, 268, 375, 290, 239, 254, and 232Ω after fresh, 10th, 25th, 50th, 75th, 175th, and 250th cycles, respectively (Fig. 7b). The NHCS electrode shows an increase in Rct and reaches its maximum in the first 25 cycles, and then dramatically decreases during the subsequently cycles. After 75 cycles, the Rct achieves a steady state. This phenomenon is interpreted as the result of subsequent stable SEI film formation throughout cycling[22]. Meanwhile, this result also suggests that the unique hollow structure helps to maintain a good electrical conduction for the charge transfer process[25].

In order to further identify the structure change of the NHCS electrode, SEM measurement was performed. Cells were disassembled and the NHCS electrode was washed with propylene carbonate to remove the residual $NaClO_4$. Fig. 8a shows the SEM images of NHCS electrode before being cycled, which is similar to those in Fig. 2a and b, suggesting that the electrode-preparing process did not damage the morphology of the NHCSs. Fig. 8b is the SEM image of NHCS electrode after 250 cycles at the current density of 100mA/g. A comparison of Fig. 8b and a shows that morphology barely change, although the cycled electrode is covered with a dense and rough layer (SEI film). The similar morphology indicates perfect structural stability during the Na^+ insertion/extraction processes.

The excellent electrochemical performance of the NHCS electrode can be attributed to the following reasons: First, low-graphitized NHCSs possess a large interlayer spacing (0.387nm), guaranteeing the insertion of Na^+. Second, the large surface area of NHCSs leads to sufficient electrode-electrolyte interface to absorb Na^+ and promote rapid charge transfer reaction. Third, the very thin carbon shell (~18nm) guarantees a very short Na^+ diffusion distance, which plays an important role in the high rate performance[25]. Finally, the introduction of nitrogen atoms can enhance the electrical conductivity and capacity of the carbonaceous anodes.

4 Conclusions

In summary, we have developed a facile modified template method to synthesize novel NHCSs. Such NHCSs exhibit a large specific surface area, appropriate nitrogen doping, and unique hollow structure. Because of their superior reversible capacity, high rate capability, and long-term cycling stability, they can be as the anode material in SIBs. This novel NHCS electrode shows a superior reversible discharge capacity of 334mA·h/g after 100 cycles at 50mA/g. Even

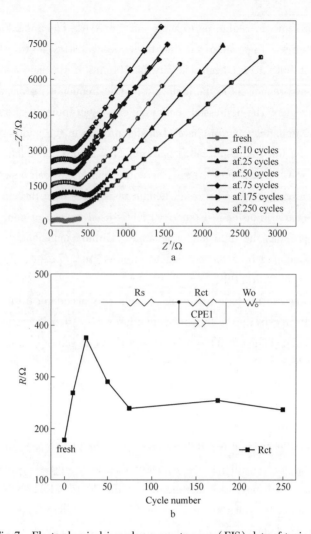

Fig. 7 Electrochemical impedance spectroscopy(EIS)plots of typical cycles for NHCS electrode cycling at the current density of 100mA/g (a)and the corresponding EIS parameters(Rct)derived from the equivalent circuit(inset)(b)(A colour version of this figure can be viewed online)

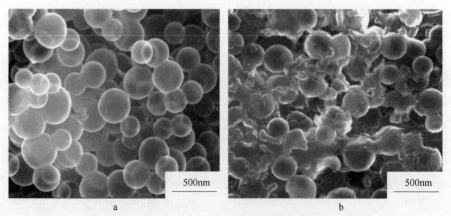

Fig. 8 SEM images of the NHCS electrode: fresh (a)and after 250 cycles at the current density of 100mA/g (b)(A colour version of this figure can be viewed online)

at an extremely high current density of 10A/g, a high discharge capacity of 114mA · h/g is obtained. Moreover, the as-prepared NHCS electrode exhibits a long-term cycling stability up to 1200 cycles even at 500mA/g. The results showed that the NHCSs are promising materials for SIB anodes. Furthermore, this study provides a novel route for NHCS synthesis, and the obtained NHCSs may find application in other fields, including lithium-sulfur batteries, lithium-air batteries, supercapacitors, catalysis, and hydrogen storage.

5 Acknowledgments

The authors acknowledge the financial support of the Teacher Research Fund of Central South University(2013JSJJ027). This study was supported by grants from the Project of Innovation-driven Plan in Central South University(2015CXS018)(2015CX001)and the State Key Laboratory of Powder Metallurgy, Central South University, Changsha, China.

References

[1] B. Dunn, H. Kamath, J. M. Tarascon. Electrical energy storage for the grid: a battery of choices, Science 334(2011):928-935.

[2] B. Scrosati, J. Hassoun, Y. K. Sun. Lithium-ion batteries. A look into the future, Energy Environ. Sci. 4 (2011):3287-3295.

[3] M. D. Slater, D. Kim, E. Lee, C. S. Johnson. Sodium-ion batteries, Adv. Funct. Mater 23(2013):947-958.

[4] S. W. Kim, D. H. Seo, X. Ma, G. Ceder, K. Kang. Electrode materials for rechargeable sodium-ion batteries: potential alternatives to current lithiumion batteries, Adv. Energy Mater 2(2012):710-721.

[5] D. A. Stevens, J. R. Dahn, The mechanisms of lithium and sodium insertion in carbon materials, J. Electrochem Soc. 148(8)(2001):A803-A811.

[6] L. Wang, Y. Lu, J. Liu, M. Xu, J. Cheng, D. Zhang, J. B. Goodenough, A superior low-cost cathode for a Na-ion battery, Angew. Chem. Int. Ed. 52(2013):1964-1967.

[7] Y. Kim, K. H. Ha, S. M. Oh, K. T. Lee. High-capacity anode materials for sodiumion batteries, Chem. Eur. J. 20(38)(2014):11980-11992.

[8] M. Dahbi, N. Yabuuchi, K. Kubota, K. Tokiwa, S. Komaba. Negative electrodes for Na-ion batteries, Phys. Chem. Chem. Phys. 16(2014):15007-15028.

[9] Y. Liu, N. Zhang, L. Jiao, Z. Tao, J. Chen. Ultrasmall Sn nanoparticles embedded in carbon as high-performance anode for sodium-ion batteries, Adv. Funct. Mater 25(2)(2014):214-220.

[10] Y. Xu, Y. Zhu, Y. Liu, C. Wang. Electrochemical performance of porous carbon/Tin composite anodes for sodium-ion and lithium-ion batteries, Adv. Energy Mater 3(2013):128-133.

[11] L. Baggetto, J. K. Keum, J. F. Browning, G. M. Veith. Germanium as negative electrode material for sodium-ion batteries, Electrochem Commun. 34(2013):41-44.

[12] P. R. Abel, Y. M. Lin, T. Souza, C. Chou, A. Gupta, J. B. Goodenough, et al. Nanocolumnar germanium thin films as a high-rate sodium-ion battery anode material, J. Phys. Chem. C 117(2013):18885-18890.

[13] L. Wu, X. Hu, J. Qian, F. Pei, F. Wu, R. Mao, et al. Sb-C nanofibers with long cycle life as an anode material for high-performance sodium-ion batteries, Energy Environ. Sci. 7(2014):323-328.

[14] Y. Zhu, X. Han, Y. Xu, Y. Liu, S. Zheng, K. Xu, et al. Electrospun Sb/C fibers for a stable and fast sodium-ion battery anode, ACS Nano 7(7)(2013):6378-6386.

[15] X. Zhou, Y. Zhong, M. Yang, M. Hu, J. Wei, Z. Zhou. Sb nanoparticles decorated N-rich carbon nanosheets as anode materials for sodium ion batteries with superior rate capability and long cycling stability,

Chem. Commun. 50(2014):12888-12891.

[16] J. Song, Z. Yu, M. Gordin, S. Hu, R. Yi, et al. Chemically bonded phosphorus/graphene hybrid as a high performance anode for sodium-ion batteries, Nano Lett. 14(2014):6329-6335.

[17] W. Li, S. Chou, J. Wang, H. Liu, S. Dou. Simply mixed commercial red phosphorus and carbon nanotube composite with exceptionally reversible sodium-ion storage, Nano Lett. 13(2013):5480-5484.

[18] J. Qian, X. Wu, Y. Cao, X. Ai, H. Yang. High capacity and rate capability of amorphous phosphorus for sodium ion batteries, Angew. Chem. 125(2013):4633-4636.

[19] Y. Kim, Y. Park, A. Choi, N.-S. Choi, J. Kim, J. Lee, et al. An amorphous red phosphorus/carbon composite as a promising anode material for sodium ion batteries, Adv. Mater 25(2013):3045-3049.

[20] M. Doeff, Y. Ma, S. Visco, L. Jonghe, Electrochemical insertion of sodium into carbon, J. Electrochem Soc. 140(12)(1993):L169-L170.

[21] Y. Cao, L. Xiao, M. Sushko, W. Wang, B. Schwenzer, J. Xiao, et al. Sodium ion insertion in hollow carbon nanowires for battery applications, Nano Lett. 12(2012):3783-3787.

[22] J. Ding, H. Wang, Z. Li, A. Kohandehghan, K. Cui, Z. Xu, et al. Carbon nanosheet frameworks derived from peat moss as high performance sodium ion battery anodes, ACS Nano 7(2013):11004-11015.

[23] W. Li, L. Zeng, Z. Yang, L. Gu, J. Wang, X. Liu, et al. Free-standing and binder-free sodium-ion electrodes with ultralong cycle life and high rate performance based on porous carbon nanofibers, Nanoscale 6(2014):693-698.

[24] H. Song, N. Li, H. Cui, C. Wang. Enhanced storage capability and kinetic processes by pores-and heteroatoms-riched carbon nanobubbles for lithium-ion and sodium-ion batteries anodes, Nano Energy 4(2014):81-87.

[25] K. Tang, L. Fu, R. White, L. Yu, M. Titirici, M. Antonietti, J. Maier. Hollow carbon nanospheres with superior rate capability for sodium-based batteries, Adv. Energy Mater 2(2012):873-877.

[26] Y. Wen, K. He, Y. Zhu, F. Han, Y. Xu, I. Matsuda, et al. Expanded graphite as superior anode for sodium-ion batteries, Nat. Commun. 5(2014). 4033, http://dx.doi.org/10.1038/ncomms5033.

[27] H. R. Byon, B. M. Gallant, S. W. Lee, Y. Shao-Horn. Role of oxygen functional groups in carbon nanotube/graphene freestanding electrodes for high performance lithium batteries, Adv. Funct. Mater 23(2013):1037-1045.

[28] H. Song, G. Yang, C. Wang. General scalable strategy toward heterogeneously doped hierarchical porous graphitic carbon bubbles for lithium-ion battery anodes, ACS Appl. Mater. Interfaces 6(2014):21661-21668.

[29] Y. Mao, H. Duan, B. Xu, L. Zhang, Y. Hu, C. Zhao, et al. Lithium storage in nitrogen-rich mesoporous carbon materials, Energy Environ. Sci. 5(2012):7950-7955.

[30] C. Yang, Y. Yin, H. Ye, K. Jiang, J. Zhang, Y. Guo. Insight into the effect of boron doping on sulfur/carbon cathode in lithium-sulfur batteries, ACS Appl. Mater Interfaces 6(2014):8789-8795.

[31] K. See, Y. Jun, J. Gerbec, J. Sprafke, F. Wudl, G. Stucky, et al. Sulfur-functionalized mesoporous carbons as sulfur hosts in Li-S batteries: increasing the affinity of polysulfide intermediates to enhance performance, ACS Appl. Mater Interfaces 6(2014):10908-10916.

[32] L. Qie, W. Chen, Z. Wang, Q. Shao, X. Li, L. Yuan, et al. Nitrogen-doped porous carbon nanofiber webs as anodes for lithium ion batteries with a super-high capacity and rate capability, Adv. Mater 24(2012):2047-2050.

[33] F. Su, C. Poh, J. Chen, G. Xu, D. Wang, Q. Li, et al. Nitrogen-containing microporous carbon nanospheres with improved capacitive properties, Energy Environ. Sci. 4(2011):717-724.

[34] L. Chen, X. Zhang, H. Liang, M. Kong, Q. Guan, P. Chen, et al. Synthesis of nitrogen-doped porous carbon nanofibers as an efficient electrode material for supercapacitors, ACS Nano 6(2012):7092-7102.

[35] Z. Song, T. Xu, M. Gordin, Y. Jiang, I. Bae, Q. Xiao, et al. Polymer-graphene nanocomposites as ultrafast-charge and -discharge cathodes for rechargeable lithium batteries, Nano Lett. 12(2012):2205-2211.

[36] H. Wang, Z. Wu, F. Meng, D. Ma, X. Huang, L. Wang, et al. Nitrogen-doped porous carbon nanosheets as low-cost, high-performance anode material for sodium-ion batteries, ChemSusChem 6(2013):56-60.

[37] J. Xu, M. Wang, N. Wickramaratne, M. Jaroniec, S. Dou, L. Dai. High-performance sodium ion batteries based on a 3D anode from nitrogen-doped graphene foams, Adv. Mater 27(2015):2042-2048.

[38] N. Jayaprakash, J. Shen, S. Moganty, A. Corona, L. Archer. Porous hollow carbon@sulfur composites for high-power lithium-sulfur batteries, Angew. Chem. Int. Ed. 50(2011):5904-5908.

[39] K. Zhang, Q. Zhao, Z. Tao, J. Chen. Composite of sulfur impregnated in porous hollow carbon spheres as the cathode of Li-S batteries with high performance, Nano Res. 6(1)(2013):38-46.

[40] N. Brun, K. Sakaushi, L. Yu, L. Giebeler, J. Eckert, M. Titirici. Hydrothermal carbon-based nanostructured hollow spheres as electrode materials for high-power lithium-sulfur batteries, Phys. Chem. Chem. Phys. 15(2013):6080-6087.

[41] S. Feng, W. Li, Q. Shi, Y. Li, J. Chen, Y. Ling, A. Asiri, D. Zhao. Synthesis of nitrogen-doped hollow carbon nanospheres for CO_2 capture, Chem. Commun. 50 (2014):329-331.

[42] N. Wickramaratne, J. Xu, M. Wang, L. Zhu, L. Dai, M. Jaroniec. Nitrogen enriched porous carbon spheres: attractive materials for supercapacitor electrodes and CO_2 adsorption, Chem. Mater 26 (2014): 2820-2828.

[43] L. Yu, N. Brun, K. Sakaushi, J. Eckert, M. Titirici. Hydrothermal nanocasting: synthesis of hierarchically porous carbon monoliths and their application in lithium-sulfur batteries, Carbon 61(2013):245-253.

[44] X. Zhou, J. Tang, J. Yang, J. Xie, B. Huang. Seaweed-like porous carbon from the decomposition of polypyrrole nanowires for application in lithium ion batteries, J. Mater Chem. A1(2013):5037-5044.

[45] L. Li, A. Manthiram. O-and N-doped carbon nanowebs as metal-free catalysts for hybrid Li-air batteries, Adv. Energy Mater 4(10)(2014). http://dx.doi.org/10.1002/aenm.201301795.

[46] D. Wang, F. Li, L. Yin, X. Lu, Z. Chen, I. Gentle, et al. Nitrogen-doped carbon monolith for alkaline supercapacitors and understanding nitrogen-induced redox transitions, Chem. Eur. J. 18(17)(2012):5345-5351.

[47] W. Zhou, X. Xiao, M. Cai, L. Yang. Polydopamine-coated, nitrogen-doped, hollow carbon-sulfur double-layered core-shell structure for improving lithium-sulfur batteries, Nano Lett. 14(9)(2014):5250-5256.

[48] A. Chen, Y. Yu, H. Lv, Y. Wang, S. Shen, Y. Hu, et al. Thin-walled, mesoporous and nitrogen-doped hollow carbon spheres using ionic liquids as precursors, J. Mater. Chem. A1(2013):1045-1047.

[49] J. Han, G. Xu, B. Ding, J. Pan, H. Dou, D. MacFarlane. Porous nitrogen-doped hollow carbon spheres derived from polyaniline for high performance supercapacitors, J. Mater. Chem. A2(2014):5352-5357.

[50] H. Wang, C. Zhang, Z. Liu, L. Wang, P. Han, H. Xu, et al. Nitrogen-doped graphene nanosheets with excellent lithium storage properties, J. Mater Chem. 21 (2011):5430-5434.

[51] W. Shin, H. Jeong, B. Kim, J. Kang, J. Choi. Nitrogen-doped multiwall carbon nanotubes for lithium storage with extremely high capacity, Nano Lett. 12(2012):2283-2288.

[52] D. Hulicova-Jurcakova, M. Seredych, G. Lu, T. Bandosz. Combined effect of nitrogen-and oxygen-containing functional groups of microporous activated carbon on its electrochemical performance in supercapacitors, Adv. Funct. Mater 19 (2009):438-447.

[53] H. Hou, C. Banks, M. Jing, Y. Zhang, X. Ji. Carbon quantum dots and their derivative 3D porous carbon frameworks for sodium-ion batteries with ultralong cycle life, Adv. Mater 27(47)(2015):7861-7866.

[54] L. Fu, K. Tang, K. Song, et al. Nitrogen doped porous carbon fibres as anode materials for sodium ion batteries with excellent rate performance, Nanoscale 6 (2014):1384-1389.

[55] Z. Wang, L. Qie, L. Yuan, W. Zhang, X. Hu, Y. Huang. Functionalized N-doped interconnected carbon nanofibers as an anode material for sodium-ion storage with excellent performance, Carbon 55(2013):

328-334.

[56] Y. Matsumura, S. Wang, J. Mondori. Mechanism leading to irreversible capacity loss in Li ion rechargeable batteries, J. Electrochem. Soc. 142(9)(1995):2914-2918.

[57] P. Thomas, D. Billaud. Electrochemical insertion of sodium into hard carbons, Electrochim Acta 47(20)(2002):3303-3307.

[58] G. Lota, K. Fic, E. Frackowiak. Carbon nanotubes and their composites in electrochemical applications, Energy Environ. Sci. 4(2011):1592-1605.

[59] Y. Matsuo, K. Ueda. Pyrolytic carbon from graphite oxide as a negative electrode of sodium-ion battery, J. Power Sources 263(2014):158-162.

[60] Y. Matsuo, K. Hashiguchi, K. Ueda, Y. Muramatsu. Electrochemical intercalation of sodium ions into thermally reduced graphite oxide, Electrochemistry 83(5)(2015):345-347.

[61] S. Komaba, W. Murata, T. Ishikawa, et al. Electrochemical Na insertion and solid electrolyte interphase for hard-carbon electrodes and application to Na-ion batteries, Adv. Funct. Mater 21(2011):3859-3867.

[62] C. Bommier, T. Surta, M. Dolgos, X. Ji. New mechanistic insights on Na-ion storage in nongraphitizable carbon, Nano Lett. 15(2015):5888-5892.

[63] Y. Yan, Y. Yin, Y. Guo, L. Wan. A sandwich-like hierarchically porous carbon/graphene composite as a high-performance anode material for sodium-ion batteries, Adv. Energy Mater (2014) 4, http://dx.doi.org/10.1002/aenm.201301584.

[64] C. Jo, Y. Park, J. Jeong. K. Lee, J. Lee. Structural effect on electrochemical performance of ordered porous carbon electrodes for Na-ion batteries, ACS Appl. Mater. Interfaces 7 (2015):11748-11754.

[65] Y. Li, S. Xu, X. Wu, J. Yu, Y. Wang, Y. Hu, et al. Amorphous monodispersed hard carbon micro-spherules derived from biomass as a high performance negative electrode material for sodium-ion batteries, J. Mater. Chem. A3(2015):71-77.

[66] H. Yan, X. Huang, H. Li, L. Chen. Electrochemical study on $LiCoO_2$ synthesized by microwave energy, Solid State Ion. 113-115(1998):11-15.

Confining Selenium in Nitrogen-containing Hierarchical Porous Carbon for High-rate Rechargeable Lithium-selenium Batteries[*]

Abstract A novel nitrogen-containing hierarchical porous carbon(NCHPC) was prepared by a simple template process and chemical activation and a selenium/carbon composite based on NCHPC was synthesized for lithium-selenium batteries by a melt-diffusion method. The Se-NCHPC composite was characterized by X-ray diffraction (XRD), field emission scanning electron microscopy (SEM), and transmission electron microscopy (TEM) measurements. It is found that the elemental selenium was dispersed inside the hierarchical pores of NCHPC. It is demonstrated from cyclic voltammetry(CV) and galvanostatic discharge-charge processes that the Se-NCHPC composite has a large reversible capacity and high rate performance as cathode materials. The Se-NCHPC composite with a selenium content of 56.2 wt% displays an initial discharge capacity of 435mA·h/g and a reversible discharge capacity of 305mA·h/g after 60 cycles at a 2 C charge-discharge rate. In particular, the Se-NCHPC composite presents a long electrochemical stability at a high rate of 5 C. The results reveal that the electrochemical reaction constrained inside the interconnected macro/meso/micropores of NCHPC would be the dominant factor for the enhancement of the high rate performance of the selenium cathode, and the nitrogen-containing hierarchical porous carbon network would be a promising carbon matrix structure for lithium-selenium batteries.

1 Introduction

Rechargeable batteries with high energy density are essential for solving imminent energy and environmental issues. Among various types of rechargeable batteries, lithium-sulfur(Li-S) batteries hold great potential due to their high theoretical specific capacity of 1675mA·h/g and specific energy of 2600W·h/kg[1]. Li-S batteries have been studied as one of the most promising systems for next generation high-energy rechargeable lithium batteries due to their high theoretical specific capacity(1675mA·h/g) and energy density(2600W·h/kg).[2-4] However, Li-S batteries suffer from the electronically and ionically insulating nature of sulfur and the solubility of reductive polysulfides in organic electrolytes during cycling.[4-7]

As a congener of sulfur, selenium being electrochemically similar to sulfur can be expected to react with lithium to generate selenides. Although the theoretical gravimetric capacity of selenium(675mA·h/g) is lower than that of sulfur(1672mA·h/g), the theoretical volumetric capacity of selenium(3253mA·h/cm^3 based on 4.82g/cm^3) is comparable to that of sulfur (3467mA·h/cm^3 based on 2.07g/cm^3). In addition, selenium has 20 orders of magnitude higher electrical conductivity than sulfur.[8-13] The advantages of selenium make it a prospective candidate for the cathode material in high energy density rechargeable batteries for specific ap-

---

[*] Copartner: Yaohui Qu, Zhian Zhang, Shaofeng Jiang, Xiwen Wang, Yanqing Lai, Jie Li. Reprinted from Journal of Materials Chemistry A, 2014, 2: 12255-12261.

plications. However, at present, research on lithium-selenium(Li-Se) batteries is still at a very early stage. Similar to sulfur, selenium cathodes also face the dissolution issue of high-order poly selenides, resulting in fast capacity fading, poor cycle performance and low Coulombic efficiency[9,10].

In attempts to improve the electrochemical performance of the selenium cathode, various strategies have been explored. One strategy is synthesizing nanoselenium with different morphologies and structures, such as nano-fibrous selenium[13] and nanoporous selenium[12]. Wu et al.[12] prepared nanoporous selenium as a cathode material with a reversible capacity of 206mA·h/g after 20 cycles at a current density of 100mA/g. The other strategy is confining selenium within various forms of porous carbon matrices exhibiting good electrochemical performance, such as multi-walled carbon nanotubes[8], porous carbon spheres[10], and ordered mesoporous carbon[11]. Abouimrane et al[8]. conducted pioneering work on the use of the Se/C composite as a cathode material. The Li-Se system sustained a reversible capacity of ~300mA·h/g at a low current density(50mA/g, ~C/12) after 100 cycles, demonstrating the excellent cycle life of this system. Guo et al.[11] synthesized the selenium/carbon composite(49.0 wt% selenium) through a facile melt-diffusion process from a ball-milled mixture of CMK-3 and Se. The Se/C cathode exhibited excellent cycling stability with a discharge capacity of 600mA·h/g at a current density of 67mA/g(0.1 C) after 50 cycles. In particular, porous carbon matrices have been proved to be effective and facile candidates to improve the selenium utilization and restrain the solubility of polyselenides on account of their excellent conductivity, large surface area, abundant porous channels and strong adsorbent properties. Herein, we were motivated to prepare a novel hierarchical porous carbon with large surface area, high pore volume and advanced pore structure to improve the utilization of selenium and the electrochemical performance of Li-Se batteries.

Additionally, recent studies have also shown that surface functional groups or doped heteroatoms play important roles in improving the performances of the carbon-based electrode materials[14-18]. The heteroatoms such as N, O, B, S or P in the porous carbon may enhance the electrical conductivity, influence the wettability, restrain the diffusion of soluble polysulfides, retard the shuttle effect and improve the cyclability of lithium-sulfur batteries[19-23]. Meanwhile, pyrolyzing nitrogenrich chemicals such as polyacrylonitrile, polypyrrole and polyaniline is the most common way to obtain nitrogen-containing porous carbon materials. Gelatin is a mixture of peptides and proteins by partial hydrolysis of collagen extracted from various animal by-products. Taking into account its low-cost and high contents of carbon/nitrogen species, gelatin is expected to be an excellent candidate precursor for producing porous carbon, especially containing nitrogen.

Therefore, in this study, we synthesize a novel nitrogencontaining hierarchical porous carbon (NCHPC) to confine selenium for lithium-selenium batteries. The NCHPC was fabricated *via* a facile template route with gelatin as a precursor and silica spheres as hard templates. The obtained NCHPC possesses a unique hierarchical porous structure with interconnected macro/meso/micropores. The NCHPC with a special interconnected porous framework could provide appropriate surfaces and channels to facilitate selenium infiltration into the interior pores, provide a continuous pathway for electron transport, and reduce the transport length of Li$^+$ ions. The selenium-NCHPC(Se-NCHPC) composite was prepared by a melt-diffusion method. Used as the cathode material in Li-Se batteries, showing the high-rate performance, the as-prepared Se-NCHPC composite(56.2 wt% Se) retained 305mA·h/g after 60 cycles at a 2 C charge-discharge rate.

2 Experimental

The nitrogen-containing hierarchical porous carbon(NCHPC) was prepared *via* a facile template route and chemical activation process. First, the colloidal silica spheres with a diameter of 130 nm were prepared by the Stöber method as reported previously[24]. Second, 4.0g gelatin, 6.0g colloidal silica spheres and 300mL distilled-deionized water were mixed to form a homogeneous solution under stirring, the solution was then transferred into a fask and kept in a water bath at 95℃ under vigorous stirring until the solvent was completely evaporated. Third, the dried solid samples were subjected to carbonization by heating at 800℃ for 2h under a nitrogen atmosphere in a tube furnace, and the silica spheres were removed by etching the products in 10 wt% HF to generate nitrogen-containing carbon. Fourth, a mixture of the nitrogen-containing carbon and KOH in the weight ratio of 1∶4 was heated in a tube furnace in a nitrogen atmosphere at 800℃ for 2h, the product was then washed with 1mol/L HCl solution and deionized water until the filtrate became neutral. Finally, the product was dried overnight at 120℃ in an oven and the resulting NCHPC was obtained.

To prepare the Se-NCHPC composite, elemental selenium(AR, Aladdin, China) and the as-prepared NCHPC with a weight ratio of 6∶4 were mixed together and placed in a sealed vessel and the mixture was heated to 260℃ for 15h under an argon atmosphere with a heating rate of 5℃/min, and then the Se-NCHPC composite was obtained. The commercial activated carbons (YP17, kuraray Co., Ltd.) are widely used in supercapacitors; as a comparison, the Se-activated carbon composite(Se-AC) was also prepared by mixing selenium with activated carbon in a weight ratio of 6∶4 and heated at 260℃ for 15h.

Field emission scanning electron microscopy(SEM, Nova NanoSEM 230) and transmission electron microscopy(TEM, Tecnai G2 20ST) were applied to characterize the materials. X-ray diffraction (XRD, Rigaku3014) measurements were carried out with CuKα radiation. Thermogravimetric analysis(TGA, SDTQ600) was conducted in determining the selenium content in the composite. N_2 adsorption/desorption measurements were performed by using a Quantachrome instrument(Quabrasorb SI-3MP) at 77K.

The selenium cathode was prepared by mixing 80 wt% active material(Se-NCHPC or Se-AC composite), 10 wt% acetylene black, and 10 wt% sodium alginate(SA) binder in deionized water solvent. The slurry was spread onto aluminum foil, and dried at 60℃ overnight, then the cathodes were cut into pellets with a diameter of 1.0cm and dried for 12h in a vacuum oven at 60℃. The typical mass loading of the active material was 0.7-1.0mg/cm^2. The electrochemical performance was performed using a CR2025 coin-type cell. CR2025-type coin cells were assembled in an argon-filled glove box(Universal 2440/750) in which oxygen and water contents were less than 1×10^{-6}. The electrolyte used was 2 M bis(trifluoromethane)sulfonamide lithium salt(LiTFSI,Sigma Aldrich) in a solvent mixture of 1,3-dioxolane and 1,2-dimethoxyethane (1:1,v/v)(Acros Organics). A lithium metal was used as the counter electrode and the reference electrode and a Celgard 2400 was used as a separator. Cyclic voltammetry(CV) measurements were conducted using a PARSTAT 2273 electrochemical measurement system. CV tests were performed at a scan rate of 0.2mV/s in the voltage range of 1.2 to 3.0V. Galvanostatic charge-discharge tests were performed in the potential range of 1.2 to 3.0V at 25℃ by using a LAND CT2001A battery-testing instrument.

3 Results and Discussion

The schematic illustration of the synthesis procedure of the Se-NCHPC is shown in Fig. 1. Gelatin was firstly dissolved in the colloidal silica spheres to form a homogeneous solution under stirring. Because of the electrostatic interaction between the colloidal silica spheres and gelatin, gelatin can be absorbed and coated on the surface of the colloidal silica spheres. During solvent evaporation, the gelatin was melted and flowed down along the shell of the colloidal silica spheres, and then the polymer/silica composite was obtained. Secondly, the polymer/silica composite was carbonized at 800℃ for 2h under an argon gas atmosphere. The silica spheres were removed by etching the products in 10 wt% HF, then the nitrogen-containing carbon was obtained. Thirdly, nitrogen-containing carbon was activated via the KOH activation process and the NCHPC was prepared. Finally, the obtained NCHPC was mixed with elemental selenium. The mixture was heated to 260℃ for 15h under an argon atmosphere, and then the Se-NCHPC composite was obtained.

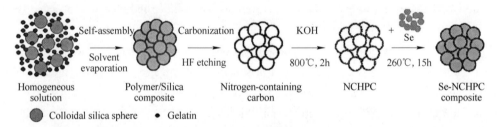

Fig. 1 Schematic illustration of the synthesis procedure of the Se-NCHPC composite

Field-emission scanning electron microscopy (SEM) and transmission electron microscopy (TEM) were used to examine the morphology and microstructure of the as-prepared NCHPC. As shown in Fig. 2a and b, it can be observed that the NCHPC is a hierarchical porous structure and the size of the interconnected pores is in the range of 120-140nm, agreed well with the size of colloidal silica spheres. The typical TEM images shown in Fig. 2c and d reveal that the NCHPC has a hollow structure with large internal void space and a thin porous wall, which indicates a hierarchical porous structure to facilitate sufficient filling of the active material[25,26]. The HRTEM image (inset in Fig. 2d) of the NCHPC indicates that the thickness of the porous wall is only about 5-7nm and reveals that the porous wall is the microporous texture. The NCHPC has a unique open ordered hierarchical porous structure; the special hierarchical porous structure could favor the elemental selenium confined within the carbon matrix and facilitate the transport of electrolyte ions[10,11].

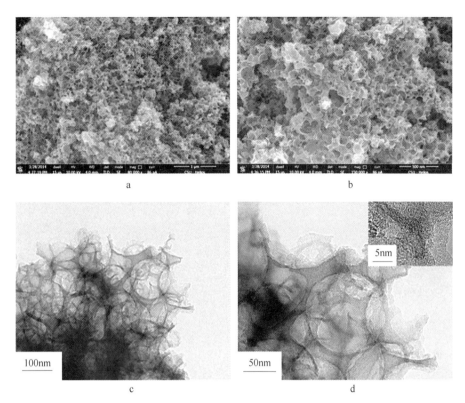

Fig. 2　SEM images of the NCHPC (a and b), TEM images of the NCHPC (c and d), and HRTEM (inset) image of the NCHPC (d)

The specific surface area analysis of the as-prepared NCHPC was performed by nitrogen BET adsorption measurements. The pore size distribution of carbon was calculated by the BJH method. As shown in Fig. 3a and b, a type IV isotherm is observed, indicating the assembly of micro-, meso-and macropore structure of the NCHPC. The total pore volume is 1.36cm^3/g. The specific BET surface area is about 2001.5m^2/g with the contribution of 1600.1m^2/g from the micropore, revealing the presence of abundant micropores. The abundant micropores are attributed to the KOH activation process, while the mesopores and macropores are contributed by the removal of the colloidal silica spheres embedded in the carbon matrix. Those results are consistent with the SEM/TEM measurements. On the basis of these results, it can be concluded that the NCHPC has a hierarchical pore structure with interconnected macro/meso/micropores, these pores can provide room for selenium loading and act as pathways for the electrolyte diffusion.

In order to confirm the chemical composition and the surface properties of NCHPC, XPS measurements were performed and the result is shown in Fig. 3. Fig. 3c shows XPS survey spectra of NCHPC, three peaks centering at 284.6, 400.5, and 531.8eV, corresponding to C1s, N1s, and O1s, respectively. As estimated by XPS results, the atom percent of nitrogen is 5.99 at% for NCHPC. The high resolution N1s peaks(Fig. 3d) can be divided into three fitted peaks with different binding energies. The peak centered at $ca.$ (398.1±0.3)eV is due to pyridinic nitrogen(N-6)[27], the one at $ca.$ (400.6±0.3)eV to pyrrolic nitrogen and/or pyridonic nitrogen(N-5)[28,29]. The contribution of the peak at *ca.* 402.5±0.3 eV derives from graphitic nitrogen.[30] Compared with the surface chemical post-modification (for example, NH_3 treatment of porous carbon at high temperature), our work provides a simple route to produce N-containing porous carbon[31-33].

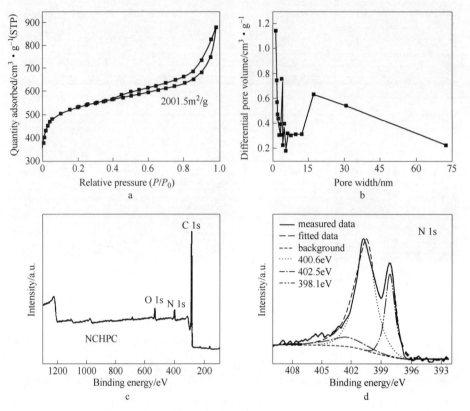

Fig. 3 The nitrogen adsorption/desorption isotherm of the NCHPC(a), the pore size distribution of the NCHPC(b), XPS spectra of the NCHPC(c), and high-resolution spectra of N1s for the NCHPC(d)

The selenium-nitrogen-containing hierarchical porous carbon (Se-NCHPC) composite was prepared *via* a simple melt-diffusion strategy.[11] Fig. 4a and b show SEM images of the Se-NCHPC composite. There is no distinguishable morphological difference between the Se-NCHPC composite and the NCHPC (Fig. 2a and b), which suggests that selenium is successfully distributed within the pores of NCHPC. Furthermore, the TEM image of the Se-NCHPC composite (Fig. 4c) illustrates that no large bulk selenium could be observed in the composite, indicating that selenium homogeneously disperses in NCHPC. Moreover, the TEM elemental mapping images of carbon, nitrogen and selenium (Fig. 4f-h) are found to have similar intensity across the Se-NCHPC composite, indicating that nitrogen is evenly distributed throughout the carbon framework and confirming that selenium has a highly dispersed state in NCHPC, which corroborates well with the SEM and TEM observation.

X-ray diffraction (XRD) patterns of elemental selenium (AR, Aladdin, China), NCHPC and the Se-NCHPC composite are given in Fig. 5a. For selenium, there are several peaks at 23.5° (100), 29.7° (101), 41.3° (110), 43.6° (102), 45.4° (111), 51.8° (201), 55.7° (112),

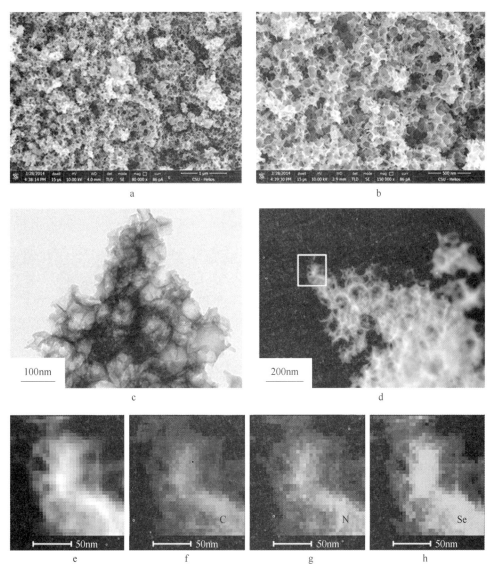

Fig. 4 SEM images of the Se-NCHPC composite(a and b), TEM images of the Se-NCHPC composite(c and d), (e) is a magnified image of the labeled position in(d). (f)-(h) are the corresponding TEM elemental mapping of carbon, nitrogen and selenium in the Se-NCHPC composite, respectively

and 61.5°(202), which are in good accordance with the diffraction peaks of the trigonal phase of selenium(JCPDS 06-0362).[34,35] The XRD pattern of NCHPC shows a broad reflection at 2θ of about 24°, which can be attributed to the amorphous characteristic of the as-prepared NCHPC. After encapsulating selenium into the NCHPC, as shown in the XRD curve of the Se-NCHPC composite, the sharp diffraction peaks of bulk crystalline selenium disappear entirely, which may be due to the dispersion of amorphous selenium at a molecular level confined in the pores of NCHPC,[11] which agrees well with the results of SEM and TEM. To determine the selenium content in the Se-NCHPC composite, this material was analyzed by thermogravimetric analysis(TGA) under a nitrogen atmosphere, as shown in Fig. 5b. The TGA result shows that the selenium content is 56.2 wt% in the Se-NCHPC composite, which is consistent with the proportions of the added amount. On the basis of these results, it can be concluded that the meltdiffusion method can successfully trap selenium inside the pores of the NCHPC, which is similar to the result of Guo et al.[11]

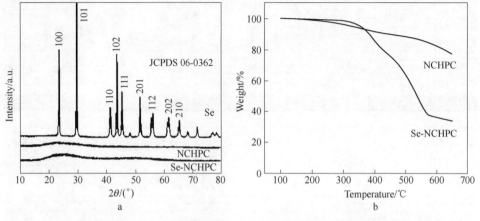

Fig. 5 XRD patterns of elemental selenium, NCHPC and Se-NCHPC composite
(a) and TGA curve of NCHPC and Se-NCHPC composite (b)

To study the electrochemical properties of the Se-NCHPC composite, the CR2025 coin cells with a metallic lithium counter electrode were assembled and evaluated, along with the comparison of the Se-AC composite. Fig. 6a shows the typical cyclic voltammogram (CV) curves of the Se-NCHPC composite cathode (56.2 wt% selenium) in the voltage range of 1.2-3.0V with a constant scan rate of 0.2mV/s. In the first cycle, two obvious cathodic peaks and one anodic peak were observed, which are consistent with the previous report.[9,36] The two remarkable reduction peaks for the Se-NCHPC composite cathode are about 1.9V and 2.1V, respectively, corresponding to the reduction of elemental selenium to soluble polyselenides and then to the insoluble Li_2Se_2 and Li_2Se.[9] In the anodic scan, only one sharp oxidation peak can be observed at about 2.23V for the Se-NCHPC composite cathode, which corresponds to the conversion of Li_2Se into high-order soluble polyselenides.[9] In the subsequent scans, the main reduction peaks are shifted to higher potentials and the oxidation peaks to lower potentials, indicating an improvement of reversibility of the cathode with cycling. In addition, the peak current is decreased slowly during the next 4 cycles, indicating that the Se-NCHPC composite cathode may have good electrochemical stability. It also indicated that the hierarchical porous carbon structure is quite effective in preventing the loss of selenium in the electrolyte and maintained high utilization of the active selenium in the redox reactions.

In order to investigate the cycling performance and rate capability of the Se-NCHPC composite (56.2 wt% selenium), galvanostatic charge-discharge cycling was performed. Fig. 6b depicts the first cycle charge and discharge curves of the Se-NCHPC composite cathode at different rates (0.5 C, 1 C, 2 C, 3 C, and 1 C is 675mA/g) and shows the typical two-plateau behavior of a selenium cathode, consistent with the result of the cyclic voltammetry measurement, which can be ascribed to the two step reaction of elemental selenium with metallic lithium during the discharge process.[9] With the increase of the discharge rate from 0.5 C to 3 C, the first discharge

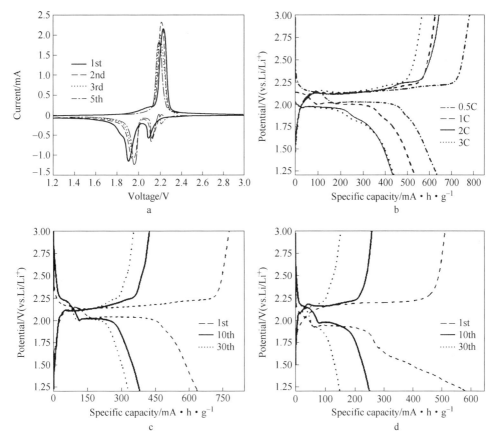

Fig. 6 Cyclic voltammograms of the Se-NCHPC composite cathode at a scanning rate of 0.2mV/s (a), charge-discharge curves of the Se-NCHPC composite cathode at different current densities between 1.2V and 3.0V(b). Charge-discharge curves of the Se-NCHPC composite cathodes at different cycles at a current density of 0.5 C(c). Charge-discharge curves of the Se-AC composite cathodes at different cycles at a current density of 0.5C (d)

capacity slightly fades, and the charge-discharge plateaus gradually rise/drop, but the typical two plateaus in the discharge curves still maintain during all the cycles even at very high current rates(2 C and 3 C), suggesting a little kinetic barrier in the electrode process and high rate capability. This could be attributed to the high quality of NCHPC and the thin porous carbon wall(only about 5-7nm), which significantly improved electronic and ionic transport at the selenium cathode.

To demonstrate the possible structural advantages of the NCHPC, the Se-AC(53.5 wt% selenium)composite was also prepared for comparison. Fig. 6c and d show charge-discharge curves of Se-NCHPC and Se-AC composite cathodes at different cycles at 0.5 C (1 C is 675mA/g). Two distinct discharge plateaus for the two cathodes are observed. The high plateau is attributed to the reduction of elemental Se into lithium polyselenide $Li_2Se_n(n \geq 4)$, while the other lower plateau is ascribed to the transition from lithium polyselenide $Li_2Se_n(n \geq 4)$ to Li_2Se_2 and Li_2Se. During the charge process, the charge plateau involves the oxidation from discharge product Li_2Se to intermediate polyselenides $Li_2Se_n(n \geq 4)$ and Se.[9] The Se-AC composite cathode shows a discharge capacity of 582mA·h/g in the first cycle, which drops to 249mA·h/g in the 10th cycle and quickly decreases to about 149mA·h/g in the 30th cycle, showing a poor cycle performance. For the Se-NCHPC composite cathode, the higher initial discharge capacity of 636mA·h/g is obtained, more importantly, the capacity remains at 330mA·h/g in the 30th cycle. It is found that the cell with the Se-NCHPC composite cathode shows more complete and stable plateaus at about 2.2V and 2.0V. The discharge plateau of the Se-AC composite cathode obviously shrinks with the increasing cycle number while the Se-NCHPC composite cathode has overlapping upper plateaus.

Cycling performance and Coulombic effciency of the Se-NCHPC and Se-AC composite cathodes are presented in Fig. 7a and b. All capacity values in this study were calculated based on selenium mass. As shown in Fig. 7a, the initial discharge capacity of the Se-AC composite cathode is 357mA·h/g at 1 C(1 C is 675mA/g) and delivers a reversible capacity of 119mA·h/g after 60 cycles. Wu et al.[12] prepared commercial Se particles as a cathode material with a reversible capacity of 60mA·h/g after 20 cycles at a current density of 100mA/g Note that the Se-NCHPC composite cathode displays higher utilization of active materials than the Se-AC composite cathode with an initial discharge capacity of 535mA·h/g. After 60 cycles, the Se-NCHPC composite cathode retains a reversible capacity of 267mA·h/g, which is higher than the prepared Se-AC composite cathode. Furthermore, at a higher rate of 2 C, as described in Fig. 7b, the initial discharge capacity of the Se-AC composite cathode is 321mA·h/g, after 60 cycles, the low discharge capacity of 103mA·h/g is retained. However, the Se-NCHPC composite cathode shows an initial discharge capacity of 435mA·h/g and still retains a capacity of 305mA·h/g after 60 cycles, almost three times higher than that of the Se-AC composite cathode. Obviously, especially at a very high current rate(2 C), the Se-NCHPC composite cathode shows higher discharge capacity than the Se-AC composite.

Moreover, the cycling performance of the Se-NCHPC composite cathode at a high constant rate(2 C) is better than the cycling performance at a low rate(1 C). This can probably be attributed to the relatively rapid charge-discharge process at a high constant rate, where polyselenides are not immediately dissolved in the electrolyte with the adsorption effect of the interconnected macro/meso/micropores in NCHPC. The shuttle effect is suppressed so that the cycling performance of the Se-NCHPC composite cathode can improve.[8,10] From Fig. 7a and b, we can also see that the Se-NCHPC and the Se-AC composite cathodes show high Columbic efficiency. The remarkable Columbic efficiency for the Se-NCHPC composite cathode should benefit from the hierarchical porous structure; the interconnected macro/meso/micropore framework can provide a physical barrier for overcharge due to alleviation of polyselenide dissolution and the shuttle effect.[10,11] The high Coulombic efficiency of the Se-AC composite cathode could be related to a larger surface area(1640m^2/g) of activated carbon, which helps to restrain polyselenide diffusion.

The cycling performance of the Se-NCHPC composite cathode at high discharge rates(3 C,4 C,5 C) was investigated. As shown in Fig. 7c, the initial discharge capacities of the Se-NCHPC composite cathode at a discharge rate of 3 C is 439mA·h/g; the cathode maintains an irreversible capacity of 240mA·h/g after 80 cycles. More importantly, at 4 C and 5 C, the dis-

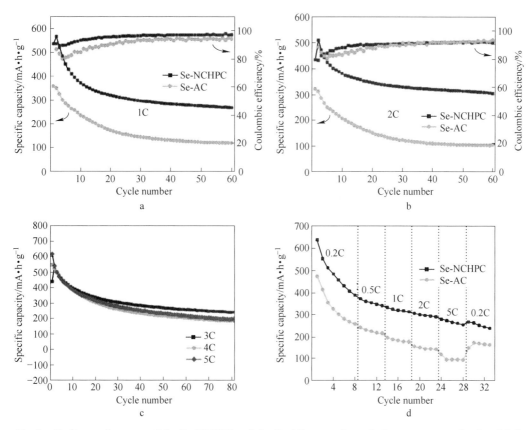

Fig. 7 Cycling performance of the Se-NCHPC and the Se-AC composite cathodes at a current density of 1 C (a). Cycling performance of the Se-NCHPC and the Se-AC composite cathodes at a current density of 2 C (b). Cycling performance of the Se-NCHPC composite cathode at different current densities (c). The rate capability of the Se-NCHPC composite cathode and the Se-AC composite cathodes (d)

charge capacity of the Se-NCHPC composite cathode is very similar after 80 cycles, suggesting a high rate capability and durable cycling stability. The rate capability of the Se-NCHPC composite cathode is shown in Fig. 7d. The discharge capacity gradually decreases as the current rate increased from 0.2 C to 5 C(1 C is 675mA/g). After 13 cycles at 0.5 C, the Se-NCHPC and Se-AC composite cathodes show the reversible capacity of 348mA · h/g and 215mA · h/g, respectively. Even after 28 cycles, a satisfactory capacity of 261mA · h/g is obtained for Se-NCHPC at 5 C, which is almost 2.8 times higher than that of the Se-AC composite cathode (94mA · h/g). The Se-NCHPC composite exhibits high-rate performance, especially under high rate conditions (2 C and 5 C), likely because of the facile electronic/ionic transport and improved reaction kinetics in the NCHPC. Wang et al.[10] prepared the Se/C composite as a cathode material for Li-Se batteries, which also shows excellent high-rate capability, but the content of selenium is lower (30.0 wt% selenium). The hierarchical porous carbon structure exhibits a similar high-rate performance in the lithium-sulfur batteries.[26,37]

The excellent high rate capability and high capacity retention of the Se-NCHPC composite cathode can be attributed to the unique nitrogen-containing hierarchical porous structure. The NCHPC with a special interconnected macro/meso/micropore framework can provide appropriate surfaces and channels to facilitate selenium infiltration into the interior pores to improve the conductivity of selenium, absorb the formed polyselenides, and retard the shuttle effect during the charge-discharge process. Moreover, the hierarchical porous with an interconnected macro/meso/micropore framework can supply facile transport channels for Li$^+$ ions during the electrochemical process. Therefore, the electrochemical reaction constrained inside the interconnected macro/meso/micropores of NCHPC proposed here would be the dominant factor for the excellent high rate capability and high capacity retention of the selenium cathode.

4 Conclusions

The novel nitrogen-containing hierarchical porous carbon (NCHPC) was prepared by a simple template process and chemical activation; the Se-NCHPC composite was synthesized for lithium-selenium batteries by the melt-diffusion method. The results demonstrated that the NCHPC is a promising candidate for lithium-selenium battery cathodes. Such NCHPC, combining a unique interconnected macro/meso/micropore, large internal void space and thin porous wall, and the special hierarchical porous framework can effectively confine selenium and suppress the diffusion of dissolved polyselenides. As a result, the Se-NCHPC composite cathode exhibits a high initial discharge capacity of 435mA·h/g and retains as high as 305mA·h/g after 60 cycles even at a high current rate of 2 C. Moreover, the Se-NCHPC composite presents a long electro-chemical stability at a high rate of 5 C. It is concluded that the high rate capability and high capacity retention of the selenium cathode can be enhanced significantly by confining selenium in the nitrogen-containing hierarchical porous carbon network nanostructure. Consequently, the NCHPC would be a promising carbon matrix to develop lithium-selenium batteries with high rate capability and high capacity.

5 Acknowledgements

The authors thank the financial support of the Teacher Research Fund of Central South University (2013JSJJ027). We also thank the support of the Engineering Research Center of Advanced Battery Materials and the Ministry of Education, China.

References

[1] P. G. Bruce, S. A. Freunberger, L. J. Hardwick, J. M. Tarascon. Nat. Mater., 2012, 11:19.

[2] J. Nelson, S. Misra, Y. Yang, A. Jackson, Y. Liu, H. Wang, H. Dai, J. C. Andrews, Y. Cui and M. F. Toney. J. Am. Chem. Soc., 2012, 134:6337.

[3] X. Ji, K. T. Lee, L. F. Nazar, Nat. Mater., 2009, 8:500.

[4] S. S. Zhang. J. Power Sources, 2013, 231:153.

[5] L. Yin, J. Wang, F. Lin, J. Yang, Y. Nuli. Energy Environ. Sci., 2012, 5:6966.

[6] N. Jayaprakash, J. Shen, S. S. Moganty, A. Corona and L. A. Archer, Angew. Chem., Int. Ed., 2011, 50:1.

[7] C. F. Zhang, H. B. Wu, C. Z. Yuan, Z. P. Guo, X. W. Lou. Confining sulfur in double-shelled hollow carbon spheres for lithium-sulfur batteries, Angew. Chem., Int. Ed., 2012, 51:9592.

[8] A. Abouimrane, D. Dambournet, K. Chapman, P. Chupas, W. Weng and K. Amine. J. Am. Chem. Soc., 2012: 134, 4505.

[9] Y. Cui, A. Abouimrane, J. Lu, T. Bolin, Y. Ren, W. Weng, C. Sun, V. Maroni, S. Heald, K. Amine. J. Am.

Chem. Soc. ,2013,135:8047.

[10] C. Luo,Y. Xu,Y. Zhu,Y. Liu,S. Zheng,Y. Liu,A. Langrock,C. Wang:ACS Nano,2013,7: 8003.

[11] C. Yang,S. Xin,Y. Yin,H. Ye,J. Zhang and Y. Guo,Angew. Chem,Int. Ed. ,2013,52:1.

[12] L. Liu,Y. Hou,S. Xiao,Z. Chang,Y. Yang,Y. P. Wu,Chem. Commun. ,2013,49:11515.

[13] D. Kundu,F. Krumeich and R. Nesper. J. Power Sources,2013,236,112.

[14] Z. Wen,X. Wang,S. Mao,Z. Bo,H. Kim,S. Cui,G. Lu,X. Feng,J. Chen. Adv. Mater. ,2012,24:5610.

[15] H. R. Byon,B. M. Gallant,S. W. Lee and Y. Shao-Horn. Adv. Funct. Mater. ,2013,23:1037.

[16] L. Zhao, L. Z. Fan, M. Q. Zhou, H. Guan, S. Qiao, M. Antonietti, M. M. Titirici. Adv. Mater. , 2010, 22:5202.

[17] F. Su,C. K. Poh,J. S. Chen,G. Xu,D. Wang,Q. Li,J. Lin,X. W. Lou. Energy Environ. Sci. ,2011,4:717.

[18] Y. Mao, H. Duan, B. Xu, L. Zhang, Y. Hu, C. Zhao, Z. Wang, L. Chen, Y. Yang. Energy Environ. Sci. , 2012,5:7950.

[19] D. Hulicova-Jurcakova,M. Seredych,G. Q. Lu and T. J. Bandosz. Adv. Funct. Mater. ,2009,19:438.

[20] L. F. Chen, X. D. Zhang, H. W. Liang, M. Kong, Q. F. Guan, P. Chen, Z. Y. Wu, S. H. Yu. ACS Nano, 2012,6:7092.

[21] F. Sun,J. Wang,H. Chen,W. Li,W. Qiao,D. H. Long,L. Ling. ACS Appl. Mater. Interfaces,2013,5:5630.

[22] X. Wang,Z. Zhang,Y. Qu,Y. Lai,J. Li. J. Power Sources,2014,256:361.

[23] J. Song, T. Xu, M. L. Gordin, P. Zhu, D. Lv, Y. Jiang, Y. Chen, Y. Duan, D. H. Wang. Adv. Funct. Mater. , 2014,24:1243.

[24] N. C. Strandwitz and G. D. Stucky. Hollow microporous cerium oxide spheres templated by colloidal silica, Chem. Mater. ,2009,21:4577.

[25] S. Wei,H. Zhang,Y. Huang,W. Wang,Y. Xia,Z. Yu. Energy Environ. Sci. ,2011,4:736.

[26] B. Ding,C. Yuan,L. Shen,G. Xu,P. Nie and X. Zhang. Chem. -Eur. J. ,2013,19:1013.

[27] X. Zhou,J. Tang,J. Yang,J. Xie and B. Huang. J. Mater. Chem. A,2013,1,5037.

[28] L. Qie,W. M. Chen,Z. H. Wang,Q. G. Shao,X. Li,L. X. Yuan,X. L. Hu,W. X. Zhang,Y. H. Huang. Adv. Mater. ,2012,24:2047.

[29] L. Li and A. Manthiram. Adv. Energy Mater. ,2014,DOI: 10. 1002/aenm. 201301795.

[30] D. W. Wang, F. Li, L. C. Yin, X. Lu, Z. G. Chen, I. R. Gentle, G. Q. Lu, H. M. Cheng. Chem. -Eur. J. , 2012,18:5345.

[31] C. Li,X. Yin,L. Chen,Q. Li and T. Wang. J. Phys. Chem. C,2009,113:13438.

[32] M. Sevilla,R. Mokaya and A. B. Fuertes. Energy Environ. Sci. ,2011,4:2930.

[33] X. Xiang,Z. Huang,E. Liu,H. Shen,Y. Tian,H. Xie,Y. Wu,Z. Wu. Electrochim. Acta,2011,56:9350.

[34] X. M. Li,Y. Li,S. Q. Li,W. W. Zhou,H. B. Chu,W. Chen,I. L. Li,Z. K. Tang. Cryst. Growth Des. ,2005, 5:911.

[35] B. Cheng and E. T. Samulski. Chem. Commun. ,2003,16:2024.

[36] Z. A. Zhang,Z. Y. Zhang,K. Zhang,X. Yang,Q. Li. RSC Adv. ,2014,4:15489.

[37] J. Huang, X. Liu, Q. Zhang, C. Chen, M. Zhao, S. Zhang, W. Zhu, W. Qian, F. Wei. Nano Energy, 2013, 2:314.

A Simple SDS-assisted Self-assembly Method for the Synthesis of Hollow Carbon Nanospheres to Encapsulate Sulfur for Advanced Lithium-sulfur Batteries[*]

Abstract The hollow carbon nanospheres(HCNSs) were prepared using a simple SDS-assisted self-assembly method, and a sulfur-carbon composite based on HCNSs was synthesized for lithium-sulfur batteries by a vapor phase infusion method. The sulfur-HCNS composite was characterized by X-ray diffraction(XRD), field emission scanning electron microscopy(SEM), transmission electron microscopy(TEM), and thermogravimetry(TG) measurements. It is found that the elemental sulfur was dispersed inside the pores of carbon spheres. It is demonstrated from the galvanostatic discharge-charge process and cyclic voltammetry(CV) that the sulfur-HCNS composite has a large reversible capacity and an excellent cycling performance as cathode materials. The sulfur-HCNS composite with a sulfur content of 47.6 wt% displays an initial discharge capacity of 1031mA·h/g and a reversible discharge capacity of 477mA·h/g after 100 cycles at 0.5 C charge-discharge rate.

1 Introduction

High energy density rechargeable batteries have received great attention in recent years because of their potential applications, such as the power source for electric vehicles, energy storage devices and smart grids.[1] Among various types of rechargeable batteries, lithium-sulfur batteries hold great potential due to their high theoretical specific capacity of 1675mA·h/g and specific energy of 2600W·h/kg.[2] In addition, sulfur as an active cathode material is non-toxic, naturally abundant, and environmentally benign. However, the practical realization of lithium-sulfur batteries has met great challenges, including low utilization of active materials and rapid capacity fading because of the insulating nature of sulfur, the solubility of polysulfides(Li_2S_x, $3 \leqslant x \leqslant 8$) as intermediary products, and the insoluble agglomerates of irreversible discharge products(Li_2S_2 and Li_2S).[2-5]

In order to overcome the above problems, different strategies have been attempted, including optimization of the organic electrolyte,[6] preparation of the sulfur-conductive polymer composite,[7] and fabrication of the sulfur-carbon composite.[8,9] In particular, the sulfur-carbon composite has been considered as one of the most promising cathode materials for advanced lithium-sulfur batteries, because carbon materials are effective to improve the sulfur utilization and restrain the solubility of lithium polysulfides, these improved characteristics are attributed to the excellent conductivity, large surface area, narrow pores, and strong adsorbent properties of carbon materials.

In recent years, various types of carbon materials have been applied, such as activated carbon

---

[*] Copartner: Yaohui Qu, Zhian Zhang, Xiwen Wang, Yanqing Lai, and Jie Li. Reprinted from Journal of Materials Chemistry A, 2013, 1, 14306-14310.

fibers,[10] porous carbon spheres,[11] hollow carbon nanofibers,[12,13] graphene oxides,[14,15] and hollow carbon spheres,[16] and have received much attention because of high conductivity and porous structure. Among them, the sulfur-carbon composite by confining sulfur in hollow carbon spheres shows outstanding electrochemical properties, including cycling performance and rate capability.[16] To date, there have been several reports on the fabrication of sulfur-carbon composites by encapsulating sulfur into the hollow carbon spheres,[16-18] however, the hollow carbon spheres were obtained *via* a hard template approach using silica particles or SnO_2 hollow spheres as the hard templates in these reports.[16-18] Moreover, the hard template method has many disadvantages, which is tedious, and the post-treatment necessary to remove the hard template adds complexity to the whole synthetic process and increases the chance of the structural deformation as well as the introduction of impurities.

In order to overcome these disadvantages, herein, we report a simple sodium dodecyl sulfate (SDS)-assisted self-assembly approach to prepare hollow carbon nanospheres(HCNSs). In this study, the self-assembly route without using any hard template is very simple, which will lead to the development of an effective and low cost fabrication process that has high potential for scaling up. Then, the sulfur-HCNS composite was prepared by the vapor phase infusion process with optimized sulfur contents. Physical properties of the HCNSs and the sulfur-HCNS composite and electrochemical performances were investigated. The sulfur-HCNS composite with a sulfur content of 47.6 wt% exhibits an initial discharge capacity of 1031mA·h/g and a reversible discharge capacity of 477mA·h/g after 100 cycles at 0.5 C charge-discharge rate, which are higher than those of the sulfur-activated carbon composite.

2 Experimental

The HCNSs were synthesized by an easy SDS-assisted self-assembly method,[19] using glucose as the precursor and sodium dodecyl sulfate(SDS) as the surfactant. First, 3.5g glucose and 0.08g SDS were dissolved in 35mL water to form a homogeneous solution under manual stirring, the solution was then transferred into a 60mL Teflon-lined stainless steel autoclave and hydrothermally treated at 160℃ for 16h. Second, the puce products were filtered and washed with absolute ethanol and distilled water several times, the samples were obtained after oven-drying at 60℃ for 12h. Finally, the dried solid samples were carbonized by heating at 800℃ for 2h under argon gas flow in a tube furnace, then we obtained the black final HCNSs.

The sulfur-HCNS composite was prepared *via* the vapor phase infusion process[13] Sublimed sulfur(AR, Aladdin, China) and HCNSs were mixed in a 3∶2 weight ratio, the mixture were put in quartz tubes that were sealed under vacuum, the sulfur impregnation of the HCNSs was obtained by heating the quartz tube at 160℃ for 10h and being further heated at 500℃ for a further 5h. As a comparison, the commercial activated carbons (YP17, kuraray Co., Ltd) were used; the sulfur-activated carbon composite (S-AC) was also prepared by mixing sulfur with activated carbon in a weight ratio of 3∶2 and heated at 155℃ for 12h.

Field emission scanning electron microscopy (SEM, Nova NanoSEM 230) and transmission electron microscopy (TEM, Tecnai G2 20ST) were applied to characterize the materials. X-ray diffraction (XRD, Rigaku3014) measurements were carried out with Cu Kα radiation. Thermogravimetric analysis (TGA, SDTQ600) was conducted to determine the sulfur content in the composites. N_2 adsorption/desorption measurements were performed by using a Quantachrome instrument (Quabrasorb SI-3 MP) at 77K.

The sulfur cathode was prepared by mixing 80 wt% active material (S-HCNS or S-AC composite), 10 wt% acetylene black, and 10 wt% polyvinylidene fluoride (PVDF) binder in the N-methyl-2-pyrrolidinone (NMP) solvent. The slurry was spread onto aluminum foil, and then dried at 60℃ overnight, then the cathodes were cut into pellets with a diameter of 1.0cm and dried for 12h in a vacuum oven at 60℃. The typical mass loading of active S was 0.8-1.0mg/cm^2. The electrochemical performance was performed using a CR2025 coin-type cell. CR2025-type coin cells were assembled in an argon-filled glove box (Universal 2440/750) in which oxygen and water contents were less than 1×10^{-6}. The electrolyte used was 2 M bis(trifluoromethane)sulfonamide lithium salt (LiTFSI, Sigma Aldrich) in a solvent mixture of 1,3-dioxolane and 1,2-dimethoxyethane (1∶1, v/v) (Acros Organics), including 0.1 M lithium nitrate ($LiNO_3$) as an additive. Lithium metal was used as the counter electrode and reference electrode and Celgard 2400 was used as the separator. Cyclic voltammetry (CV) measurements were conducted using a PARSTAT 2273 electrochemical measurement system. CV tests were performed at a scan rate of 0.2mV/s in the voltage range of 1.5 to 3.0V. Galvanostatic charge-discharge tests were performed in the potential range of 1.5 to 3.0V at 25℃ by using a LAND CT2001A battery-testing instrument.

3 Results and Discussion

The schematic illustration of a self-assembled process of the HCNSs is shown in Fig. 1. Glucose solution can produce a lot of bubbles under the hydrothermal conditions, mainly CO_2, CO, and H_2,[20,21] and these bubbles might serve as templates for construction of hollow spheres.[22,23] In aqueous solution, the SDS molecules are easily adsorbed onto the interfaces of the bubbles; the hydrophilic group of the SDS, namely, the sulfonic group, points to the outer surface of the bubbles, while the hydrophobic end points to the inner. Because of the electrostatic interaction of the sulfonic group and the glucose, the outer surface of the bubbles is occupied by a lot of glucose. Then the absorbed glucose will carbonize in this region during the hydrothermal process,[20] and the obtained C will gradually deposit on the surface of the bubbles to form the HCNSs, after hydrothermal carbonization the particles were further carbonized at 800℃ for 2h in an argon gas flow in a tube furnace.

Field-emission scanning electron microscopy (SEM) and transmission electron microscopy (TEM) were used to examine the morphology and microstructure of the as-prepared HCNSs. As shown in Fig. 2a and b, the majority of the products exhibit spherical and hemispherical morphology with diameters ranging from 200 to 500nm. A typical TEM image shown in Fig. 2c and d reveals that the spheres have a hollow structure with large internal void space and thin shells, which is similar to the results reported by Archer et al.[16] and Lou et al.,[17] moreover, it can be seen in Fig. 2d, the spheres have a unique hollow hemispherical structure compared with the previous reports,[16-18] and the special open structure could favor the elemental sulfur confined within the carbon matrix.

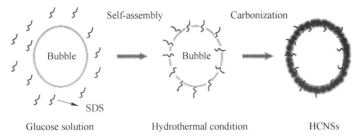

Fig. 1 Schematic illustration of the assembled process of the HCNSs

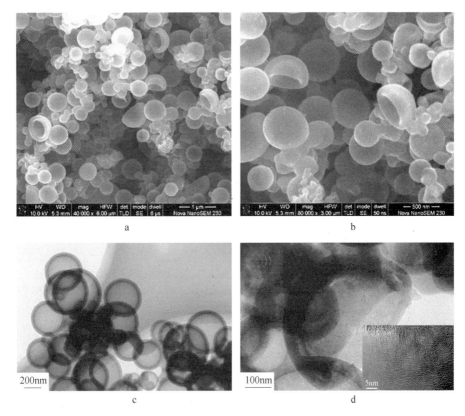

Fig. 2 SEM images of the HCNSs(a and b), TEM images of the HCNSs(c), TEM and HRTEM(inset) of the HCNSs(d)

The XRD pattern of the as-prepared HCNSs(Fig. 3a) shows the presence of reflections characteristic of the carbon hexagonal phase, the broad diffraction peak appears at about 24° and 44° which can be indexed to(002) and(100)/(101) planes, respectively, suggesting the presence of partially graphitized structures of HCNSs. Moreover, the partially graphitized structures of the HCNSs could be further supported by Raman spectroscopy(Fig. 3b). Two characteristic peaks around 1440 and 1490cm^{-1} could be ascribed to the D band arising from the defects and disorder in carbonaceous solid and the G band from the stretching mode of C-C bonds of typical graphite, respectively.[24] The intensity ratio(I_D/I_G) of the two bands is about 0.96, further verifying the partial graphitization degree of the as-obtained HCNSs. Because the electrical conductivity of graphitic carbon is substantially higher than that of amorphous carbon, the partially graphitized HCNSs may ensure good electronic conductivity, and facilitate transport of electrons from the poorly conducting sulfur, so it is expected that the HCNSs might exhibit excellent electrochemical performances.

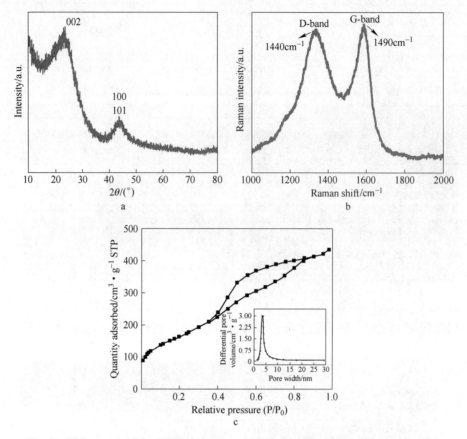

Fig. 3　XRD pattern of the HCNSs(a), Raman spectrum of the HCNSs(b), the nitrogen adsorption/desorption isotherm and the inset is the pore size distribution(c)

　　The specific surface area analysis of the as-prepared HCNSs was performed by nitrogen BET adsorption measurements. The pore size distribution of carbon was calculated by the BJH method. As shown in Fig. 3c, a type Ⅳ isotherm was observed, indicating the characteristic mesoporous structure of the HCNSs. The specific BET surface area is 580.1m^2/g, and the pores of the HCNSs with a narrow size distribution near 3-4nm, well consistent with the TEM results, which is similar to the result reported by Titirici et al.[25] These pores can act as pathways for the impregnation of sulfur into the interior when the sulfur-carbon composite is formed.[4]

　　The sulfur-HCNS composite was prepared *via* the vapor phase infusion process. Fig. 4a is the SEM of the S-HCNS composite. There is no distinguishable morphological difference between the S-HCNS composite and the HCNSs(Fig. 2b), which suggests that the homogeneous distribution of sulfur in the pores of HCNSs. Furthermore, the TEM of the S-HCNS composite (Fig. 4b), clearly shows the uniform distribution of the sulfur in its internal hollow structure. Elemental composition of the S-HCNS composite analyzed by energy-dispersive X-ray (EDX)microanalysis is shown in Fig. 4c. EDX spectra collected from different locations within the S-HCNS composite also indicates the presence of sulfur throughout the HCNSs. This can be further confirmed by the BET analysis that the specific surface area, average pore diameter and pore volume of the S-HCNS composite are reduced to $18.2m^2/g$, $1.3nm$ and $0.05cm^3/g$ from the initial $580.1m^2/g$, $3.4nm$ and $0.67cm^3/g$ of HCNSs, respectively. The results indicate that the sulfur is filled inside the pore structure of carbon spheres, which is similar to the reported literature.[4,16]

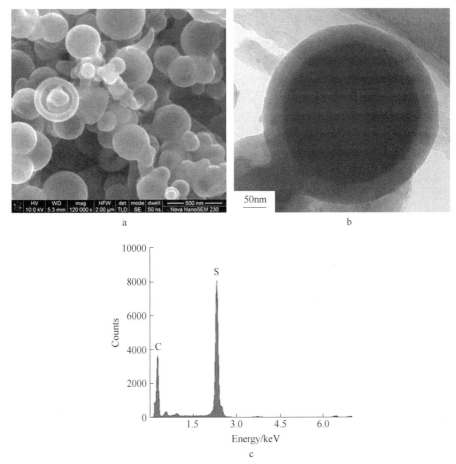

Fig. 4 SEM image of the S-HCNS composite(a), TEM image of the
S-HCNS composite(b), and EDX analysis of the S-HCNS composite(c)

X-ray diffraction(XRD) patterns of sublimed sulfur(AR, Aladdin, China) and the S-HCNS composite are given in Fig. 5a. The sharp diffraction peaks of crystalline sulfur are retained in the sublimed sulfur, meanwhile, the sulfur in the S-HCNS composite exists in a crystallized form, however, the broadened diffraction peaks with much reduced intensity compared with that of sublimed sulfur suggest that most of the sulfur penetrated into the pores of the HCNSs.[10,18] The content of sulfur was measured by thermogravimetric analysis(TGA) under a N_2 atmosphere. The TGA results(Fig. 5b) show that the sulfur contents are 47.6 wt% and 53.3 wt% in the S-HCNS composite and the S-AC composite. In the preparation of the composite, the weight ratio S to HCNS or AC is 3 : 2, which is larger than that in the product. The lower content of sulfur in the resulting composite(47.6 wt% or 53.3 wt%) can be ascribed to the inevitable loss of S during preparation. Moreover, compared with the S-AC composite, the sulfur component in the S-HCNS composite evaporates at a slightly elevated temperature. This is likely due to the strong interaction between sulfur and HCNSs, and the excellent encapsulation capability of the HCNSs.[9]

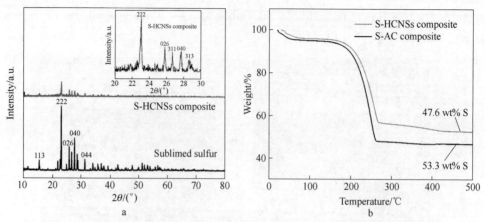

Fig. 5 XRD patterns of the sublimed sulfur and the S-HCNS composite, and the inset is the enlargement of the 20°-30° (2θ) region of the S-HCNS composite(a), and TGA curves of the S-HCNS composite and the S-AC composite(b)

Fig. 6 shows the cyclic voltammogram(CV) curves of the S-HCNS composite cathode(47.6 wt% sulfur) at the first, fifth, and tenth cycles. At the first cycle, there are two obvious reduction peaks at about 1.9V and 2.2V. The peak at 2.2V associates with the conversion of elemental sulfur to soluble lithium polysulfide(Li_2S_n, $4 \leqslant n \leqslant 8$), and the peak at about 1.9V is related to the reduction of lithium polysulfides to insoluble Li_2S_2 and Li_2S, respectively.[3,26,27] These two changes correspond to the two discharge plateaus in the discharge curve. Two expected oxidation peaks overlap to form one large peak at about 2.5-2.7V, which corresponds to the conversion of Li_2S into high-order soluble polysulfides.[28,29] According to the result of CV, the shapes of the CV curves are close to each other and the CV peak positions are not charged after first ten cycles, indicating that the S-HCNS composite cathode may have a similar reversibility during scanning.

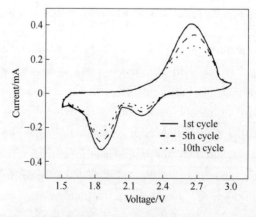

Fig. 6 Cyclic voltammograms of the S-HCNS composite cathode at a scanning rate of 0.2mV·s^{-1}

Fig. 7a and b show charge-discharge curves of S-HCNS and S-AC composite cathodes at different cycles. There exist two distinct discharge plateaus for the two cathodes. The high plateau is attributed to the reduction of S_8 into $Li_2S_n(4 \leqslant n \leqslant 8)$, while the other lower plateau is ascribed to the transition from high-order lithium polysulfide to insoluble lithium sulfide(Li_2S_2 or Li_2S). During the discharge process, polysulfide is formed. The charge plateau involves the oxidation from loworder lithium polysulfide to high-order lithium polysulfide. The S-AC composite cathode shows a discharge capacity of 669mA · h/g in the first cycle, which drops to 349mA · h/g in the 10th cycle and quickly decreases to about 185mA · h/g in the 50th cycle, which is similar to the result of the sulfur-carbon nanotube,[30] and the carbon black-sulfur.[17] For the S-HCNS composite cathode, the higher initial discharge capacity of 1031mA · h/g is obtained, more importantly, the capacity remains at 598mA · h/g in the 50th cycle.

Cycling performances of the S-HCNS and the S-AC composite cathodes are presented in Fig. 7c and d. All capacity values in this study were calculated based on sulfur mass with a constant current density of 0.5 C. As shown in Fig. 7c and d, the initial discharge capacity of the S-AC composite cathode is 669mA · h/g, and delivers a reversible capacity of 99mA · h/g after 100 cycles. The S-HCNS composite cathode exhibits an initial discharge capacity of 1031mA · h/g, and still maintains a capacity of 477mA · h/g after 100 cycles, which is higher than that of the

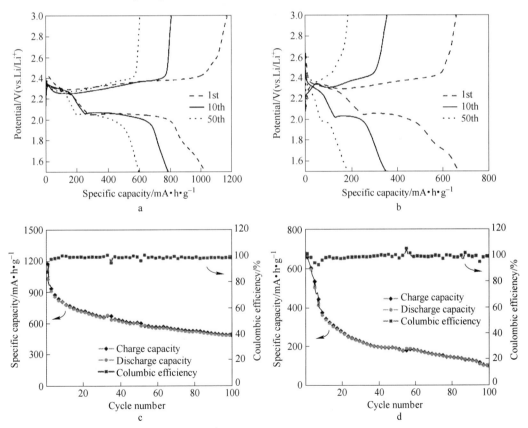

Fig. 7 Charge-discharge curves of the S-HCNS(a) and the S-AC
(b) composite cathodes at different cycles at a current density of 0.5
C. Cycling performance of the S-HCNS(c) and the S-AC (d) composite cathodes at a current density of 0.5 C

sulfur impregnated in mesoporous carbon,[31] and even better than the sulfur-carbon nanotube @ meso C.[30] The cyclability is significantly improved for the S-HCNS composite cathode, which can be attributed to the porous hollow carbon structure. The HCNSs can confine sulfur into the porous hollow framework to improve the conductivity of insulate sulfur, which can accommodate the formed polysulfides, retard the shuttle effect, and restrain the volume change during charge-discharge process.[9,16] Moreover, the porous hollow structure and the thin shells can facilitate transport of Li$^+$ ions in the electrolyte due to the existence of large amount of pores.

From Fig. 7c and d, we can also see that the S-HCNS and the S-AC composite cathodes show high coulombic efficiency. In our case, the coulombic efficiency for the S-HCNS and the S-AC composite cathodes is ~100%. The remarkable coulombic efficiency for the S-HCNS composite cathode should benefit from the porous hollow structure, the porous framework can provide a physical barrier for overcharge due to alleviation of polysulfide dissolution and the shuttle effect.[16,32] The high columbic efficiency of the S-AC composite cathode could be related to a larger surface area (1640m^2/g) of activated carbon, which helps to restrain polysulfides diffusion.[33]

4 Conclusions

Novel HCNSs were prepared *via* a simple SDS-assisted selfassembly method, the S-HCNS composite was synthesized for lithium-sulfur batteries by the vapor phase infusion method. The results show that the highly conductive HCNSs matrix allow fast transport of electrons and Li$^+$ ions, provide a porous hollow structure to encapsulate sulfur and absorb polysulfides, and suppresses the shuttle effect during the charge-discharge process. Therefore, the S-HCNS composite cathode exhibits better active material utilization and electrochemical reversibility than the S-AC composite cathode. The sulfur-HCNS composite cathode with a sulfur content of 47.6 wt% displays an initial discharge capacity of 1031mA · h/g and a reversible discharge capacity of 477mA · h/g after 100 cycles at 0.5 C charge-discharge rate. Consequently, the HCNSs would be a promising carbon matrix for Li-S batteries and the simple synthetic method indicates great potential for mass production in the future.

Acknowledgements

Financial support by "the Strategic Emerging Industries Program of Shenzhen, China (JCYJ201-20618164543322)" is gratefully acknowledged.

References

[1] J. M. Tarascon and M. Armand, Nature, 2001, 414, 359.
[2] P. G. Bruce, S. A. Freunberger, L. J. Hardwick and J. M. Tarascon, Nat. Mater., 2012, 11, 19.
[3] X. L. Ji and L. F. Nazar, J. Mater. Chem., 2010, 20, 9821.
[4] X. L. Ji, K. T. Lee and L. F. Nazar, Nat. Mater., 2009, 8, 500.
[5] S. -E. Cheon, S. -S. Choi, J. -S. Han, Y. -S. Choi, B. -H. Jung and H. S. Lim, J. Electrochem. Soc., 2004, 151, A2067.
[6] Z. Lin, Z. C. Liu, W. J. Fu, N. J. Dudney and C. D. Liang, Adv. Funct. Mater., 2013, 23, 1064.

[7] F. Wu, J. Z. Chen, R. J. Chen, S. X. Wu, L. Li, S. Chen and T. Zhao, J. Phys. Chem. C, 2011, 115, 6057.

[8] J. Wang, S. Y. Chew, Z. W. Zhao, S. Ashraf, D. Wexler, J. Chen, S. H. Ng, S. L. Chou and H. K. Liu, Carbon, 2008, 46, 229.

[9] C. D. Liang, N. J. Dudney and J. Y. Howe, Chem. Mater., 2009, 21, 4724.

[10] R. Elazari, G. Salitra, A. Garsuch, A. Panchenko and D. Aurbach, Adv. Mater., 2011, 23, 5641.

[11] J. Kim, D. -J. Lee, H. -G. Jung, Y. -K. Sun, J. Hassoun and B. Scrosati, Adv. Funct. Mater., 2013, 23, 1076.

[12] G. Y. Zheng, Y. Yang, J. J. Cha, S. S. Hong and Y. Cui, Nano Lett., 2011, 11, 4462.

[13] J. C. Guo, Y. H. Xu and C. S. Wang, Nano Lett., 2011, 11, 4288.

[14] L. W. Ji, M. M. Rao, H. M. Zheng, L. Zhang, Y. C. Li, W. H. Duan, J. H. Guo, E. J. Cairns and Y. G. Zhang, J. Am. Chem. Soc., 2011, 133, 18522.

[15] H. L. Wang, Y. Yang, Y. Y. Liang, J. T. Robinson, Y. G. Li, A. Jackson, Y. Cui and H. J. Dai, Nano Lett., 2011, 11, 2644.

[16] N. Jayaprakash, J. Shen, S. S. Moganty, A. Corona and L. A. Archer, Angew. Chem., Int. Ed., 2011, 50, 5904.

[17] C. F. Zhang, H. B. Wu, C. Z. Yuan, Z. P. Guo and X. W. Lou, Angew. Chem., Int. Ed., 2012, 51, 1.

[18] K. Zhang, Q. Zhao, Z. L. Tao and J. Chen, Nano Res., 2013, 6, 38.

[19] X. M. Sun and Y. D. Li, J. Colloid Interface Sci., 2005, 291, 7.

[20] G. C. A. Luijkx, F. Rantwijk, H. van Bekkum and M. J. Antal Jr, Carbohydr. Res., 1995, 272, 191.

[21] R. D. Cortright, R. R. Davda and J. A. Dumesic, Nature, 2002, 418, 964.

[22] Q. Peng, Y. J. Dong and Y. D. Li, Angew. Chem., Int. Ed., 2003, 42, 3027.

[23] V. Bajpai, P. G. He and L. M. Dai, Adv. Funct. Mater., 2004, 14, 145.

[24] J. Robertson, Mater. Sci. Eng., R, 2002, 37, 129.

[25] N. Brun, K. Sakaushi, L. H. Yu, L. Giebeler, J. Eckert and M. M. Titirici, Phys. Chem. Chem. Phys., 2013, 15, 6080.

[26] Y. J. Jung and S. Kim, Electrochem. Commun., 2007, 9, 249.

[27] Y. J. Li, H. Zhan, S. Q. Liu, K. L. Huang and Y. H. Zhou, J. Power Sources, 2010, 195, 2945.

[28] S. E. Cheon, K. S. Ko, J. H. Cho, S. W. Kim, E. Y. Chin and H. T. Kim, J. Electrochem. Soc., 2003, 150, A796.

[29] J. R. Akridge, Y. V. Mikhaylik and N. White, Solid State Ionics, 2004, 175, 243.

[30] D. L. Wang, Y. C. Yu, W. D. Zhou, H. Chen, F. J. Disalvo, D. A. Muller and H. D. Abruna, Phys. Chem. Chem. Phys., 2013, 15, 9051.

[31] D. Li, F. Han, S. Wang, F. Cheng, Q. Sun and W. C. Li, ACS Appl. Mater. Interfaces, 2013, 5, 2208.

[32] X. L. Ji, S. Evers, R. Black and L. F. Nazar, Nat. Commun., 2011, 2, 325.

[33] Z. F. Deng, Z. A. Zhang, Y. Q. Lai, J. Liu, Y. X. Liu and J. Li, Solid State Ionics, 2013, 238, 44.

Electrochemical Impedance Spectroscopy Study of A Lithium/Sulfur Battery: Modeling and Analysis of Capacity Fading *

Abstract The electrochemical behavior of a lithium/sulfur(Li/S) battery was studied by electrochemical impedance spectroscopy(EIS). An impedance model based on the analysis of EIS spectra as a function of temperature and depth of discharge was developed. Then, by monitoring the evolution of impedance during the cycling process, the capacity fading mechanism of Li/S battery was investigated. The results show that the semicircle at the middle frequency of the EIS spectra is ascribed to the charge-transfer process and the semicircle at high frequency is related to the interphase contact resistance. Furthermore, electrolyte resistance, interphase contact resistance, and charge-transfer resistance vary with cycle number in different manners, and the charge-transfer resistance is the key factor contributing to the capacity fading of Li/S battery.

The lithium/sulfur(Li/S) battery has been recognized as one of the most promising energy storage systems for next-generation lithium batteries due to its high theoretical specific capacity (1675mA · h/g) and high energy density(2600Wh/kg). [1,2] In addition, sulfur as an active cathode material is non-toxic, naturally abundant, and environmentally benign. However, the practical realization of Li/S batteries has met huge challenges, including low utilization of active materials and rapid capacity fading because of the insulating nature of sulfur, the solubility of polysulfides(Li_2S_x, $3 \leqslant x \leqslant 8$) as intermediary products, and the insoluble agglomerates of irreversible discharge products(Li_2S_2 and Li_2S). [1-4] To overcome these problems effectively, studies of the electrochemical process and capacity fading mechanism of Li/S batteries are of great importance.

Electrochemical impedance spectroscopy(EIS), as one of the most powerful electroanalytical tools, has been widely used for studying kinetics in various electrochemical systems. Usually, a circuit model is required to analyze EIS data, such as assigning different segments of the measured EIS spectra to different physical/chemical processes and evaluating kinetic parameters. Recently, EIS was used to study the electrochemical behaviors of Li/S batteries. [5-9] A typical EIS spectrum is normally composed of two depressed semicircles and an inclined line. However, assignments of semicircles are controversial. It was accepted in some reports [8,10] that the semicircle in the high-frequency(HF) region is related to the charge-transfer resistance and the semicircle in the middle-frequency(MF) range is caused by formation of the solid film of Li_2S and Li_2S_2. Other authors [5,7] have used models which associated the semicircle at HF with solid film resistance and assigned the semicircle at MF to the charge-transfer resistance. Hence, the

---

* Copartner: Zhaofeng Deng, Zhian Zhang, Yanqing Lai, Jin Liu, Jie Li. Reprinted from Journal of The Electrochemical Society, 2013, 160(4): A553-A558.

reconsideration of the EIS model for the Li/S battery becomes necessary.

On the other hand, previous EIS studies mainly focused on the investigation of the electrochemical process and kinetics of Li/S bateries during a specific cycle. Kolosnitsyn et al.[5] studied the influence of polysulfides on impedance as a function of the state of charge, suggesting that the conductivity of the electrolyte has a significant impact on the rates of electrochemical processes. Barchasz et al.[6] conducted EIS measurements for sulfur cathode at different voltages in a specific cycle and analyzed the change in the shape of EIS spectra. Ahn et al.[7] prepared sulfur multi-walled carbon nanotube composites and investigated the EIS variation of the composite cathode as a function of the depth of discharge. Wang et al.[9] employed EIS to study the sulfurized polyacrylonitrile composite cathode during the first cycle. In addition, Zhang et al.[11] synthesized a sulfur-carbon sphere composite and conducted EIS measurements for the composite cathode in the 1st, 20th, 50th, and 100th cycle, respectively. The result indicated that the composite showing better cyclability has more moderate impedance variation during cycling. Nevertheless, to our knowledge, no systematical study of the capacity fading mechanism of Li/S batteries has been performed using the EIS technique.

In this paper, we propose a model for the impedance of a Li/S battery based on analyzes of EIS spectra at various temperatures and discharge depths. Then, the capacity fading mechanism of the Li/S battery is investigated by monitoring impedance variation as a function of cycle number. The kinetic parameters in the first discharge process and during progressive cycling are also discussed.

1 Experimental

Sulfur cathode slurry was prepared by mixing 60 wt% sublimed sulfur, 30 wt% carbon blacks, and 10 wt% polyvinylidene fluoride (PVDF) binder in N-methyl-2-pyrrolidinone (NMP) solvent. The slurry was spread onto aluminum foil (20μm), and then dried at 60℃ under vacuum for 24h. Coin-type electrodes were cut at a diameter of 10 mm and sulfur loading was 1.0mg/cm^2.

CR2025-type coin cells were assembled in an argon-filled glove box (Universal 2440/750), in which the oxygen and water contents were less than 1×10^{-6}. The electrolyte was 1.5 M bis(trifluoromethane) sulfonamide lithium salt (LiTFSI, Sigma Aldrich) in 1,3-dioxolane (DOL) and 1,2-dimethoxyethane (DME) (1 : 1, v/v) (Acros Organics). Metallic lithium foil was used as the counter electrode and Celgard 2400 was used as the separator. Galvanostatic charge/discharge tests were carried out at a constant current density of 168mA/g(0.1 C) in the potential range of 1.5 to 3.0V at 25℃ on a LAND CT2001A charge-discharge system. EIS measurements were carried out using a PARSTAT 2273 electrochemical measurement system (PerkinElmer Instrument USA). The impedance spectra were recorded on cells in the frequency range between 100kHz and 10mHz with a perturbation amplitude of 5mV. A high-low temperature test-chamber (GDH-2005C) was used to provide a constant temperature environment for EIS test. The obtained EIS data were fitted using ZView software (Scribner and Associates).

2 Results and Discussion

Variation of impedance with temperature. - To investigate the attribution of each semicircle, EIS measurements for Li/S cells were performed at −10, 0, 10, 25, and 40℃ because resistance parameters exhibit different sensitivity to temperature.[12-15] The cells were measured at the full charged state after 20 cycles with open-circuit potential relaxed to 2.33V. Note that the open-circuit potential of the cells declined to 2.18V when the temperature was increased to 40℃, which is much lower than the value at the other four temperatures owing to obvious self-discharge behavior at high temperature.

The Nyquist plots of a Li/S cell at different temperatures are presented in Fig. 1. As shown, the Nyquist plots are composed of a depressed semicircle in the HF region (100-1kHz), a depressed semicircle in MF region (1kHz-1Hz), and a straight slopping line in the low-frequency (LF) region (1Hz-10mHz), and the two semicircles are partially overlapped. It can be clearly observed that the magnitude of the MF semicircle increases dramatically with the de-clining of temperature.

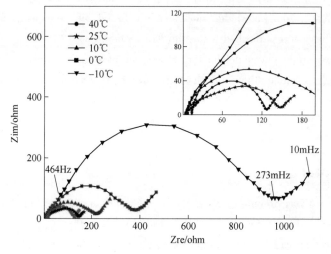

Fig. 1 (Color online) Nyquist plots of a lithium/sulfur cell at the fully charged state after 20 cycles for temperatures of −10, 0, 10, 25 and 40℃; Inset is the close-up of the high-frequency range

To get a better understanding of the change of impedance parameters, the Nyquist plots were analyzed by Zview software. Based on the characteristic of the Nyquist plots, we here propose a simplified circuit for the EIS spectra, as presented in Fig. 2. The R_1 corresponds to the resistance of the electrolyte.[5,7-9] The semicircle at HF is represented by $R_2//CPE_1$, which is assigned to the resistance and capacitance in this frequency region. Constant phase element (CPE) is used in the model in place of a capacitor to compensate for non-ideal behavior of electrode, for example, rough or porous electrode surface.[16,17] The semicircle in the MF region can be modeled using $R_3//CPE_2$. For the LF region, we choose to model the inclined line by a CPE. As reported in some literatures,[7,8] Warburg impedance (W_o) is used to represent diffusion of the ions within the cathode. However, it is found that either W_o-R or W_o-T must be set as "Fixed" to get rid of fitting failure at the fitting process. Therefore, the CPE is used in this equivalent circuit to produce an infinite length Warburg element. As presented in Fig. 2, by using this equivalent circuit, the fitted impedance plots agree well with the actual impedance spectra.

Fitted values of R_1, R_2, and R_3 regarding the spectra in Fig. 1 are plotted as a function of

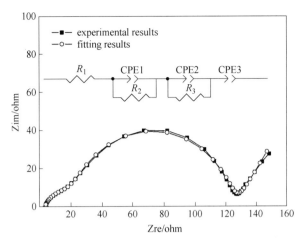

Fig. 2　The equivalent circuit for analysis of the Nyquist plots in Fig. 1

temperature. As shown in Fig. 3, R_1 increases slightly with a decrease of temperature because of the diminution of the ionic conductivity of electrolyte in accordance with some literature on lithium batteries.[12,13,15] R_3 increases dramatically with a decline of temperature, rising from 106Ω at 40℃ to 854Ω at −10℃. This behavior reveals that the semicircle at MF is associated with the charge-transfer process, which is strongly temperature dependent.[12-15] According to References 15 and 18, when $T^{[-1]}$ is very small the temperature dependence of the charge-transfer resistance (R_{ct}) can be expressed as

$$\ln R_{ct} = \ln\left(\frac{R}{z^2 F^2 c_T A_f [\text{Li}^+]^{1-\alpha}(1-x)^{1-\alpha} x^\alpha}\right) + \frac{\Delta G - R}{RT} + 1$$

where c_T is the maximum lithiation concentration of active material; A_f is the pre-exponential factor; [Li$^+$] is Li$^+$ concentration at the electrode surface; α is the symmetry factor of the electrode reaction; x is the intercalation level of the active material. When x is unchanged, there will be a linear relationship between $\ln R_{ct}$ and T^{-1}. Considering the consistency in x, all data but that obtained at 40℃ were analyzed. As shown in the inset of Fig. 3, the $\ln R_{ct}$ – T^{-1} relationship is approximately linear.

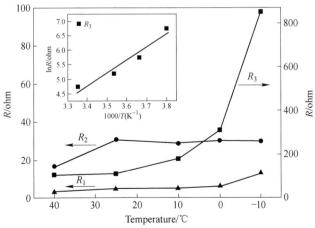

Fig. 3　(Color online) Temperature dependence of resistance parameters (R_1, R_2, and R_3) for lithium/sulfur cell after 20 cycles; Inset is the linear fitting of $\ln R_3$-T^{-1}

For R_2, little change is observed in the temperature range from −10 to 25℃. If the semicircle at HF is closely related to the ion diffusion through the solid film on the cathode and lithium anode, this means that R_2 will change with temperature,[12] which has not been observed in our experiment. The above results indicate that the semicircle in HF could be linked to other elements rather than the simplex ionic conduction process through solid film on electrodes. Therefore, in the next step, variation of impedance with depth of discharge(DOD) was studied.

Variation of impedance with depth of discharge. Fig. 4 presents the discharge cure of a typical Li/S cell at the first cycle. The discharge curve shows two potential plateaus. The upper plateau corresponds to the reduction of insulating sulfur to soluble high-order lithium polysulfides, and the lower plateau is related to the reduction of soluble lithium polysulfides to insoluble reduction products(Li_2S_2 and Li_2S).[1,2]

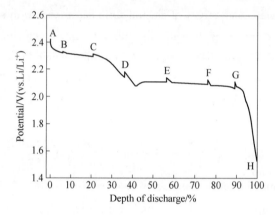

Fig. 4 (Color online)The discharge curve of lithium/sulfur cell during the first cycle

Fig. 5 illustrates the Nyquist plots obtained from the Li/S cell under different DOD. The cell delivers an initial discharge capacity of 910mA·h/g-S, corresponding to 100% DOD. As shown in Fig. 5a, the impedance spectra at 21% DOD (point C) and 36% DOD (point D) clearly exhibit two semicircles followed by an inclined line, which is similar to the result reported by Ahn et al.[7] Because the insoluble discharge products would not generate before the appearance of lower voltage plateau,[1,2,19,20] the HF semicircle could not be linked to the formation of a solid film of insoluble discharge products on cathode, which supports the analysis in temperature dependence of EIS.

On the other hand, in some studies of lithium batteries[16,21-24] the HF semicircle is interpreted as an interphase contact resistance in the electrode bulk. The behavior of the impedance spectra in Fig. 5 is in agreement with this interpretation. The consumption of sulfur makes for a better interphase electronic contact between particles, resulting in smaller HF semicircle, and the production of insulating reduction products on particle surface could increase the interphase contact resistance and enlarge the HF semicircle moderately. The interface impedance of the lithium anode also contributes to the HF semicircle.[5,7] But, the contribution of the anode im-

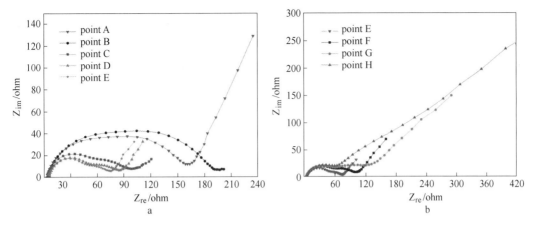

Fig. 5 (Color online) Nyquist plots of lithium/sulfur cell at different depths of discharge in the first cycle

pedance to the HF semicircle is neglected in the following discussion because the anode impedance in an electrolyte with polysulfides is small[25] and plays a minor role to the change of the HF semicircle. Based on the above discussion, we here propose a circuit model as shown in Fig. 6 and possible pictures for physical/chemical process of sulfur electrode are presented as well. Fig. 6a shows the schematic when sulfur particles exist in the cathode, and Fig. 6b shows when sulfur is fully reduced and solid products generate. In the proposed circuit model, R_e is the resistance of electrolyte. $R_{int}//CPE_{int}$ is the interphase contact resistance and its related capacitance in the sulfur electrode bulk, which simulates the process of electron conduction from the current collector to reaction sites. $R_{ct}//CPE_{dl}$ is the charge-transfer resistance and its related capacitance, which reflects the charge-transfer process at the interface between the conductive agent and the electrolyte. CPE_{dif} is diffusion impedance that probably represents Li-ion diffusion process.

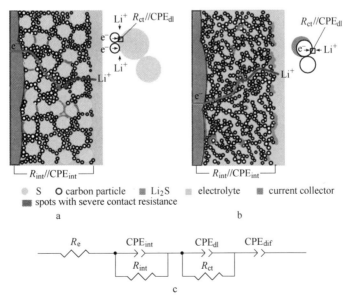

Fig. 6 (Color online) Schematic presentation of physical/chemical processes involved in the sulfur cathode: sulfur particles exist(a). sulfur is reduced(b) and the proposed equivalent circuit for lithium/sulfur cell(c)

To better understand the main phenomena occurring during the first discharge process, the Nyquist plots in Fig. 5 were analyzed using the equivalent circuit presented in Fig. 6. Table 1 lists the fitted values for all of the equivalent circuit elements. Note that CPE_{dl} elements (Y_2 and n_2) vary significantly during the discharge process. The CPE_{dl} may be related to the roughness of the electrode surface. The consumption of sulfur contributes to a rough and porous electrode surface, causing an increase of the value of Y_2 and a decline of n_2; the generation of the solid product layer results in a smooth electrode surface, and therefore the value of Y_2 decreases sharply and n_2 rises. The CPE_{dif} elements (Y_3 and n_3) also vary considerably with the state of discharge.

Table 1 Fitted values for the equivalent circuit elements in Fig. 6 by simulation of impedance spectra in Fig. 5

Point	R_e/Ω	CPE_{int}*		R_{int}/Ω	CPE_{dl}		R_{ct}/Ω	CPE_{dif}	
		$Y_1/\Omega^{-1}\cdot s^n$	n_1		$Y_2/\Omega^{-1}\cdot s^n$	n_2		$Y_3/\Omega^{-1}\cdot s^n$	n_3
A	2.52	9.29E-06	0.842	50.4	1.16E-04	0.698	107	3.26E-02	0.661
B	3.88	8.44E-06	0.842	56.6	1.10E-04	0.689	125	8.86E-02	0.320
C	3.73	5.43E-06	0.868	41.2	4.35E-04	0.627	47.1	6.05E-02	0.304
D	5.10	6.98E-06	0.840	39.1	1.03E-03	0.575	40.9	9.25E-02	0.567
E	3.35	11.8E-06	0.801	40.5	1.14E-03	0.607	30.6	9.20E-02	0.564
F	3.54	11.8E-06	0.796	46.0	6.98E-05	0.606	51.9	5.37E-02	0.567
G	2.77	8.65E-06	0.823	46.3	6.49E-05	0.620	65.7	1.60E-02	0.466
H	2.75	8.81E-06	0.818	49.0	—	—	~477	—	—

* The CPE impedance is expressed as: $Z_{CPE} = Y^{-1}(j\omega)^{-n}$.

DOD dependences of the R_e, R_{int}, and R_{ct} are plotted in Fig. 7. As shown in Fig. 7a, R_e increases from point A(0% DOD) to point D(36% DOD), and then decreases until the end of discharge. Similar results were reported in other references.[5,7] As discharge proceeds, sulfur transforms to soluble polysulfides, and thus the viscosity of electrolyte increases, resulting in an increase in electrolyte resistance. The following decline of electrolyte resistance is linked with the decreased viscosity of the electrolyte because of the reduction of polysulfides to insoluble Li_2S and Li_2S_2.

Fig. 7b illustrates the values of R_{int} at different depths of discharge. The R_{int} increases at the beginning of discharge(point B). This behavior is possibly attributed to the rearrangement of particles as electrochemical reaction proceeds. Afterward, there is an obvious decrease of R_{int} from point B(6% DOD) to point C(21% DOD). Then, after a little fluctuation the R_{int} slowly increases. As mentioned above, the obvious decrease of R_{int} could be ascribed to the reduction of insulating sulfur and the following increase of R_{int} is probably due to the production of nonconductive reduction products.

The variation of R_{ct} as a function of DOD is shown in Fig. 7c. Similar to R_{int}, the value of R_{ct}

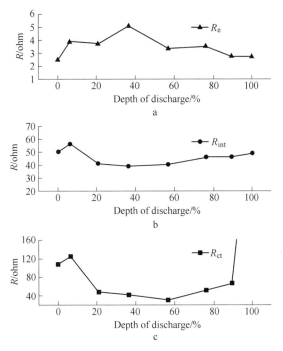

Fig. 7 (Color online) Plots of electrolyte resistance(R_e) (a), interphase contact resistance(R_{int}) (b) and charge-transfer resistance(R_{ct}) (c) against the depth of discharge in the first cycle

increases at the beginning of the discharge. After that, the R_{ct} obviously decreases and reaches the minimum value (30.6Ω) at point E. Subsequently, the R_{ct} increases and gets to 65.7Ω at point G (89% DOD). The significant decrease of R_{ct} from point B to point C is associated with the improved electrochemical accessibility resulting from the consumption of insulating sulfur; the increase of R_{ct} could be ascribed to the generation and accumulation of insulating Li_2S and Li_2S_2 and fewer soluble polysulfides. In particular, for the Nyquist plot obtained at point H, at which the discharge process completes, the MF semicircle is replaced with an arc, and the LF region is characteristic of a large arc. This can be explained by the fact that a sulfur electrode surface fully covered with insulating precipitations behaves as a blocking interface, resulting in blocking of the charge transfer.[6] These results indicate that the transfer reaction kinetic is closely related to electronic conductivity of sulfur cathode.

Variation of impedance with cycle number. — Capacity fading, generally observed in Li/S batteries with prolonged cycling, is a vital problem of concern. Fig. 8 displays the discharge/charge curves of a Li/S cell. Note that the discharge capacity decreases obviously during the first few cycles and reaches a steady state around 13 cycles, after which very moderate capacity fading is observed. Many studies on sulfur cathodes showed similar cyclability.[26-28] The solubility of polysulfides is one of the capacity fading factors, as reported in the literature.[1,11] It will lead to a polysulfide shuttle phenomenon that delays the end of the charge process, decreases the active materials utilization in the discharge process, and contributes to capacity fading. Another important factor of capacity fading is the irreversible precipitates (Li_2S and Li_2S_2) on the cathode, which is insoluble and electrochemically inaccessible.[1,11] Some reports[26,29] with no visible shuttle phenomenon still present similar cyclability. In this study, EIS measurements were carried out to study the capacity fading mechanism of sulfur cathode by monitoring the impedance evolution during progressive cycling. All measurements were conducted with cells that charged to 3.0V at a 0.1 C and then kept at 3.0V for 1h.

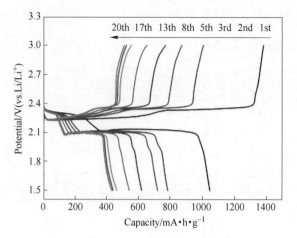

Fig. 8 (Color online) Discharge/charge curves of lithium/sulfur cell at a current density of $168mA \cdot g^{-1}$ (0.1 C)

Fig. 9 shows the Nyquist plots of sulfur electrode recorded at various cycles. The magnitude of the MF semicircle increases obviously in the first 8 cycles, which suggests a close relationship between impedance and capacity fading. To explore this relationship, the Nyquist plots were fitted by using the circuit model illustrated in Fig. 6 and all fitted elements are presented in Table 2. Note that the value of Y_2 declines moderately for the first 13 cycles and then it slightly increases. The decline of Y_2 probably indicates a decrease of porosity of cathode surface because of the accumulation of residual Li_2S and Li_2S_2. There is no regular change of CPE_{int} with the increase of cycle number and the value of Y_1 after the first cycle is the smallest.

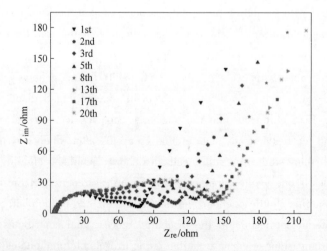

Fig. 9 (Color online) Nyquist plots of lithium/sulfur cell after charge in a series of cycles

Table 2 Fitted values for the equivalent circuit elements in Fig. 6 by simulation of impedance spectra in Fig. 9

Cycle number	R_e/Ω	CPE_{int}		R_{int}/Ω	CPE_{dl}		R_{ct}/Ω	CPE_{dif}	
		$Y_1/\Omega^{-1}\cdot s^n$	n_1		$Y_2/\Omega^{-1}\cdot s^n$	n_2		$Y_3/\Omega^{-1}\cdot s^n$	n_3
1	2.55	8.63E-06	0.868	40.1	6.70E-04	0.698	33.7	4.56E-02	0.718
2	2.34	9.53E-06	0.864	42.3	5.46E-04	0.693	46.1	6.56E-02	0.640
3	2.66	13.9E-06	0.834	38.8	4.33E-04	0.684	68.5	4.37E-02	0.751
5	2.32	13.5E-06	0.843	37.3	3.67E-04	0.680	88.7	3.00E-02	0.786
8	2.29	11.9E-06	0.857	38.3	3.54E-04	0.683	100	3.40E-02	0.781
13	3.38	19.5E-06	0.801	39.1	3.38E-04	0.698	102	5.22E-02	0.747
17	3.96	13.3E-06	0.839	42.5	3.80E-04	0.685	96.2	6.01E-02	0.723
20	3.66	18.7E-06	0.798	46.4	3.91E-04	0.685	102	4.48E-02	0.775

Fig. 10 displays the plots of R_e, R_{int}, and R_{ct} as a function of cycle number. Note that the change trends of R_e, R_{int}, and R_{ct} with cycle number are different. The R_{ct} increases rapidly during the first 8 cycles, and then slowly increases. The R_e remains at similar values during all cycles. As to the R_{int}, there is a little fluctuation with the values from the 3rd cycle to the 13th cycle, and then it slightly increases, which could be attributed to the appearance of cracks within the cathode after several cycles.[30] Comparing the change rules of three resistances (R_e, R_{int}, and R_{ct}) with the change of discharge capacity that showed in Fig. 8, we conclude that the R_{ct} plays a crucial role in the capacity fading of the Li/S battery.

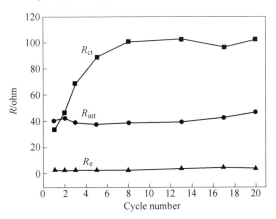

Fig. 10 (Color online) Plots of electrolyte resistance (R_e), interphase contact resistance (R_{int}) and charge-transfer resistance (R_{ct}) as a function of cycle number

The degradation of the cathode surface is an important factor contributing to dramatic increase of R_{ct}. During the charge process, insulating discharge products (Li_2S, Li_2S_2) are reoxidized to soluble polysulfides, but few of them are left even at the fully charged state. Thus, insulating agglomerates of residual Li_2S and Li_2S_2 are formed on the surface of cathode with increasing cycle. Elazari et al.[30] investigated the surface conductivity of a sulfur electrode and confirmed that the electrode surfaces become less conductive as cycling proceeds. Furthermore, it is well known that the charge-transfer reactions occur at the interface of conductive additive and electrolyte. Hence, the obvious increase of R_{ct} may be closely related to the cumulative agglomerates, which decrease the electrically conductive area on the cathode surface and hinder the transportation of ions toward the inside of the cathode, resulting in slow transfer reaction kinetic and capacity fading.

The following smooth increase of R_{ct} may be related to the cracks, which appear within the sulfur cathode after several cycles, as mentioned above. Although surface conductivity is poor, the cathode material exposed inside the cracks is still electrochemically accessible. In addition, the cracks allow a good contact between solution species and conductive carbon materials. The above results suggest that R_{ct} is closely related to the surface properties of sulfur cathode.

3 Conclusions

In summary, we propose an impedance model for a Li/S battery by analyzing the impedance variations as functions of temperature and state of discharge. Then, EIS measurements are performed at a series of cycle numbers to investigate the capacity fading mechanism. The results indicate that the signature of the middle frequency semicircle is the charge-transfer resistance and its relative capacitance, and the semicircle at high frequency may be related to the interphase contact resistance. Moreover, the increase of the charge-transfer resistance, which is strongly affected by the conductivity of sulfur cathode and the cathode surface properties, is an important factor in the capacity fading of Li/S battery. Therefore, it is important to design efficient cathode materials which help to minimize irreversible agglomerates on cathode surface to decrease the charge-transfer resistance, and to improve the cycling stability of Li/S batteries.

Acknowledgments

The authors acknowledge the financial support of the Strategic Emerging Industries Program of Shenzhen, China (JCYJ20120618164543322) and the National Natural Science Foundation of China (20803095). We also appreciate the support of the Engineering Research Center of Advanced Battery Materials, the Ministry of Education, China.

References

[1] X. Ji and L. F. Nazar. J. Mater. Chem., 20, 9821 (2010).
[2] P. G. Bruce, S. A. Freunberger, L. J. Hardwick, and J. M. Tarascon, Nat. Mater., 11, 19 (2012).
[3] X. Ji, K. T. Lee, and L. F. Nazar, Nat. Mater., 8, 500 (2009).
[4] S. E. Cheon, S. S. Choi, J. S. Han, Y. S. Choi, B. H. Jung, and H. S. Lim, J. Electrochem. Soc., 151, A2067 (2004).
[5] V. S. Kolosnitsyn, E. V. Kuzmina, E. V. Karaseva, and S. E. Mochalov, J. Power Sources, 196, 1478 (2011).
[6] C. Barchasz, J. C. Leprêtre, F. Alloin, and S. Patoux, J. Power Sources, 199, 322 (2012).
[7] W. Ahn, K. B. Kim, K. N. Jung, K. H. Shin, and C. S. Jin, J. Power Sources, 202, 394 (2012).
[8] L. Yuan, X. Qiu, L. Chen, and W. Zhu, J. Power Sources, 189, 127 (2009).
[9] L. Wang, J. Zhao, X. He, and C. Wan, Electrochim. Acta, 56, 5252 (2011).
[10] M. He, L. X. Yuan, W. X. Zhang, X. L. Hu, and Y. H. Huang, J. Phys. Chem. C, 115, 15703 (2011).
[11] B. Zhang, X. Qin, G. R. Li, and X. P. Gao, Energy Environ. Sci., 3, 1531 (2010).

[12] S. S. Zhang, K. Xu, and T. R. Jow, Electrochim. Acta, 49, 1057(2004).

[13] S. S. Zhang, K. Xu, and T. R. Jow, J. Power Sources, 115, 137(2003).

[14] S. S. Zhang, K. Xu, and T. R. Jow, Electrochem. Commun., 4, 928(2002).

[15] J. Li, C. F. Yuan, Z. H. Guo, Z. A. Zhang, Y. Q. Lai, and J. Liu, Electrochim. Acta, 59, 69(2012).

[16] M. Holzapfel, A. Martinent, F. Alloin, B. Le Gorrec, R. Yazami, and C. Montella, J. Electroanal. Chem., 546, 41(2003).

[17] T. Pajkossy, Solid State Ionics, 176, 1997(2005).

[18] Q. C. Zhuang, G. Z. Wei, J. M. Xu, X. Y. Fan, Q. F. Dong, and S. G. Sun, Acta Chim. Sin., 66, 722(2008).

[19] C. Barchasz, F. Molton, C. Duboc, J. C. Leprêtre, S. Patoux, and F. Alloin, Anal. Chem., 84, 3973(2012).

[20] Y. Li, H. Zhan, S. Liu, K. Huang, and Y. Zhou, J. Power Sources, 195, 2945(2010).

[21] J. Illig, M. Ender, T. Chrobak, J. P. Schmidt, D. Klotz, and E. Ivers-Tiffée, J. Electrochem. Soc., 159, A952 (2012).

[22] M. Gaberscek, J. Moskon, B. Erjavec, R. Dominko, and J. Jamnik, Electrochem. Solid-State Lett., 11, A170 (2008).

[23] J. Guo, A. Sun, X. Chen, C. Wang, and A. Manivannan, Electrochim. Acta, 56, 3981(2011).

[24] J. M. Atebamba, J. Moskon, S. Pejovnik, and M. Gaberscek, J. Electrochem. Soc., 157, A1218(2010).

[25] D. Aurbach, E. Pollak, R. Elazari, G. Salitra, C. S. Kelley, and J. Affinito, J. Electrochem. Soc., 156, A694 (2009).

[26] X. Li, Y. Cao, W. Qi, L. V. Saraf, J. Xiao, Z. Nie, J. Mietek, J. G. Zhang, B. Schwenzer, and J. Liu, J. Mater. Chem., 21, 16603(2011).

[27] J. Schuster, G. He, B. Mandlmeier, T. Yim, K. T. Lee, T. Bein, and L. F. Nazar, Angew. Chem. Int. Ed., 51, 1(2012).

[28] H. Wang, Y. Yang, Y. Liang, J. T. Robinson, Y. Li, A. Jackson, Y. Cui, and H. Dai, Nano Lett., 11, 2644 (2011).

[29] J. Chen, Q. Zhang, Y. Shi, L. Qin, Y. Cao, M. Zheng, and Q. Dong, Phys. Chem. Chem. Phys., 14, 5376 (2012).

[30] R. Elazari, G. Salitra, Y. Talyosef, J. Grinblat, C. S. Kelley, A. Xiao, J. Affinito, and D. Aurbach, J. Electrochem. Soc., 157, A1131(2010).

An Electrochemical-thermal Model Based on Dynamic Responses for Lithium Iron Phosphate Battery *

Abstract An electrochemical-thermal model is developed to predict electrochemical and thermal behaviors of commercial LiFePO$_4$ battery during a discharging process. A series of temperatures and lithium ion concentrations dependent parameters relevant to the reaction rate and Li$^+$ transport are employed in this model. A non-negligible contribution of current collectors to the average heat generation of the battery is considered. Simulation results on rate capability and temperature performance show good agreement with the literature data. The behavior of Li$^+$ distribution at pulse-relaxation discharge, the variation of electrochemical reaction rate and thermal behavior at a constant current discharge are studied. Results of pulse-relaxation discharge describe the dynamic change of Li$^+$ concentration distribution in liquid and solid phases, which is helpful to analysis the polarization of the battery. In constant current discharge processes, the electrochemical reaction rate of positive electrode has a regular change with the time and the position in the electrode. When discharge finished, there is still a part of the LiFePO$_4$ material has not been adequately utilized. At low rate, the discharge process accompanies endothermic and exothermic processes. With the rate increasing, the endothermic process disappears gradually, and only exothermic process left at high rate.

1 Introduction

Lithium ion battery is nowadays one of the most popular energy storage devices due to high energy, power density and cycle life characteristics[1,2]. It has been known that the overall performance of batteries not only depends on electrolyte and electrode materials, but also depends on operation conditions and choice of physical parameters[3]. Designers need an understanding on thermodynamic and kinetic characteristics of batteries that is costly and time-consuming by experimental methods. Conversely, numerical modeling and simulation for batteries are economic, which can provide guidelines for design in a short time[4], and information during the electrochemical and transport process. For example, the local electric potential and ion concentrations throughout the porous electrodes can be calculated while the data cannot be obtained experimentally[5,6].

The accuracy of numerical modeling and simulation of electrochemical and thermal behavior relies on the model construction and the parameters applied during simulation[7]. The most famous and practical model for lithium ion battery is the porous electrode model[8,9], which was based on the porous electrode theory containing charge transfer kinetics at reaction sites, species and charge conservations. It was combined with an energy conservation equation by Newman and Pals[10,11], and was developed as an electro-thermal model, which made it possibly simulate

---

* Copartner: Jie Li, Yun Cheng, Ming Jia, Yiwei Tang, Yue Lin, Zhian Zhang. Reprinted from Journal of Power Sources. 2013, 255: 130-143.

and predict the interaction between temperature and electrochemical reaction of batteries. Previously, the electrochemical models and electrochemical thermal models were built for simulating the cathode materials of lithium ion battery, such as $LiCoO_2$, $LiMn_2O_4$ or $Li(NiCoMn)O_2$[12-18]. Later, based on the observation of a phase change occurred in the LiFePO$_4$ cathode during the lithiated and unlithiated process[19,20], Srinivasan[21] developed a model that accounted for the phase change with the shrinking core, and investigated the cause for the low power capability of the materials. The other models without special features about the two-phase process were also founded for further qualitative analysis[22-31]. For instance, Wang[22] analyzed the effect of local current density on electrode design for the LiFePO$_4$ battery. Ye[23] developed an electro-thermal cycle life model by incorporating the dominant capacity fading mechanism to investigate the capacity fading effect on the performance. Gerver[24] and Christian Hellwig[25] gave a multidimensional modeling framework for simulating coupled thermal and electrochemical phenomena that are critical for safety, durability and design optimization studies. These models demonstrated that the porous electrode theory can be used to simulate a LiFePO$_4$ battery system without considering the phase change.

In the previous work[8-31], the computational domains were divided into three parts, named negative electrode, separator and positive electrode, but the current collectors of a battery were neglected. Because a rate of the heat generation is an average value obtained through dividing the total heat generation by the cell volume, about 10% of the cell volume comes from the current collectors, indicating that the contribution of the collectors to the heat generation rate cannot be neglected.

Besides, the material properties treated as parameters also have an important influence on the accuracy of the simulation[7]. During the discharge process, lithium ions deintercalate from the negative porous electrode, then transfer through separator and intercalate into the positive electrode, resulting in changes of temperature and lithium ion concentrations in the solid and liquid phases that may affect the diffusion coefficient, ionic conductivity, ion transference number as well as the reaction rate constant[24,32-34]. Therefore, it is necessary to rectify dynamically these parameters during a simulation.

In this study, we develop an electrochemical thermal model for a LiFePO$_4$ battery by considering the current collectors into the computational domain, dynamic responses in lithium ion concentration, and temperature as parameters during the discharge process. By comparing experimental results with simulation at different operating temperatures and discharge rates, this model can be used to study the dynamic evolution for pulses, relaxation behavior, electrochemical reaction and thermal behavior at a constant discharge rate in lithium iron phosphate battery.

2 Model Development

2.1 Model assumption and simulation domain

This electrochemical-thermal model for a LiFePO$_4$ battery is developed based on the porous electrode model[8,9]. The active materials of solid electrodes are treated as homogenous media, and are comprised with spherical particles. Since the current collectors play an important role in electron and heat transferring, the current collectors are considered as a computing domain.

Fig. 1 shows a schematic computational domain of onedimensional (1D) battery model[8,9]. Because both positive and negative current collectors have two sides contacting and reacting with electrolyte, we take a half of their thickness into simulation domain in order to accurately calculate the current flow and the heat generation. There are six boundaries in the model. Four inner boundaries (anode current collector/anode interface boundaries 2, anode/separator interface boundaries 3, separator/cathode interface boundaries 4 and cathode/cathode current collector interface boundaries 5) and two external boundaries (anode current collector external-face boundary 1 and cathode current collector external-face boundary 6) are shown in Fig. 1.

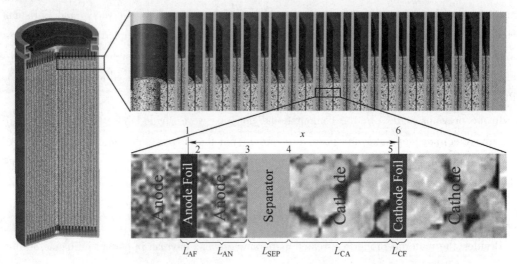

Fig. 1　Schematic diagram of lithium iron phosphate battery and computational domain

2.2　Electrochemical part

2.2.1　Electronic charge conservation

(a) Solid phase. There is a double layer capacitance at the interface between the active materials and electrolyte[35]. The capacitance can store additional energy and smooth the abrupt change of the cell voltage caused by the short-time pulses passing through a cell. However, the capacitance was seldom considered in the previous models[35]. Small modification in the electronic charge conservation is required to account of the double layer. Electronic charge conservation for solid phase can be expressed as follows:

$$\nabla \cdot (-\sigma_c \nabla \phi_c) = -J_i \tag{1}$$

$$\nabla \cdot (-k_1^{\text{eff}} \nabla \phi_1) = -S_{a,i}\left[j_{\text{loc},i} + C_{\text{dl}}\left(\frac{\partial \phi_1}{\partial t} - \frac{\partial \phi_2}{\partial t}\right)\right] \tag{2}$$

$$S_{a,i} = \frac{3\varepsilon_{1,i}}{r_{p,1}}; k_1^{\text{eff}} = k_1 \varepsilon_1^{\gamma_1} \tag{3}$$

where ϕ_c is arbitrarily set to zero at boundary 1; at boundary 6, the charge flux is set to equal the average current density of the battery, which is expressed as Eq. (4); at boundary 3 and boundary 4, there is no charge flux, which is expressed as Eq. (5) and the boundary condition is set as isolation.

$$\phi_c\big|_{\mathcal{X}=0}=0;-\sigma_c\nabla\phi_c\big|_{\mathcal{X}=L_{ncc}+L_n+L_s+L_p+L_{pcc}}=-I_{app} \quad (4)$$

$$-k_1^{eff}\nabla\phi_1\big|_{\mathcal{X}=L_{ncc}+L_n}=0;-k_1^{eff}\nabla\phi_1\big|_{\mathcal{X}=L_{ncc}+L_n+L_s}=0 \quad (5)$$

(b) Solution phase. According to the concentrated solution theory, the governing equation for electronic charge conservation in solution phase is expressed as:

$$\nabla\cdot\left\{k_2^{eff}\left[-\nabla\phi_2+\frac{2RT}{F}\left(1+\frac{\partial\ln f}{\partial\ln c_2}\right)(1-t_+)\frac{\nabla c_2}{c_2}\right]\right\}=S_{a,i}j_{loc,i} \quad (6)$$

$$k_2^{eff}=k_2\varepsilon_2^{\gamma_2} \quad (7)$$

Liquid-junction potential is introduced in Eq. (6) with expression

$$K_{junc}=\frac{2RT}{F}\left(1+\frac{\partial\ln f}{\partial\ln c_2}\right)(1-t_+)=\frac{2RT}{F}\nu \quad (8)$$

The parameter ν is the thermodynamic factor relating to electrolyte activity, and it is concentration and temperature dependent.

There is no flux at external boundaries (boundary 1 and boundary 6), which is expressed as Eq. (9); the parameter φ_2 is taken to be continuous at boundary 2 and boundary 5.

$$\frac{\partial\phi_2}{\partial\mathcal{X}}\bigg|_{\mathcal{X}=L_{ncc}}=\frac{\partial\phi_2}{\partial\mathcal{X}}\bigg|_{\mathcal{X}=L_{ncc}+L_n+L_s+L_p}=0 \quad (9)$$

2.2.2 Mass conservation

(a) Solid phase. The mass conservation of lithium ions in an intercalation particle of the electrode active material is described by Fick's law. The parameter r is the distance from the center of solid sphere, the mass transport within solid phase in spherical coordinates can be described as:

$$\frac{\partial c_{1,i}}{\partial t}=\frac{D_{1,i}}{r^2}\left[\frac{\partial}{\partial r}\left(r^2\frac{\partial c_{1,i}}{\partial r}\right)\right]=D_{1,i}\left(\frac{2}{r}\frac{\partial c_{1,i}}{\partial r}+\frac{\partial^2 c_{1,i}}{\partial r^2}\right) \quad (10)$$

Let the variable τ equal to r/R_i, the Eq. (10) can be described as

$$\tau^2 R_i\frac{\partial c_{1,i}}{\partial t}=\frac{D_{1,i}}{R_i}\left[\frac{\partial}{\partial\tau}\left(\tau^2\frac{\partial c_{1,i}}{\partial\tau}\right)\right] \quad (11)$$

The parameter R_i is the particle radius. There is no species source at the center of sphere, so $\partial c_{1,i}/\partial\tau\big|_{\tau=0}=0$. The Li$^+$ concentration on the surface of the particles is coupled to the concentration and flux in the 1D model for the charge and material transport in the electrolyte.

(b) Solution phase. Solution phase mass conservation for LiF$_6$ dissolved in the liquid phase

$$\varepsilon_2\frac{dc_2}{dt}+\nabla\cdot\{-D_2^{eff}\nabla c_2\}=\frac{S_{a,i}j_{loc,i}}{F}(1-t_+) \quad (12)$$

$$D_2^{eff}=D_2\varepsilon_2^{\gamma_2} \quad (13)$$

The flux of liquid species is arbitrarily set to zero at boundary 2 and boundary 5, liquid species flux and species concentration at boundary 3 and boundary 4 are taken to be continuous.

2.2.3 Electrochemical kinetics

The local current per active material area is calculated using the Butler-Volmer equation:

$$j_{loc,i} = j_{0,i}\left[\exp\left(\frac{\alpha_{a,i}\eta_i F}{RT}\right) - \exp\left(\frac{-\alpha_{c,i}\eta_i F}{RT}\right)\right] \quad (14)$$

The parameter $j_{loc,i}$ is driven by over-potential, the parameter η_i is defined as the difference between solid and electrolyte phase potentials minus U_i.

$$\eta_i = \phi_{1,i} - \phi_{2,i} - U_i \quad (15)$$

The parameter U_i is the thermodynamic equilibrium potential of the solid phase and is taken to be a function of the local SOC on the surface of active particles.

The temperature-dependent open circuit potentials of positive and negative electrodes are approximated by Taylor's first order expansion around a reference temperature:

$$U_i = U_{ref,i} + (T - T_{ref})\frac{dU_i}{dT} \quad (16)$$

The parameter $U_{ref,i}$ is the open circuit potential under the reference temperature.

In Eq. (14), exchange current density, $j_{0,i}$, connects concentrations in both solid and liquid phase:

$$j_{0,i} = Fk_i c_2^{\alpha_{a,i}}(c_{1,max,i} - c_{1,surf,i})^{\alpha_{a,i}} c_{1,surf,i}^{\alpha_{c,i}} \quad (17)$$

The parameter k_i is the reaction rate, considered temperature dependent in this paper. The parameters α_a and α_c are the anodic and cathodic transfer coefficients, respectively.

The outputs of the model are the cell potential, current density distribution, species and concentrations distributions, and the cell potential is derived by following expression:

$$E = \phi_1\Big|_{x = L_{ncc}+L_n+L_s+L_p+L_{pcc}} - \phi_1\Big|_{x=0} \quad (18)$$

2.3 Energy conservation

The heat generation is modeled with the local heat generation model of Rao and Newman[36] using the formulation of Gu and Wang[37]. According to the position of heat generation, the total heat generation is the summation of heat generated in the two electrodes, separator and current collectors. The summary of the heat generation mechanisms is listed in Table 1.

Table 1 Heat generation mechanism in a cell.

Heat generation mechanism		Equation	
Porous electrode	(1) Electrochemical interface	$q_{i,(1)} = S_a j_{loc,i}\left(T\dfrac{dU_i}{dT} + \phi_{1,i} - \phi_{2,i} - U_i\right)$	(19)
	(2) Electrical ohmic heat	$q_{i,(2)} = \sigma_i^{eff} \nabla\phi_{1,i} \cdot \nabla\phi_{1,i}$	(20)
	(3) Ionic ohmic heat	$q_{i,(3)} = k_2^{eff} \nabla\phi_{2,i} \cdot \nabla\phi_{2,i} + \dfrac{2RTk_2^{eff}}{F}(t_+ - 1)\left(1 + \dfrac{\partial \ln f}{\partial \ln c_{2,i}}\right) \cdot \nabla(\ln c_{2,i}) \cdot \nabla\phi_{2,i}$	(21)
Separator	Ionic ohmic heat	$q_s = k_2^{eff} \nabla\phi_2 \cdot \nabla\phi_2 + \dfrac{2RTk_2^{eff}}{F}(t_+ - 1)\left(1 + \dfrac{\partial \ln f}{\partial \ln c_2}\right) \cdot \nabla(\ln c_2) \cdot \nabla\phi_2$	(22)
	Current collector Electrical ohmic heat	$q_{i,c} = \sigma_{i,c}\nabla\phi_{i,c} \cdot \nabla\phi_{i,c}$	(23)

Subscript $i = n$ or p for a negative or a positive electrode.

According to the type of the heat generation, there are three parts of heat sources during the charge or discharge processes, including reaction heat, Q_{rea}, due to entropy change during the discharge; ohmic heat, Q_{ohm}, due to the ohmic potential drop and active polarization heat, Q_{act}, due to the electrochemical reaction polarization between active material particle surface and the electrolyte.

Reaction heat generation is:

$$Q_{rea} = S_{a,i} j_{loc,i} T \frac{dU_i}{dT} \tag{24}$$

Ohmic heat generation is:

$$Q_{ohm} = q_{i,(2)} + q_{i,(3)} + q_s + q_{i,c} \tag{25}$$

Active heat generation is:

$$Q_{act} = S_{a,i} j_{loc,i} (\phi_{1,i} - \phi_{2,i} - U_i) \tag{26}$$

Active heat Q_{act} and ohmic Q_{ohm} are irreversible, while reaction heat Q_{ura} is reversible.

$$Q_{irr} = Q_{ohm} + Q_{act} \tag{27}$$

$$Q_{re} = Q_{rea} \tag{28}$$

The energy conservation in lithium ion battery is shown as follows:

$$\rho_i C_{p,i} \frac{\partial T}{\partial t} + \nabla \cdot (-k_i \nabla T) = Q_{irr} + Q_{re} \tag{29}$$

The parameters ρ_i, $C_{p,i}$ and k_i are density, heat capacity and thermal conductivity, respectively.

According to Newton's cooling law and radiation law, the boundary condition for energy conservation is expressed as:

$$-\lambda \nabla T = -h(T_{amb} - T) - \varepsilon \sigma (T_{amb}^4 - T^4) \tag{30}$$

The parameter λ is thermal conductivity of stainless steel case, h is natural convection heat transfer coefficient ($h = 7.17 W/(m^2 \cdot K)^{-1[38]}$), T_{amb} is ambient temperature, and ε is the blackness of the battery surface ($\varepsilon = 0.8^{[38]}$).

3 Model Parameters and Model Validation

3.1 Battery parameters and thermal properties

The physical properties of battery components and battery design parameters are summarized in Tables 2 and 3, respectively.

Table 2 Thermal properties of battery components

Materials	Density/kg · m^{-3}	Heat capacity /J · (kg · K)$^{-1}$	Thermal conductivity /W · (m · K)$^{-1}$	Electrical conductivity /s · m^{-1}
Negative electrode	2223	641	1.04	100
Positive electrode	1500	800	1.48	0.5
Separator	900	1883	0.5	—
Electrolyte	1210	1518	0.099	Eqs. (42) and (43)

Continued Table 2

Materials	Density/kg·m^{-3}	Heat capacity /J·(kg·K)$^{-1}$	Thermal conductivity /W·(m·K)$^{-1}$	Electrical conductivity /s·m^{-1}
Copper foil	8700	396	398	$-0.04889T^3+54.65T^2$ $-218.00T+3.52\times10^6$ (s·cm^{-1})
Aluminum foil	2700	897	237	$-0.0325T^3+37.07T^2$ $-15,000T+2.408$ $\times10^6$(s·cm^{-1})
Stainless steel case	7500	460	14	—

Table 3 Critical battery parameters used in baseline for a 2.3Ah LiFePO$_4$ cylindrical 26650 type battery

Quantity	Negative electrode	Positive electrode	Separator	Electrolyte	Copper foil	Aluminum foil
Design specifications	(geometry and volume fractions)					
A_{cell}/m^2	0.1694					
$\varepsilon_{1,i}$	0.55	0.43	—			
$\varepsilon_{2,i}$	0.33	0.332	0.54			
L_i/μm	34	70	25		6.2(half thickness)	10(half thickness)
R_i/μm	0.0365	3.5				
Lithium ion concentrations						
$c_{ini,2}$/mol·m^{-3}				1200		
c_{max}/mol·m^{-3}	31.370	22.806				
$c_{ini,1}$/mol·m^{-3}	31 370 * 0.86	22 860 * 0.022				
Kinetic and transport properties						
$a_{a,i}, a_{c,i}$	0.5	0.5				
γ_i	1.5	1.5	1.5			
D_2/m^2·s^{-1}				Eq. (40)		
C_{dl}/F·m^{-2}	0.2	0.2				
$D_{1,i}$/m^2·s^{-1}	Eq. (33)	Eq. (32)				
k_i/m$^{2.5}$·mol$^{-0.5}$·s^{-1}	Eq. (34)	Eq. (35)				
$E_{a,k,i}$/J·mol^{-1}	20,000	30,000				
k_1/S·m^{-1}	100	0.5				
k_2/S·m^{-1}				Eqs. (42) and (43)		
σ_c						
t^+				Eq. (44)		
ν				Eq. (41)		
Constant quantity						
T_{ref}/T	298.15					
F/C·mol^{-1}	96.487					

3.2 Dynamic parameters

The dynamic response of battery properties is attributed to the change of temperature and lithium-ion concentration during the discharge and charge processes. It is considered for eight sets of physical properties. These are (1) $k_{0,i}$, the reaction rate constant; (2) $D_{1,i}$, the diffusion coefficient of Li$^+$ in the solid active particles; (3) open circuit potential (OCP) of the electrodes; (4) D_2, the diffusion coefficient of Li$^+$ in the electrolyte; (5) k_2, the electrolyte ionic conductivity; (6) ν, the thermodynamic factor relating to electrolyte activity; (7) t^+, the Li$^+$ transference number; (8) σ_c, current collector electrical conductivity.

3.2.1 Dynamic variables related to electrodes

3.2.1.1 Electrode kinetics properties.

The dynamic temperature dependences of $D_{1,i}$ follow the Arrhenius equation. However, the LFP electrode exhibited significantly higher utilization on charging than on discharging at the same current density[39]. This asymmetry was ascribed to the different transport limitation between the fully charged and discharged states[40]. To account for this asymmetry, an empirical equation is used for the solid-state Li$^+$ diffusion coefficient that is dependent on the average lithium ion concentration of particles[27]:

$$D_{1,p} = \frac{1.18 \times 10^{-18}}{(1+y)^{1.6}} \tag{31}$$

In this study, we take a combination of the Arrhenius equation and Eq. (31) into the formula for the $D_{1,p}$, which is expressed as:

$$D_{1,p} = \frac{1.18 \times 10^{-18}}{(1+y)^{1.6}} \exp\left[-\frac{E_{1,D,p}}{R}\left(\frac{1}{T} - \frac{1}{298.15}\right)\right] \tag{32}$$

The state of charge (SOC) and temperature also affect $k_{0,i}$ and $D_{1,n}$, but duo to lack of data, the SOC is not considered. $k_{1,i}$ and $D_{1,n}$ are expressed by Eqs. (33)-(35) followed the Arrhenius formula [41].

$$D_{1,n} = 3.9 \times 10^{-14} \exp\left[-\frac{E_{1,D,n}}{R}\left(\frac{1}{T} - \frac{1}{298.15}\right)\right] \tag{33}$$

$$k_{1,n} = 3 \times 10^{-11} \exp\left[-\frac{E_{1,k,n}}{R}\left(\frac{1}{T} - \frac{1}{298.15}\right)\right] \tag{34}$$

$$k_{1,p} = 1.4 \times 10^{-12} \exp(-3y) \exp\left[-\frac{E_{1,k,p}}{R}\left(\frac{1}{T} - \frac{1}{298.15}\right)\right] \tag{35}$$

The parameters $E_{1,D,p}$ and $E_{1,D,n}$ are the activation energy for diffusion of lithium respectively in the positive and negative electrodes, ($E_{1,D,p} = 35\text{kJ/mol}$[23,28], $E_{1,D,n} = 35\text{kJ/mol}$[23,42]). The parameters $E_{1,k,p}$ and $E_{1,k,n}$ are the activation energy for the rate constant of the positive and negative electrodes, respectively ($E_{1,k,p} = 20\text{kJ/mol}$[23,43], $E_{1,k,n} = 30\text{kJ/mol}$[23,28]).

3.2.1.2 Electrode thermodynamics properties.

The open circuit potentials of the positive electrode U_p and the negative electrode U_n are a function of local SOC on the surface of active particles. U_p and U_n come from Ref.[28], which are expressed by Eqs. (36) and (37).

$$U_{ref} = 3.4323 - 0.4828\exp\left[-80.2493(1-y)^{1.3198}\right]$$

$$-3.2474\times10^{-6}\exp[20.2645(1-y)^{3.8003}]$$
$$+3.2482\times10^{-6}\exp[20.2646(1-y)^{3.7995}] \tag{36}$$

$$U_{ref} = 0.6379 + 0.5416\exp(-305.5309x) +$$
$$0.044\tanh\left(-\frac{x-0.1958}{0.1088}\right) - 0.1978\tanh\left(\frac{x-1.0571}{0.0854}\right) -$$
$$0.6875\tanh\left(\frac{x-0.0117}{0.0529}\right) - 0.0175\tanh\left(\frac{x-0.5692}{0.0875}\right) \tag{37}$$

Entropy change in electrodes is $\Delta S = nF(dU/dT)$. dU_p/dT and dU_n/dT are the entropy changes of lithium iron phosphate positive electrode and the negative electrode, respectively. Their curves[44,45] are as shown in Fig. 2, and are expressed by eqs. (38) and (39).

$$\frac{dU_p}{dT} = -0.35376y^8 + 1.3902y^7 - 2.2585y^6 + 1.9635y^5 -$$
$$0.98716y^4 + 0.28857y^3 - 0.046272y^2 + 0.0032158y -$$
$$1.9186\times10^{-5} \tag{38}$$

$$\frac{dU_n}{dT} = 344.1347148\times$$
$$\frac{\exp(-32.9633287x + 8.316711484)}{1 + 749.0756003\exp(-34.79099646x + 8.887143624)} -$$
$$0.8520278805x + 0.362299229x^2 + 0.2698001697 \tag{39}$$

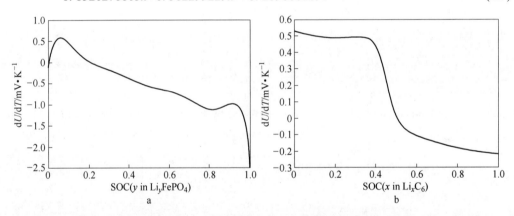

Fig. 2 The entropy change of the positive electrode and negative electrode as a function of SOC

3.2.2 Dynamic variables related to lithium ion transport in electrolyte

Due to lack of the proportion of the electrolyte components, the dynamic variables dependent on temperature of lithium ion transport in $LiPF_6$ are assumed to follow the properties of similar electrolyte systems. Valøen[32] investigated a full set of transport properties for $LiPF_6$ in PC/EC/DMC(10∶27∶63 by volume) experimentally, showing the k_2, D_2, and ν as functions of temperature and $LiPF_6$ concentration.

$$D_2 = 1\times10^{-4}\times10^{-4.43\frac{54.0}{T-229.0-0.05C}-2.2\times10^{-4}C} \tag{40}$$

$$\nu = 0.601 - 0.24\sqrt{10^{-3}C} +$$
$$0.982[1 - 0.0052(T-294.0)\sqrt{10^{-9}C^3}] \tag{41}$$

$$k_2 = 1\times10^{-4}(-10.5 + 0.074T - 6.69\times10^{-5}T^2 + 6.68\times$$

$$10^{-4}C - 1.78 \times 10^{-5}CT + 2.8 \times 10^{-8}CT^2 +$$
$$4.49 \times 10^{-7}C^2 - 8.86 \times 10^{-10}C^2T) \tag{42}$$

By combined the parameters of $LiPF_6$ in EC/DMC(2:1 by volume) at 25℃ [46] with that in the PC/EC/DMC system [32], the following expression is obtained:

$$k_2 = 1.12 \times 10^{-4}(-8.2488 + 0.053248T - 2.9871 \times 10^{-5}T^2 +$$
$$0.26235C - 9.3063 \times 10^{-3}CT + 8.069 \times$$
$$10^{-6}CT^2 + 0.22002C^2 - 1.765 \times 10^{-4}C^2T) \tag{43}$$

The Li^+ transference number, t^+, is considered as a constant, such as $t^+ = 0.363$ [14]. However, t^+ also appears to be dependent of Li^+ concentration and temperature [47,48]. Hence, t^+ is adopted from the published literature [47,48]. that is a function of temperature and Li^+ concentration is this study.

$$t^+ = 2.67 \times 10^{-4} \exp\left(\frac{833}{T}\right)\left(\frac{C}{1000}\right) + 3.09 \times$$
$$10^{-3} \exp\left(\frac{653}{T}\right)\left(\frac{C}{1000}\right) + 0.517 \exp\left(-\frac{49.6}{T}\right) \tag{44}$$

It should be noted that the solvent mixture for the commercial battery is unknown. We assume that Eqs.(40)-(44) are applicable in commercial batteries, so, these data are used in this simulation.

3.3 Model validation

The electrochemical-thermal model is validated from electrochemical performance. Experiment data are from the introduction document of ANR26650m1-a lithium ion cylindrical cell (A123 company, US) [49].

Firstly, the experiment data [49] of battery potential against the discharge capacity are compared with simulation results that are in fairly good agreement with the experimental results (see Fig. 3). In order to further validate the electrochemical performance at different discharge rates, the simulation results are compared with the experiment data [25] at different discharge rates (C/10, 1/2C, 1C, 2C) at 25℃ in Fig. 4. The results indicate this electrochemical-thermal model based dynamic response is reliable to simulate the discharge performance of lithium iron phosphate battery at different discharge rates.

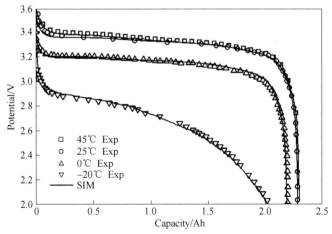

Fig. 3 -20℃, 0℃, 25℃, 45℃, 1C discharge validations

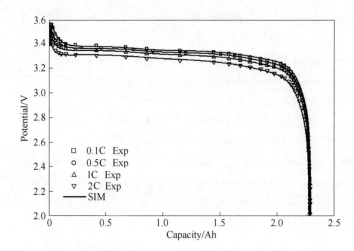

Fig. 4 Different discharge rates(0.1C, 0.5C, 1C, 2C) validation at 25℃

4 Results and Discussion

This model is mainly used to predict the electrochemical performance and thermal behavior of lithium ion batteries.

4.1 Dynamic evolution for pulse behavior

Lithium ion batteries have been used in many aspects which work intermittently, for example, assisting the engine of hybrid electric vehicles during vehicle acceleration. Because these actions are of short duration, on the order of seconds, the pulse and relaxation behaviors are considered important aspects to analyze polarization in battery. In this study, pulse test with different discharge rates was chosen to obtain Li$^+$ concentration distribution. During simulated operation, the battery is discharged at 0.2C, 1C, or 2C constant current during 0-200s and is relaxed by cutting off current during 200-1000s. The Boundaries 2, 3, 4 and 5 are used to examine the spatial and temporal distribution of lithium ion concentration in liquid and solid phases.

The spatial and temporal distribution of lithium ion concentration in electrolyte (c_2) during the pulse test at different discharge rates is depicted in Fig. 5. At the beginning of the discharge, a distribution of lithium ion rapidly builds up inside the battery, c_2 decreases from Boundary 2 to Boundary 5, forming a gradient which drives lithium ions along the battery direction from the negative electrode to positive electrode. At a high discharge rate (1C or 2C), c_2 increases in the negative electrode and decreases in the positive electrode during the very early period (about 80s), and then stays steady from 80s to 200s. On the contrary, at a low discharge rate, c_2 changes from 0s to 200s. When the current is cut down at 200s, c_2 changes rapidly in a timescale of millisecond. Then a self-adjusting period driven by diffusion is taken to attain equilibrium. The greater discharge rate is, the longer self-adjusting period. Due to the employment of double layer capacitance of the active materials and electrolyte interface, c_2 in equilibrium shows an increase compared to the initial level (1200 mol · m^{-3}). At 0.2, 1 and 2C discharge rates, the augmentation is 0.6, 8.4, 9.9mol/m, respectively, showing a rising trend along with an increase of the discharge rates. However, this rising is not obvious at high discharge rates.

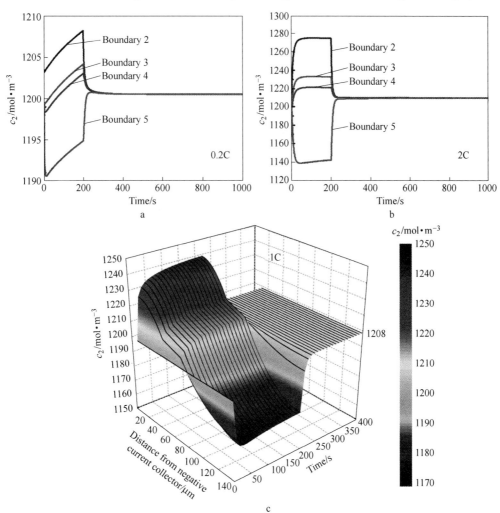

Fig. 5 Spatial and temporal distribution of Li$^+$ concentration in electrolyte

Lithium ion concentration gradient in solid phase can be described as an indicator of polarization in a particle. Large concentration gradient in a solid particle means high polarization, which may lead to high diffusion induced stress (DIS), cause particle fracture and reduce battery capacity and battery power[50]. As introduced in the section of model development, active materials (LiFePO$_4$ and Graphite) in solid electrodes are considered to be homogenous, and they are composed of spherical particles. In order to study their interior lithium ion concentration gradient, four special particles at Boundary 2, 3, 4 and 5, respectively, are selected, and they divided into 10 parts along the radial direction. The center and surface of the particles are located at $r/R = 0$ and 1, respectively. Fig. 6 gives lithium ion concentration distribution throughout the particles.

At all of the boundaries, lithium ion concentrations of the particles show great changes at pulse time of 0-200s. They decrease at Boundary 2 and 3 in the negative electrode with dis-

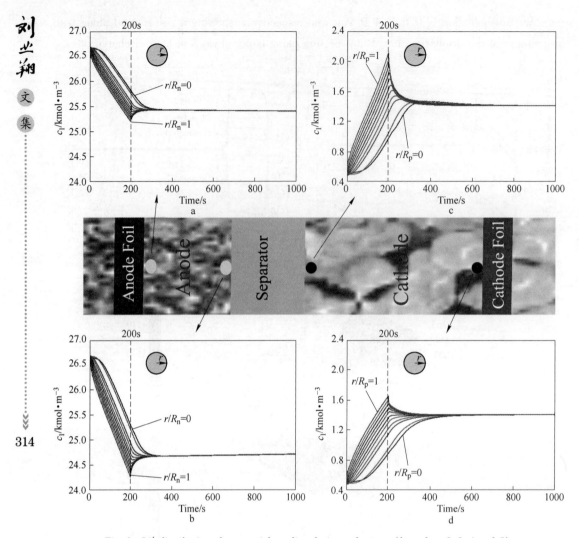

Fig. 6 Li$^+$ distribution along particle radius during pulse tests (boundary 2, 3, 4 and 5)

charge time, as shown in Fig. 6a and b. However, an increase with discharging time at Boundary 4 and 5 in the positive electrode is seen in Fig. 6c and d. At 200s, the surface concentrations of particles at Boundary 2 and 3 decrease to 25.2kmol/m^3 and 24.3kmol/m^3, from initial concentration (26.98kmol/m^3), and at Boundary 4 and 5 the concentrations increase to 2.3kmol/m^3 and 1.7kmol/m^3 from initial concentration (0.5kmol/m^3). These indicate the lithium ion distribution is in location dependent.

After the current is cut down at 200s, the battery is in relaxation. In the relaxation period (200~1000s), the variation of lithium ion concentration in the negative electrode including Boundary 2 and 3 appears the similar tendency. At r/R_n = 0, 0.1, 0.2, 0.3, 0.4, the concentrations decrease gradually until unchanged. And at r/R_n = 0.5, 0.6, 0.7, 0.8, 0.9, 1.0 the concentrations turn to increase until unchanged. As for the positive electrode, the variation of interior lithium ion concentration of the particle at Boundary 5 shows a contrary location dependent with Boundary 2 and 3. At r/R_p = 0, 0.1, 0.3, 0.2, 0.4, the concentrations continue increasing gradually until unchanged, and at r/R_p = 0.5, 0.6, 0.7, 0.8, 0.9, 1.0 the concentrations turn to decrease until unchanged. But for Boundary 4, lithium ion concentrations of the particle decrease gradually until unchanged, appearing no location dependence that is quite different from at other boundaries.

In the relaxation period, the interior concentrations remain unchanged ultimately, which is in equilibrium. It is well known that the cells of a battery require a certain time to reach a steady state in terms of charge, concentration and temperature. As introduced in literature[51], relaxation time factor (R^2/D) denotes the theory time needed for interior concentrations to obtain equilibrium after pulse discharge, which reflects the depolarization ability of the battery. In theory, at 25℃, $R_n^2/D_n = 136s$, and $R_p^2/D_p = 1129s$. With the discharge proceeding, battery temperature will increase that leads to D increase (see Eqs. (32) and (33)). So R^2/D at discharge process is lower than that at 25℃. That explains why the relaxation times of negative and positive electrodes concluded from Fig. 6 (about 120s and 370s) are lower than theoretical values. Increasing lithium ion diffusivity in solid phase may decrease the polarization and improve the relaxation performance of battery.

Besides, Fig. 7 shows the potential in solid phase also changes at pulse test. The potential decreases with time as a typical saw-tooth behavior, corresponding to the increase of lithium ion concentration on the surface of positive particle (Fig. 6c and d). Within 200s, the voltage decrease at boundary 5 that is larger than that at boundary 4 due to the ohmic overvoltage contributed from the thickness of positive electrode. The relaxation of the potential after current interruption occurs in two stages. When the current is cut down at 200s, the decrease is also shut down immediately. And the potential increases by 0.04V quickly with a timescale less than 0.2s, which is caused by the disappearance of ohmic polarization. Duo to the diffusion of lithium ions, the potential increases slowly until equilibrium is attained.

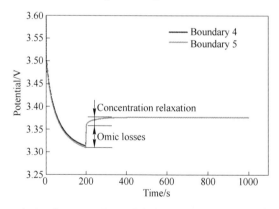

Fig. 7 Dynamic evolution for potential in solid phase during pulse tests (boundary 4 and 5)

4.2 Dynamic evolution for electrochemical reaction

In order to observe the dynamic evolution in the electrodes, the local current density, surface concentration and the state of charge are simulated. The curves of these properties at different time during discharge are shown in Figs. 8 and 9. In the negative electrode (Fig. 8a), the local current density does not change in regularity. For the positive electrode, the local current density at Boundary 4 increases severely at the beginning of discharge. A peak is seen at 1220s in the local current density curve, and it moves from the separator-positive electrode interface to the positive electrode-current collector interface. This indicates large electrochemical reaction rate near the separator-negative electrode interface. This result is different from the report of Wang and his co-works[22] in which the peak moved from the current collector to the cathode-separator interface as discharge proceeded. This is because the peak location depends upon the comparison between the effective conductivity of the solid phase (k_1^{eff}) and that of the liquid (k_2^{eff})[51]. When $k_1^{eff} < k_2^{eff}$, the peak appears near the current collector and moves from the current collector to the cathodesepartor interface as discharge proceeds. On the contrary, when $k_1^{eff} > k_2^{eff}$, the peak appears near the cathode-separator interface and moves from the cathode-separator interface to the current collector as discharge proceeds. In our work, k_1^{eff} is larger than k_2^{eff}, therefore, the surface concentration near the separator-positive electrode is increased rapidly to the maximum concentration of 22,806mol/m³ as shown in Fig. 9a, and the positive particles approximately reach the fully state of charge in a short time (about 1800s) (Fig. 9b).

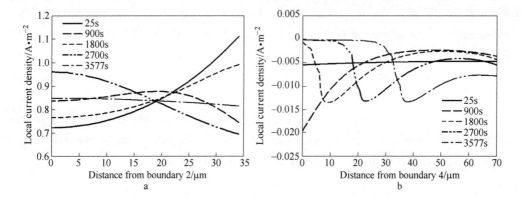

Fig. 8 The local current density distributions at different time during the discharge at 1C rate

a—The negative electrode side; b—The positive electrode side

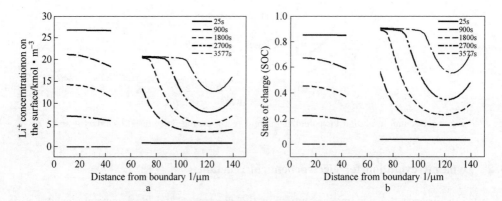

Fig. 9 The state of charge distribution at different time during the discharge at 1C discharge

The state of charge (SOC) is defined by the quotient of the surface concentration divided by the maximum concentration of 22,806mol/m³. Therefore, the surface concentration and the state of charge distributions at different time represent the same as shown in Fig. 9. The particles near to the separator-positive electrode interface firstly approach the fully discharge state. As a consequence, the utilization of active materials is not the same. As the curve shown in Fig. 8b, the peak of the local current density at the end of discharge has not reached the positive electrode-current collector boundary at $t = 3577$s. Correspondingly, the surface concentration in the region close to the current collector also has not reached the maximum concentration. In addition, the local state of discharge is far less than 1, which indicates that the active material is not adequately utilized. That is because the stop condition is determined by the cut-off voltage in both simulation and practice. When the cut-off voltage is reached, the utilization of active materials may not be 100%. To optimize battery design and achieve better performance, it is essential to choose appropriate values for some design parameters.

4.3 Heat generation and thermal behavior

Understanding the thermal behavior of these heat generation sources would offer much valuable information to develop the battery thermal management strategies. There are several heat generation mechanisms, as shown in Fig. 10. The overall heat generation of the battery discharge is consisted of electrochemical reaction heat, ohmic heat and active polarization heat produced in the negative electrode, electrolyte and positive electrode. The heat generations of negative and positive electrodes are complex, as shown in Table 1. The ionic ohmic and electrochemical interface heats are dominant contributor, nearly 100% of the total heat generation. The ionic ohmic heat from negative electrode is positive and nearly constant through the whole discharge except 0-25s, which is mainly due to the rapid change of the ionic ohmic at early discharge. An S-shape change of the electrochemical interface heat is observed from negative to positive as a function of time. Its shape resembles a mirror image of the entropy change versus SOC curve of the Li_xC_6 anode in Fig. 2a. Due to the electrochemical interface heat, the total heat generation rate of negative electrode also exhibits an S shape.

As shown in Fig. 10b, the value of ionic ohmic heat from positive electrode is also positive through the whole discharge progress. But it appears a large range, which illustrates that the ionic ohmic of positive electrode is more variable than the negative electrode. With the combined effects of ionic ohmic heat and electrochemical interface heat, the total heat of the positive electrode does not exhibit the shape as the entropy of the positive electrode change versus SOC curve of the $LiFePO_4$ cathode in Fig. 2b. Compared to the heat generation of porous electrodes, the heat generation of the current collectors and separator are tiny and can be neglected, as shown in Fig. 10c and d.

As introduced in the part of energy balance, the heat generation can be divided into two types, irreversible heat and reversible heat. The heat generation of the current collectors and separator shown in Fig. 10c and d is irreversible joule heat. Fig. 11 plots the irreversible heat, reversible heat and the total heat of porous electrodes and the battery system. Comparing Eqs. (19), (24) with Eq. (26), we obtain the relation as Eq. (45), which means that the electrochemical interface heat of porous electrodes comprises reversible heat generation and active heat generation. The curve of the reversible heat in Fig. 11a almost has the same shape with the

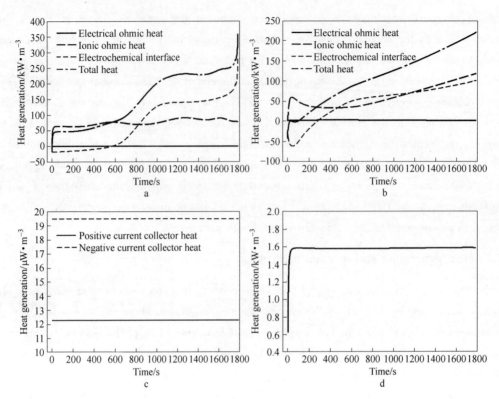

Fig. 10 Time history of heat generation by various mechanisms during discharge at 2C rate
a—Negative electrode; b—Positive electrode; c—Current collectors; d—Separator

curve of electrochemical interface heat in Fig. 10a, except the time range from 1450s to discharge end. From the value and shape difference, we can deduce that the active heat generation of negative electrode is about $5 \times 10^4 \mathrm{W \cdot m^{-3}}$ and have a server increase from 1450 s to the discharge end. And the active heat generation of the positive electrode remains a stable value of $2 \times 10^4 \mathrm{W \cdot m^{-3}}$. The active heat generation, also called polarization heat generation, is directly dependent on the overpotential, so we can conclude that the variation of the overpotential in negative electrode is severer than in positive electrode, especially at the last stage of discharge. The negative electrode plays an important role for the contribution to the overpotential in a complete battery. The total heat generation of the negative electrode is slightly higher than that of the positive electrode. While the thickness of the positive electrode (70μm) is more than twice that of the negative electrode (34μm), so the heat generation of a complete battery come mostly from the positive electrode, which is the reason why the shape of the total heat generation is similar to that of the positive electrode, as shown in Fig. 11b and c.

$$q_{i,(1)} = S_{a,j_{loc,i}} \left(T \frac{dU_i}{dT} + \phi_{1,i} - \phi_{2,i} - U_i \right)$$

$$= \left(Q_{rea} = S_{a,j_{loc,i}} T \frac{dU_i}{dT} \right) + \left[Q_{act} = S_{a,j_{loc,i}} (\phi_{1,i} - \phi_{2,i} - U_i) \right] \tag{45}$$

As shown in Fig. 11, the reversible heat accompanied endothermic process and exothermic process, and the irreversible heat is exothermic progress and increases with increasing the dis-

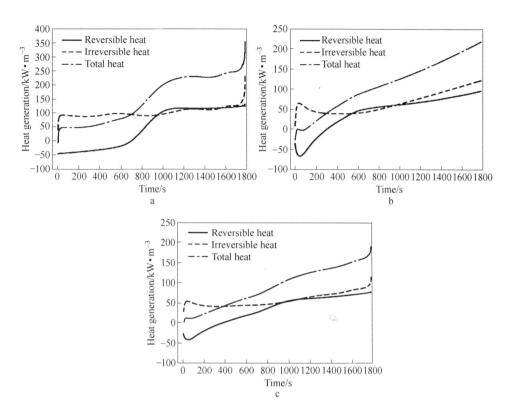

Fig. 11 Time history of reversible and irreversible heat generation during discharge at 2C rate
a—Negative electrode; b—Positive electrode; c—Complete battery

charge rate[52]. Their combination determines whether the discharge process is exothermic or not. Fig. 12 shows the total heat generation of complete battery at different discharge rate. The blue (in the web version) area stands for the endothermic heat, the red area represents the exothermic heat with an equal amount of the blue area, and the yellow area represents the net heat. At low discharge rate like 0.5C, the discharge electric current is so small that the polarization is not serious, which leads smaller exothermic amount of irreversible heat than the endothermic amount of reversible heat. Hence, there are three colored areas existing in Fig. 12a, indicating the discharge process of the LiFePO$_4$ battery is accompanied exothermic and endothermic processes at low rates. The same phenomena still exists in 1C discharge process. But when the rate is increased to 2C and above, only yellow area is left, indicating only exothermic phenomena existing. This phenomenon is in a good agreement with the experimental work of Liubin Song and his coworkers[53,54]. Fig. 13a shows evolutions of the average battery temperature during 1C discharging. The battery is surrounded by natural air at 25℃. At the initial stage of discharge, the average temperature increases slowly. But, the average temperature increases to about 37℃ at the end of discharge. Fig. 13b is the contour plot of the battery temperature distribution at the end of 1C discharging. Hot area is distributed at the axial core of the battery. Due to larger thermal conductivity in the axial direction, the temperature distribution in this direction is uniform. The main temperature variation is in the radial direction with a value about 1.5℃. Basing on the comparison between the heat generation of 5C and 1C in Fig. 12 and the temperature contour plot at 1C in Fig. 13, the battery temperature will reach more than 60℃ at 5C discharging rate. Appropriate rates can be used to avoid the internal heat accumulation of batteries.

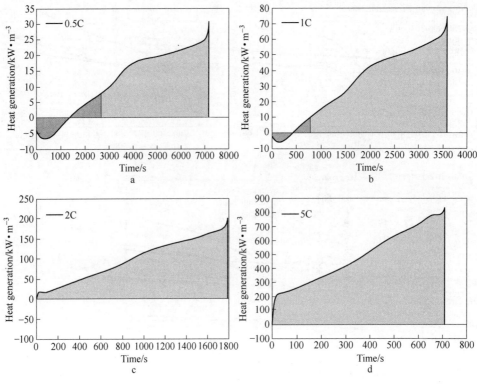

Fig. 12 The total heat generation of complete battery at different discharge rate

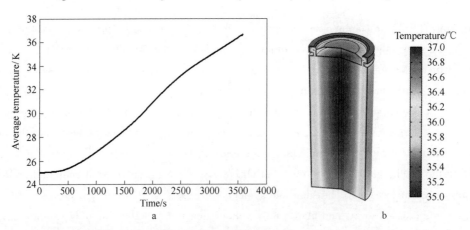

Fig. 13 Evolutions of the average battery temperature during 1C discharging(a) and the contour plot of the battery temperature distribution at the end of 1C discharging(b)

All computations are carried out on a DELL PRECISION T1650 Workstation with two quad-core processors(Intel Core i7-3770, 3.4GHz, with a total of eight processor cores) and a total of 16 GB random access memory(RAM).

5 Conclusion

In this paper, an electrochemical-thermal model based dynamic materials response for lithium iron phosphate battery is developed by employing the comprehensive dynamic parameters in thermodynamics and kinetics. The current collectors are considered in the model. This model is validated in aspects of electrochemical performance, thermal performance, which is in a good agreement between the simulated results and experimental results.

The pulse tests show that the self-adjusting period needed for lithium ion concentration in electrolyte to attain equilibrium after relaxation is affected by the discharge rate. The greater discharge rate is, the longer self-adjusting period. Because the rate of lithium ions deintercalating from the negative porous electrode is larger than that of intercalating into the positive electrode, after enough relaxation, the lithium ion concentration in electrolyte shows an augmentation compared to initial level. And the self-adjusting period needed for lithium ion concentration in particle to attain equilibrium is longer than that for electrolyte. The existence of critical makes the dynamic evolution for lithium ion concentration in solid phase much complicated. Moreover, in constant current discharge processes, the electrochemical reaction rate is location-related, and the maximum rate moves from the separator-positive electrode interface to the positive electrode-current collector interface, resulting the utilization of active material is not uniform. At the end of discharge, there is still a part of the active material has not been adequately utilized. The discharge process of LiFePO$_4$ battery accompanied exothermic process and endothermic process at a low rate. With the rate increasing, the endothermic phenomena disappears gradually, the heat production rate and the enthalpy change during the discharge process of LiFePO$_4$ battery increase. When the rate is increased to 2C, only exothermic phenomena exists. These suggest that appropriate rates for batteries should be chosen in order to avoid safety problem.

Acknowledgment

This work is supported by funds from the National Natural Science Foundation of China (No. 51204211), the China Postdoctoral Science Foundation (No. 2012M521543) and the Specialized Research Fund for the Doctoral Program of Higher Education of China (No. 20120162120089). The assistance from Prof. Jin Liu in manuscript polishing is also greatly appreciated.

Nomenclature

list of symbols

A_{cell}	area of the positive electrode (both sides) (m^2)
$c_{1,i}$	lithium in active material (mol · m^{-3})
$C_{1,max,i}$	maximum concentration (mol · m^{-3})
$c_{1,surf,i}$	Li$^+$ concentration on the surface of active material particles (mol · m^{-3})
$C_{p,i}$	heat capacity (J · (kg · K)$^{-1}$)
$D_{1,i}$	solid phase diffusivity (m^2 · s^{-1})
$D_{10,i}$	reference solid phase diffusivity (m^2 · s^{-1})
$E_{a,D,i}$	diffusion active energy (kJ · mol^{-1})
$E_{a,k,i}$	reaction active energy (kJ · mol^{-1})

h		heat transfer coefficient($W(m^{-2} \cdot K)^{-1}$)
I		current(A)
I_{app}		cell current density related to A_{cell}($A \cdot m^{-2}$)
$j_{0,i}$		exchange current density($A \cdot m^{-2}$)
$j_{loc,i}$		local current density($A \cdot m^{-2}$)
$k_{0,i}$		reaction rate constant($m^{2.5} \cdot mol^{-0.5} s^{-1}$)
k_i		thermal conductivity($W \cdot (m\ K)^{-1}$)
L_i		thickness of component(m)
Q_{act}		active heat generation($J \cdot m^{-3}$)
Q_{ohm}		ohmic heat generation($J \cdot m^{-3}$)
Q_{rea}		reaction heat generation($J \cdot m^{-3}$)
r		radius distance variable of particle(m)
R_i		characteristic radius of electrode particles(m)
$S_{a,i}$		specific surface area(m^{-1})
$SOC_{0,i}$		initial state of charge
t		time(s)
t^+		Li^+ transference number
T		absolute temperature(K)
T_{amb}		ambient temperature(K)
U_i		open circuit voltage(V)
ν		thermodynamic factor relating to electrolyte activity
x		distance from half negative foil along negative-positive direction(m)

Greek letters

$\alpha_{a,i}$		transfer coefficient for anodic current
$\alpha_{c,i}$		transfer coefficient for cathodic current
$\varepsilon_{1,i}$		active material volume fraction
$\varepsilon_{2,i}$		volume fraction
φ_i		electric potential(V)
γ_i		Bruggeman exponent
k		ionic or electronic conductivity($S \cdot m^{-1}$)
ρ_i		density($kg \cdot m^{-3}$)
σ_i		solid phase conductivity($S \cdot m^{-1}$)

Subscripts and superscripts

0		initial or equilibrated state
1		solid phase
2		liquid phase
amb		ambient temperature
n		negative electrode

p	positive electrode
irr	irreversible
re	reversible
s	separator

References

[1] B. Scrosati, J. Garche, J. Power Sources 195(2010) 2419-2430.

[2] J. Wang, X. Sun, Energy Environ. Sci. 5(2012) 5163-5185.

[3] T. Reddy, Linden's Handbook of Batteries, 2011.

[4] K. -J. Lee, K. Smith, A. Pesaran, G. -H. Kim, J. Power Sources 241(2013) 20-32.

[5] Y. Tang, M. Jia, Y. Cheng, K. Zhang, H. Zhang, J. Li, Acta Phys. Sin. 62(2013) 158201-1-158201-10.

[6] A. A. Franco, RSC Adv. 3(2013) 13027-13058.

[7] W. B. J. Zimmerman, Process Modelling and Simulation with Finite Element Methods, 2004.

[8] M. Doyle, T. F. Fuller, J. Newman, J. Electrochem. Soc. 140(1993) 1526-1533.

[9] T. F. Fuller, M. Doyle, J. Newman, J. Electrochem. Soc. 141(1994) 1-10.

[10] C. R. Pals, J. Newman, J. Electrochem. Soc. 142(1995) 3274-3281.

[11] C. R. Pals, J. Newman, J. Electrochem. Soc. 142(1995) 3282-3288.

[12] J. Yi, U. S. Kim, C. B. Shin, T. Han, S. Park, J. Power Sources 244(2013) 143-148.

[13] L. Cai, Y. Dai, M. Nicholson, R. E. White, K. Jagannathan, G. Bhatia, J. Power Sources 221(2013) 191-200.

[14] Y. Ye, Y. Shi, N. Cai, J. Lee, X. He, J. Power Sources 199(2012) 227-238.

[15] K. Somasundaram, E. Birgersson, A. S. Mujumdar, J. Power Sources 203(2012) 84-96.

[16] S. Elul, Y. Cohen, D. Aurbach, J. Electroanal. Chem. 682(2012) 53-65.

[17] S. Santhanagopalan, Q. Guo, P. Ramadass, R. E. White, J. Power Sources 156(2006) 620-628.

[18] L. Cai, R. E. White, J. Power Sources 196(2011) 5985-5989.

[19] A. K. Padhi, K. S. Nanjundaswamy, J. B. Goodenough, J. Electrochem. Soc. 144(1997) 1188-1194.

[20] A. Yamada, Y. Kudo, K. -Y. Liu, J. Electrochem. Soc. 148(2001) A1153-A1158.

[21] V. Srinivasan, J. Newman, J. Electrochem. Soc. 151(2004) A1517-A1529.

[22] M. Wang, J. Li, X. He, H. Wu, C. Wan, J. Power Sources 207(2012) 127-133.

[23] Y. Ye, Y. Shi, A. A. O. Tay, J. Power Sources 217(2012) 509-518.

[24] R. E. Gerver, J. P. Meyers, J. Electrochem. Soc. 158(2011) A835-A843.

[25] C. Hellwig, S. Sorgel, W. G. Bessler, ECS Trans. 35(2011) 215-228.

[26] V. Ramadesigan, P. W. C. Northrop, S. De, S. Santhanagopalan, R. D. Braatz, V. R. Subramanian, J. Electrochem. Soc. 159(2012) R31-R45.

[27] M. Safari, C. Delacourt, J. Electrochem. Soc. 158(2011) A63-A73.

[28] M. Safari, C. Delacourt, J. Electrochem. Soc. 158(2011) A562-A571.

[29] M. Safari, C. Delacourt, J. Electrochem. Soc. 158(2011) A1436-A1447.

[30] M. Safari, C. Delacourt, J. Electrochem. Soc. 158(2011) A1123-A1135.

[31] C. Delacourt, M. Safari, J. Electrochem. Soc. 159(2012) A1283-A1291.

[32] L. O. Valøen, J. N. Reimers, J. Electrochem. Soc. 152(2005) A882-A891.

[33] W. Wu, X. Xiao, X. Huang, Electrochimica Acta 83(2012) 227-240.

[34] M. Guo, G. -H. Kim, R. E. White, J. Power Sources 240(2013) 80-94.

[35] I. J. Ong, J. Newman, J. Electrochem. Soc. 146(1999) 4360-4365.

[36] L. Rao, J. Newman, J. Electrochem. Soc. 144(1997) 2697-2704.

[37] W. B. Gu, C. Y. Wang, J. Electrochem. Soc. 147(2000) 2910-2922.

[38] G. -H. Kim, A. Pesaran, R. Spotnitz. J. Power Sources 170(2007)476-489.

[39] V. Srinivasan, J. Newman. Electrochem. Solid-State Lett. 9 A110-A2006.

[40] D. Morgan, A. Van der Ven, G. Ceder. Electrochem. Solid-State Lett. 7(2004)A30-A32.

[41] D. Bernardi, E. Pawlikowski. J. Newman, J. Electrochem. Soc. 132(1985)5-12.

[42] O. Y. Egorkina, A. M. Skundin. J. Solid State Electrochem. 2(1998)216-220.

[43] T. L. Kulova, A. M. Skundin, E. A. Nizhnikovskii, A. V. Fesenko. Russ. J. Electrochem. 42(2006)259-262.

[44] V. Srinivasan, C. Y. Wang. J. Electrochem. Soc. 150(2003)A98-A106.

[45] J. L. Dodd, Ph. D. Dissertation, California Institute of Technology, 2007.

[46] J. Newman, W. Tiedemann, AIChE J. 21(1975) 25-41.

[47] G. -H. Kim, K. Smith. J. Ireland, A. Pesaran, J. Power Sources 210(2012)243-253.

[48] M. Guo, R. E. White. J. Power Sources 221(2013) 334-344.

[49] http://www.akukeskus.ee/anr26650m1a_datasheet_april_2009.pdf.

[50] C. Lim, B. Yan, L. Yin, L. Zhu. Electrochimica Acta 75(2012)279-287.

[51] D. M. Bernardi, J. -Y. Go. J. Power Sources 196(2011)412-427.

[52] V. V. Viswanathan, D. Choi, D. Wang, W. Xu, S. Towne. R. E. Williford, J. -G. Zhang, J. Liu, Z. Yang, J. Power Sources 195(2010)3720-3729.

[53] L. Song, X. Li, Z. Wang, H. Guo, Z. Xiao, F. Zhang. S. Peng. Electrochimica Acta 90 (2013) 461-467.

[54] L. Song, X. Li, Z. Wang, X. Xiong, Z. Xiao, F. Zhang. Int. J. Electrochem. Sci. 7(2012) 6571-6579.

Numerical Analysis of Distribution and Evolution of Reaction Current Density in Discharge Process of Lithium-ion Power Battery[*]

Abstract The reaction current density is an important process parameter of lithium-ion battery, significantly influencing its electrochemical performance. In this study, aimed at the discharge process of lithium-ion power battery, an electrochemical-thermal model was established to analyze the distribution of the reaction current density at various parts of the cathode and its evolution with the time of discharge, and to probe into the causes of distribution and evolution. The investigation revealed that the electrochemical-thermal coupled model showed more accurate compared to the single electrochemical model, which was more obvious in high rate discharge. The results demonstrated that the conductivity of solid and liquid-phases was an important factor affecting the distribution of the reaction current density. Moreover, the uniformity of the distribution of the current density was related to the rate of utilization of the active materials in the electrodes. By optimizing the porosity and thickness of the electrode, not only the distribution of the current density was improved, but also the rate of utilization of the active materials in the electrodes and the energy density of batteries were significantly enhanced.

Lithium-ion battery is low maintenance with a series of advantages such as high voltage, high energy density, long cycle life, and no "memory effect"[1-3]; therefore, it has been extensively employed in portable electronic products and become a preferred battery of choice for electric vehicles and hybrid electric vehicles.[4] The study of Li-ion battery is of significant scientific and technological interest. In recent years, the new energy vehicle market has developed rapidly; thus, to satisfy the needs for the Li-ion battery with higher single capacity and specific energy, the battery manufacturers are encouraged to improve their electrode design. Therefore, it is extremely important and urgent to analyze in depth the dynamics of Li-ion power battery because it is inevitably necessary to increase the battery power.

The research conducted on the Li-ion batteries is based on the experimental approach; thus, to gain insight into the characteristics of the battery and to examine its performance, different experiments would have to be performed. Operating mechanism of the battery can be well understood by the visual data obtained from the experiments and by summarizing the related criterion for electrode design by comparing the effect of different electrode designs on the battery performance. However, the Li-ion battery is a closed chemical system with complex internal structure and components; therefore, it is difficult to acquire directly the distribution of its internal physical quantities from the experiments in real time, which significantly affects the understanding for the battery operation. Instead, real time management of the electrochemical

---

  [*] Copartner: Yiwei Tang, Ming Jia, Jie Li, Yanqing Lai, Yun Cheng. Reprinted from Journal of The Electrochemical Society. 2014. 161(8): E3021-E3027.

process can be effectively studied by applying the computer numerical simulation technology to establish mathematical models on the basis of a strict electrode dynamics theory framework and enormous amount of accumulated battery data. The mathematical models across multiple scales were widely used in understanding and describing behaviors of Li-ion battery,[5-8] it forms the core of systems engineering approach for the optimal design of Li-ion battery.[9] Newman et al.[10-17] applied Butler-Volmer equation to describe the electrochemical process occurring between the interface of electrode and electrolyte based on the porous electrode theory. Fick's law was used to describe the intercalation and deintercalation of the Li-ions inside the active-material particles, and the mass transfer process of the Li-ion in electrolyte was described using concentrated solution theory. Moreover, the changes in the concentration distribution, electrochemical potential, and exchange current density of the battery at various parts with time of discharge were obtained from the calculation. Wang[18] et al. applied the abovementioned model to study the distribution and changes in local reaction current density during the discharge process and to probe into the relationship between the reaction current density and electrode design. However, this study ignored the effect of temperature on the electrochemical process revealing that a large error could be produced by the increase in temperature due to a high-powered discharge of the battery.[9] Incorporation of the energy conservation in the electrochemical process would be helpful to improve the accuracy of model.[19] Smith[20] et al. and Ye[21] et al. utilized the electrochemical-thermal coupled model to study the relationships of the temperature with the electrochemical reaction, as well as with the key parameters such as the diffusion coefficients of the solid and liquid-phases, indicating that the influence of temperature change on the electrochemical parameters could not be neglected; and verifying the accuracy of the model via experimental method. The reaction current density is an important parameter in the operating process of the Li-ion battery and significantly influencing its electrochemical performance. To analyze the dynamics during operation and to acquire a deeper understanding of the battery, real-time and quantitive analysis should be conducted to study the parameters and the factors influencing them.

In allusion to the Li-ion power battery; this study established a one-dimensional electrochemical-thermal coupled model capable of investigating the distribution and evolution of the local reaction current density during discharge process by considering anode as an example. The model was useful in analyzing the causes of distribution and evolution; thus, further providing guidance for the design of Li-ion power battery.

1 Model development and experimental

Taking into account the coupling relationship between electrochemical reaction and heat, an electrochemical thermal coupling model was established to investigate the electrochemical process of lithium ion battery. The schematic of the battery modeled in this study is shown in Fig. 1. The complete electrochemical system is composed of five media, namely negative current collector, negative electrode, separator, positive electrode and positive current collector. The active materials of solid electrodes are treated as homogenous media, and are comprised with spherical particles.

Electrochemical model. —The model developed in this paper considers porous electrode theory, Ohm's law, concentrated solution theory, intercalation/deintercalation kinetics and transport in solid phase and electrolyte phase. The main governing equations and boundary conditions required in this model are as follows:

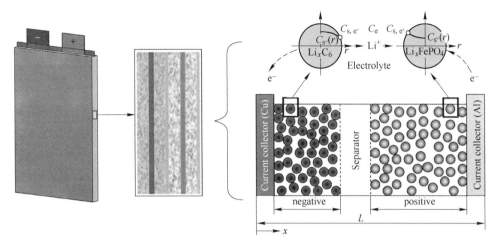

Fig. 1 Schematic diagram of lithium iron phosphate battery model

For the mass balance of lithium in an intercalation particle of electrode active material is described by Fick's second law in spherical coordinates can be described as:

$$\frac{\partial c_s}{\partial t} = \frac{D_s}{r^2}\left[\frac{\partial}{\partial r}\left(r^2 \frac{\partial c_s}{\partial r}\right)\right] = D_s\left(\frac{2}{r}\frac{\partial c_s}{\partial r} + \frac{\partial^2 c_s}{\partial r^2}\right) \quad (1)$$

The parameter r is the distance from the center of solid sphere. There is no species source at the center of sphere, so $\left.\frac{\partial c_s}{\partial r}\right|_{r=0} = 0$. The Li$^+$ concentration on the surface of the particles is coupled to the concentration and flux which can be described as: $\left.\frac{\partial c_s}{\partial r}\right|_{r=R_s} = \frac{j^{Li}}{a_s F}$. Under ideally close packing condition, the specific interfacial area of the porous electrode is calculated by

$$a_s = \frac{3\varepsilon_s}{r_s} \quad (2)$$

Where ε_s represent the volume fraction of solid active material in the electrode regions.

For the mass balance in the electrolyte phase,

$$\varepsilon_e \frac{dc_e}{dt} + \nabla \cdot \{-D_e^{eff} \nabla c_e\} = \frac{j^{Li}}{F}(1-t_+) \quad (3)$$

$$D_e^{eff} = D_e \varepsilon_e^{\gamma} \quad (4)$$

The flux of liquid species is arbitrarily set to zero at current collector/electrode interface, liquid species flux and species concentration at separator/electrode interface are taken to be continuous.

For the Ohm's law in the solid phase, there is a double layer capacitance at the interface between the active materials and electrolyte.[22] The capacitance can store additional energy and smooth the abrupt change of the cell voltage caused by the short-time pulses passing through a cell. However, the capacitance was seldom considered in the previous models.[22] Small modification in the electronic charge conservation is required to account of the double layer. Electronic charge conservation for solid phase can be expressed as follows:

$$\nabla \cdot (-\sigma_c \nabla \varphi_c) = -J \tag{5}$$

$$\nabla \cdot (-k_s^{eff} \nabla \varphi_s) = -a_s \left(\frac{j^{Li}}{a_s} + C_{dt} \left(\frac{\partial \varphi_s}{\partial t} - \frac{\partial \varphi_e}{\partial t} \right) \right) \tag{6}$$

where φ_c is arbitrarily set to zero at anode current collector external-face; at cathode current collector external-face the charge flux is set to equal the average current density of the battery, which is expressed as Eq. 7.

$$\varphi_c \big|_{x=0} = 0; \; -\sigma_c \nabla \varphi_c \big|_{x=L_{ncc}+L_n+L_s+L_p+L_{pcc}} = -I_{app} \tag{7}$$

For the charge conversation in the solution phase, according to the concentrated solution theory, the governing equation for electronic charge conservation in solution phase is expressed as:

$$\nabla \cdot \left\{ k_e^{eff} \left[-\nabla \varphi_e + \frac{2RT}{F} \left(1 + \frac{\partial \ln f}{\partial \ln c_e} \right) (1-t_+) \frac{\nabla c_e}{c_e} \right] \right\} = j^{Li} \tag{8}$$

Liquid-junction potential is introduced in Eq. 10 with expression

$$K_{junc} = \frac{2RT}{F} \left[1 + \frac{\partial \ln f}{\partial \ln c_e} \right] (1-t_+) = \frac{2RT}{F} \nu \tag{9}$$

The parameter ν is the thermodynamic factor relating to electrolyte activity, and it is concentration and temperature dependent.

Current collectors present an impermeable wall to the electrolyte, therefore the lithium ion flux is null in these boundaries.

$$\frac{\partial \varphi_e}{\partial x} \bigg|_{x=0} = \frac{\partial \varphi_e}{\partial x} \bigg|_{x=L} = 0 \tag{10}$$

The reaction current density is calculated using the Bulter-Volmer equation:

$$j^{Li} = a_s i_0 \left[\exp\left(\frac{\alpha_a F}{RT} \eta \right) - \exp\left(-\frac{\alpha_c F}{RT} \eta \right) \right] \tag{11}$$

$$\eta = \varphi_s - \varphi_e - U - j^{Li} R_{sei} \tag{12}$$

Thermal model.—The heat generation is modeled with the local heat generation model of Rao and Newman[23] using the formulation of Gu and Wang.[24] According to the position of heat generation, the total heat generation is the summation of heat generated in the two electrodes, separator and current collectors. The summary of the heat generation mechanisms is listed in Table 1.

Table 1 Heat generation mechanism in a cell

	Heat generation mechanism	Equation	
Porous electrode	(1) Electrochemical interface	$q_{(1)} = j^{Li} \left(T \frac{dU}{dT} + \varphi_s - \varphi_e - U \right)$	(13)
	(2) Electrical ohmic heat	$q_{(2)} = \sigma^{eff} \nabla \varphi_s \cdot \nabla \varphi_s$	(14)
	(3) Ionic ohmic heat	$q_{(3)} = k_e^{eff} \nabla \varphi_e \cdot \nabla \varphi_e + \frac{2RTk_e^{eff}}{F}(t_+ - 1)\left(1 + \frac{\partial \ln f}{\partial \ln c_e}\right) \cdot \nabla(\ln c_e) \cdot \nabla \varphi_e$	(15)
Separator	Ionic ohmic heat	$q_{sep} = k_e^{eff} \nabla \varphi_e \cdot \nabla \varphi_e + \frac{2RTk_e^{eff}}{F}(t_+ - 1)\left(1 + \frac{\partial \ln f}{\partial \ln c_e}\right) \cdot \nabla(\ln c_e) \cdot \nabla \varphi_e$	(16)
	Current collector Electrical ohmic heat	$q_c = \sigma_c \nabla \varphi_c \cdot \nabla \varphi_c$	(17)

According to the type of the heat generation, there are three parts of heat sources during the charge or discharge processes, including reaction heat, Q_{rea}, due to entropy change during the discharge; ohmic heat, Q_{ohm}, due to the ohmic potential drop and active polarization heat, Q_{act}, due to the electrochemical reaction polarization between active material particle surface and the electrolyte.

Reaction heat is:
$$Q_{rea}=j^{Li}T\frac{dU}{dT} \tag{18}$$

Ohmic heat is:
$$Q_{ohm}=q_{(2)}+q_{(3)}+q_{sep}+q_c \tag{19}$$

Active heat is:
$$Q_{act}=j^{Li}(\varphi_s-\varphi_e-U) \tag{20}$$

Active heat Q_{act} and ohmic Q_{ohm} are irreversible, while reaction heat Q_{rea} is reversible.
$$Q_{irr}=Q_{ohm}+Q_{act} \tag{21}$$
$$Q_{re}=Q_{rea} \tag{22}$$

The following equation governs energy conservation in lithium ion battery.
$$\rho C_p\frac{\partial T}{\partial t}+\nabla\cdot(-k\nabla T)=Q_{irr}+Q_{re} \tag{23}$$

The parameters ρ, C_p, and k are density, heat capacity and thermal conductivity, respectively. Physiochemical property values are made temperature dependent, coupling the 1D electrochemical model to the thermal model. An Arrhenius equation defines the temperature sensitivity of a general physiochemical property, ψ, as

$$\psi=\psi_{ref}\exp\left[\frac{E_{act}^{\psi}}{R}\left(\frac{1}{T_{ref}}-\frac{1}{T}\right)\right] \tag{24}$$

Where E_{act}^{ψ} is the activation energy, controls the temperature sensitivity of each individual property, ψ. ψ_{ref} is the property value defined at reference temperature $T_{ref}=25\text{℃}$.

Model parameters. —The physical properties of battery components and battery design parameters are summarized in Table 2 and Table 3, respectively.

Table 2 Critical battery parameters used in electrochemical model

Quantity	Negative electrode	Positive electrode	Separator	Electrolyte	Copper foil	Aluminum foil
Design specifications (geometry and volume fractions)						
ε_s	0.55	0.43	—			
ε_e	0.33	0.332	0.54			
$L(\mu m)$	34	70	25		14	20
$R(\mu m)$	3.5	0.08				
Lithium ion concentrations						
$c_{ini,e}(\text{mol}\cdot\text{m}^{-3})$				1200		
$c_{max,s}(\text{mol}\cdot\text{m}^{-3})$	31,370	22,806				
$c_{ini,s}(\text{mol}\cdot\text{m}^{-3})$	31.370×0.85	22.806×0.015				
Kinetic and transport properties						
a_a, a_c	0.5	0.5				

Continued Table 2

Quantity	Negative electrode	Positive electrode	Separator	Electrolyte	Copper foil	Aluminum foil
γ	1.5	1.5	1.5			
$D_e(m^2 \cdot s^{-1})$				Eq. 25		
$C_{dl}(F \cdot m^{-2})$	0.2	0.2				
$D_s(m^2 \cdot s^{-1})$	3.9×10^{-14}	9×10^{-18}				
$k_s(S \cdot m^{-1})$	100	0.5				
$k_c(S \cdot m^{-1})$				Eq. 26		

Table 3 Thermal properties of battery components

Materials	Density/kg·m^{-3}	Heat capacity/J·(kg·K)$^{-1}$	Thermal conductivity/W·(m·K)$^{-1}$
Negative electrode	2223	641	1.04
Positive electrode	1500	800	1.48
Separator	900	1883	0.5
Electrolyte	1210	1518	0.099
Copper foil	8700	396	398
Aluminum foil	2700	897	237

Valøen[25] investigated a full set of transport properties for LiPF$_6$ in PC/EC/DMC experimentally, showing the k_e, D_e as functions of temperature and LiPF$_6$ concentration.

$$D_e = 1 \times 10^{-4} \times 10^{-4.43 - \frac{54.0}{T - 229.0 - 0.05C} - 2.2 \times 10^{-4}C} \tag{25}$$

$$k_e = 1 \times 10^{-4}(-10.5 + 0.074T - 6.69 \times 10^{-5}T^2 + 6.68 \times 10^{-4}C - 1.78 \times 10^{-5}CT + 2.8 \times 10^{-8}CT^2 + 4.49 \times 10^{-7}C^2 - 8.86 \times 10^{-10}C^2T) \tag{26}$$

The open circuit potentials of the positive electrode U_p and the negative electrode U_n are a function of local SOC on the surface of active particles. U_p and U_n come from Ref. [26], which are expressed by Eq. 27 and Eq. 28.

$$U_p = 3.4323 - 0.4828\exp[-80.2493(1-y)^{1.3198}] - 3.2474 \times 10^{-6}\exp[20.2645(1-y)^{3.8003}] + 3.2482 \times 10^{-6}\exp[20.2646(1-y)^{3.7995}] \tag{27}$$

$$U_n = 0.6379 + 0.5416\exp(-305.5309x) + 0.044\tanh\left(-\frac{x-0.1958}{0.1088}\right) - 0.1978\tanh\left(\frac{x-1.0571}{0.0854}\right) - 0.6875\tanh\left(\frac{x+0.0117}{0.0529}\right) - 0.0175\tanh\left(\frac{x-0.5692}{0.0875}\right) \tag{28}$$

The x and y are dimensionless solid-state Li concentration at the surface of graphite and LiFePO$_4$ particles, respectively.

Experimental.—Experimental studies were done on a lithiumion pouch cells(10mm×140mm×105mm, capacity 10.0Ah, LiC$_6$ anode material, LiFePO$_4$ cathode material, 1.2mol·L^{-1} LiPF$_6$ in PC/EC/DMC solvent). A battery test system(HT-V5C200D200-4, China) was used to monitor the charge-discharge current and battery voltage. Real-time graph is created and stored in computer. The experimental discharge data used in this work were obtained at 25℃ with following steps: (1) the battery is fully charged to 3.65V with a constant current of 0.5C(5.0A) and with constant voltage to 3.65V till charge current declined to 0.05C(0.5A); (2) After 10min rest, battery was discharged with some constant current(1C,5C) until the cell voltage reached 2.0 V.

2 Results and Discussion

The electrochemical process of the Li-ion battery is a continuous multiphase and multistep behavior; involving the solid and liquid phases mass transfer, and chemical reactions at the interface. The resistances of the solid and liquid phases mass transfers varied because of certain thickness of the electrode; thus, the electrochemical reactions at various sites could not be uniformly realized. Besides, the distribution of local current density of electrode/solution interface had a direct bearing on the rate of utilization of the active materials inside the battery. Fig. 2 shows the distribution of reaction current density in the direction of the positive electrode at 3C discharge rate. In the initial stage of discharge, the peak value of the current density appeared in the electrode zone near the separator, indicating that the rate of electrode reaction in this zone was greater than that in the other sites.

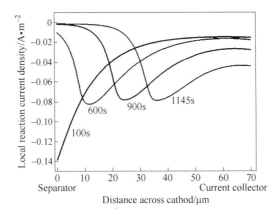

Fig. 2 Distribution of reaction current densities in the positive electrode direction at different periods during 3C discharge rate

The Li-ions were deintercalated from the cathode and then intercalated into the anode during the discharge process; therefore, the concentration of the Li-ions on the surface of the active material particles near the separator increased dramatically, as shown in Fig. 3. The surface of the particle in the zone reached the maximum Li intercalation concentration with the progress in the discharge process, further leading to a difficulty for Li-ions to continue their intercalation; thus, resulting in a reduced reaction current density (however, such density could not be promptly dropped to zero). The abovementioned phenomenon could be explained by assuming the active electrode materials being made up of spherical particles with certain radius. The spherical surface and the center of the sphere relied on diffusion to realize the mass balance between them. Thus, when the Li-ions on the surface of the particles reached the maximum concentration and made it difficult to continue the Li intercalation, the Li-ions were delivered into the interior of the particles by diffusion. In this state, the Li-ions obtained from the particle surface through the reaction were equal to the Li-ions lost during diffusion; thus, the reaction current was equal to the diffusion current. When the overall particles reached the maximum Li intercalation concentration, the reaction current density dropped to zero. The adopted method of constant-current discharge states that according to the charge conservation, when the reaction current density near separator was reduced, the current density in the other zones could be increased correspondingly, thus making the reaction current peak move to the current collector till the completion of the discharge process.

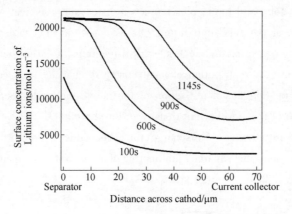

Fig. 3 Distribution of reaction current densities in the anode direction at different periods during 3C discharge rate

Fig. 3 demonstrates that the reaction zone initially located in the surface layer of the electrode near the separator, and gradually moved inward with time of discharge, follows a reverse trend compared to the study of Wang[18] et al. because of the difference in the effective conductivity ratios of the solid-phase to the liquid-phase. The electrode of the Li-ion battery was made up of a porous structure. Based on the porous electrode theory, the electrode contained the electronic conductive network constituted by solid-phase conductive particles and ion transportation network with electrolyte as the carrier. Moreover, the transfer processes of electrons and ions were respectively related to the volume fraction and refraction coefficient of the pore structure in the solid and liquid phases. Considering the complexity of the pore structure of the real electrode, it was very difficult to obtain precisely the effective conductivity in the solid and liquid phases. In general, Formulas 29 and 30 are used to calculate the effective ionic conductivity ratios in the solid and liquid phases.

$$\kappa_e^{\text{eff}} = \kappa_e \varepsilon_e^{\gamma} \tag{29}$$

$$\kappa_s^{\text{eff}} = \kappa_s \varepsilon_s^{\gamma} \tag{30}$$

Fig. 4 exhibits that under different conductivity ratios in the solid and liquid phases, the electrochemical-thermal coupled model is used to obtain the distribution curve of the reaction current density when the 3C was discharged for 5s.

When $\kappa_e^{\text{eff}} > \kappa_s^{\text{eff}}$, the initial site of the reaction peak was near the current collector; when $\kappa_e^{\text{eff}} = \kappa_s^{\text{eff}}$, the initial reaction peak was on both the ends of the electrodes; however, when $\kappa_e^{\text{eff}} < \kappa_s^{\text{eff}}$, the initial site of the reaction peak was near the separator. To investigate the influence of the site of initial reaction peak on the discharge capacity of the battery, the solid-phase conductivity was kept constant, and liquid-phase conductivity was increased and then decreased to change the

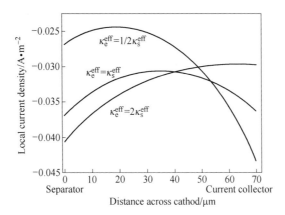

Fig. 4 Distribution of reaction current density in the anode direction when
3C was discharged for 5s under different conductivity ratios in the solid and liquid phases

site of the initial reaction peak and to simulate the 3C discharge process of the battery according to the aforementioned conditions. Fig. 5 reveals the influence of the changed liquid-phase conductivity on the discharge voltage and capacity. When the liquid-phase conductivity was increased by tenfold compared to the previous level, the discharge capacity increased by 0.35%; however, when the liquid-phase conductivity ratio reduced to one-tenth of its previous level, the discharge capacity was decreased by 0.94%. The significant increase or decrease in the liquid-phase conductivity led to a small change in the discharge capacity indicating that the conductivity in the solid and liquid phases were not the rate determining steps. Under this condition, the site of the initial reaction peak would exert a small influence on the discharge capacity of the battery; simultaneously revealing that, during the electrode design, it was necessary to keep the conductivity in both the solid and liquid phases identical. This would generate a limited influence on improving the battery performance by increasing the conductivity of only one phase.

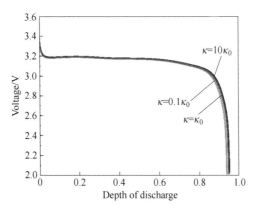

Fig. 5 Voltage change during 3C discharge process with the variation in depths
of discharge under different liquid-phase specific resistances

According to formula 24, the conductivity in the solid and liquid phases change with the variation in the temperature. During the discharge process, the internal temperature of the Li-ion battery would be changed because of the accumulation of heat and due to the change in thermogenesis rate, further leading to a corresponding change in the conductivity in the solid and liquid phases. Fig. 6 reveals the change in the conductivity of the liquid-phase with the variation in depths of discharge at 1C, 3C, and 5C discharge rates, respectively.

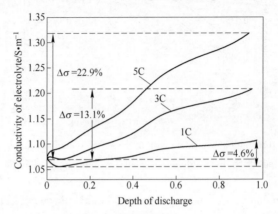

Fig. 6　Change in liquid conductivity ratio with the variation in depths of discharge during different discharge rates

As shown in Fig. 6, during 1C discharge process, the conductivity barely changed; however, at the end of 3C and 5C discharge processes, the conductivity increased respectively by 13.1% and 22.9% compared to the initial value. Fig. 6 explains that during a high-powered discharge process, the change in temperature exerts a significant influence on the liquid-phase conductivity because under the similar cooling condition, the average temperature of the battery would increase with the increase in the current rate.[3,27] Moreover, a drastic change in temperature resulted in the change in liquid-phase conductivity. Thus, the electrochemical-thermal coupled model is considered as a realistic reflection to study the distribution of local current density of Li-ion power battery during the discharge process. Fig. 7 exhibits the comparison chart of simulated battery discharge curves and experimental results at the discharge rates of 1C and 5C. The calculated results of the electrochemical-thermal coupled model showed a smaller error compared to the experimental value, which was more obvious in 5C discharge. This study considered the spatial variation of temperature, and calculated the heat generation of cathode, anode and current collector, respectively. However, due to the thermal resistance is very small, the spatial variation is extremely minor, a lumped type analysis can be used for higher computational efficiency.

Figs. 2 and 3 show that, at the end of the discharge process, the Li-ion concentration on the surface of the active particles at current collector has not reached its maximum level, indicating that during the operation, the interior and exterior surfaces of the electrode did not play identical effective roles. Moreover, the polarizations on various electrodes/solution interfaces were not uniform; thus, generating a negative effect on the electrode performance.

To provide maximum power and energy output, it was necessary to optimize the electrode. Multiple factors influencing the design of the electrode were taken into account. When the material system is determined, according to the formulas 29 and 30, the porosity of the electrode

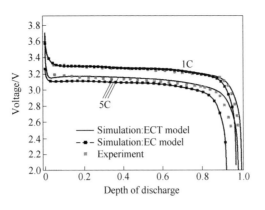

Fig. 7 Comparison chart of the simulated battery discharge curve and experimental result at different discharge rates. ECT: electrochemical and thermal coupling model; EC: electrochemical model

is regarded as a key parameter. Moreover, the thickness of the electrode determined the length of the transmission route. Therefore, the designed value of the thickness of the electrode was equal to the "critical thickness" of its effective reaction layer.[18] Thus, at the end of the discharge, the active materials of the electrodes were completely utilized. If the electrode thickness was greater than the "critical thickness", a part of the active materials could not attain the maximum Li intercalation concentration; on the contrary, with the insufficient amount of the active materials of the electrodes, the volume energy density was small and the spatial utilization rate was low. The porosity was related to the degree of the electrode infiltrated by the electrolyte; specifically, if the electrode was insufficiently infiltrated, the reaction resistance of the electrode was high.

Fig. 8 shows the depths of discharge of the electrode with different porosity and thickness, in different sites at the end of the 3C discharge process. When the active materials in the electrodes were in a certain proportion, the change in the thickness of the electrode led to a change in the content of the active materials. To assure that the electrodes of different thickness were discharged at the same rate, the discharge current in the simulated experiment was correspondingly adjusted with the change in the thickness.

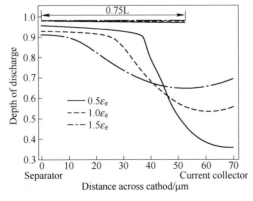

Fig. 8 Depths of discharge at various positions of electrodes at the end of 3C discharge in different porosity and thickness

Thus, the comparison reveals that, the increase in the volume fraction of electrolyte in the same electrode thickness contributes to a uniformly distributed reaction current density, a smaller difference in depth of discharge of the active materials at various positions of the electrodes at the end of the discharge process, and an improved rate of utilization of the material. When the volume fractions of electrolyte were identical, with the increase in the electrode thickness, the reaction current density was distributed non-uniformly, resulting in a large difference in the depth of discharge at various positions of the electrodes and a low rate of utilization of some active materials. As shown in Fig. 8, when the electrode thickness was 52.5μm (0.75L), the depths of discharge of the active substances at various positions were distributed uniformly and the value was approximately one, indicating that such a thickness was almost similar to the "critical thickness", at this condition, the smaller volume fraction cannot exert a significant influence on the uniformity of the distribution of depths of discharge. Therefore, during the electrode design, when the discharge rate was determined, to promote the rate of utilization and energy density of the active substances, it was required to make the value of the thickness closer to the "critical thickness" as far as possible.

3 Conclusions

In this study, an electrochemical-thermal model was established to analyze the distribution of the reaction current density at various parts of the electrode and its evolution with the time of discharge, and probed into the causes of distribution and evolution.

The investigation revealed that the conductivity of solid and liquid-phases was an important factor affecting the distribution of the reaction current density. The results demonstrated the following phenomenon: when the conductivity of the solid-phase was less than that of the liquid-phase, the initial reaction zone was near the current collector; when the conductivities of the solid and liquid-phases were equal, the initial reaction zone was on both the ends of the electrode; however, when the conductivity of the solid-phase was greater than that of the liquid-phase, the initial reaction zone was near the separator. Moreover, the uniformity of the distribution of the current density was related to the rate of utilization of the active materials in the electrodes. Therefore, by optimizing the porosity and thickness of the electrode, not only the distribution of the current density was improved, but also the rate of utilization of the active materials in the electrodes and the energy density of batteries were significantly enhanced. This work are helpful to an analysis the dynamics during operation and acquire a deeper understanding of the battery, which are significant and crucial for proper electrode designing.

Acknowledgments

This work is supported by funds from the National Natural Science Foundation of China (No. 51204211 and No. 51222403) and the China Postdoctoral Science Foundation (No. 2012M521543), which are greatly appreciated.

Nomenclature

List of symbols

a_s	active surface area per electrode unit volume(m^{-1})
c_s	lithium in active material($mol \cdot m^{-3}$)
$c_{max,s}$	maximum concentration($mol \cdot m^{-3}$)
C_p	heat capacity($J/(kg \cdot K)^{-1}$)
D_s	solid phase diffusivity($m^2 \cdot s^{-1}$)
h	heat transfer coefficient($W \cdot m^{-2} \cdot K^{-1}$)
I	current(A)
I_{app}	cell current density($A \cdot m^{-2}$)
i_0	exchange current density($A \cdot m^{-2}$)
j^{Li}	reaction current density($A \cdot m^{-2}$)
k	thermal conductivity($W/(m \cdot K)^{-1}$)
L	thickness of component(m)
Q_{act}	active heat generation($J \cdot m^{-3}$)
Q_{ohm}	ohmic heat generation($J \cdot m^{-3}$)
Q_{rea}	reaction heat generation($J \cdot m^{-3}$)
R	radius distance variable of particle(m)
R_{sei}	film resistance on an electrode surface($\Omega \cdot m^2$)
soc	state of charge
t	time(s)
t^+	Li^+ transference number
T	absolute temperature(K)
U	open circuit voltage(V)
V	thermodynamic factor relating to electrolyte activity

Greek letters

α_a	transfer coefficient for anodic current
α_c	transfer coefficient for cathodic current
ε	active material volume fraction or porosity of a phase
φ	potential(V)
γ	Bruggeman exponent
κ	ionic or electronic conductivity($S \cdot m^{-1}$)
ρ	density($kg \cdot m^{-3}$)
σ	solid phase conductivity($S \cdot m^{-1}$)

Subscripts

e	electrolyte phase
max	maximum value
ref	with respect to a reference state
s	solid phase
sep	separator

References

[1] R. E. Gerver and J. P. Meyers, J. Electrochem. Soc. , 158, A835(2011).

[2] J. Wang and X. Sun, Energy and Environmental Science. , 5, 5163(2012).

[3] K. -J. Lee, K. Smith, A. Pesaran, and G. -H. Kim, J. Power Sources. , 241, 20(2013).

[4] L. Cai and R. E. White, J. Power Sources. , 196, 5985(2011).

[5] P. Gomadam, R. E. White, and J. W. Weidner, ECS Transactions. , 19, 1(2009).

[6] P. Gomadam, R. E. White, and J. W. Weidner, J. Electrochem. Soc. , 150, A1339(2003).

[7] X. C. Zhang, W. Shyy, and A. M. Sastry, J. Electrochem. Soc. , 154, S21(2007).

[8] P. C. Northrop, V. Ramadesigan, S. De, and V. R. Subramanian, J. Electrochem. Soc. , 159, S5(2012).

[9] V. Ramadesigan, P. C. Northrop, S. De, S. Santhanagopalan, R. Braatz, and V. R. Subramanian, J. Electrochem. Soc. , 159, R31(2012).

[10] V. Srinivasan and J. Newman, J. Electrochem. Soc. , 151, A1517(2004).

[11] M. Doyle, T. F. Fuller, and J. Newman, J. Electrochem. Soc. , 140, 1526(1993).

[12] C. R. Pals and J. Newman, J. Electrochem. Soc. , 142, 3274(1995).

[13] M. Safari and C. Delacourt, J. Electrochem. Soc. , 158, A63(2011).

[14] M. Safari and C. Delacourt, J. Electrochem. Soc. , 158, A1436(2011).

[15] T. F. Fuller, M. Doyle, and J. Newman, J. Electrochem. Soc. , 141, 1(1994).

[16] M. Safari and C. Delacourt, J. Electrochem. Soc. , 158, A1123(2011).

[17] G. Botte and R. E. White, J. Electrochem. Soc. , 148, 54(2001).

[18] W. Ming, L. J. Jun, H. X. Ming, W. Han, and W. C. Rong, J. Power Sources. , 207, 127(2012).

[19] J. Park, J. H. Seo, G. Plett, W. Lu, and A. M. Sastry, Electrochemical and solid-State letters. , 14, A14(2011).

[20] K. Smith and W. C. Yang, J. Power Sources. , 160, 662(2006).

[21] Y. Ye, Y. Shi, and A. A. O. Tay, J. Power Sources. , 217, 509(2012).

[22] I. J. Ong and J. Newman, J. Electrochem. Soc. , 146, 4360(1999).

[23] L. Rao and J. Newman, J. Electrochem. Soc. , 144, 2697(1997).

[24] W. B. Gu and C. Y. Wang, J. Electrochem. Soc. , 147, 2910(2000).

[25] L. O. Valøen and J. N. Reimers, J. Electrochem. Soc. , 152, A882(2005).

[26] M. Safari and C. Delacourt, J. Electrochem. Soc. , 158, A562(2011).

[27] Y. W. Tang, M. Jia, Y. Cheng, K. Zhang, H. Zhang, and J. Li, Acta Phys. Sin. , 62, 158201(2013).

Unique Starch Polymer Electrolyte for High Capacity All-solid-state Lithium Sulfur Battery*

Abstract Solid polymer electrolyte (SPE)-based lithium sulfur battery offers high energy and safety for new energy vehicles and storage. However, the low room temperature ionic conductivity of the existing SPE limits the battery performance. Herein, a novel SPE film using food grade starch as a host was fabricated. This electrolyte provides exceptional lithium ion transportability with an ionic conductivity of 3.39×10^{-4} S/cm and lithium ion transference number of 0.80 at 25℃. The application potential of this starch hosted electrolyte was demonstrated by all-solid-state lithium sulfur battery systems presenting the initial discharge capacity of 1442mA · h/g, an average discharge capacity of 864±16mA · h/g at 0.1 C for 100 cycles, 562±118mA · h/g at 0.5 C for 1000 cycles at room temperature, and 388±138mA · h/g for 2000 cycles at 2 C and 45℃. This opens a bright route towards realizing energy power and safety with low cost and high sustainability.

1 Introduction

The development of alternative transportation, such as fully electric or hybrid vehicles, has become a key need for a sustainable long-term solution of climate change and non-renewable resources.[1] Lithium batteries with high energy and long life have been the core power source technology for new energy vehicles.[2-5] Despite the classical figures of merit, a number of issues have been recognized such as severe safety problems present in lithium battery operating on liquid organic electrolyte, environmental concern and restricted battery design.[6-8] A challenge in the very near future is to find a safer, cheaper and more sustainable technology that would enable electric vehicles an extended driving range. All-solid-state lithium sulfur batteries satisfy the requirements. However, the performance of batteries depends greatly on the lithium ionic conductivities of the electrolyte. Therefore, the best path is entrenched in the design of solid electrolytes using natural products that are comparable to, or exceed the performance of today's materials.

Solid polymer electrolyte (SPE) has the flexible properties, which can facilitate various designs for batteries, while further improving the energy performance at high safety levels. General improvement of the ionic conductivity for SPE is based largely on modification of the polyethylene oxide (PEO) electrolyte.[9-13] PEO consisting of ether oxygen (-C-O-C-) functional groups can dissolve lithium salts and provide paths for lithium ion transport, however, such an electrolyte faces barriers in practical applicability. For instance, the low room temperature ionic conductivities of the electrolyte still inhibit the performance of lithium batteries. Thus, there is an urgent necessity for new hosts to achieve high lithium ion transport efficiency.

---

* Copartner: Yue Lin, Jie Li, Kathy Liu, Jin Liu and Xuming Wang. Reprinted from Green Chemistry, 2016.

Turning to nature, we can find numerous resources with -C-O-C- functional groups similar to the PEO structure such as polysaccharides, especially starch materials. Starch, a form of energy storage in plants, is the largest percentage carbohydrate in the human diet and is found in substantial quantities in staple foods such as potatoes, wheat, corn, and rice.[14] It is composed of repeating glucose monomers joined by glycosidic linkages forming a stable helix structure,[15,16] which would allow the reversible transportation of ions. In this sense, we became interested in starch as a solid electrolyte host for lithium batteries. The operating performance of the starch hosted electrolyte, including ionic conductivity and electrochemical stability, as well as capacity and cycling of assembled lithium batteries, has been evaluated. This electrolyte demonstrates excellent lithium ion transportability at room temperature. Based on this electrolyte, all-solid-state lithium sulfur batteries achieve near the theoretical discharge capacity and long cycling performance, showing great potential for the use of natural products in new energy materials and devices. With a sustainable and vastly safer design, this also paves a way for biocompatible batteries used in organisms.

2 Experimental

2.1 Materials

Food grade corn starch was purchased from a supermarket without purification. γ-(2,3-Epoxypropoxy)propyltrimethoxy-silane(KH560) with a purity of 97% was purchased from Aladdin and used as a crosslinking reagent. Lithium bis-(trifluoromethanesulfonyl)imide(LiTFSI) was obtained from Sigma-Aldrich and stored in an argon-filled box before use. All solvents and other reagents were of analytical grade and used without further purification.

2.2 Synthesis of the starch host

0.5000g corn starch was dissolved into 10.000g dimethyl sulfoxide(DMSO) and stirred at 90℃ to form a uniform solution. A certain amount of KH560 was added to the solution according to the molar ratio of glucose monomer in starch/KH560 from 0.25∶1 to 1.75∶1. After the reaction in an argon atmosphere at 90℃ for 12 hours, a light yellow viscous solution was obtained. The solution was dried at 60℃ for 10 hours to obtain the synthesis product(starch host). Solid ^{29}Si NMR(BRUKER, AVANCE III 400MHz) of the product with glucose monomer in a starch/KH560 molar ratio of 1∶1 and ^{29}Si NMR of the coupling reagent was performed to examine the extent of the crosslinking reaction. Gel permeation chromatography(GPC) of the product solution was carried out using an Agilent 1100 chromatograph and N,N-dimethylformamide as the eluent. Standardization of the GPC was accomplished using poly(methyl methacrylate).

2.3 Preparation and characterizations of the starch hosted electrolyte

To prepare the starch-hosted electrolyte film, the starch host solution was mixed directly with LiTFSI in an argon-filled glove box and dried at 60℃ for 10 hours. The weight contents of LiTFSI varied from 20% to 60%. The ionic conductivities of the films were evaluated from the complex plane impedance plots between 25℃ and 100℃ (the temperature controlled by an oven with temperature accuracy of ±1℃) with an impedance analyzer (PARSTAT 4000, Princeton Applied Research). Each film (about 20mg) was sandwiched between a stainless steel (SS) disc ($d=1.6$cm) and the positive shell of a 2025 coin cell blocking electrode to form a symmetrical SS/electrolyte/SS cell. The cell was sealed to prevent contamination from oxygen, moisture, and other substances. The scanning frequency ranged from 500kHz to 10Hz with a perturbation voltage of 10mV.

The ionic conductivity was calculated from the electrolyte resistance (R_b) obtained from the intercept of the AC impedance spectra with the real axis, the film thickness (l) (~260μm), and the electrode area (S, 2cm^2) using the equation:

$$\sigma = \frac{l}{SR_b} \tag{1}$$

The Li$^+$ ion transference number (t^+) of the electrolyte film was tested in a symmetric cell (Li/Starch+LiTFSI/Li) using PARSTAT 4000. When a constant polarization voltage of 10mV was applied to the cell, a current that was fulfilled from the initial value to a steady-state value was measured. The AC impedance plots of the film before and after polarization were obtained. The frequency range was from 500kHz to 10Hz and the signal amplitude was 10mV. Three cells containing the same conformation from three starch samples were measured using the same method to obtain the averaged t^+. The t^+ is given by the following expression:[17]

$$t^+ = \frac{I_s(V - I_0 R_0)}{I_0(V - I_s R_s)} \tag{2}$$

where V is the DC voltage applied to the cell; R_0 and R_s are the initial and steady-state resistances of the passivating layer, respectively; I_0 and I_s are the initial and steady-state current, respectively.

The electrochemical windows of the electrolyte film were measured by cyclic voltammetry using Li/Starch+LiTFSI/SS cells with the scan route in 2.8 → -0.8 → 6V at a scan rate of 10mV/s. Thermogravimetric analysis (TGA) was conducted by NETZSCH STA 449C to investigate the thermal stability of the electrolyte. The morphology of the starch-hosted electrolyte film was investigated by scanning electron microscopy (SEM, Sirion 200).

The stability of the electrolyte with lithium during cycling was measured using the Li/Starch+LiTFSI/Li cells. The diameter of lithium foil was 15.6mm. The cycling (one-hour charge and one-hour discharge) current was 0.01, 0.02, and 0.03mA, respectively, for every successive 30 cycles and then back to 0.01mA for the last 30 cycles. The cell was directly tested in a room temperature environment, while the cell was maintained in oven for 45℃ tests.

2.4 Battery tests

The starch hosted electrolyte based lithium sulfur cell laboratory prototype was assembled by contacting in sequence a lithium metal disk anode ($d=15.6$mm, thickness=0.3mm), a layer of the starch hosted electrolyte film and a disk of a macro-structural sulfur cathode ($d=10$mm). The sulfur cathode was fabricated according to the previous report.[18] Conductive carbon black (Super P) and sublimed sulfur were ball-milled to obtain the S/C composite. Aniline was then polymerized on the mixture by *in situ* chemical polymerization method.[18-20] The obtained precipitate was filtered and washed with distilled water, dried in a vacuum oven and then treated at 280℃ to gain the macro-structural sulfur materials. The sulfur content in the materials was 43.0% by weight. The cathode was made from slurry of the macrostructural sulfur, Super P and binder(70 : 20 : 10 by weight). The slurry was then coated onto aluminium foil using a doctor blade and dried at 50℃ to form a working cathode. Therefore, the sulfur content in the cathode is about 30.1%(43.0%×70%).

The electrode-electrolyte was housed and sealed in a 2025 coin cell to build a Li/Starch+LiTFSI/sulfur battery. All the cells were assembled in an argon filled glove box. Galvanostatic charge/discharge cycling at 45℃ was performed in a LAND CT2001A battery test system and the battery was held in an oven with a temperature accuracy of ±1℃. The battery was directly tested at 25℃ to test the room temperature cycling performance. The cutoff voltage of the discharge and charge process was 1.2 and 3.0V, respectively.

3 Results and Discussion

3.1 Synthesis and structure of starch host

Fig. 1 shows the process of food grade corn starch used to realize the concept of natural materials as the electrolyte host. The starch was crosslinked with KH560 chiefly to improve stretchable properties of the electrolyte film (Fig. 1a). The molecular weight distribution and molecular structure of the starch host were evaluated by GPC(Fig. 1b) and NMR(Fig. 1c).

The results from GPC analysis in Fig. 1b show that the starch host has a wide molecular weight distribution. It can be divided into two fractions: one is the molecular weight from 10^6 to 10^{12} corresponding to the evolution time from 0.5 to 10.0min, another is lower than 80 000 with the evolution time from 11.4 to 19.0min. The former can be considered due to the formation of the crosslinked starch polymer. The latter contains three peaks with a molecular weight of 2300, 236 and 78 that may come from crosslinked starch small molecules, unreacted silicon coupling reagent and dimethyl sulfoxide solvent, respectively.

Fig. 1c shows the ^{29}Si NMR spectrum of the reactant of KH560(green) and solid ^{29}Si NMR spectrum of the product of the starch host (orange). For KH560, only one peak at −41.7 ppm from -Si-(OCH_3)$_3$ is observed.[21] For the product, three new shifts at −51.1, −59.7 and −67.4ppm are allocated to the three different substitutions of -Si-(OCH_3)$_3$, which are Si-(OCH_3)$_2$Sub$_1$, -Si-OCH_3Sub$_2$ and -Si-Sub$_3$. Because starch is composed of repeating glucose monomers joined by glycosidic linkages, the three -OH groups in per unit from starch may react with -Si-(OCH_3)$_3$, which would bring about changes in the chemical shifts. The shift at −42.8 ppm may be assigned to the unreacted -Si-(OCH_3)$_3$ or small molecular starch connected to KH560. According to the GPC and NMR results, the starch host was synthesized and a possible structure of the host is suggested in Fig. 1d.

3.2 Lithium ions transportability of starch hosted electrolyte

Ionic conductivity is a key parameter to describe the ion transport ability of an electrolyte.

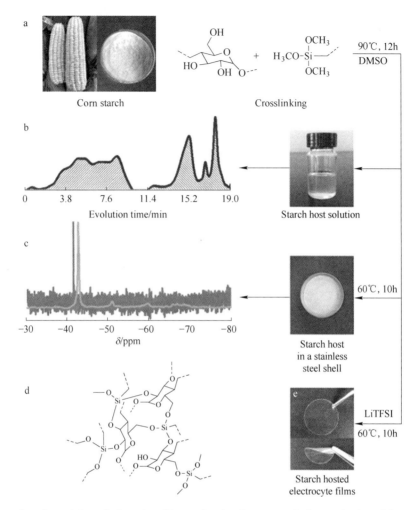

Fig. 1 Preparation of starch hosted electrolyte films and molecular structural characterization of the starch host

a—Scheme of starch host synthesis; b—GPC result of the starch host solution with DMSO solvent, ▨(evolution time from 0.5 to 10.0min) represents molecular weight from 10^6 to 10^{12}, ▩(evolution time from 11.4 to 19.0min) represents molecular weight lower than 80000; c—Solid ^{29}Si NMR spectrum of the starch host after vaporizing DMSO at 60℃ for 10 hours(orange) and ^{29}Si NMR spectrum of the reactant of KH560(green); d—Suggested structure of the starch host; e—Images of the transparent and flexible electrolyte films(Starch+LiTFSI) after mixing with lithium salt (LiTFSI) and then vaporizing DMSO at 60℃ for 10 hours

Therefore, the starch host solution was fabricated into the electrolyte film by mixing with lithium salt(LiTFSI) and then vaporizing the solvent(see Fig. 1e) and ionic transportability was tested.

Optimizations of the ionic conductivity were performed by adjusting the molar ratio of Starch/KH560 in the host and the lithium salt concentration in the starch hosted electrolyte(Starch+LiTFSI), which can be found in Fig. 2a and b. The highest ionic conductivity is attained with 40 wt% LiTFSI and at the Starch/KH560 molar ratio of 1∶1. The optimized conditions were maintained in the following measurements.

Fig. 2c shows the temperature dependence of the ionic conductivity of this starch-hosted electrolyte in the temperature range of 25-100℃. The ionic conductivity increased steadily with in-

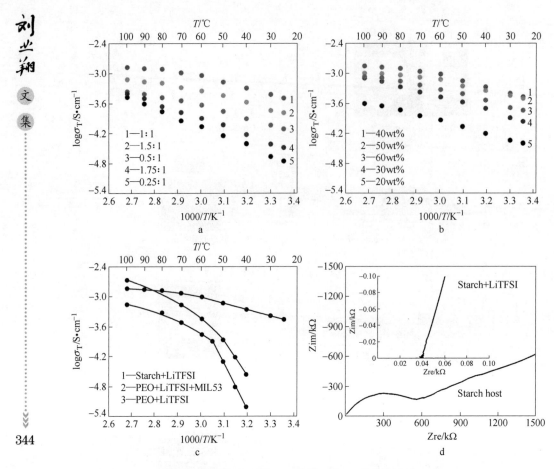

Fig. 2 Lithium ion transportability of starch hosted electrolyte

a— Ionic conductivities of the starch hosted electrolyte films with different molar ratios of glucose monomer of Starch/KH560 at 40 wt% LiTFSI as a function of temperature; b—Ionic conductivities of the starch hosted electrolyte films with different concentrations of LiTFSI at the Starch/KH560 molar ratio of 1 : 1 as a function of temperature; c—Temperature dependency of ionic conductivities of the Starch+LiTFSI electrolyte at the Starch/KH560 molar ratio of 1 : 1 and 40 wt% LiTFSI. The ionic conductivities of the PEO+LiTFSI and PEO+LiTFSI+MIL53 electrolytes are shown for comparison; d—AC impedance spectra of the starch hosted film without LiTFSI(Starch host) and the starch hosted electrolyte film(Starch+LiTFSI) at 25 ℃

creasing temperature. No turning points were observed. At 100 ℃, the ionic conductivity was 1.41×10^{-3} S·cm^{-1}, which is only 2.6 times of 5.37×10^{-4} S·cm^{-1} at 40 ℃, and 4.2 times of 3.39×10^{-4} S·cm^{-1} at 25 ℃. In general, the room temperature ionic conductivity of polymer-based electrolytes ranges from 10^{-8} to 10^{-4} S·cm^{-1};[22-28] thus, this starch hosted electrolyte shows distinctive ionic conductibility.

As a comparison, the ionic conductivity for the PEO electrolyte(PEO+LiTFSI, EO : Li molar ratio of 15 : 1) and the PEO based electrolyte(PEO+LiTFSI+MIL53, EO : Li molar ratio of 15 : 1, MIL53 concentration of 10 wt%) are also shown in Fig. 2c.[11] Both PEO+LiTFSI and PEO+LiTFSI+MIL53 have a sharp decrease in ionic conductivity below 60 ℃. At 40 ℃, the ionic conductivities were 6.03×10^{-6} and 2.69×10^{-5} S/cm, respectively, for PEO+LiTFSI and PEO+LiTFSI+MIL53, which are a thousand fold lower than those at 100 ℃. Previous experience has shown that these types of batteries assembled using PEO-based electrolytes only work at temperatures higher than 60℃.[11,18] Therefore, the temperature-insensitive behavior of the starch-hosted electrolyte can maintain the battery capacities in a wide temperature range.

In addition, a control experiment of the ionic conductivity measurements for the starch host film without LiTFSI was conducted (see Fig. 2d). At 25℃, its ionic conductivity was 4.15×10^{-9} S/cm, which is about 10^5 lower than the Starch+LiTFSI electrolyte, demonstrating that the high room temperature ionic conductivity is not due to the starch host itself, but the host acting as a "solid solvent" for the quick dissolution of LiTFSI.

Apart from the high room temperature ionic conductivity, the starch hosted electrolyte also offers a very high lithium ion transference number (t^+), as shown in Fig. 3. The t^+ was measured using electrochemical impedance spectroscopy combined with potential polarization.[17] When the t^+ equals 1, it means only Li$^+$ ions transfer in the electrolyte. An average t^+ of 0.80±0.13 was obtained for three starch electrolyte samples, which is remarkably higher than previous SPEs (where the t^+ is generally less than 0.5).[11,26,29] This value is also higher than 0.2~0.4 of the liquid electrolytes with commercial separators.[30,31] Although single ion conductors were reported to have t^+ higher than 0.80, the room temperature ionic conductivities were dramatically low, even lower than 10^{-8} S/cm.[32,33] Therefore, the starch-hosted electrolyte uniquely demonstrates high room temperature conductivity combined with high t^+, which is excellent for use in lithium batteries.

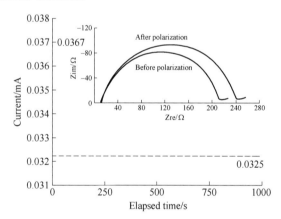

Fig. 3　Chronoamperometry of the Li/Starch+LiTFSI/Li cell at a potential step of 10mV and at 25℃. Inset: the AC impedance spectra of the same cell before polarization and after the steady-state current

This enhancement in the lithium ions transportability may be attributed to the high dielectric constant of the starch host. As opposed to the PEO base, the corn starch with cyclic polar units has a dielectric constant of 50,[34] which is ten times higher than that of PEO with straight ether oxygen chains.[35] The Li$^+$ ions would be trapped in the cyclic groups of the starch, assisting in the dissociation of LiTFSI. Another reason may be attributed to the silane portion. Crosslinking by silane obviously decreases the crystallinity of starch, greatly advantaging Li$^+$ ion movement. Consequently, the Li$^+$ ion transportability is increased.

3.3　Electrochemical stability of the starch hosted electrolyte

In addition to the high Li$^+$ ionic conductivity, compatibility of the electrolyte with the electrodes is also crucial for applications in batteries. This property can be evaluated by the electrochemical stability. First, the electrochemical decomposition voltage was used as a rule for the stability at the cathode/electrolyte interface. Cyclic voltammogram measurements were carried out with SS as the working electrode and lithium as the reference and counter electrodes. The results are displayed in Fig. 4a. It is observed that the starch-hosted electrolyte is stable up to 4.80V and 4.60V, respectively, at 25℃ and 45℃, both of which are higher than the decomposition voltage of commercial liquid organic electrolytes (~4.2V). They are also significantly higher than lithium insertion/extraction voltages of most cathode active materials such as $LiFePO_4$, $LiCoO_2$, $LiMnO_2$, $LiNi_xCo_yMn_zO_2$ and sulfur.

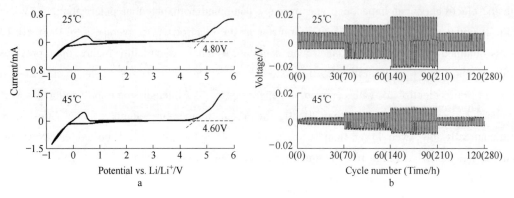

Fig. 4 Electrochemical stability of the starch hosted electrolyte

a—Cyclic voltammetry of SS/Starch+LiTFSI/Li cells at the scan route at 2.8V→ -0.8V → 6.0V and at a rate of 10mV·s^{-1}; b—Polarization tests of Li/Starch+LiTFSI/Li cells at 25 and 45℃ at 0.01mA for the first 30 cycles, 0.02mA for the 31st to 60th cycles, 0.03mA for the 61st to 90th cycles, and 0.01mA for the last 30 cycles. During each cycle, the battery was rested for 10min, charged for 1h, rested for 10min and then discharged for 1h

The cathodic current onsets observed at potentials of -0.22V at 25℃ and -0.15V at 45℃ correspond to the electrochemical deposition of lithium. The lithium stripping peaks appear at 0.46V(25℃) and 0.39V(45℃) that reveal a reversible Li plating/stripping process on the stainless steel electrode. Therefore, the starch hosted electrolyte will effectively satisfy high potential and high capacity cathodes.

The anodic stability was investigated by assembling the electrolyte into a symmetric Li/Starch+LiTFSI/Li cell. Onehour charge and one-hour discharge cycling at a specific current density was designed to mimic the lithium stripping/plating process. The galvanostatic cycling curves are shown in Fig. 4b and the polarization potentials are listed in Table 1.

Table 1 Polarization potentials of the Li/Starch + LiTFSI/Li cells at different currents at 25℃ and 45℃

Current/mA	Polarization potential/mV	
	25℃	45℃
0.01(1st–30th cycles)	5.74±0.07	2.37±0.19
0.02(31st–60th cycles)	11.9±0.13	5.55±0.13
0.03(61st–90th cycles)	17.9±0.15	8.64±0.10
0.01(91st–120th cycles)	6.25±0.13	2.71±0.04

In this test, lithium is stripped continuously from one electrode and plated on the other and the process reverses when the current direction is reversed. Because the potentials can reflect the resistance needed for the Li$^+$ ion reaction at the interface, this resistance will express the lithium ion transportability in the electrolyte and the reaction rate at the interface. The averaged resistances are 574±7.0Ω, 595±6.5Ω and 596±5.0Ω at 25℃, and 237±19.0Ω, 278±6.5Ω and 288±3.3Ω at 45℃ at 0.01mA, 0.02mA, and 0.03mA, which are very comparable with those of SPEs at a high temperature of 90℃[36,37] and the liquid electrolyte at room temperature.[38] It is also noticed that the resistance is 625Ω and 271Ω at room temperature and 45℃, respectively, after 280 hours polarization. An increase of only 51Ω (625-574Ω) and 34Ω (271-237Ω) was observed, which further reveals that a compatible lithium anode/electrolyte interface is formed. However, the slight increases in potentials observed after 280 hours may be due to poorer contact between the metal lithium and the electrolyte when the surface of the lithium foil would become rougher after longtime stripping/plating.[36]

3.4 Thermal stability and morphology of the starch hosted electrolyte

TGA of the starch hosted electrolyte film was conducted to check the thermal stability and the results are shown in Fig. 5. The film was prepared by mixing the starch host with LiTFSI and then vaporizing DMSO at 60℃ under an argon flow for 5, 10, and 24 hours. The three samples have similar curves in the temperature range of 150-465℃. The weight loss for the two main pe-

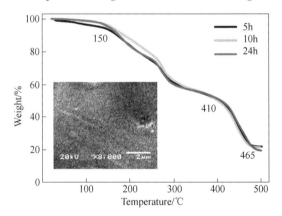

Fig. 5 Thermal analysis of the starch-hosted electrolytes after evaporation for 5, 10 and 24 hours at 60℃ and under an argon flow. Inset: SEM image of the film on top view

riods of 150-410℃ and 410-465℃ is about 44% and 32%, which is assigned to the decomposition of starch host and LiTFSI,[39] respectively. The contents of LiTFSI and the starch host used for preparation of the electrolyte film are 40 and 60 wt%, respectively. The residuals above 465℃ are about 22% that are from the decomposition products of LiTFSI and the starch host containing Li, C, Si and O. A slight difference was observed at temperatures lower than 150℃, which may come from water absorbed during the sample transfer process and/or unevaporated solvent such as DMSO. When the heating time was 5h, the weight loss was about 6%. When it was 10 or 24h, the weight loss was within 2%, indicating that DMSO can be removed almost completely under the conditions of 60℃ and 10h in an argon atmosphere. Therefore, the starch hosted electrolyte is thermally stable up to 150℃.

SEM image of the film on the top view is shown in the inset of Fig. 5. A smooth surface was observed at a magnification of 8000, showing the good film-forming property of the starch hosted electrolyte. This may also benefit lithium ion transportation.

3.5 Cycling performance of starch hosted electrolyte lithium sulfur battery

The starch hosted electrolyte with 40% LiTFSI was used to fabricate high capacity all-solid-state lithium sulfur battery (lithium sulfur battery theoretical capacity is 1672mA · h/g). The all-solid-state macro-structural sulfur battery with MIL53 modified PEO electrolyte was studied in our previous study, which delivered the initial discharge capacity of 1520mA · h/g, the initial charge capacity of 920mA · h/g, and stable capacity higher than 800mA · h/g during 60 cycles at 0.2 C and 80℃.[18] In this study, the same cathode materials and Li anode were used, but a different solid electrolyte, the starch hosted electrolyte, was used to investigate the electro-chemical properties and performance of the lithium sulfur battery. The cycling test was conducted, as shown in Fig. 6.

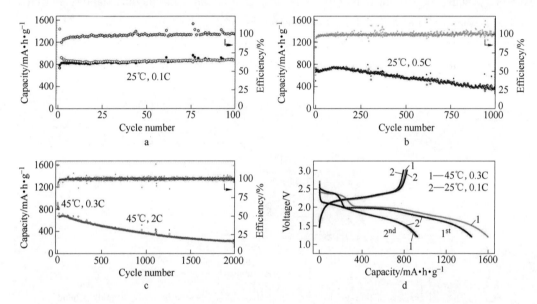

Fig. 6 Cycling performance of the starch hosted electrolyte based all-solid-state lithium sulfur battery, where the anode is metal Li and the cathode is macro-structural sulfur with the same formulation as in a previous report.[18] The discharge capacity (hollow), charge capacity (solid) and efficiency (charge capacity divided by discharge capacity, black) at (a) 25℃ and 0.1C, (b) 25℃ and 0.5C, (c) 45℃, 0.3C from the first cycle to the 17th cycle and 2 C from 18th cycle to the 2000th cycle. (d) The initial and second charge/discharge profile of the battery at 45℃, 0.3 C and 25℃, 0.1 C

At 25℃ and 0.1 C (Fig. 6a), the initial discharge capacity was 1442mA · h/g, the initial charge capacity was 790mA · h/g, and the second discharge capacity was 926mA · h/g. In the following cycles, the battery presents stable discharge capacities with an averaged value of 864±16mA · h/g in the 100 discharge/charge cycles with 93% retention compared to the second discharge capacity, and close to 100% efficiency for each cycle.

Fig. 6b is the cycling performance of the battery at 0.5℃ and at 25℃. The initial discharge capacity of 1110mA·h/g and an averaged discharge capacity of 562±118mA·h/g during 1000 cycles were obtained with approximately 100% efficiency for each cycle.

The higher temperature cycling performance of the battery was also investigated at 45℃. At 0.3 C, the initial discharge capacity was 1597mA·h/g and the stable discharge capacity is 842mA·h/g for the following 16 cycles. To further test the faster charge/discharge ability of the battery, a test at an increase in the current rate of 2 C was performed and the result is also shown in Fig. 6c. The battery delivered a discharge capacity of 719mA·h/g and a charge capacity of 662mA·h/g. After 2000 cycles, the battery still has a discharge capacity of 221mA·h/g with an average capacity of 388±138mA·h/g. The higher discharge capacity at the higher temperature is the result of the better compatibility of the electrolyte with electrodes and faster lithium ion transfer of the electrolyte, which is in accordance with the above stability test.

The initial and second charge/discharge profile of the battery cycled at 25℃, 0.1 C and 45℃, 0.3 C can be found in Fig. 6d. At both temperatures, two distinguishable plateaus at 2.5-2.3 V and 2.0-1.5 V were observed in the initial discharge process. The first plateau was associated with high polymerized sulfur linking to the macro structure reduced to $-S_4^{2-}$ and the second plateau is related to the $-S_4^{2-}$ transformed to $-S^{2-}$.[18] While during the initial charging process, only one plateau at 2.1-2.4V was present, which shows that the $-S^{2-}$ is oxidized to $-S_4^{2-}$. In the following discharge/charge process, there was only one pair of reduction/oxidation plateaus, corresponding to the redox reaction of $-S_4^{2-} \rightleftharpoons -S^{2-}$. This change from macro-structural sulfur to intermediate $-S_4^{2-}$ would cause the initial charge capacity loss. Because the macrostructural intermediate is highly electrochemically reversible, the coulombic efficiencies for all other cycles(except for the 1st cycle) are close to 100%.

It is worth noting that this electrochemical behavior of the starch hosted electrolyte lithium sulfur battery is different from common lithium sulfur batteries,[5,40-42] but quite similar to the MIL53-modified PEO electrolyte battery,[18] which has much lower working temperature of 25℃, a more than 35℃ decrease. Compared to all-solid-state lithium sulfur batteries reported,[18,43,44] the operating temperature is also lowered significantly. This proves that the starch-hosted electrolyte affords all-solid-state lithium sulfur batteries better capacity performance at lower temperatures.

4 Conclusions

We demonstrated great potential to use food grade starch for advanced flexible electrolyte. Using the starch hosted electrolyte, all-solid-state lithium sulfur batteries were developed that possess impressive gains in capacity, cycling, temperature and safety performance. This electrolyte is widely available, more economically advantageous than most other solid polymer electrolytes developed to date and is environmentally sustainable. This opens a bright route towards realizing energy in green transportation at low cost and making micro-scale batteries for devices in organisms in future research.

Acknowledgements

We acknowledge support from the National Natural Science Foundation of China (No. 51274239), the program of China Scholarships Council and the Fundamental Research Funds for the Central Universities of Central South University (No. 2015zzts036).

References

[1] J. Tollefson, Nature, 2008, 456, 436-440.

[2] E. C. Evarts, Nature, 2015, 526, S93-S95.

[3] J. Motavalli, Nature, 2015, 526, S96-S97.

[4] Y. Su, S. Li, D. Wu, F. Zhang, H. Liang, P. Gao, C. Cheng and X. Feng, ACS Nano, 2012, 6, 8349-8356.

[5] J. -Q. Huang, Q. Zhang, H. -J. Peng, X. -Y. Liu, W. -Z. Qian and F. Wei, Energy Environ. Sci., 2014, 7, 347-353.

[6] Y. J. Nam, S. J. Cho, D. Y. Oh, J. M. Lim, S. Y. Kim, J. H. Song, Y. G. Lee, S. Y. Lee and Y. S. Jung, Nano Lett., 2015, 15, 3317-3323.

[7] M. Armand and J. M. Tarascon, Nature, 2008, 451, 652-657.

[8] D. Larcher and J. M. Tarascon, Nat. Chem., 2015, 7, 19-29.

[9] F. Croce, G. B. Appetecchi, L. Persi and B. Scrosati, Nature, 1998, 394, 456-458.

[10] E. Staunton, Y. G. Andreev and P. G. Bruce, J. Am. Chem. Soc., 2005, 127, 12176-12177.

[11] K. Zhu, Y. Liu and J. Liu, RSC Adv., 2014, 4, 42278-42284.

[12] Z. Zhang, L. J. Lyons, K. Amine and R. West, Macro-molecules, 2005, 38, 5714-5720.

[13] C. Yuan, J. Li, P. Han, Y. Lai, Z. Zhang and J. Liu, J. Power Sources, 2013, 240, 653-658.

[14] S. Jobling, Curr. Opin. Plant Biol., 2004, 7, 210-218.

[15] S. M. Southall, P. J. Simpson, H. J. Gilbert, G. Williamson and M. P. Williamson, FEBS Lett., 1999, 447, 58-60.

[16] K. Sorimachi, M. -F. L. Gal-Coëffet, G. Williamson, D. B. Archer and M. P. Williamson, Structure, 1997, 5, 647-661.

[17] J. Evans, C. A. Vincent and P. G. Bruce, Polymer, 1987, 28, 2324-2328.

[18] C. Zhang, Y. Lin and J. Liu, J. Mater. Chem. A, 2015, 3, 10760-10766.

[19] S. Li, D. Wu, C. Cheng, J. Wang, F. Zhang, Y. Su and X. Feng, Angew. Chem., Int. Ed., 2013, 52, 12105-12109.

[20] S. Li, D. Wu, H. Liang, J. Wang, X. Zhuang, Y. Mai, Y. Su and X. Feng, ChemSusChem, 2014, 7, 3002-3006.

[21] M. Tian, W. Liang, G. Rao, L. Zhang and C. Guo, Compos. Sci. Technol., 2005, 65, 1129-1138.

[22] E. Quartarone and P. Mustarelli, Chem. Soc. Rev., 2011, 40, 2525-2540.

[23] D. Lin, W. Liu, Y. Liu, H. R. Lee, P. -C. Hsu, K. Liu and Y. Cui, Nano Lett., 2016, 16, 459-465.

[24] J. Zhang, L. Yue, P. Hu, Z. Liu, B. Qin, B. Zhang, Q. Wang, G. Ding, C. Zhang, X. Zhou, J. Yao, G. Cui and L. Chen, Sci. Rep., 2014, 4, 6272.

[25] Z. Zhu, M. Hong, D. Guo, J. Shi, Z. Tao and J. Chen, J. Am. Chem. Soc., 2014, 136, 16461-16464.

[26] J. Li, Y. Lin, H. Yao, C. Yuan and J. Liu, ChemSusChem, 2014, 7, 1901-1908.

[27] Z. Xue, D. He and X. Xie, J. Mater. Chem. A, 2015, 3, 19218-19253.

[28] Z. Zhang, D. Sherlock, R. West, R. West, K. Amine and L. J. Lyons, Macromolecules, 2003, 36, 9176-9180.

[29] M. -K. Song, J. -Y. Cho, B. W. Cho and H. -W. Rhee, J. Power Sources, 2002, 110, 209-215.

[30] D. Zhou, Y. -B. He, R. Liu, M. Liu, H. Du, B. Li, Q. Cai, Q. -H. Yang and F. Kang, Adv. Energy Mater.,

2015, 5, 1500353.

[31] S. Zugmann, M. Fleischmann, M. Amereller, R. M. Gschwind, H. D. Wiemhöfer and H. J. Gores, Electrochim. Acta, 2011, 56, 3926-3933.

[32] R. Bouchet, S. Maria, R. Meziane, A. Aboulaich, L. Lienafa, J.-P. Bonnet, T. N. T. Phan, D. Bertin, D. Gigmes, D. Devaux, R. Denoyel and M. Armand, Nat. Mater., 2013, 12, 452-457.

[33] Q. Ma, H. Zhang, C. Zhou, L. Zheng, P. Cheng, J. Nie, W. Feng, Y.-S. Hu, H. Li, X. Huang, L. Chen, M. Armand and Z. Zhou, Angew. Chem., Int. Ed., 2016, 55, 2521-2525.

[34] M. K. Ndife, G. Şumnu and L. Bayindirli, Food Res. Int., 1998, 31, 43-52.

[35] A. Das, S. Pisana, B. Chakraborty, S. Piscanec, S. K. Saha, U. V. Waghmare, K. S. Novoselov, H. R. Krishnamurthy, A. K. Geim, A. C. Ferrari and A. K. Sood, Nat. Nanotechnol., 2008, 3, 210-215.

[36] Q. Pan, D. M. Smith, H. Qi, S. Wang and C. Y. Li, Adv. Mater., 2015, 27, 5995-6001.

[37] G. M. Stone, S. A. Mullin, A. A. Teran, D. T. Hallinan, A. M. Minor, A. Hexemer and N. P. Balsara, J. Electrochem. Soc., 2012, 159, A222-A227.

[38] Y. Lu, K. Korf, Y. Kambe, Z. Tu and L. A. Archer, Angew. Chem., Int. Ed., 2014, 53, 488-492.

[39] Z. Lu, L. Yang and Y. Guo, J. Power Sources, 2006, 156, 555-559.

[40] J. Song, H. Noh, H. Lee, J.-N. Lee, D. J. Lee, Y. Lee, C. H. Kim, Y. M. Lee, J.-K. Park and H.-T. Kim, J. Mater. Chem. A, 2015, 3, 323-330.

[41] J.-Q. Huang, T.-Z. Zhuang, Q. Zhang, H.-J. Peng, C.-M. Chen and F. Wei, ACS Nano, 2015, 9, 3002-3011.

[42] C.-H. Chang, S.-H. Chung and A. Manthiram, J. Mater. Chem. A, 2015, 3, 18829-18834.

[43] J. Hassoun and B. Scrosati, Adv. Mater., 2010, 22, 5198-5201.

[44] J. H. Shin, K. W. Kim, H. J. Ahn and J. H. Ahn, Mater. Sci. Eng., B, 2002, 95, 148-156.

A Fast Charging/Discharging All-solid-state Lithium Ion Battery Based on PEO-MIL-53(Al)-LiTFSI Thin Film Electrolyte[*]

Abstract Metal-organic framework aluminum 1,4-benzenedicarboxylate(MIL-53(Al)) is used as a filler for a polyethylene oxide(PEO) based thin film electrolyte. With the participation of MIL-53 (Al), the ionic conductivity of this electrolyte is increased from 9.66×10^{-4} S·cm^{-1} to 3.39×10^{-3} S·cm^{-1} at 120℃ and the oxidation potential is raised from 4.99V to 5.10V. In addition, an all-solid-state LiFePO$_4$/Li button battery based on the electrolyte is fabricated. At 5 C and 120℃, the battery delivers the discharge capacity of 136.4mA·h·g^{-1} in the initial cycle, 129.2mA·h·g^{-1} in the 300th cycle, and 83.5mA·h·g^{-1} in the 1400th cycle. At 10 C and 120℃, its discharge capacity is 116.2mA·h·g^{-1} in the initial cycle and 103.5mA·h·g^{-1} in the 110th cycle. The results indicate that this metal-organic framework(MIL-53(Al)) is a novel structural modifier for solid polymer electrolytes in fast charging/discharging lithium ion batteries.

1 Introduction

All-solid-state polymer lithium ion batteries (LIBs) have better safety than conventional LIBs since the fammable liquid electrolyte is replaced by a solid electrolyte.[1-5] However, the cycle performance of batteries assembled by such solid polymer electrolytes does not have a strong appeal when batteries are charged/discharged at high current densities. Generally, the fast charging/discharging capability of a battery is determined by subjecting the battery to high charging/discharging rates where about 40% of the charge/discharge state must be obtained within 15min(about 1.6 C rate).[6] This requires the ability of conductible ions in the electrolyte to transport quickly between two electrodes.[7] Thus, various studies were focused on the improvement of ionic conductivities of solid polymer electrolytes,[8,9] especially on the modifcation of polyethylene oxide(PEO) based electrolytes.[10-18] For example, the ionic conductivity was increased to 2.0×10^{-4} S·cm^{-1} at 80℃ by adding silane-treated Al$_2$O$_3$.[16] The LiFePO$_4$/Li battery assembled by this electrolyte showed a discharge capacity of 140mA·h·g^{-1} in the 100th cycle at 0.2 C and 90℃. In 2013, we reported the PEO based electrolyte with metal-organic-framework-5 (MOF-5) as a filler for the all-solid-state LiFePO$_4$/Li battery.[17] The ionic conductivity increased to 7.9×10^{-4} S·cm^{-1} from 3.9×10^{-4} S·cm^{-1} at 75℃. The battery attained the initial discharge capacity of 151mA·h·g^{-1} and 45% capacity retention in the 100th cycle at 1 C and 80℃. However, it did not charge/discharge at higher rates. One reason for this would be the strong absorption capacity of MOF-5 for water and small molecules, resulting in instability of the electrolyte. Recently, another MOF, aluminum(Ⅲ)-1,3,5-benzenetricar-

---

[*] Copartner: Kai Zhu, Jin Liu. Reprinted from RSC Advances, 2014, 4, 42278-42284.

boxylate(Al-BTC), was reported as a filler to add to PEO based electrolytes.[18] An ionic conductivity of $\sim 7 \times 10^{-4}$ S·cm^{-1} at 70℃ was measured. The LiFePO$_4$/Li battery with the electrolyte delivered a specifc capacity of about 100 mA·h·g^{-1} at 2 C and 70℃. At 5 C, the battery had a low specifc capacity of 40 mA·h·g^{-1}.

MOFs are inorganic-organic hybrid materials that are primarily applied in the fields of drug delivery, optoelectronics, sensing and catalysis.[19-22] Aluminum 1,4-benzenedicarboxylate (MIL-53(Al)) is one of the MOFs that is built up of corner-sharing AlO$_4$(OH)$_2$ octahedra.[23] It has remarkable water and oxygen stability compared to MOF-5.[24] In this study, the PEO based electrolyte was prepared by using MIL-53(Al) as a filler and lithium bis(trifuoro-methanesulfonyl)imide(LiTFSI) as the lithium salt. The synthesized electrolyte's ionic conductivity is 3.39×10^{-3} S·cm^{-1} at 120℃, which is higher than the 9.66×10^{-4} S·cm^{-1} of that without MIL-53(Al). The LiFePO$_4$/Li battery based on this electrolyte has an average discharge capacity of 103.5 mA·h·g^{-1} for 110 cycles at a high rate of 10 C and 120℃, demonstrating the potential of the modifed PEO electrolyte based lithium ion battery in faster charging/discharging applications.

2 Experimental

2.1 Preparation of MIL-53(Al) nanoparticles

Aluminum 1,4-benzenedicarboxylate(MIL-53(Al)) nanoparticles were synthesized under hydrothermal conditions by treating 1,4-benzenedicarboxylic acid(H$_2$BDC, 99%) and aluminum nitrate nonahydrate(Al(NO$_3$)$_3$·9H$_2$O, 98%) with N,N-dimethylformamide(DMF, +99.9%) and deionized water.[23] The materials were stirred for 2 h, and then transferred to a Tefonlined steel autoclave(50 mL). The autoclave was heated in an oil bath at 160℃ for 50 h. The free H$_2$BDC and water in the sample were removed by washing with absolute alcohol and filtering 3 times. The sample was heated in a vacuum at 120℃ for 12 h to obtain a white powder product.

2.2 Preparation of thin film electrolytes

The polyethylene oxide(PEO, M_W 4×10^6, 99.9%) was thoroughly dried at 50℃ for 12 h, and the lithium bis(trifuoro-methanesulfonyl)imide(LiN(SO$_2$CF$_3$)$_2$, LiTFSI, +99.5%) was dried at 100℃ in a vacuum for 24 h before use. Firstly, LiTFSI was added to acetonitrile(CH$_3$CN, AR grade) and stirred for 2 h. Then, MIL-53(Al) nanoparticles were added to the solution and stirred for 10 min. After that, PEO was dispersed in the solution and stirred for 36 h. The result was a homogenized colloidal solution. Finally, the solution was cast and dried into a thin film at 80℃ for 24 h in an argon-filled glove box, resulting in the formation of a PEO-MIL-53(Al)-LiTFSI thin film electrolyte. The thickness of the solid electrolyte thin film was about 60 μm and the diameter of the solid electrolyte thin film was 20 mm. Using this electrolyte, all-solid-state LiFePO$_4$/Li batteries were fabricated for all evaluations.

2.3 Characterization and instruments

The surface morphologies of the MIL-53(Al) nanoparticle and the thin film electrolyte were observed with a scanning electron microscope (SEM, sirion 200). The X-ray diffraction (XRD, RINT-2000, Rigaku) patterns were collected from a Rigaku/TTR-Ⅲ powder diffractometer equipped with Cu-Kα radiation ($\lambda = 1.5418$Å) at 25℃. The diffraction pattern was recorded from 5° to 60° with a step size of $10°\text{min}^{-1}$. The zeta electric potential of the MIL-53(Al) nanoparticles was measured with a Brookhaven Zeta Plus instrument (USA). Thermogravimetric analysis (TGA) was carried out on Perkin-Elmer Pyris-1. The thin film electrolytes were loaded in hermetically sealed aluminum pans, and measurements were carried out at a heating rate of $10℃ \cdot \text{min}^{-1}$ from 48℃ to 500℃. The mechanical strength of the thin film electrolytes was measured with a Shimadzu TA Q800-1706 instrument with a tensile speed of $1\text{N} \cdot \text{min}^{-1}$.

The electrochemical properties of the electrolytes were measured by using a PARSTAT 2273 system (PerkinElmer Instrument, USA). The electrochemical window was determined using blocking stainless steel electrode/electrolyte/Li batteries at a scan rate of $10\text{mV} \cdot \text{s}^{-1}$ from 2.5V to 6.5V. The ionic conductivities were determined by AC impedance spectroscopy, which was carried out in the 500kHz to 10Hz frequency range. The equation for calculating the conductivity is

$$\sigma = \frac{d}{R_p S} \quad (1)$$

where d is the thickness of the electrolyte; R_p is the resistance of the electrolyte and the intercept of the x-axis of the straight line; S is the area of the blocking stainless steel electrodes.

The lithium ion transference number of the electrolyte was tested in a symmetric cell (Li/electrolytes/Li) by using a PAR-STAT 2273. The symmetric Li/electrolyte/Li battery was polarized at a small voltage of 10mV. AC impedance plots of the cell before and after polarization were obtained.

The cycling performance of the all-solid-state LiFePO$_4$/electrolyte/Li batteries was investigated by using the LiFePO$_4$/electrolyte/Li batteries. Button cells (2025) with diameter 20mm and thickness 2.5mm were used to fabricate the batteries. The thickness of the LiFePO$_4$ cathode was about 30μm. Charge/discharge cycles of the batteries were performed on a Land instrument (Wuhan Land Electronic Co., Ltd. China), with the tests using cut-off voltages of 4.0V (charge) and 2.5V (discharge). All the batteries were assembled in an argon-filled UNILAB glove box.

3 Results and Discussions

3.1 Characteristics of MIL-53(Al) nanoparticles and the PEO-MIL-53(Al)-LiTFSI thin film electrolyte

Fig. 1a shows the XRD pattern of MIL-53(Al), in which the characteristic peaks of the sample are identical to the pattern simulated by the lattice parameters of MIL-53(Al).[23] The surface morphologies of MIL-53(Al) and the PEO-MIL-53(Al)-LiTFSI thin film electrolyte were investigated by SEM. Cylindrical nanoparticles in the MIL-53(Al) sample and a uniform surface for the electrolyte were observed in Fig. 1b and c respectively, suggesting that the cylindrical particles would not damage the architecture of the thin film.

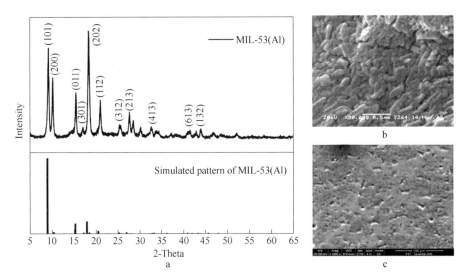

Fig. 1 (a) XRD pattern of MIL-53(Al) and the simulated pattern based on the crystallographic data of MIL-53(Al);[23] SEM images of (b) MIL-53(Al) and (c) the PEO-MIL-53(Al)-LiTFSI thin film electrolyte

3.2 Lithium ionic conductivities of thin film electrolytes

The ionic conductivities of the thin film electrolytes with different EO(ethylene oxide in PEO): Li molar ratios and different MIL-53(Al) concentrations at 30 ℃, 60 ℃ and 80 ℃ were measured, with the results listed in Table 1. At a fxed MIL-53(Al) concentration of 10 wt%, the electrolyte with the EO : Li ratio of 15 : 1 had the highest ionic conductivities at all three temperatures compared to the other ratios of 10 : 1, 20 : 1 and 25 : 1. At the constant EO : Li ratio of 15 : 1 and various MIL-53(Al) concentrations of 0, 10 and 20 wt%, there were the highest ionic conductivities at 10 wt% MIL-53(Al) and at all three temperatures. Hence, the EO:Li ratio of 15 : 1 and the MIL-53(Al) concentration of 10 wt% were selected as optimized parameters for the preparation of the electrolyte.

Table 1 Ionic conductivities at 30 ℃, 60 ℃, and 80 ℃ for the PEO-X wt%MIL-53(Al)-LiTFSI thin film electrolytes with different EO : Li ratios

EO : Li	X	Ionic conductivities/S · cm^{-1}		
		30 ℃	60 ℃	80 ℃
10 : 1	10	1.29×10^{-5}	4.30×10^{-4}	9.03×10^{-4}
15 : 1	10	1.62×10^{-5}	4.48×10^{-4}	9.71×10^{-4}
20 : 1	10	3.36×10^{-6}	8.41×10^{-5}	1.68×10^{-4}
25 : 1	10	1.14×10^{-6}	6.60×10^{-5}	1.65×10^{-4}
15 : 1	0	6.35×10^{-7}	4.70×10^{-5}	9.24×10^{-5}
15 : 1	20	1.13×10^{-6}	2.63×10^{-5}	2.41×10^{-4}

Previous studies found that the ionic conduction behavior is different in the crystalline phase and amorphous phase of the PEO based electrolyte, and the ionic conductivity is higher in the amorphous phase than in the crystalline phase.[25] In order to determine the phase transition temperature and find the applicable operating temperature range of the PEO-MIL-53(Al)-LiTFSI electrolyte, ionic conductivities at 40, 45, 50, 55, 60, 65, 70, 80, 100, 120 and 150℃ were measured (see Fig. 2). The ionic conductivity of the electrolyte is 3.39×10^{-3} S·cm^{-1} at 120℃, 3.5 times greater than that of the PEO-LiTFSI electrolyte without MIL-53(Al) (9.66×10^{-4} S·cm^{-1}) at the same EO∶Li ratio of 15∶1. Meanwhile, the log σ shows a linear relation with $1/T$ at low temperatures, while at high temperatures a nonlinear relation is presented. Thus, by fitting the data with the Arrhenius equation $\left(\sigma = A \exp\dfrac{E_a}{RT}\right)$[26] at low temperatures and the Vogel-Tamman-Fulcher (VTF) equation $\left(\sigma = AT^{-0.5}\exp\left[-\dfrac{B}{k(T-T_0)}\right]\right)$[27] at high temperatures, the phase transition temperatures of 50.3℃ for the PEO-MIL-53(Al)-LiTFSI electrolyte and 56.9℃ for the PEO-LiTFSI electrolyte are obtained, where σ is the ionic conductivity, A is the pre-exponential factor, E_a is the activation energy for ionic transport, R is the gas constant, k is the Boltzmann constant, B is the pseudo activation energy, T is the test absolute temperature, and T_0 is the equilibrium glass transition temperature. There is a decrease of 6.6℃ in the phase transition temperature. This is ascribed to the addition of MIL-53(Al), leading to the higher ionic conductivities of the PEO-MIL-53(Al)-LiTFSI electrolyte. In addition, above the phase transition temperatures, both the electrolytes are amorphous and the ionic conductivities are higher. These results indicate that the ionic conduction behavior is indeed related to the phase structure.

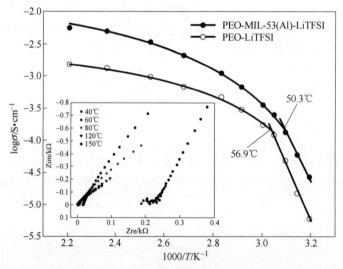

Fig. 2 Temperature-dependent conductivities for the PEO-10 wt% MIL-53(Al)-LiTFSI electrolyte and the PEO-LiTFSI electrolyte at the same EO∶Li ratio of 15∶1. The circles are the experimental data and the lines are the fitted results.[28,29] Inset: the AC impedance spectra of the PEO-MIL-53(Al)-LiTFSI electrolyte at selected temperatures of 40, 60, 80, 120 and 150℃

The lithium ion transference number, t^+, is another desirable parameter of electrolytes.[30,31] The measurement of t^+ was carried out by using AC impedance spectroscopy and chronoamperometry (CA). Fig. 3 shows the relation between time and current crossing a symmetric Li/PEO-MIL-53(Al)-LiTFSI/Li battery polarized by a small voltage of 10mV at 80℃. The inset is the

AC impedance spectra of the same battery before and after the polarization. Then, the t^+ can be calculated by the following equation:[32]

$$t^+ = \frac{I_{ss} \Delta V - I_0 R_0}{I_0 \Delta V - I_{ss} R_{ss}} \quad (2)$$

where I_0 and I_{ss} are the initial current and the steady-state current, respectively; ΔV is the voltage; R_0 and R_{ss} are the initial and steady-state interfacial resistances, respectively. From the figure inset, the first interception of the data at high frequency is related to the bulk resistance of about 16Ω. The diameters of the semicircles in the medium frequency range represent the initial interfacial resistance (33Ω) and the steady-state interfacial resistance (53Ω). The t^+ of the PEO-MIL-53(Al)-LiTFSI electrolyte is 0.343. Without the filler, the t^+ of the PEO-LiTFSI electrolyte is 0.252. This indicates that the mobility of Li$^+$ ions is enhanced due to the function of the metal-organic framework.

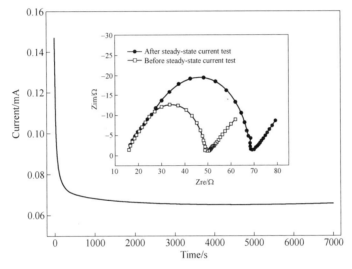

Fig. 3 Chronoamperometry of the Li/PEO-MIL-53(Al)-LiTFSI/Li cells at a potential of 10mV at 80℃.
Inset: the AC impedance spectra of the same battery before and after the polarization

3.3 Lithium ion conduction mechanism for the PEO-MIL-53(Al)-LiTFSI electrolyte

The mechanism for the enhancement of lithium ion transfer properties has been preliminarily explored. The zeta electric potential of the MIL-53(Al) nanoparticles was measured and the result is shown in Fig. 4. The point of zero charge (PZC) for the particles is at pH 9.1. When pH is lower than 9.1, the surface exhibits a Lewis acidic property. Previous studies found that the ionic conductivity and the lithium ion transference number were increased when adding Lewis acidic ceramic nanoparticles.[33-36] In this study, the pH of the PEO-MIL-53(Al)-LiTFSI electrolyte is about 7, thus, the particles have strong Lewis acidic properties. Under this condition, the N(SO$_2$CF$_3$)$_2$-anions of the LiTFSI salt may reside on the Lewis acidic surfaces of MIL-53(Al) nanoparticles, while the Li$^+$ ions are released from the salt and approach the ether oxygen atoms in the PEO chains. This interaction not only disturbs the crystallization of PEO, but also increases the lithium salt dissolution. This experiment demonstrates again that the Lewis acidic nanoparticles can improve the lithium ionic conductivity in solid polymer electrolytes.

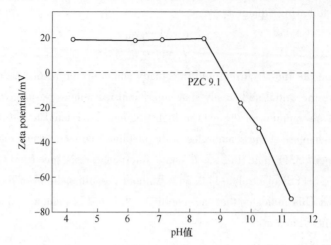

Fig. 4 The zeta electric potential measurement of MIL-53(Al) nanoparticles

3.4 Electrochemical stability of the electrolytes

The electrochemical stability of the PEO-LiTFSI electrolyte and the PEO-MIL-53(Al)-LiTFSI electrolyte at 80℃ and 120℃ were measured by using linear sweep voltammograms(LSV) in the potential range of 2.5V to 6.5V(vs. Li$^+$/Li)(see Fig. 5). The curves represent oxidation decomposition at the potentials of 5.31V at 80℃ and 5.10V at 120℃ for the PEO-MIL-53(Al)-LiTFSI electrolyte. For a comparison, the electrochemical stability of the PEO-LiTFSI electrolyte without MIL-53(Al) was investigated and the lower oxidation potentials of 5.15V at 80℃ and 4.99V at 120℃ were measured. It is noticed that there is a small current increase of about 0.2mA at 4.5V and 120℃. Here, this small current increase is ignored in the determination of the decomposition potential.

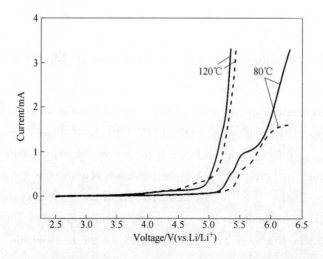

Fig. 5 Linear sweep voltammograms of SS/PEO-LiTFSI/Li(solid line) and SS/PEO-MIL-53(Al)-LiTFSI/Li (dotted line) batteries at 80℃ and 120℃. The electrolytes were swept in the potential range from 2.5V to 6.5V(vs. Li/Li$^+$) at a rate of 10mV · s^{-1}

3.5 Thermal stability and mechanical characteristics of the electrolytes

The thermal stability of electrolytes is a key factor that determines the safety performance of batteries. Thermogravimetric analysis(TGA) of the PEO-MIL-53(Al)-LiTFSI electrolyte and the PEO-LiTFSI electrolyte were carried out(results are shown in Fig. 6a). By differential processing of the TGA data, the differential thermogravimetric(DTG) curves for the electrolytes are obtained and are presented in Fig. 6b. Similar three-step degradations in the temperature range of 48-500 ℃ for both electrolytes are observed. In the PEO-MIL-53(Al)-LiTFSI electrolyte, the first degradation beginning from 195 ℃ corresponds to the decomposition of PEO.[37] The second degradation occurring at 375 ℃ is mainly due to the decomposition of LiTFSI.[38] The third is also caused by the decomposition of PEO. The residual at 500 ℃ (about 6%) for the PEO-MIL-53(Al)-LiTFSI electrolyte is from partially carbonized MIL-53(Al).[23] The decomposition peaks in the DTG curve for the PEO-MIL-53(Al)-LiTFSI electrolyte shift right compared to the PEO-LiTFSI electrolyte, with the LiTFSI decomposition temperature 15 degrees higher. The approximate 1.3% weight loss at 48 ℃ for both electrolytes is due to water absorbed during the sample transfer process.

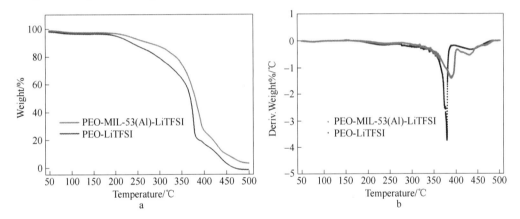

Fig. 6 (a) Thermogravimetric analysis of the PEO-LiTFSI and PEO-MIL-53(Al)-LiTFSI electrolytes and (b) DTG curves of the PEO-LiTFSI and PEO-MIL-53(Al)-LiTFSI electrolytes

In order to characterize the mechanical strength, dynamic mechanical analysis(DMA) of the PEO-MIL-53(Al)-LiTFSI and PEO-LiTFSI electrolytes was performed. From the stress-strain curves in Fig. 7, the stress of the PEO-MIL-53(Al)-LiTFSI electrolyte is obviously higher than the PEO-LiTFSI electrolyte. The enhancement in the thermal stability and mechanical strength can be attributed to the addition of MIL-53(Al) nanoparticles, which act as crossing-linking centers for PEO, thus constructing a robust network. These results further demonstrate that the MIL-53(Al) nanoparticles with Lewis acidic surfaces are beneficial for the improvement of solid polymer electrolyte performance.

3.6 Cycling performance of all-solid-state LiFePO$_4$/PEO-MIL-53(Al)-LiTFSI/Li batteries

Fig. 8 shows the cycling performance of the LiFePO$_4$/PEO-MIL-53(Al)-LiTFSI/Li batteries at

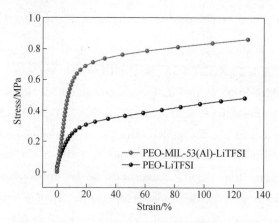

Fig. 7 Stress-strain curves of the PEO-MIL-53(Al)-LiTFSI and PEO-LiTFSI electrolytes

5 C, at 80℃ and 120℃. For the purpose of retaining high discharge capacities at fast charging/discharging, the batteries were pre-charged/discharged at 1 C for about ten cycles (the data are not included in Fig. 8). The initial discharge capacity at 5 C is 127.1mA·h·g^{-1} at 80℃ and 136.4mA·h·g^{-1} at 120℃. After 300 cycles, the discharge capacity is 116.0mA·h·g^{-1} at 80℃ and 129.2mA·h·g^{-1} at 120℃. In the inset of Fig. 8, when the battery ran 610 cycles at 80℃, the retention ratio of the discharge capacity is 80%; when cycled at 120℃, the battery has a discharge capacity of 109.3mA·h·g^{-1} in the 625th cycle with fading of 20%. The batteries were cycled for 1400 cycles at the same rate and the same temperatures, still demonstrating discharge capacities of 65.9mA·h·g^{-1} and 82.9mA·h·g^{-1} at 80℃ and 120℃, and retention ratios of 52.4% and 61.3%, respectively.

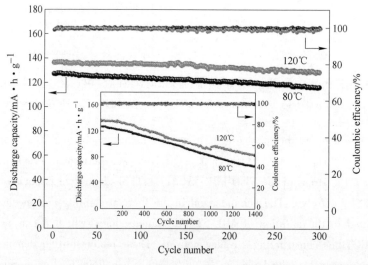

Fig. 8 The cycling performance and coulombic efficiencies of the LiFePO$_4$/PEO-MIL-53(Al)-LiTFSI/Li batteries at 5 C, at 80℃ and 120℃. Inset: the cycling performance of the same batteries for 1400 cycles. At 120℃, an increase in the capacity of about 4.3mA·h·g^{-1} in the 1003rd cycle is seen since the power system of the computer was broken for 18h by accident, but the battery was stored in the oven that was kept working at 120℃

In order to further evaluate fast charging/discharging performance, the battery was cycled at 10 C and 120℃ (the result is shown in Fig. 9). When cycling at 10 C (about 4-5min for charging or discharging), the battery has the discharge capacity of 116.2mA · h · g^{-1} in the first cycle and 103.5mA · h · g^{-1} in the 110th cycle with a retention ratio of 90%. At the same temperature and 1 C, another battery can run 530 cycles with the discharge capacity of 129.8mA · h · g^{-1}, retaining 90% of the highest discharge capacity (see Fig. 9). Additionally, the inset of Fig. 9 also shows the typical potential vs. time profiles at 1 C, 10C and 120℃. Flat voltage plateaus reflecting the insertion/removal of Li$^+$ ions at the LiFePO$_4$ cathode are seen. Even in the 110th cycle at 10 C, the plateaus still exist. Compared to all-solid-state lithium ion batteries with PEO based electrolyte reported,[10-18] the LiFePO$_4$/PEO-MIL-53(Al)-LiTFSI/Li battery exhibits better cycling performance at high rates and temperatures.

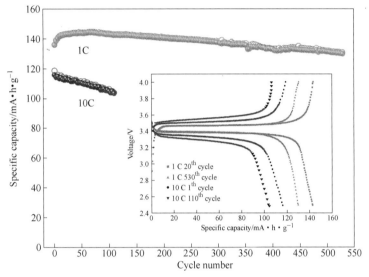

Fig. 9 The charge (hollow circle) and discharge (solid circle) cycling performance of the LiFePO$_4$/PEO-MIL-53(Al)-LiTFSI/Li batteries at 1 C, 10 C at 120℃. The cycle number cut off when the testing batteries delivered 90% or less of the highest discharge capacity. Inset: the typical potential vs. time profiles at 1 C, 10 C and at 120℃

4 Conclusions

A PEO-MIL-53(Al)-LiTFSI electrolyte was prepared by using the metal-organic framework MIL-53(Al) as a filler for all-solid-state LIBs. The MIL-53(Al) nanoparticles have Lewis acidic surfaces and interact with $N(SO_2CF_3)_2^-$ anions that promote the dissolution of LiTFSI, thus increasing the lithium ionic conductivity. The particles also act as crossing-linking centers for PEO, leading to the increase in thermal stability and mechanical strength from the robust network built. The battery with the electrolyte displays fast charging/discharging abilities. Such electrolytes, which may replace both the separator and liquid electrolyte in conventional batteries, can be used directly for fast charging/discharging all-solid-state lithium ion batteries. This is an important step in expanding applications of lithium ion batteries.

Acknowledgements

This work was supported by the National Natural Science Foundation of China (No. 51274239) and Central South University, which is greatly appreciated.

References

[1] H. J. Zhang, S. Kulkarni and S. L. Wunder, J. Phys. Chem. B, 2007, 111, 3583-3590.

[2] J. F. M. Oudenhoven, L. Baggetto and P. H. L. Notten, Adv. Energy Mater., 2011, 1, 10-33.

[3] S. Lee, M. Schomer, H. Peng, K. A. Page, D. Wilms, H. Frey, C. L. Soles and D. Y. Yoon, Chem. Mater., 2011, 23, 2685-2688.

[4] M. Patel, M. U. M. Patel and A. J. Bhattacharyya, ChemSusChem, 2010, 3, 1371-1374.

[5] J. Cao, L. Wang, X. M. He, M. Fang, J. Gao, J. J. Li, L. F. Deng, H. Chen, G. Y. Tian, J. L. Wang and S. S. Fan, J. Mater. Chem. A, 2013, 1, 5955-5961.

[6] United States Advanced Battery Consortium, Electric Vehicle Battery Test Procedures Manual, Revision 2, Southfeld, MI, 1996.

[7] F. Croce, G. B. Appetecchi, L. Persi and B. Scrosati, Nature, 1998, 394, 456-458.

[8] Y. Lin, J. Li, Y. Q. Lai, C. F. Yuan, Y. Cheng and J. Liu, RSC Adv., 2013, 3, 10722-10730.

[9] J. Li, Y. Lin, H. H. Yao, C. F. Yuan and J. Liu, ChemSusChem, 2014, 7, 1901-1908.

[10] S. T. Ren, H. F. Chang, L. J. He, X. F. Dang, Y. Y. Fang, Y. Zhang, H. Y. Li, Y. L. Hu and Y. Lin, J. Appl. Polym. Sci., 2013, 129, 1131-1142.

[11] Y. F. Tong, L. Chen, X. H. He and Y. W. Chen, Electrochim. Acta, 2014, 118, 33-40.

[12] Y. H. Li, X. L. Wu, J. H. Kim, S. Xin, J. Su, Y. Yan, J. S. Lee and Y. G. Guo, J. Power Sources, 2013, 244, 234-239.

[13] J. S. Syzdek, M. B. Armand, P. Falkowski, M. Gizowska, M. Karzowicz, L. Lukaszuk, M. L. Marcinek, A. Zalewska, M. Szafran, C. Masquelier, J. M. Tarascon, W. G. Wieczorek and G. Zukowska, Chem. Mater., 2011, 23, 1785-1797.

[14] J. L. Schaefer, D. A. Yanga and L. A. Archer, Chem. Mater., 2013, 25, 834-839.

[15] R. R. Madathingal and S. L. Wunder, Macromolecules, 2011, 44, 2873-2882.

[16] B. W. Zewde, S. Admassie, J. Zimmermann, C. S. Isfort, B. Scrosati and J. Hassoun, ChemSusChem, 2013, 6, 1400-1405.

[17] C. F. Yuan, J. Li, P. F. Han, Y. Q. Lai, Z. A. Zhang and J. Liu, J. Power Sources, 2013, 240, 653-658.

[18] C. Gerbaldi, J. R. Nair, M. A. Kulandainathan, R. S. Kumar, c. Ferrara, P. Mustarelli and A. M. Stephan, J. Mater. Chem. A, 2014, 2, 9948-9954.

[19] W. J. Rieter, K. M. Pott, K. M. L. Taylor and W. Lin, J. Am. Chem. Soc., 2008, 130, 11584-11585.

[20] K. A. White, D. A. Chengelis, K. A. Gogick, J. Stehman, N. L. Rosi and S. Petoud, J. Am. Chem. Soc., 2009, 131, 18069-18071.

[21] A. J. Lan, K. H. Li, H. H. Wu, D. H. Olson, T. J. Emge, W. Ki, M. C. Hong and J. Li, Angew. Chem., Int. Ed., 2009, 48, 2334-2338.

[22] S. J. Garibay and S. M. Cohen, Chem. Commun., 2010, 46, 7700-7702.

[23] T. Loiseau, C. Serre, C. Huguenard, G. Fink, F. Taulelle, M. Henry, T. Bataille and G. Férey, Chem. -Eur. J., 2004, 10, 1373-1382.

[24] J. A. Greathouse and M. D. Allendorf, J. Am. Chem. Soc., 2006, 128, 10678-10679.

[25] B. Kumar, S. J. Rodrigues and S. Koka, Electrochim. Acta, 2002, 47, 4125-4131.

[26] Y. H. Li, X. L. Wu, J. H. Kim, S. Xin, J. Su, Y. Yan, J. S. Lee and Y. G. Guo, J. Power Sources, 2013, 244,

234-239.
- [27] R. S. Bobadea, S. V. Pakade and S. P. Yawale, J. Non-Cryst. Solids, 2009, 355, 2410-2414.
- [28] F. B. Dias, L. Plomp and J. J. B. Veldhuis, J. Power Sources, 2000, 88, 169-191.
- [29] C. D. Robitaille and D. Fauteux, J. Electrochem. Soc., 1986, 133, 315-325.
- [30] M. Doyle, T. F. Fuller and J. Newman, Electrochim. Acta, 1994, 39, 2073-2081.
- [31] K. E. Thomas, S. E. Sloop, J. B. Kerr and J. Newman, J. Power Sources, 2000, 89, 132-138.
- [32] J. Evans, C. A. Vincent and P. G. Bruce, Polymer, 1987, 28, 2324-2328.
- [33] F. Croce, L. Persi, B. Scrosati, F. S. Fiory, E. Plichta and M. A. Hendrickson, Electrochim. Acta, 2001, 46, 2457-2461.
- [34] W. Wieczorek, Z. Florjancyk and J. R. Stevens, Electrochim. Acta, 1995, 40, 2251.
- [35] F. Croce, R. Curini, A. Martinelli, L. Persi, F. Ronci and B. Scrosati, J. Phys. Chem., 1999, 103, 10632-10638.
- [36] N. Angulakshmi, K. S. Nahm, J. R. Nair, C. Gerbaldi, R. Bongiovanni, N. Penazzi and A. M. Stephan, Electrochim. Acta, 2013, 90, 179-185.
- [37] T. Shodai, B. B. Owens, T. Ostsuka and J. Yamaki, J. Electrochem. Soc., 1994, 141, 2978-2981.
- [38] M. Echeverri, N. Kim and T. Kyu, Macromolecules, 2012, 45, 6068-6077.

Effect of Ni-doping on Electrochemical Capacitance of MnO$_2$ Electrode Materials*

Abstract Mn/Ni composite oxides as active electrode materials for supercapacitors were prepared by solid-state reaction through the reduction of KMnO$_4$ with manganese acetate and nickel acetate at low temperature. The products were characterized by X-ray diffractometry (XRD) and transmission electron microscopy (TEM). The electrochemical characterizations were performed by cyclic voltammetry (CV) and constant current charge-discharge in a three-electrode system. The effects of different potential windows, scan rates, and cycle numbers on the capacitance behavior of Mn$_{0.8}$Ni$_{0.2}$O$_x$ composite oxide were also investigated. The results show that the composite oxides are of nano-size and amorphous structure. With increasing the molar ratio of Ni, the specific capacitance goes through a maximum at molar fraction of Ni of 20%. The specific capacitance of Mn$_{0.8}$Ni$_{0.2}$O$_x$ composite oxide is 194.5F/g at constant current discharge of 5mA.

1 Introduction

Supercapacitor is a promising energy storage device for meeting the highpower electric market[1-4]. Complementary to battery, supercapacitor can provide superior power density and cyclability, thus it is regarded as an intermediate device between traditional capacitor and battery. Supercapacitors are generally categorized into two types based on the charge-discharge mechanism. The electric double-layer capacitance (EDLC) arises from electrostatics separation of charges at the interface between electrode and electrolyte. On the other hand, pseudocapacitance, typically 10 times greater than EDLC, results from either rapid adsorption/desorption superficial reaction or multi-electron-transfer faradic reactions with fast charge/discharge properties. Many transition metal oxides have been shown to be excellent electrode materials for supercapacitors whose charge storage mechanism is primarily based on pseudocapacitance. RuO$_2$, for example, delivers a high capacity up to 700F/g and an excellent cyclability. But RuO$_2$-based supercapacitors are too expensive to be commercially attractive[5,6]. Nonnoble oxides such as NiO, Co$_3$O$_4$, MnO$_2$ are very promising candidates for electrode materials in supercapacitors[7-13]. These oxides show specific capacitance varying between 100 and 200 F/g. Recent research is focused on increasing the specific capacitance of the oxides by introducing other oxides technology[14,15].

Zhang et al[16] reported that MnO$_2$ prepared by solid-state reaction at low temperature shows ideal capacitive behavior in Na$_2$SO$_4$ solution in the potential range from 0.2 to 0.8V (vs. SCE). However, the relatively low specific capacitance (150F/g) needs to be improved for supercapacitor application. In this study, Ni-doped MnO$_2$ composite oxides as electrode materials were pre-

---

* Copartner: Zhang Zhian, Lai Yanqing, Li Jie. Reprinted from J of Central South University of Technology, 2007, 14(5): 638-642.

pared by solid-state reaction through the reduction of KMnO$_4$ with manganese acetate and nickel acetate at low temperature. The physical properties of the oxides, such as crystallinity and particle size were characterized by X-ray diffractometry(XRD) and transmission electron microscopy (TEM). The capacitive characteristics of the composite oxides as the active electrode materials for supercapacitors were also investigated by cyclic voltammetry and chronopotentiometry.

2 EXperimental

2.1 Preparation of electrode materials

The electrode materials were synthesized by reduction of KMnO$_4$ with manganese acetate and nickel acetate. First, a controlled amount of Mn(Ac)$_2 \cdot$ 4H$_2$O and Ni(Ac)$_2 \cdot$ 3H$_2$O were mixed and ground for 5min in an agate mortar before mixing with the powder of KMnO$_4$; then, the mixture was ground at room temperature for 0.5h; next, it was heated in a water bath at 65℃ for 6h to make the reaction proceed completely; finally, the as-prepared powder was washed and filtered several times. with deionized water and annealed in a vacuum oven at 200℃ overnight. The composite oxides(in molar ratio, A: $n(Mn)/n(Ni) = 10:0$, B: $n(Mn)/n(Ni) = 9:1$, C: $n(Mn)/n(Ni) = 8:2$, D: $n(Mn)/n(Ni) = 7:3$, E: $n(Mn)/n(Ni) = 6:4$) can be obtained.

2.2 Preparation of electrode

Electrodes were prepared by mixing 75%(mass fraction) of the composite oxide powder as active material with 20% acetylene black and 5% polytetrafluoroethylene(PTFE) binder. The former two constituents were firstly mixed together to obtain a homogeneous black powder. The binder was then added with a few drops of ethanol. The mixture was ground completely to form a rubberlike paste, andpressed with a hand oil press under 10MPa onto a Ni-foam current collector as a working electrode, followed by drying at 80℃ under vacuum for 12h.

2.3 Material characteristics

The X-ray diffraction patterns of the products were recorded using a Philips X'pert Pro with Cu K$_\alpha$ radiation ($\lambda = 0.154056$nm) operated at 40kV and 40mA. The morphology and particle size of the products were characterized by transmission electron microscope(TEM, JEM-100CX JEOL, Japan).

2.4 Electrochemical measurement

A Ps-14 potentiostat/galvanostat was used for all electrochemical measurements. A beaker type electrochemical cell equipped with the working electrode, an SCE electrode reference and 20mm×20mm Pt film counter electrode were used. A Luggin capillary, whose tip was set at a distance of 1~2mm from the surface of the working electrode, was used to minimize errors due to IR drop in the electrolyte. Aqueous solutions of 1mol/L Na$_2$SO$_4$, employed as the electrolyte, were degassed with purified argon before measurement. The solution temperature was maintained at 25℃. The cyclic voltammogram(CV) and the galvanostatic charge-discharge curves of the electrode were recorded after 20 cycles. All the specific capacitances are reported per gram of active material unless otherwise specified. The specific capacitance in F/g evaluated by chrono-potentiometry can be calculated

$$C_s = \frac{dQ}{mdV} = \frac{dQ}{mdt}\frac{dt}{dV} \approx \frac{i\Delta t}{m\Delta V} \quad (1)$$

where Q is the charge on the electrode; V is the potential, i is the discharge current; t is the discharge time; m is the mass of the active material in an electrode.

3 Results and Discussion

3.1 Structure characteristics of material

The XRD patterns of the undoped MnO_2 and the Ni-doped MnO_2 composite oxides are shown in Fig. 1. For the pure MnO_2, the obvious peaks are obtained, indicating excellent crystalline. However, for the Ni-doped MnO_2 composite oxides, the peaks are broadened and lowered. When the molar ratio of Mn to Ni($n(Mn)/n(Ni)$) is 9 : 1, the width of the peaks starts to increase, indicating the presence of amorphous phase. When $n(Mn)/n(Ni)$ is 8 : 2, the main peak appears at $2\theta = 37$℃ and other peaks are broadened and lowered, indicating that the structure changes to amorphous. When $n(Mn)/n(Ni)$ is 7 : 3 or 6 : 4, no obvious peaks are observed. Previous research showed that amorphous nanostructured oxides exhibiting pseudocapacitance were considered to be promising materials for high-energy-density application[12-13].

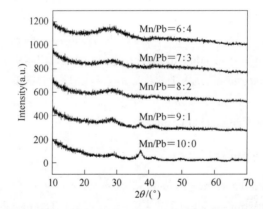

Fig. 1 XRD patterns of pure MnO_2 and Mn/Ni composite oxides

The morphologies of the undoped MnO_2 and the Ni-doped MnO_2 composite oxides were examined by TEM. Typical images of the pure MnO_2 and of the Mn/Ni composite oxides are shown in Fig. 2. For the pure MnO_2, the particle size is not larger than 40nm. The typical morphology is irregularly spherical. When $n(Mn)/n(Ni)$ is 8 : 2, the particle size of the Mn/Ni composite oxide is reduced to 15nm and the composite oxide disperses well. The morphology is spherical or needle. When $n(Mn)/n(Ni)$ is 6 : 4, some agglomerations are found.

3.2 Effect of Ni-doping on specific capacitance of composite oxides

To investigate the performance of the composite oxide electrode, all of these electrodes were charged/discharged at different discharge currents of 5, 10 and 20mA using a potential window

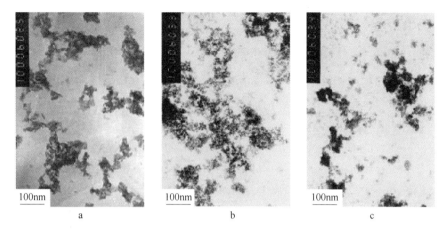

Fig. 2 TEM images of pure MnO_2 and Mn/Ni composite oxides

a—Mn/Pb=10∶0 b—Mn/Pb=8∶2 c—Mn/Pb=6∶4

between −0.2 and 0.8V(vs. SCE). Fig. 3 shows the dependence of the specific capacitance on molar ratio of Ni. By 5mA constant current charges-discharge, the pure MnO_2 has the specific capacitance of 150F/g. With increasing the molar ratio of Ni, the specific capacitance goes through a maximum at approximately 20% molar fraction of Ni. The specific capacitance reaches 194.5F/g for the $Mn_{0.8}Ni_{0.2}O_x$ composite oxide. With further increase of Ni content of the composite oxide, the specific capacitance starts to decrease gradually.

From the above results, it can be inferred that the specific capacitance depends on the material microstructure of the composite oxide. For the pure MnO_2, the particle size is relatively large, and the surface area is relatively low. When $n(Mn)/n(Ni)$ is 8∶2, the oxide exhibits an amorphous structure, the particle size decreases, the surface area increases, and the active site area and the utilization of the oxide are enlarged, which results in an increase of the capacitance of the composite oxide when compared with the capacitance of the pure MnO_2. And, when Ni content increases, agglomerations will lead to a decrease of the surface area and the oxide utilization.

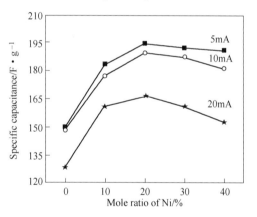

Fig. 3 Specific capacitances of different Mn/Ni composite oxides at different constant discharges currents

Moreover, for all the composite oxides, the specific capacitance of the nanocomposite oxide decreases with the increase of discharge current, which can be explained by considering utilization of the active electrode material. When the charge-discharge current is low, the time of charge-discharge process is long, and the reversible faradic reaction is carried out thoroughly, which results in high utilization of the active electrode material. When charge-discharge current is increased, the time of charge-discharge process is shortened, and the utilization of the active electrode material is lowered. Thus, the specific capacitance of the nanocomopiste decreases with increasing charge-discharge current.

3.3 Discharge curves of composite oxides

Fig. 4 shows the discharge curves of the pure MnO_2 and the Ni-doped composite oxide at the discharge currents of 5,10 and 20mA. It can be found that the discharge curve is approximately linear at 5mA. When the discharge current is 10mA, the inflexion appears at the potential of+ 0.22V (vs. SCE), corresponding to a short discharge platform. When the discharge current is 20mA, the inflexion also appears at the potential of+0.18V(vs. SCE), corresponding to a short discharge platform.

It can be found that the discharge curves for $Mn_{0.8}Ni_{0.2}O_x$ oxide are approximately linear at 5,10 and 20mA. From Fig. 4, it can be seen that the discharge time of $Mn_{0.8}Ni_{0.2}O_x$ oxide is longer than that of the pure MnO_2. Hence, the composite is a promising electrode material in the application of supercapacitors.

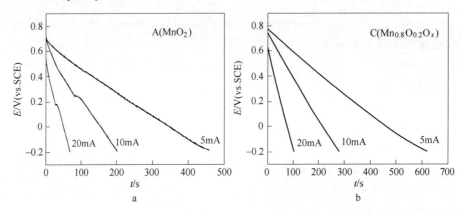

Fig. 4 Discharge curves of MnO_2(a) and $Mn_{0.8}Ni_{0.2}O_x$(b) at different constant discharge currents

3.4 Effect of Ni-doped on CV curves

Fig. 5 shows CV curves of the pure MnO_2 and the Ni-doped composite oxide at the different scan rates in 1mol/L Na_2SO_4.

For the pure MnO_2, the shape of CV curves is rectangle at the scan rate of 5mV/s, exhibiting excellent capacitance behavior of the oxide. With the increase of the scan rate, the curves of CV show distorted. The responding current fluctuates with the potential. When the scan rates are 10 and 20mV/s, respectively, the reduction current appears and fluctuates at the potential of 0.2V (vs. SCE), which corresponds to the previous results, whereas for the $Mn_{0.8}Ni_{0.2}O_x$, all the CV curves show no obvious redox peaks. It can be seen that the currents change fast when the direction of scan is changed, exhibiting no obvious electrochemical polarization.

3.5 CV characteristics of $Mn_{0.8}Ni_{0.2}O_x$ oxide at different potential windows

The electrochemical reversibility of Mn/Ni composite oxide is usually examined by varying the upper potential limit of CV. Typical CV curves of $Mn_{0.8}Ni_{0.2}O_x$ composite oxide electrode between -0.2 and 1.0V(vs. SCE), measured at 5mV/s in 1mol/L Na_2SO_4, are shown in Fig. 6. When

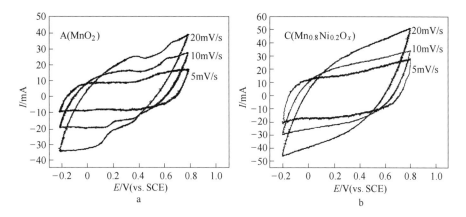

Fig. 5　CV curves of MnO_2(a) and $Mn_{0.8}Ni_{0.2}O_x$(b) at 5, 10 and 20mV/s

the potential window is between -0.2 and 1.0V (vs. SCE), the response currents vary with the potential. When the potential window range is between -0.2 and 0.9V (vs. SCE), all the curves show rectangle-like shape and no obvious redox peaks. Note that all voltammetric currents approximately follow the same trace on the positive sweeps of all CV curves in this figure. The currents on all CV curves reaching the plateau values are very fast when the direction of potential sweep is just changed. This shows typical capacitive-like characteristics. These results indicate good electrochemical reversibility

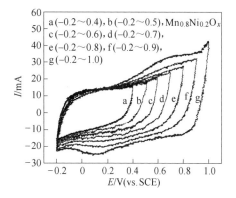

Fig. 6　Effect of potential windows on CV curves for $Mn_{0.8}Ni_{0.2}O_x$ at scan rate of 5mV/s

of the composite oxide in 1mol/L Na_2SO_4, resulting in high-power characteristics of the composite oxide.

3.6　Cycle life

The cycle life stability of the active material in the electrolyte was tested using cyclic voltammetry at the scan rate of 5mV/s. The CVs of the nanocomposite $Mn_{0.8}Ni_{0.2}O_x$ oxide measured at the 30th, 100th and 200th cycle are shown in Fig. 7. When the cycle number is 30, the corresponding current varies with the potential; when the cycle number is 100, the corresponding current is stable, no significant change of CV is observed after 200 cycles for the $Mn_{0.8}Ni_{0.2}O_x$ oxide electrode, indicating good electrochemical stability, high electrochemical re-

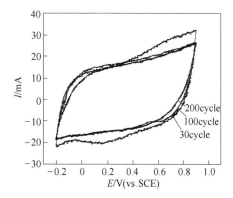

Fig. 7　Effect of cycle number on CV curves for $Mn_{0.8}Ni_{0.2}O_x$ at scan rate of 5mV/s

versibility and long cycle life. Thus a very promising electrode material for supercapacitors has been prepared using the solid-state reaction described.

4 Conclusions

(1) Mn/Ni composite oxides as active electrode materials for supercapacitor were synthesized by using a method based on solid-state reaction of $KMnO_4$ with $Mn(Ac)_2 \cdot 4H_2O$ and $Ni(Ac)_2 \cdot 3H_2O$ at low temperature.

(2) The particles of the composites are nano size, and the size decreases with increasing Ni content. With increasing the molar ratio of Ni, the specific capacitance goes through a maximum at molar fraction of Ni of 20%. The specific capacitance of $Mn_{0.8}Ni_{0.2}O_x$ composite oxide is 194.5 F/g at constant discharge current of 5mA. Therefore, the Mn/Ni composite oxide is a promising electrode material for supercapacitor.

References

[1] Kötz R, Carlen M. Principles and applications of electrochemical capacitors [J]. Electrochimica Acta, 2000, 45: 2483-2498.

[2] Burke A. Ultracapacitors: Why, how and where is the technology[J]. J Power Sources, 2000, 91: 37-50.

[3] Vol'fkovich Y M, Serdyuk T M. Electrochemical capacitors [J]. Russian J Electrochem, 2002, 38: 935-958.

[4] Lai Yanqing, Li Jing, Li Jie, et al. Preparation and electrochemical characterization of C/PANI composite electrode materials [J]. Journal of Central South University of Technology, 2006, 13(4): 353-359.

[5] Zheng J P, Cygan, P J, Jow T R. Hydrous ruthenium oxide as an electrode material for. electrochemical capacitors[J]. J Electrochem Soc, 1995, 142: 2699-2703.

[6] Barbieri O, Hahn M, Foelske A, et al. Effect of electronic resistance and water content on the performance of RuO_2 for supercapacitors[J]. J Electrochem Soc, 2006, 153: A2049. -A2054.

[7] Srinivasan V, Weidner J W. Studies on the capacitance of nickel oxide films: Effects of heating temperature and electrolyte concentration[J]. J Electrochem Soc, 2000, 147: 880-885.

[8] Wu Mengqiang, Gao Jiahui, Zhang Shuren, et al. Comparative studies of nickel oxide films on different substrates for electrochemical supercapacitors [J]. J Power Sources, 2006, 159: 365-369.

[9] LIN C, RITTER J A, POPOV B N. Characterization of sol-gel-derived cobalt oxide xerogels as electrochemical capacitors[J]. J Electrochem Soc, 1998, 145: 4097-4103.

[10] Broughton J N, Brett M J. Investigation of thin sputtered Mn films for electrochemical capacitors [J]. Electrochim Acta, 2004, 49: 4439-4446.

[11] Luo S L, Wu N L. Investigation of pseudocapacitive charge-storage reaction of $MnO_2 \cdot nH_2O$ supercapacitor in aqueous electrolytes[J]. J Electrochem Soc, 2006, 153: A1317-A1324.

[12] Devaraj S, Munichandraiah N. Electrochemical supercapacitor studies of nanostructured α-MnO_2 synthesized by microemulsion method and the effect of annealing [J]. J Electrochem Soc, 2007, 154: A80-A88.

[13] Zolfaghari A, Ataherian F, Ghaemi M, et al. Capacitive behavior of nanostructured MnO_2 prepared by sonochemistry method[J]. Electrochimica Acta, 2007, 52: 2806-2814.

[14] Wang Yonggang, Zhang Xiaogang. Preparation and electrochemical capacitance of RuO_2/TiO_2 nanotubes

composites[J]. Electrochimica Acta,2004,49:1957-1962.

[15] Kuo S L,Lee J F,Wu N L. Study on pseudocapacitance mechanism of aqueous $MnFe_2O_4$ supercapacitor [J]. J Electrochem Soc,2007,154:A34-A38.

[16] Zhang Zhian,Yang Bangchao,Hu Yongda. Electrochemical investigation of Mn-Pb nanocomposite oxide for supercapacitor electrode material [J]. J Functional Material,2006,37(2):315-318.

Preparation and Properties of Pitch Carbon Based Supercapacitor*

Abstract Using the mesophase pitch as precursor, KOH and CO_2 as activated agents, the activated carbon electrode material was fabricated by physical-chemical combined activated technique for supercapacitor. The influence of activated process on the pore structure of activated carbon was analyzed and 14F supercapacitor with working voltage of 2.5V was prepared. The charge and discharge behaviors, the properties of cyclic voltammetry, specific capacitance, equivalent serials resistance (ESR), cycle properties, and temperature properties of prepared supercapacitor were examined. The cyclic voltammetry curve results indicate that the carbon based supercapacitor using the self-made activated carbon as electrode materials shows the desired capacitance properties. In 1mol/L Et_4NBF_4/AN electrolyte, the capacitance and ESR of the supercapacitor are 14.7F and 60mΩ, respectively. The specific capacitance of activated carbon electrode materials is 99.6F/g; its energy density can reach 2.96W·h/kg under the large current discharge condition. There is no obvious capacitance decay that can be observed after 5000 cycles. The leakage current is below 0.2mA after keeping the voltage at 2.5V for 1h. Meanwhile, the supercapacitor shows desired temperature property; it can be operated normally in the temperature ranging from −40℃ to 70℃.

1 Introduction

Supercapacitor is a kind of new energy storage device, which can fill the gap between the conventional capacitor and the battery[1-3]. Supercapacitor has many advantages, such as good pulse charge and discharge character, rapid discharge ability, long life span, no environment pollution[4,5]. Supercapacitors are now utilized in many fields, such as space industry, national defense, war industry, electrical vehicle, wireless communication, and consumptive electronics. It is estimated that the total market value would be about one billion by 2015, in the commercial, industrial and automotive applications[6].

According to the energy storage principle, supercapacitor can be categorized into carbon based supercapacitors that store energy by electric double layer (Helmholtz layer) and the metal-oxide based and the polymer based supercapacitors that store energy by pseudo capacitance[7-10]. Carbon based supercapacitor is also called electric double layer supercapacitor (EDLC).

There are many types of carbon materials that can be used for carbon based supercapacitor, such as carbon aero gel[11], carbon nanotube[12], carbon fiber[13], activated carbon powder[14]. The attributes of carbon (surface area, pore structure and specific capacitance) are the key factors to determine the properties (capacitance, equivalent serials resistance (ESR) and energy

---

* Copartner: Li Jing, Lai Yanqing, Song Haisheng, Zhang Zhian. Reprinted from Journal of Central South University of Technology, 2007, 14(5): 601-606.

density, power density) of carbon based supercapacitor. Because activated carbon electrode materials enjoy many advantages of the stable electrochemical properties, wide electrochemical window, perfect cycle property and lower cost, activated carbon is now the main commercially utilized supercapacitor electrode material.

In this experiment, low-cost mesophase pitch was selected as raw materials and physical-chemical combined activated technique was utilized to fabricate the activated carbon powder that has both the high surface area and ideal pore structure. This activated carbon powder was used as electrode material and activated electrode was prepared. Meanwhile, 1mol/L Et_4NBF_4 was used as electrolyte and cylindrical type supercapacitor was also prepared, the size of prepared supercapacitor was d 12.5mm×28mm. Properties of prepared supercapacitor, including capacitance properties, capacitance, ESR, current leakage, cycle property, discharge at different power density, temperature properties were discussed.

2 Experimental

2.1 Preparation of activated carbon electrode materials

When coal tar-pitch (soft point 83℃, provided by Wuhan Ferrous Company) was carbonized at 500℃, the easily emitted compound with small molecule was removed and the mesophase pitch was obtained. The mesophase pitch was ground and sieved in order to collect the grains whose average size was in the range of 100~200μm. Sieved mesophase pitch was mixed with KOH (analytical grade, provided by Shanghai Reagent Factory) in nickel crucible, and the mass ratio of alkali to carbon was 3∶1. In the activation process, the samples were first heated at 800℃ in argon atmosphere, and argon flow was kept at 60mL/min, the heating rate was 5℃/min, and heating time was 1h. After this process, CO_2 was led into and the time of combined activated process was 2h. After activated process, the activated carbon was washed first by 1mol/L HCl, then distilled water was used to wash it until its pH value reached 6.5. After rinsing process, the activated carbon was dried in an oven at 110℃ for 10h. The characterization of the porous texture of the activated carbons was conducted using physical adsorption of N_2 at 77K in the Autosorb-6 apparatus. The morphologies of active carbon powder and electrode were examined by JSM-5600LV scanning electron microscope.

2.2 Preparation of activated carbon electrode and cylindrical type supercapacitor

80% (mass fraction) self-made active carbon, 10% acetylene black, 10% adherent reagent were mixed and stirred in the agate mortar until homogeneous black slurry was achieved, which then was spread out onto a aluminum film of 30μm. In the vacuum condition, electrode film was dried at 100℃ for 10h; the roller was then used to roll the electrode film until its thickness was less than 190μm; and the lead wire was dot welted on the electrode films. Two electrode films were separated by a separator, and the three films were wrapped in the certain wrapping machine. The wrapped electrode films were put into electrolyte in the inert atmosphere and were put onto rubber packing. At last the whole was put into aluminum case with the size of d 12.5mm×28mm, and sealed in the self-made sealer.

2.3 Examination of pitch carbon based super-capacitor

The specific capacitance, ESR, the galvanostatic measurement and the cyclic voltammeter(CV) measurements were performed in the 273A EG&G Princeton Applied Research Potentiosat. As shown in Eqns. (1) and (2), the ESR was calculated according to voltage drop of the charge and discharge curve. Specific capacitance of activated carbon was calculated by $E(t)$ slope[15]:

$$R = \Delta E/I \tag{1}$$
$$C = (2It)/(m\Delta E) \tag{2}$$

where I is the discharge current; t is the discharge time; m is the mass of carbon on a electrode; ΔE is the voltage drop in discharge, excluding the portion of IR drop; the factor of "2" comes from the fact that the total capacitance measured from the test cells is the addition of two equivalent single-electrode capacitances.

3 Results and Discussion

3.1 Influence of activated technique on physical property of activated carbon

The activated methods can be described as physical activation(CO_2, H_2O used as activated regent) and chemical activation (KOH, NaOH, $ZnCl_2$ used as activated regent). The activated effect was determined by the method of activation, the attributes of precursor and the concrete activated techniques, such as activated temperature, the rate of heating, ratio of activated regent, and so on. Chemical activation, especially KOH used as activated regent, has the advantage of low activated temperature and high activated efficiency[16,17]. But when KOH was used as activated regent, the pore structure was formed by the corrosion of micro-graphic layer, and the amount of microspore of activated carbon was high and it was difficult to control the pore structure. When the physical-chemical activated technique was used at first, the graphic layer in carbon precursor was permeated into by melted KOH and a great amount of microspores were formed. When CO_2 was led in the activated process, the microspore on the surface may have the priority to be activated and widened, because in the CO_2 activated process, the pore structure was formed by the oxidation of active point of precursor, in the pore widening process, the mesopore may be formed, and the tendency of microspore formation may be wakened. Meanwhile, KOH on the surface of carbon might prevent the pore structure from collapsing. Compared with the KOH activated process, the physical-chemical combined technique might effectively enhance the amount of mesopores. The physical parameters of activated carbon are listed in Table 1, where the surface area of activated carbon is $1959m^2/g$, pore volume is $1.059cm^3/g$, and volume fraction of mesopore is 19.96%, average pore diameter is 2.224nm.

Fig. 1 shows the N_2 adsorption isotherms of the activated carbons prepared by physical-chemical combined technique. It can be seen from Fig. 1 that the N_2 adsorption isotherms of prepared activated carbon are almost type I according to BET classification, which indicates that microspores occupy most proportion of the activated carbon pore structure. Meanwhile, it can also be seen that the plateau has the shape of a small gradual increase along the pressure axis, which indicates that this activated carbon has a significant mesopore volume.

Table 1 Physical parameters of activated carbon

Sample	BET surface area /m² · g⁻¹	Total pore volume (N₂)/cm³ · g⁻¹	Mesopore volume /cm³ · g⁻¹
C083020	1959	1.059	0.2114

Sample	Volume fraction of mesopore/%	Average pore diameter/nm	Density/g · cm⁻³
C083020	19.96	2.224	0.478

Fig. 2 shows the pore size variation of microspores for the prepared activated carbon. It can be seen from Fig. 2 that there is a pore diameter concentration plateau between 1.250 and 1.559nm, which indicates the activated carbons with larger proportion of large microspore.

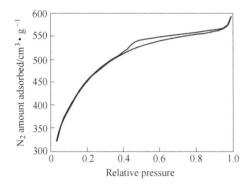

Fig. 1 Isotherms of activated carbon   Fig. 2 Microspore distribution of activated carbon

Fig. 3a and Fig. 3b show SEM images of the activated carbon powder and the electrode of activated carbon, respectively. It can be found that the activated carbon has the irregular shape; the diameter of activated carbon ranges from several μm to several tens μm.

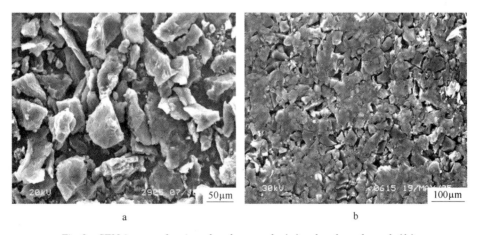

Fig. 3 SEM images of activated carbon powder(a) and carbon electrode(b)

There is no obvious pore structure to be found on the surface of activated carbon, which indicates that there is large proportion of micropore in the activated carbon. It can also be seen from SEM image of activated carbon electrode that the activated carbon powder is homogeneously mixed with acetylene black to form the net-work structure, and there are obvious pores between carbon particles, which may contribute to the accessibility of electrolyte.

3.2 Cyclic voltammograms examination of pitch carbon based supercapacitor

Fig. 4 shows the cyclic voltammograms(CV) property of pitch carbon supercapacitor at different scan rates ranging from 2 to 20mV/s, and the voltage window range of CV curve is from 0 to 2.5V. According to the CV curve, the reversibility of the supercapacitor electrode can be analyzed. For ideal supercapacitor, its CV curve may show rectangle shape, but in the real examination process, the electrolyte ion may be prevented by the migration force and the polarized resistance is produced to make the CV curve of supercapacitor a little difference from the ideal rectangle. It can be seen from Fig. 4a that the CV curves of supercapacitor show the regular rectangle-like shape, and there is no obvious oxidation or reduction peak in these curves, indicating that the energy is stored on the basis of the electric double layer. Fig. 4b shows the relationship of peak current and scan rate, from which it can be seen that in the stable electrochemical window of 0~2.5V, with the increase of scan rate, the peak current increases linearly, which indicates that the scan rate has no influence on the capacitance of carbon electrode, and the supercapacitor shows the ideal reversibility and the capacitance behavior.

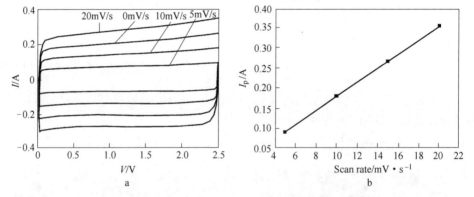

Fig. 4 CV properties of pitch carbon supercapacitor at different scan rates
a—CV curves at different scan rates; b—Relationship between peak current(I_p) and scan rate

3.3 Charge and discharge process of pitch carbon based supercapacitor

Fig. 5 shows the charge and discharge curve of pitch carbon supercapacitor at the current of 0.6A, and the maximum voltage is 2.5V. It can be seen from Fig. 5 that the voltage may change linearly with the change of charging time on the condition of keeping the current stable, which indicates that the reaction on the surface is mainly reaction of the movement of ion. The mirror-like curve indicates the ideal capacitance property of prepared supercapacitor, which is in accordance with the result observed from the CV curve, and the capacitance of carbon-based supercapacitor calculated by the slope of the curve is 14.7F, and the ESR is 60mΩ, the specific capacitance of activated carbon is 99.6F/g.

Fig. 6 shows the discharge property of pitch carbon supercapacitor at different current densities. It can be seen from Fig. 6a that the curves of the discharge still show the obvious linear

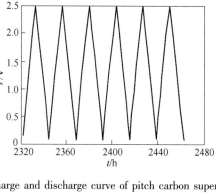

Fig. 5 Charge and discharge curve of pitch carbon supercapacitor

shape when the current increases by eight times, which indicates that the increase of current does not cause the polarized phenomenon of electrode. Meanwhile, as shown in Fig. 6b, when the discharge current increases from 0.2 to 1.6A, there is no obvious capacitance decay. The energy density of this supercapacitor is 2.96W·h/kg at the current of 1.6A, which indicates that the supercapacitor has an ideal power property and is suitable for the large current discharge process.

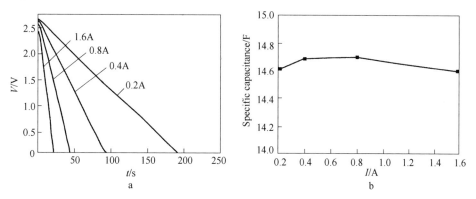

Fig. 6 Discharge properties of pitch carbon supercapacitor at different current densities
a—Discharge curve at different current densities; b—Capacitance decay at different charge current

3.4 Examination of cycle property and current leakage of pitch carbon based supercapacitor

Fig. 7 illustrates the charge and discharge curve of activated carbon electrode over 5000 cycles, and the charge and discharge current is 200mA. In the first 500 cycles, the specific capacitance of active carbon electrode shows a little decay behavior. After 500 cycles, the decay rate of specific capacitance is very slow. Compared with the capacitance of 500th cycle, the capacitance decay rate is less than 3% because energy, different from the rechargeable lithium ion battery, is stored by electric double layer. There is not the intercalation of ion into the electrode materials that may cause the volume change, the supercapacitor has the advantage of desired cycle property.

Leakage current is an important parameter to analyze the property of supercapacitor. There are many reasons for the current leakage: the reaction of impurity ion on the electrode, the micro-contact of the electrode, and the diffusion of electrolyte ion on the electrode surface because of the concentration gradient. Fig. 8 shows the curve of leakage current at 2.5V, from which it can be seen that the current leakage is very low and the current leakage value is less than 0.2mA after keeping the voltage at 2.5V for 1h.

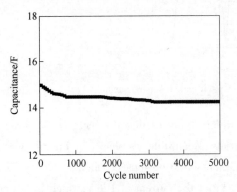
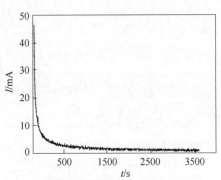

Fig. 7 Charge and discharge curve of activated carbon electrode

Fig. 8 Examination of leakage current of supercapacitor

3.5 Examination of temperature property of pitch carbon based supercapacitor

Temperature property is also an important property for supercapacitor. In the operating process of supercapacitor, the environment temperature may vary greatly, which requires that the supercapacitor has the good electric property in the different temperature conditions. Fig. 9 shows the change of capacitance and ESR of pitch carbon supercapacitor with the variation of temperature. It can be seen from Fig. 9a that the capacitance keeps stable in the range from $-30°C$ to $50°C$. The capacitance has the tendency to decay when the temperature rise to $70°C$. This is because with the increase of temperature, the electrolyte ion of electric double layer has the tendency to move from the electric double layer. For the same reason, at the low temperature, because the heating movement becomes slow, and the electric double layer is relatively stable, the capacitance of pitch carbon supercapacitor at low temperature is relatively high.

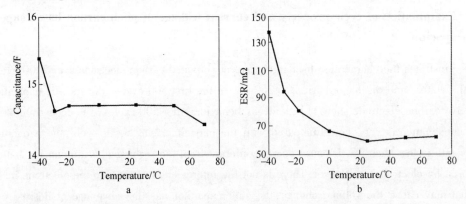

Fig. 9 Temperature properties of pitch carbon supercapacitor

a—Specific capacitance as function of temperature; b—Equivalent series resistance as function of temperature

It can be seen from Fig. 9b that the ESR decreases with the increase of the temperature in the temperature range from −40℃ to 25℃ because the movement of ion may speed up at high temperature and the polarized resistance of electrode is reduced. But with the further increase of temperature, the heating movement of electrolyte ion in disorder direction may influence the effective movement of electrolyte ion under the force of the electric field, and reduce the effect of low ion polarized resistance caused by enhancing ion movement rate. Therefore, in the temperature ranging from 25℃ to 70℃, the ESR of pitch carbon supercapacitor decreases a little with the increase of temperature. But in the temperature ranging from −40℃ to 70℃, the pitch carbon supercapacitor shows obvious capacitance behavior, the shape of supercapacitor is not changed, and there is also no damage or electrolyte leakage phenomenon during the cycle process.

4 Conclusions

(1) Using the mesophase pitch as precursor, KOH and CO_2 as activated agent, the activated carbon electrode material was fabricated by physical-chemical combined activated technique for supercapacitor. The carbon materials have the high surface area and ideal pore structure, which is suitable for engineering preparation of supercapacitor.

(2) Using the pitch carbon as electrode material, the cylinder supercapacitor was prepared, and the size of case is d 12.5mm×28mm, the capacitance of supercapacitor is 14.7 F in the electrolyte of $BF_4(CN)_4/AN$, and ESR is 60mΩ. The specific capacitance of activated carbon material is 99.6F/g and the prepared pitch carbon supercapacitor shows ideal capacitance behavior.

(3) In the current ranging from 0.2 to 1.6A, pitch carbon supercapacitor shows ideal power discharge behavior, and its capacitance is almost unchanged, the energy density of this supercapacitor is 2.96W·h/kg at the current of 1.6A. The current leakage is less than 0.2mA after keeping stable voltage of 2.5V for 1h. After 5000 cycles, the capacitance decay is less than 3%, compared with the capacitance of 500th cycle.

(4) The prepared supercapacitor shows ideal temperature properties, which can be operated at the temperature range of −40℃ to 70℃. The capacitance of supercapacitor may increase with the increase of temperature, and the ESR is relatively high at low temperature.

References

[1] Kötz R, Carlen M. Principle and application of electrochemical capacitors [J]. Electrochimica Acta, 2000, 54(11): 2483-2498.

[2] Bonnefoi L, Simon P, Fauvarque J F, et al. Electrode optimisation for carbon power supercapacitors [J]. Power Source, 1999, 87(2): 1113-1119.

[3] Wang Xiaofeng, Ruan Dianbo, Wang Dazhi. Hybrid electrochemical supercapacitors based on polyaniline and activated carbon electrodes[J]. Acta Phys Chim Sin, 2005, 21(3): 261-266.

[4] Lai Yanqing, Li Jing, Li Jie, et al. Preparation and electrochemical characterization of C/PANI composite electrode materials [J]. Journal of Central South University of Technology, 2006, 13(4): 353-359.

[5] Li Jing, Li Jie, Lai Yanqing, et al. Influence of KOH activation techniques on pore structure and electrochemical property of carbon electrode materials[J]. Journal of Central South University of Technology,

2006,13(4):360-366.

[6] Vix-guterl C,Saadallah S. Supercapacitor electrodes from new ordered porous carbon materials obtained by a template procedure [J]. Materials Science and Engineering B,2004,B108(2):148-155.

[7] Wang Yonggang, Zhang Xiao-gang. Preparation and electrochemical capacitance of RuO_2/TiO_2 nanotubes composites [J]. Electrochemical Acta,2004,49(12): 1957-1962.

[8] Pedro G R,Malgorzata C,Karina C G,et al. Hybrid organic inorganic nanocomposite materials for application in solid-state electrochemical supercapacitors [J]. Electrochemistry Communications, 2003, 5(2): 149-153.

[9] Conway B F. Transition from "supercapacitor" to "battery" behavior in electrochemical energy storage [J]. Journal of Electrochemical Society,1991,6(1):1439-1448.

[10] Passerini V,Vidakovic T,Dekanski A,et al. The properties of carbon supported hydrous ruthenium oxide obtained from RuO_xH_y sol[J]. Electrochimica Acta,2003,48(25/26):3805-3813.

[11] Li Wencui,Reichenauer G,Fracke J,et al. Carbon aerogel derived from cresol- resorcinol-formaldehyde for supercapacitor[J]. Carbon,2002,40(12):2955-2959.

[12] Frackowiak E,Jurewicz K,Delpeux S,et al. Nanotubular materials for supercapacitor[J]. Journal of Power Source,2001,98(1):822-825.

[13] Kim Y J,Horie Y,Matsuzawa Y,et al. Structure features necessary to obtain a high specific capacitance in electric double layer capacitor [J]. Carbon,2004,42(12/13):2423-2432.

[14] Qu De-yang. Studies of the activated carbon used in double-layer supercapacitors [J]. Journal of Power Source,2002,109(1):403-411.

[15] Portet C,Taberna P L,Simon P,et al. Modification of Al current collector surface by sol-gel deposit for carbon-carbon supercapacitor applications[J]. Electrochimica Acta,2004,49(9):905-912.

[16] Lozano-castell D,Lillo-R Denas M A,Cazorla-amorós D,et al. Preparation of activated carbons from Spanish anthracite Ⅰ:Activation by KOH[J]. Carbon,2001,39(5):741-749.

[17] Lillo M A,Juanjuan J,Cazorlaamorós D,et al. About reaction occurring during chemical activation with hydroxides[J]. Carbon,2004,42(7):1365-1369.

Cyclic Voltammetry Study of Electrodeposition Cu(In, Ga)Se$_2$ Thin Films[*]

Abstract The electrodeposition of Cu(In, Ga)Se$_2$ has been investigated by cyclic voltammetry (CV) in a DMF-aqueous solution containing citrate as complexing agent. The effects of the citrate ion on the reduction potentials of Cu^{2+}, In^{3+}, Ga^{3+} and H$_2$SeO$_3$ were examined in unitary system. For better understanding the electrodeposition behaviour, cyclic voltammetry study was performed in ternary Cu-In-Se, quaternary Cu-In-Ga-Se system, and binary Cu-Se, In-Se and Ga-Se system as well. It has been shown that the insertion of In and Ga into solid phase may proceed by underpotential deposition mechanism involving two different routes: In^{3+} and Ga^{3+} reduction by surface-induced effect of Cu$_3$Se$_2$, and/or reaction with H$_2$Se.

1 Introduction

Cu(In, Ga)Se$_2$(CIGS) photovoltaic material is one of the most promising candidates in developing low-cost and high-efficiency thin film solar cells[1]. Cu(In, Ga)Se$_2$ has an adjustable band gap about in the range 1.05~1.67eV, which is within the maximum solar absorption region, and a large optical absorption coefficient(10^5cm^{-1}), which results from a direct energy gap and permits thin films with the thickness of only 1~2μm[2]. CIGS solar cells have already surpassed the conversion efficiency of 19.5% based on a multistep process using physical vapor deposition(PVD)[3]. The PVD technology is excellent for good quality film growth, but difficult to scale up. Currently, a great deal of effort is directed to a large-scale, high-quality and low-cost technology for preparing CIGS thin films. Electrodeposition is highly suitable to achieve that goal[4]. The record efficiency of 11.3% is reported for a cell using electrodeposition technique[5].

Cu(In, Ga)Se$_2$ thin films have been electrodeposited mainly employing various process routes: one-step electrodeposition of In-Ga[6], Cu-In-Ga[7] or Cu-In-Ga-Se[8-13], and sequential electrodeposition of individual metal[14] or alloy[15] films. As-deposited films often need a post-annealing under reactive or insert atmosphere to drive CIGS formation reactions and film recrystallization. In order to simplify film manufacture and avoid pollution, the most desirable way is one step co-electrodeposition of Cu-In-Ga-Se simultaneously, which can offer the possibility of post-annealing in a atmosphere without any harmful reagents like H$_2$Se or Se. However, the co-electrodeposition of quaternary Cu-In-Ga-Se is rather difficult due to the wide difference in deposition potential for each constituent. Only adjusting the pH and electrolyte concentration can

---

[*] Copartner: Lai Yanqing, Liu Fangyang, Zhang Zhian, Liu Jun, Li Yi, Kuang Sanshuang, Li Jie. Reprinted from Electrochimica Acta, 2009, 54: 3004-3010.

not change the deposition potentials greatly, but may introduce significant concentration polarization. Adding complexing agent in the bath is one of the most effective methods to overcome the bottleneck by bringing the potentials of the four elements closer. The most attractive complexing agents is citrate ion[16], which is found to form complexes with both Cu^{2+} and $HSeO_2^-$ ions, but has no effect on In^{3+} in the electrodeposition of $CuInSe_2$. In addition, citrate is a non-toxic and easily available compound. L. Zhang et al.[17] have reported the co-electrodeposited of CIGS precursors in a citrate bath containing as high as 23 atom% Ga. K. Bouabid et al. electrodeposited CIGS films from a citric acid bath with the band gap varying between 1.01 and 1.26eV by increasing the Ga content[18] and similar morphology and structure to those prepared by selenization of sputtered alloy films[19]. D. L. Xia et al.[20] have also reported recently the co-electrodeposition of CIGS films with the ratio of Ga/(Ga+In) from 0.24 to 0.35 in a citrate bath.

Although several groups have investigated the electrodeposition mechanism of $CuInSe_2$ in citrate bath[21,22], there are no reports yet on electrodeposition mechanism of $Cu(In,Ga)Se_2$ electrodeposition in this bath system, as far as the authors know. For better understanding the electrochemical behavior during the co-deposition of Cu, In, Ga and Se in a citrate bath, a systematic cyclic voltammetric study was undertaken on a SnO_2-coated soda-lime glass electrode. The well-known Mo-coated soda-lime glass electrode for device fabrication is not adopted here due to the oxidation of Mo when CV measurement was carried out at the positive potential[21,23]. In order to assist identifying the reaction mechanisms, the composition, structure and morphology of deposited films under different conditions were also studied.

2 Experimental

Electrodeposition was carried out potentiostatically in a three-electrode cell configuration where the reference electrode was a saturated calomel electrode (SCE), the counter electrode was a platinum plate, and the working electrode was SnO_2-coated soda-lime glass substrates. All substrates were cleaned with acetone and deinonized water ultrasonically and dried. The electrolyte bath contained 2~5mM $CuCl_2$, 20mM $InCl_3$, 30mM $GaCl_3$, 6~10mM H_2SeO_3, 200mM Na-citrate($Na_3C_6H_5O_7$), 100mM KCl and 100mM NH_4Cl dissolved in 50 vol% deionized water and 50 vol% N,N-dimethylformamide(C_3H_7NO, DMF) solvent mixtures. The pH 2.0 of the solution was adjusted by adding drops of concentrated HCl. Oxygen in the bath was removed by N_2 bubbling for 20min prior to deposition.

An EG&G Princeton Applied Research potentiostat/galvanostat Model 273A was used for the film depositions and cyclic voltammetry studies. The typical cyclic voltammograms were measured at a scan rate of 10mV/s and were scanned first to the negative direction. All film depositions and cyclic voltammogram measurements were performed at room temperature and without stirring.

The surface morphology, chemical composition and crystalline properties of the electrodeposited films were characterized by scanning electron microscopy (SEM, JSM-6360LV), energy dispersive X-ray spectroscopy (EDS, EDAX-GENSIS60S) and X-ray diffraction (XRD, Rigaku3014), respectively.

3 Results and Discussion

3.1 Cyclic voltammograms of unitary Cu, In, Ga and Se system

Fig. 1 illustrates the effect of DMF on the potentials of hydrogen evolution by comparing the cyclic voltammograms of the base solutions of water solvent and 50 vol% water+50 vol% DMF mixture solvent. Both base solutions were adjusted to pH=2.0, and contained 100mM KCl and 100mM NH_4Cl. As seen in the figure, the peaks of H^+ reduction to H_2 are at −0.64V and −0.74V in water and 50 vol% water+50 vol% DMF base solution respectively. This variation is due to the fact that adding of DMF, which is more inert and has larger molecule volume than H_2O, brings on the decrease of concentration of H^+ in electric double layer. It also means that the 50 vol% water+50 vol% DMF base solution has wider electrochemical window, which can weaken the destructive effect on electrodeposited films caused by hydrogen evolution when applying more negative deposition potential for Ga electrochemical doping. So below, all of the electrodepositions and cyclic voltammetry studies is carried out in 50 vol% water+50 vol% DMF base solution.

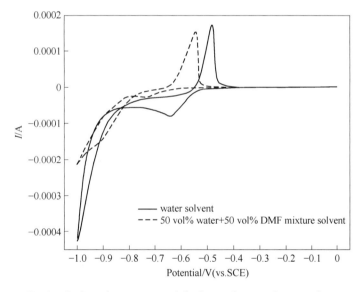

Fig. 1 Cyclic voltammograms of the base solutions of water solvent and 50 vol% water+50 vol% DMF mixture solvent

Fig. 2 shows the changes in cyclic voltammograms when 200mM Na-citrate is added to a $CuCl_2$ solution. There is a negative shift from −0.03V to −0.13V of the Cu^{2+} reduction peak (Eq. (1)) and a decrease of its maximum value. This result indicates that the citrate anion and copper cation form complex compound which results in the drop of copper cation activity in solution. It also shows that citrate can increase the hydrogen evolution overpotential slightly by shifting H^+ reduction peak towards negative direction. Moreover, it's important to remark that the peak potential of hydrogen evolution is more positive than that in Fig. 1, which suggests hydrogen evolution overpotential on SnO_2 electrode surface is higher than on electrode surface covered with Cu nucleation sites.

$$Cu^{2+} + 2e \longrightarrow Cu \tag{1}$$

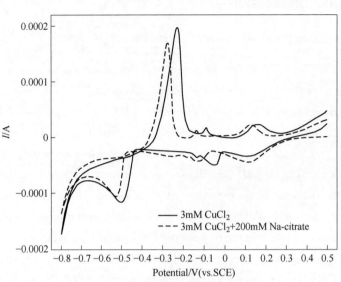

Fig. 2 Cyclic voltammograms of 3mM $CuCl_2$
and 3mM $CuCl_2$ + 200mM Na-citrate

Fig. 3 compares the cyclic voltammograms of $InCl_3$ in solution with and without 200mM Na-citrate. As seen in the figure, the cyclic voltammogram of $InCl_3$ is not significantly affected by Na-citrate. The reduction potential of In^{3+} is almost the same in both solution. The first reduction peak observed between $-0.65V$ and $-0.75V$ is attributed to the reduction of H^+ to H_2. The second peak between $-0.75V$ and $-0.88V$ probably corresponds to the In^{3+} reduction to In according to Eq. (2). Citrate does not shift the In^{3+} reduction potential obviously due to a weak complexation of citrate and In^{3+}.

$$In^{3+} + 3e \longrightarrow In \tag{2}$$

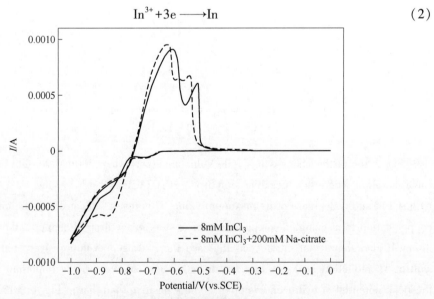

Fig. 3 Cyclic voltammograms of 8mM $InCl_3$
and 8mM $InCl_3$ + 200mM Na-citrate

In Fig. 4 cyclic voltammograms of GaCl$_3$ obtained in solution with and without 200mM Na-citrate are presented. It is seen that the reduction peak of H$^+$ to H$_2$ is between -0.75V and -0.85V. A small negative shift of H$^+$ reduction peak is observed again after adding 200mM Na-citrate. No reduction peak of Ga^{3+} to Ga is found throughout the potential scanning region in both cyclic voltammograms. There is also no indication of gallium electrodeposition on electrode surface in the course of both cyclic voltammetry measurements. It means that the electrochemical insertion of gallium is very difficult to achieve in this electrolyte bath due to so negative Ga reduction potential. This may be why the formation of the quaternary CIGS phase by one step electrodeposition is not reported for long time.

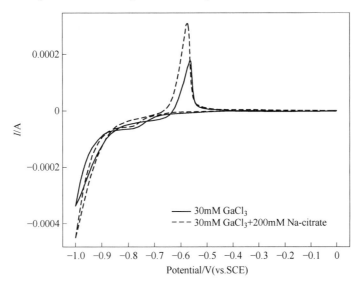

Fig. 4 Cyclic voltammograms of 8mM InCl$_3$
and 8mM InCl$_3$+200mM Na-citrate

Fig. 5 compares the cyclic voltammograms of H$_2$SeO$_3$ in solution with and without 200mM Na-citrate. The reduction peak of H$_2$SeO$_3$ to Se(Eq. (3)) is too weak to distinguish. But it can be inferred that citrate can shift the reduction of H$_2$SeO$_3$ significantly to the negative direction by formation of complex compound from the fact that the reduction of H$_2$SeO$_3$ to Se begins at about -0.2V and -0.4V in solution without and with 200mM Na-citrate, respectively. In addition, the peaks at about -0.6V and -0.8V in solution without and with 200mM Na-citrate respectively correspond to the reduction of H$_2$Se(Eq. (4)) probably. This difference again indicates the complexing effect of citrate on H$_2$SeO$_3$. H$_2$Se then reacts with the dissolved H$_2$SeO$_3$ to form a red Se suspension according to Eq. (5), as previously proposed by S. Massaccesi et al.[24] and K. K. Mishra et al.[25]. The oxidation peak of Se is at about the same potentials in both solutions during the reverse scan, but much weaker in solution with 200mM Na-citrate than in solution without Na-citrate, which shows that Se films electrodeposited in solution with 200mM Na-citrate is much more intense.

$$H_2SeO_3+4H^++4e \longrightarrow Se+H_2O \quad (3)$$
$$H_2SeO_3+6H^++6e \longrightarrow H_2Se+3H_2O \quad (4)$$

$$H_2SeO_3 + 2H_2Se + 6e \longrightarrow Se + 3H_2O \tag{5}$$

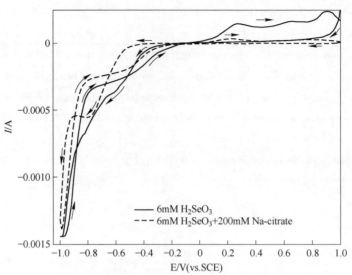

Fig. 5 Cyclic voltammograms of 6mM H_2SeO_3
and 6mM H_2SeO_3+200mM Na-citrate

By comparing the cyclic voltammograms in Fig. 2 to Fig. 5, the reduction order is seen clearly as follows: Cu^{2+} first, then H_2SeO_3, In^{3+} third, and Ga^{3+} finally. Individual reduction of Ga^{3+} is very difficult to achieve in this solution system. It is also evident that citrate shift the reduction potential of Cu^{2+} and H_2SeO_3 significantly to the negative direction, but have no obvious effect on reduction of In^{3+} and Ga^{3+}.

3.2 Cyclic voltammograms of binary Cu-Se, In-Se and Ga-Se system

Fig. 6 illustrates the typical cyclic voltammograms of binary $CuCl_2+H_2SeO_3$, $GaCl_3+H_2SeO_3$ and $InCl_3+H_2SeO_3$ in solution with 200mM Na-citrate. The cyclic voltammogram of unitary H_2SeO_3 solution is also appended into the figure as the control. Comparing with cyclic voltammograms of unitary $CuCl_2$ and H_2SeO_3 in separate citrate solutions, there are several modifications in the cyclic voltammograms of binary Cu-Se system: (1) Comparing with one anodic peak in unitary H_2SeO_3 solution, the two anodic peak in binary solution at about 0.1V and 0.2V are clearly related to the formation of copper selenides. But it's not possible to confirm all phases by XRD due to the poor film crystallinity. (2) A small positive shift of Cu^{2+} reduction peak from −0.13V in unitary $CuCl_2$ solution to −0.05V in binary solution that may be attributed to a decrease of the nucleation overpotential on a previously deposited copper selenide films. (3) Peak of H_2SeO_3 reduction to Se is at about −0.14V in binary solution while the reduction does not begin until at about −0.4V in unitary H_2SeO_3 solution. This sharp positive shift implies that the previously deposited Cu induces H_2SeO_3 reduction to Se or lowers the nucleation overpotential of Se deposition greatly. These modifications indicate that the formation of copper selenides occurs by underpotential deposition. Copper selenides is very likely to be Cu_3Se_2 supported by XRD analysis in the below section (Fig. 9). Furthermore, The large reduction peak at about −0.59V may be assigned to the H$_2$SeO$_3$ reduction to copper selenides(Eq. (6)). Whatever the copper selenide stoichiometry is, Cu$_3$Se$_2$ is still presented here as copper selenides for simplicity. Another reduction peak at about −0.63V may corresponds to the copper selenides reduction to H$_2$Se according to Eq. (7).

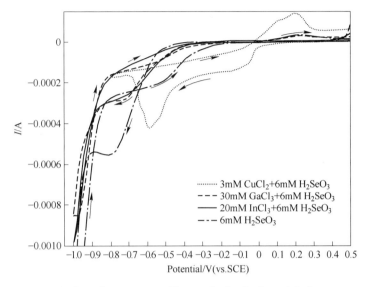

Fig. 6 Cyclic voltammograms of binary Cu-Se, Ga-Se and In-Se system

$$2H_2SeO_3+3Cu^{2+}+8H^++14e \longrightarrow Cu_3Se_2+6H_2O \qquad (6)$$
$$Cu_3Se_2+4H^++4e \longrightarrow 3Cu+2H_2Se \qquad (7)$$

Comparing the cyclic voltammograms of unitary H$_2$SeO$_3$ solution with binary GaCl$_3$+H$_2$SeO$_3$ solution, it is observed that the reduction of H$_2$SeO$_3$ to Se begins a little earlier in the latter, which reveals that the state of electrode surface has changed due to the formation of gallium selenides in previous several repetitious measurements to obtain the typical and reproducible cyclic voltammogram, and the modified surface may lower the overpotential of H$_2$SeO$_3$ reduction. Deposits prepared at potential of −0.4V containing Se but no Ga indicates that Ga^{3+} is difficult to react with deposited Se to form gallium selenides at this potential. However, Ga^{3+} is favorable to H$_2$SeO$_3$ reduction to H$_2$Se as evidenced by positive shift of its reduction peak in binary solution. This may be attributed to the fact that the consumption of H$_2$Se by reaction with Ga^{3+}(Eq. (8)) promotes the reduction of H$_2$SeO$_3$ to H$_2$Se. Combining Eq. (4) and Eq. (8) to Eq. (9), it is more clear that Ga^{3+} has promoting effect on H$_2$SeO$_3$ reduction due to large free energy of formation of gallium selenides (for instance, −418kJ/mol for Ga$_2$Se$_3$). Gallium selenides also may be formed at more negative potential according to Eq. (10), preceded by Eq. (3). In any cases the underpotential deposition of gallium as gallium selenides is apparent. From another point of view, the observation of a grey deposit with red reflection, which indicates the probable formation of a Ga-Se compound containing red Se, and the film composition of GaSe$_{6.7}$ examined by EDS(Fig. 7a) confirm formation of gallium selenides by underpotential deposition. It should be pointed out that peaks of Si and Sn etc. in EDS pattern can only be from the substrate which is also the main source of oxygen element. But it does not exclude the possibility that farthing oxygen element may exist in deposit as gallium oxide or hydroxide. The morphology of the films, from the SEM micrograph in Fig. 7b, shows that overall surface is covered by cauliflower-like grains and large-sized clusters. However, it was not possible to determine the phases of deposit by XRD due to its amorphous nature.

$$3H_2Se + 2Ga^{3+} \longrightarrow Ga_2Se_3 + 6H^+ \tag{8}$$

$$3H_2SeO_3 + 2Ga^{3+} + 12H^+ + 18e^+ \longrightarrow Ga_2Se_3 + 9H_2O \tag{9}$$

$$3Se + 2Ga^{3+} + 6e \longrightarrow Ga_2Se_3 \tag{10}$$

A similar electrochemical behavior is observed from the cyclic voltammograms of binary $InCl_3$ + H_2SeO_3 solution due to the similar solution systems and properties of the third main group In and Ga element. Again the formation of indium selenides proceeds by underpotential deposition mechanism.

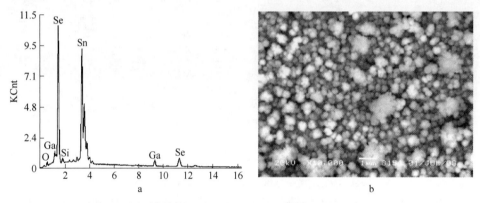

Fig. 7 Typical EDS spectra(a) and SEM morphology(b) of the film deposited at potential of −0.8V for 1800s in binary 30mM $GaCl_3$ + 6mM H_2SeO_3 solution

3.3 Cyclic voltammograms of ternary Cu-In-Se and quaternary Cu-In-Ga-Se system

Fig. 8 shows the typical cyclic voltammograms of ternary $CuCl_2$ + $InCl_3$ + H_2SeO_3 and quaternary $CuCl_2$ + $InCl_3$ + $GaCl_3$ + H_2SeO_3 in solution with 200mM Na-citrate.

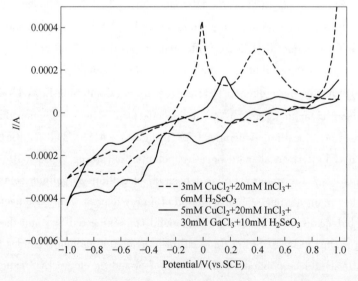

Fig. 8 Cyclic voltammograms of Cu-In-Se and Cu-In-Ga-Se system

From the cyclic voltammogram of ternary Cu-In-Se system and the change of film composition examined by EDS analysis as seen in Table 1, it is inferred that peaks at about −0.07V, −0.26V and −0.42V probably correspond to the reduction reaction of Cu, Se and In, respectively. The co-electrodeposition of Cu, In and Se occurs at more positive potential than where In^{3+} alone reduces. The formation of different phases under various potentials of electrodeposition is confirmed by XRD (Fig. 9). From Table 1, the films deposited at −0.1 ~ −0.3V contain more than 50 atom % Se, however, XRD indicates only the presence of Cu_3Se_2 crystal phase. The discrepancy in film composition may be due to the presence of amorphous elemental Se or other copper selenides. At potentials more negative than −0.3V, the insertion of In is observed by EDS, and with the potentials moving to negative, In level shows a significant increase. From XRD pattern in Fig. 9, it is observed that (101) peak intensity of Cu_3Se_2 phase decreases gradually beginning from −0.4V with the insertion of In and a weak (220) peak of $CuInSe_2$ appears at potential of −0.6V. The evolvement of (112) peak of $CuInSe_2$ can not be distinguished due to poor crystallinity of as-deposited $CuInSe_2$ film and the similar XRD peak positions to (110) peak of SnO_2. These results reveal one possible CIS deposition route which is similar to that reported by M. E. Calixto et al.[27]: the pre-deposited Cu_3Se_2 phase induces the reduction of In^{3+} to In, which is called surface-reduced phenomenon by M. C. F. Oliveira et al.[22]. The newly generated In may be assimilated into the formation of $CuInSe_2$ directly or converted to indium selenides by reaction with excess Se, which will be quickly absorbed into the formation of more stable $CuInSe_2$ on the basis of formation free energy calculation[28]. The formation of $CuInSe_2$ may proceed according to one of the following apparent reactions:

$$Cu_3Se_2 + 3In^{3+} + 4H_2SeO_3 + 16H^+ + 25e \longrightarrow 3CuInSe_2 + 12H_2O \quad (11)$$

$$Cu_3Se_2 + 3In^{3+} + 4Se + 9e \longrightarrow 3CuInSe_2 \quad (12)$$

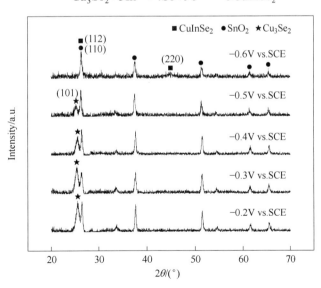

Fig. 9 XRD patterns of CIS films deposited at various potentials

Based on the analysis in unitary and binary systems, the peak at about −0.6V in Fig. 8 may be assigned to H_2SeO_3 reduction to H_2Se. The formed H_2Se_3 is expected to form In_2Se_3 at the electrode surface by reaction with In^{3+} (Eq. (13)) inferred from CV study in In-Se binary system.

$$3H_2Se + 2In^{3+} \longrightarrow In_2Se_3 + 6H \quad (13)$$

The generated In_2Se_3 also will be rapidly transformed into more stable $CuInSe_2$. Thus, another $CuInSe_2$ formation route which involves H_2Se and/or In_2Se_3 may be in existence according to one of the following reactions:

$$3In^{3+} + 4H_2Se + Cu_3Se_2 + e \longrightarrow 3CuInSe_2 + 8H^+ \quad (14)$$

$$2Cu^{2+} + In_2Se_3 + H_2SeO_3 + 4H^+ + 8e \longrightarrow 2CuInSe_2 + 3H_2O \quad (15)$$

$$2Cu^{2+} + In_2Se_3 + Se + 4e \longrightarrow 2CuInSe_2 \quad (16)$$

$$2Cu^{2+} + In_2Se_3 + H_2Se + 2e \longrightarrow 2CuInSe_2 + 2H^+ \quad (17)$$

$$2Cu_3Se_2 + 3In_2Se_3 \longrightarrow 6CuInSe_2 + Se \quad (18)$$

In both cases, the underpotential deposition mechanism of $CuInSe_2$ is definitive. In addition, it should be emphasized that it is very difficult to confirm the actual reaction pathway only by CV due to the complexity of the ternary system. Although many groups have studied the mechanism of $CuInSe_2$ formation, there is no widely accepted conclusion formed.

Following the peak at about $-0.6V$, a flat limiting current not affected by potential markedly is observed at potentials from $-0.6V$ to $-0.9V$ which indicates electrode process is mainly controlled by diffusion. Thus, the potential range of $-0.6V$ to $-0.9V$ is suitable to obtain films of stable and controllable in stoichiometry which is also supported by film composition presented in Table 1. In this potential range, the atomic ratio of Cu/In is about 1, and there is little fluctuation in relative proportion of Cu, In and Se with potential variation. At more negative potential, the cathodic current rises rapidly which indicates hydrogen reduction becomes the main reaction. Hydrogen evolution destroys the stabilization of electric double layer on electrode surface and leads to the large composition fluctuation.

Table 1 The EDS composition analysis of CIS films electrodeposited at various potential

Potential /V vs. SCE	Atomic percent/%			Cu/In ratio	Se/(Cu+In) ratio	stoichiometry
	Cu	In	Se			
-0.2	34.59	0	65.41	—	1.89	$CuSe_{1.89}$
-0.3	26.3	0	73.68	—	2.80	$CuSe_{2.80}$
-0.4	24.75	7.47	67.8	3.31	2.10	$CuIn_{0.30}Se_{2.74}$
-0.5	23.94	16.05	60.01	1.49	1.50	$CuIn_{0.67}Se_{2.51}$
-0.6	19.38	16.00	64.62	1.21	1.83	$CuIn_{0.83}Se_{3.33}$
-0.7	17.68	19.10	63.23	0.93	1.72	$CuIn_{1.08}Se_{3.58}$
-0.8	18.06	16.23	65.70	1.11	1.92	$CuIn_{0.90}Se_{3.64}$
-0.9	19.73	20.26	60.01	0.97	1.50	$CuIn_{1.03}Se_{3.04}$
-1.0	24.81	15.59	59.60	1.59	1.48	$CuIn_{0.63}Se_{2.40}$
-1.1	25.74	15.83	58.43	1.63	1.41	$CuIn_{0.61}Se_{2.27}$
-1.2	24.62	11.91	63.47	2.07	1.74	$CuIn_{0.48}Se_{2.58}$

In the cyclic voltammogram of quaternary Cu-In-Ga-Se system, a large peak between 0.2V to $-0.1V$ is associated with the simultaneous reduction of $CuCl_2$ and H_2SeO_3 to Cu_3Se_2

probably. Furthermore, the typical morphology of the deposit at −0.2V shows the needle-like structure, as seen in Fig. 10a, which is in agreement with that reported in the literature for Cu_3Se_2[26]. The rapid increase in cathodic current from −0.3V should be attributed to the beginning reduction of In^{3+} and Ga^{3+}, in combination with film composition analysis in Table 2. The peak at −0.35V and −0.45V may correspond to the reduction of In^{3+} and Ga^{3+} respectively. The co-electrodeposition of Cu, In, Ga and Se occurs at more positive potential than where In^{3+} or Ga^{3+} alone reduces. The underpotential of In and Ga is undoubted. Fig. 10b shows the deposited CIGS precursor morphology consisted of isolated grain with uniform size and well-defined boundaries. Due to the similarity of the systems, and the similarity of In^{3+} and Ga^{3+} chemical properties, it may be expected that the electrochemical mechanism of $Cu(In, Ga)Se_2$ electrodeposition will be similar to that of $CuInSe_2$ electrodeposition[12]. The pre-deposited Cu_3Se_2 phase also induces the underpotential deposition of Ga, which leads to the formation of $CuGaSe_2$. Similarly, the formation of $CuGaSe_2$ may proceed according to one of the following reactions:

$$Cu_3Se_2 + 3Ga^{3+} + 4Se + 9e \longrightarrow 3CuGaSe_2 \qquad (19)$$

$$Cu_3Se_2 + 3Ga^{3+} + 4H_2SeO_3 + 16H^+ + 25e \longrightarrow 3CuGaSe_2 + 12H_2O \qquad (20)$$

Table 2 The EDS composition analysis of CIGS films electrodeposited at various potential

Potential /V vs. SCE	Atomic percent/%				Cu/(In+Ga) ratio	Se/(Cu+In+Ga) ratio	Stoichiometry
	Cu	In	Ga	Se			
−0.1	47.4	—	—	52.6	—	1.11	$CuSe_{1.11}$
−0.2	35.71	—	—	64.29	—	1.80	$CuSe_{1.80}$
−0.3	32.67	6.86	3.27	57.19	3.22	1.33	$CuIn_{0.21}Ga_{0.10}Se_{1.75}$
−0.4	33.9	13.56	4.74	47.8	1.85	0.92	$CuIn_{0.40}Ga_{0.14}Se_{1.41}$
−0.5	37.31	13.06	4.48	45.15	2.13	0.82	$CuIn_{0.35}Ga_{0.12}Se_{1.21}$
−0.6	35.33	14.85	6.0	43.82	1.69	0.78	$CuIn_{0.42}Ga_{0.17}Se_{1.24}$
−0.7	34.13	14.68	6.14	45.05	1.64	0.82	$CuIn_{0.43}Ga_{0.18}Se_{1.32}$
−0.8	34.36	16.15	7.57	41.92	1.45	0.72	$CuIn_{0.47}Ga_{0.22}Se_{1.22}$
−0.9	32.9	20.39	7.24	39.47	1.19	0.65	$CuIn_{0.62}Ga_{0.22}Se_{1.20}$
−1.0	30.77	17.54	4.92	46.77	1.37	0.88	$CuIn_{0.57}Ga_{0.16}Se_{1.52}$

It is important to remark that reactions of Eq. (19) and Eq. (20) may occur more easily than reactions of Eq. (11) and Eq. (12) respectively, due to the larger formation energy of $CuGaSe_2$ (−224kJ/mol) than that of $CuInSe_2$ (−218kJ/mol)[29]. This may be one of the reasons why Ga insertion can be achieved at a relative positive potential of −0.3V. The formed $CuGaSe_2$ and $CuInSe_2$ are miscible in all proportions but their mixture has a positive heat of mixing that can make the alloy up to 3.3kJ/mol less stable than the ideal mixture[30]. This means that the deposit may contains $CuGaSe_2$ and $CuInSe_2$ mixture, but formation of $Cu(In, Ga)Se_2$ is unlikely.

Similar to Cu-In-Se system, the peak at about −0.7V may be assigned to H_2SeO_3 reduction to H_2Se. The generated H_2Se is expected to form Ga_2Se_3 with Ga^{3+} prior to indium selenides or copper

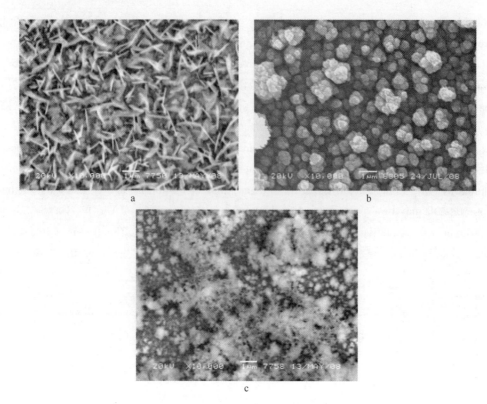

Fig. 10 Morphology of deposited CIGS precursor at various deposition potentials
a—0.2V;b—0.6V;c—0.9V

selenides on the electrode surface due to the largest formation energy of Ga_2Se_3[28]. Similarly, the formed Ga_2Se_3 trends to be quickly assimilated into the formation of more stable $CuGaSe_2$. Thus, $CuGaSe_2$ formation mechanism which involves H_2Se or Ga_2Se_3 also may be in existence similar to Eq. (14) to Eq. (18), supported by the similarities of the Cu-Ga-Se and Cu-In-Se systems. Moreover, Ga_2Se_3 may also react with $CuInSe_2$ or In_2Se_3 to form $Cu(In,Ga)Se_2$, but there is not enough evidence to support this conjecture.

These results allow us to conclude that the insertion of In and Ga proceeds by underpotential deposition mechanism which is achieved by two different routes: In^{3+} and Ga^{3+} surface-induced reduction by Cu_3Se_2 and/or reaction with H_2Se. But it is impossible to determine the reaction pathway only by cyclic voltammetry in a so complex system and confirm the formation of $Cu(In,Ga)Se_2$ by XRD due to the poor crystallinity of as-deposited films.

Similar to ternary system, the limiting current plateau appears in the range −0.77V to −0.95V where it is suitable to obtain $Cu(In,Ga)Se_2$ films of stable and controllable in stoichiometry. However, it must be noted that, when deposited under limiting current in this solution system, a dendrite structure and worsening morphology of films will occur due to the high electrode reaction rate and excessive concentration polarization at the quite negative potential, as seen in Fig. 10c.

4 Conclusions

One step electrodeposition of $Cu(In,Ga)Se_2$ thin films has been investigated by cyclic voltammetry. It is found that citrate ions presents the complexing effect on Cu^{2+} and H_2SeO_2 but no obvious effect on In^{3+} and Ga^{3+}. It is also shown that Cu_3Se_2 is formed in Cu-Se binary system by underpotential deposition where Cu is first reduced and then induces co-deposition of Se. In Ga-Se and In-Se binary system, the formation of indium and gallium selenides can proceed through reactions of Ga^{3+} and In^{3+} with H_2Se respectively. It is concluded that the insertion of In and Ga by underpotential depostion mechanism involves two different routes. One is the pre-deposited Cu_3Se_2 induces In^{3+} and Ga^{3+} reduction, leading to the formation of $CuInSe_2$ and $CuGaSe_2$. Another is that In^{3+} and Ga^{3+} may react with H_2Se leading to the formation of indium and gallium selenides which will be rapidly assimilated into the formation more stable $CuInSe_2$ and $CuGaSe_2$ respectively. However, whether $Cu(In,Ga)Se_2$ phase is formed throughout electrodeposition process cannot be confirmed. The actual formation mechanism of $Cu(In,Ga)Se_2$ will be a subject of further studies.

5 Acknowledgements

This work was supported by Key Technologies R&D Program of Hunan Province in China under grant No. 2007FJ4108.

References

[1] A. M. Hermann, R. Westfall, R. Wind, Sol. Energy Mater. Sol. Cells, 1982, 52: 355.
[2] R. N. Bhattacharya, W. Batchelor, J. E. Granata, F. Hasoon, K. Wiesner, K. Ramanathan, J. Keane, R. N. Noufi, Sol. Energy Mater. Sol. Cells, 1998, 55: 83.
[3] M. A. Contreras, M. J. Romero, R. Noufi, Thin Solid Films, 2006, 511-512: 51.
[4] R. N. Bhattacharya, J. F. Hiltner, W. Batchelor, M. A. Contreras, R. N. Noufi, J. R. Sites, Thin Solid Films, 2000, 361-362: 396.
[5] D. Lincot, J. F. Guillemoles, S. Taunier, D. Guimard, J. Sicx-Kurdi, A. Chaumont, O. Roussel, O. Ramdani, C. Hubert, J. P. Fauvarque, N. Bodereau, L. Parissi, P. Panheleux, P. Fanouillere, N. Naghavi, P. P. Grand, M. Benfarah, P. Mogensen, O. Kerrec, Solar Energy, 2004, 77: 725.
[6] J. Zank, M. Mehlin, H. P. Fritz, Thin Solid Films, 1996, 286: 259.
[7] M. Ganchev, J. Kois, M. Kaelin, S. Bereznev, E. Tzvetkova, O. Volobujeva, N. Stratieva, A. Tiwari, Thin Solid Films, 2006, 511-512: 325.
[8] T. Matsuoka, Y. Nagahori, S. Endo, Jpn. J. Appl. Phys. 1994, 33: 6105.
[9] A. Kampmann, V. Sittinger, J. Rechid, R. Reineke-Koch, Thin Solid Films, 2000, 361-362: 309.
[10] M. Fahoume, H. Boudrainel, M. Aggourl, F. Chra bi, A. Ennaoui, J. L. Delplancke, J. Phys. IV France, 2005, 123: 75.
[11] N. B. Chaure, A. P. Samantilleke, R. P. Burton, J. Young, I. M. Dharmadasa, Thin Solid Films, 2005, 472: 212.
[12] M. E. Calixto, K. D. Dobson, B. E. McCandless, R. W. Birkmire, J. Electoro. Soc., 2006, 153: G521.
[13] A. M. Fernandez, R. N. Bhattacharya, Thin Solid Films, 2005, 474: 10.
[14] A. Kampmann, J. Rechid, A. Raitzig, S. Wulff, M. Mihhailova, R. Thyen, K. Kalberlah, Mat. Res. Soc. Symp. Proc., 2003, 763: 323.
[15] R. Friedfeld, R. P. Raffaelle, J. G. Mantovani, Sol. Energy Mater. Sol. Cells, 1999, 58: 375.
[16] F. Chraibi, M. Fahoume, A. Ennaoui, J. L. Delplancke, phys. stat. sol. (a), 2001, 186: 373.
[17] L. Zhang, F. D. Jiang, J. Y. Feng, Sol. Energy Mater. Sol. Cells, 2003, 80: 483.

[18] K. Bouabid, A. Ihlal, A. Manar, A. Outzourhit, E. L. Ameziane, Thin Solid Films, 2005, 488: 62.

[19] A. Ihlal, K. Bouabid, D. Soubane, M. Nya, O. Ait-Taleb-Ali, Y. Amira, A. Outzourhit, G. Nouet, Thin Solid Films, 2007, 515: 5852.

[20] D. I. Xia, J. Z. Li, M. Xu, X. J. Zhao, J. Non-cryst. Solids, 2008, 354: 1447.

[21] L. Thouin, S. Massaccesi, S. Sanchez, J. Vedel, J. Electroanal. Chem., 1994, 374: 81.

[22] M. C. F. Oliveira, M. Azevedo, A. Cunha, Thin Solid Films, 2002, 405: 129.

[23] M. Kemell, H. Saloniemi, M. Ritala, M. Leskela, J. Electoro. Soc., 2001, 148: C110.

[24] S. Massaccesi, S. Sanchez, J. Vedel, J. Electroanal. Chem., 1996, 412: 95.

[25] K. K. Mishra, K. Rajeshwar, J. Electroanal. Chem. 1989, 271: 279.

[26] A. N. Molin, A. I. Dikusar, G. A. Kiosse, P. A. Petrenko, A. I. Sokolovsky, Yu. G. Saltanovsky, Thin Solid Films, 1994, 437: 66.

[27] M. E. Calixto, K. D. Dobson, B. E. McCandless, R. W. Birkmire, Mat. Res. Soc. Symp. Proc., 2005, 865: 431.

[28] J. F. Guillemoles, Thin Solid Films, 2000, 361-362: 338.

[29] J. P. Ao, G. Z. Sun, L. Yan, F. Kang, L. Yang, Q. He, Z. Q. Zhou, F. Y. Li, Y. Sun, Acta Phys.-Chim. Sin., 2008, 24: 1073.

[30] I. V. Bodnar, A. P. Bologa, Cryst. Res. Technol., 1982, 17: 339.

In Situ Growth of Cu$_2$ZnSnS$_4$ Thin Films by Reactive Magnetron Co-sputtering*

Abstract High-quality Cu$_2$ZnSnS$_4$(CZTS) thin films were first *in situ* grown by a reactive magnetron co-sputtering technique. Raman examination and XRD analysis indicate that the grown film shows a single CZTS phase with good crystallinity and strong preferential orientation along (112) plane. SEM analysis reveals a homogeneous, compact surface morphology and large columnar grains throughout thickness for the film. The *in situ* grown CZTS film demonstrates an optical absorption coefficient of higher than 10^4 cm^{-1} and an optical band gap of (1.52±0.01) eV. The carrier concentration, resistivity and mobility of the CZTS film are $3.9×10^{16}$ cm^{-3}, 5.4 Ω·cm, and 30 cm^2·V^{-1}·s^{-1}, respectively and the conduction type is p-type. These optical and electrical properties are suitable for a thin film solar cell fabrication.

Key words Cu$_2$ZnSnS$_4$; *in situ* growth; reactive magnetron co-sputtering; solar cell

1 Introduction

Cu$_2$ZnSnS$_4$(CZTS) is a promising new material for thin film solar cell application that contains earth-abundant elements and has a near-optimum direct band gap energy of about 1.5eV and a large absorption coefficient($>10^4$ cm^{-1})[1]. Solar cells based on CZTS absorber have achieved conversion efficiencies as high as 6.77% under AM1.5G illumination[2].

Up to now, all Cu$_2$ZnSnS$_4$ thin films reported are prepared by thermally activated processes like evaporation(including multi-stage[3] and co-evaporation[4,5]) which is very difficult to scale up, or sequential processing consisting of the deposition of the alloy precursors followed by post-annealing, where the precursors can be deposited by various methods such as atom beam sputtering[6], electron beam evaporation[7,8], RF sputtering[9,10], hybrid sputtering[11], pulsed laser deposition[12], photo-chemical deposition[13], sol-gel[14], spray pyrolysis[15-17], electrodeposition[18,19], and soft-chemistry[20]. However, for large scale production of thin film solar cells a direct one-step deposition process is needed. Fortunately, reactive magnetron co-sputtering is a well suitable technique for this task. Reactive magnetron co-sputtering can simply provide control on film composition at a relatively low cost by adjusting the ratio of the powers applied to the targets and is suitable for large-area, continuous and multicomponent film deposition. This method is already an established technique for the preparation of thin films for magnetic, optical, and contact applications[21,22].

In this work, therefore, we first present a reactive magnetron co-sputtering technique for the *in situ* growth of Cu$_2$ZnSnS$_4$ thin films and the compositional, structural, optical, and electrical properties of the films have been characterized.

---

* Copartner: Fangyang Liu, Yi Li, Kun Zhang, Bo Wang, Chang Yan, Yanqing Lai, Zhian Zhang, Jie Li. Reprinted from solar Enger Materials and solar cells. 2018.

2 Experimental

The depositions of the Cu_2ZnSnS_4 films on soda-lime glass substrates have been performed in a reactive magnetron co-sputtering system with three sputtering targets of 60mm in diameter and 5mm in thickness. Prior to film deposition, the sputter chamber was evacuated to approximately 10^{-4} Pa. The substrates were rinsed in acetone, methanol and deionized water ultrasonically, then dried with nitrogen gas and loaded in the reaction chamber. The distance between target and substrate was about 120mm and the substrate holder rotated at a frequency of 4rev./min. H_2S (purity:98.0%) was solely introduced as both reactive and working gas during the sputter deposition with a flow of 40 sccm, and the sputtering pressure was controlled to 1 Pa. The sputtering source was operated by using metallic copper, zinc and tin targets and the magnetron plasmas were excited independently by three dc power supplies, respectively. During the deposition, the substrate was heated to temperatures of 500℃. The growth time was 30min.

The surface morphology, chemical composition and crystalline properties of the films were analyzed by scanning electron microscopy(SEM, JSM-6360LV), energy dispersive X-ray spectroscopy(EDS, EDAX-GENSIS60S) and X-ray diffraction(XRD, Rigaku3014), respectively. The optical properties of the films were determined by Shimadzu UV-2450 spectrophotometer. Electrical resistivity, carrier concentration and carrier mobility of the CZTS films were measured using van der Pauw geometry and Hall effect measurement(HMS-3000/0.55T) at room temperature using indium ohmic contacts. The Raman spectra were taken by using a Jobin-Yvon LabRAM HR800-Horiba spectrometer coupled with an Olympus metallographic microscope. Excitation was provided with the 632.8nm line from a He-Ne laser due to the minimization of thermal effects achieved with this wavelength. The laser spot on the sample is about 1μm and the penetration depth of this light on CZTS is estimated below 200nm.

3 Results and Discussion

The chemical composition of the deposited film determined by EDS analysis is the following: 24.69 at% Cu, 13.16 at% Zn, 12.09 at% Sn and 50.05 at% S. This almost stoichiometric but slightly Zn-rich and Cu-poor composition can lead to good optoelectronic properties according to previous reports[8,23].

Fig. 1a illustrates the typical Raman spectrum of the CZTS film *in situ* grown on soda-lime glass by reactive magnetron co-sputtering. From the spectrum, one can evidently observe a single peak at about 336cm^{-1} which corresponds to quaternary CZTS[24]. This result is in good agreement with the Raman spectrum of CZTS film reported by P. A. Fernandes et al[25]. In addition, it is evident that there are no additional peaks related to the presence of other compounds, which means that the single phase CZTS film was obtained.

To obtain further insight into the crystalline quality, the film was characterized by XRD, as shown in Fig. 1b. The sharp peak at $2\theta = 28.48°$ can be attributed to the diffraction of (112) plane of kesterite or stannite structure CZTS in combination with Raman analysis in Fig. 1a and its peak intensity was observed to be much greater than other peaks, revealing strong preferential orientation along (112) plane. However, it is very difficult to distinguish the

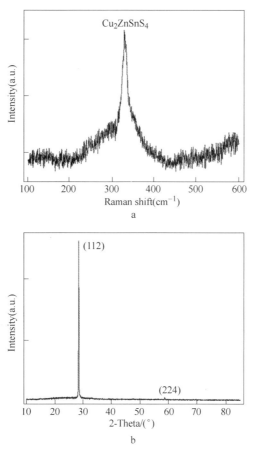

Fig. 1 Raman spectrum of the CZTS film *in situ* grown by reactive magnetron co-sputtering(a) and XRD pattern of the CZTS film *in situ* grown by reactive magnetron co-sputtering (b)

structure between kesterite and stannite structure due to the fact that the XRD patterns of these two structures differ only slightly in the splitting of high order peaks, such as (220)/(204) and (116)/(312) resulted from a slightly different tetragonal distortion ($c/2a$)[26]. Moreover, the full width at half maximum (FWHM) of the diffraction peak is very small, which indicates that the film crystallinity is fairly good. Also, no extra phases of other compounds such as binary sulfides were observed within the detection limits of XRD analysis.

Fig. 2a and b show the SEM micrographs of the surface morphology and cross-section of the CZTS film respectively. In Fig. 2a, the film shows a homogeneous, polycrystalline and extremely dense morphology without any voids, and consists of grains with uniform size of about 100nm. From the cross-section of the film as shown in Fig. 2b, it can be seen that film with a thickness of 800nm and excellent crystallinity and compactness has been grown on substrate. It should be remarked that columnar grains extending from the bottom to the top of the CZTS layer were observed. This feature is similar to that of $Cu(In,Ga)Se_2$(CIGS) films prepared by three-stage co-evaporation leading to the highest conversion efficiency of 19.9%[27], and beneficial to decreasing the minority carrier recombination during transport process when applied to solar cell.

Fig. 3a shows the optical absorption coefficient (α) of the CZTS thin films as a function of

Fig. 2　SEM images of the CZTS film *in situ* grown by reactive magnetron co-sputtering showing top view(a) and cross section(b)

photon energy ($h\nu$). The absorption coefficient is larger than $10^4 cm^{-1}$ in the visible region, which is consistent with those reported in earlier published results[6,11,14,17,19,28], and therefore, the film is considered to be a suitable material for photovoltaic solar energy conversion. Based on the allowed direct interband transition, the band gap is determined to be (1.52±0.01)eV by extrapolating the linear$(\alpha h\nu)^2$ vs. $h\nu$ plots to $(\alpha h\nu)^2 = 0$, as depicted in Fig. 3b, which is also in agreement with band gaps reported for CZTS films by other authors[6,11,14,19,28-30]. This value is quite close to the theoretical optimal value for a single-junction solar cell. Above optical characteristics indicate that the CZTS film *in situ* grown by reactive magnetron co-sputtering can be applied to the absorber layer of thin film solar cells.

The hot-probe measurement indicates that the CZTS thin film exhibits p-type conductivity. The Hall coefficient, carrier concentration, resistivity and mobility of the CZTS film tested by using Van der Paw method are +1.6×10^2 cm^3·C^{-1}, +3.9×10^{16} cm^{-3}, 5.4Ω·cm, and 30cm^2·V^{-1}·s^{-1}, respectively. The result of Hall coefficient confirms the p-type conductivity of CZTS films again. The value of carrier concentration is comparable with values for device quality CIGS[31,32] and similar to those reported by Scragg[18,30] and Katagiri[23] (10^{16} cm^{-3}), but lower than those found by Ito[6], Tanaka[11] and Zhang[33] ($10^{18} \sim 10^{19}$ cm^{-3}), who reported that the high carrier concentration might be due to the presence of Cu_xS second phase leading to the conversion film into useless degenerate semiconductor, because no second phase except for CZTS was found in

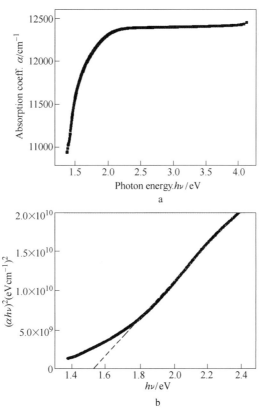

Fig. 3 Optical absorption coefficient (α) of the CZTS film *in situ* grown by reactive magnetron co-sputtering(a) and $(\alpha h\nu)^2$ as a function of photon energy($h\nu$) for CZTS film *in situ* grown by reactive magnetron co-sputtering(b)

our film from Raman and XRD data. The film shows a resistivity of a few $\Omega \cdot cm$, in agreement with the Cu-poor and Zn-rich CZTS films prepared by sulfurization of metallic precursors[34] or spray pyrolysis[15,35]. Mobility of close to $30 cm^2 \cdot V^{-1} \cdot s^{-1}$ is higher than those reported by others[6,11,33], which may be attributed to better structural quality(such as large grain throughout the film thickness, and lower defect and dislocation density accordingly). The electrical properties of the film can meet the requirements for the applications in thin film solar cells.

4 Conclusions

Cu_2ZnSnS_4(CZTS) thin films were *in situ* grown on soda-lime glass substrates by a reactive magnetron co-sputtering technique and characterized by Raman, XRD, SEM, EDS, optical transmittance and electrical measurement. The film shows a stoichiometric composition and a single CZTS phase with good crystallinity and strong preferentially oriented along(112) plane. It also has a homogeneous and dense surface morphology, large columnar grains throughout the film thickness, and an optical band gap of (1.52 ± 0.01) eV. The carrier concentration, resistivity and mobility of the CZTS film are $3.889 \times 10^{16} cm^{-3}$, $5.4 \Omega \cdot cm$, and $29.75 cm^2 \cdot V^{-1} \cdot s^{-1}$, respectively and the conduction type is p-type. These optical and electrical properties are suitable for a thin film solar cell fabrication.

References

[1] H. Katagiri, K. Saitoh, T. Washio, H. Shinohara, T. Kurumadani, S. Miyajima, Development of thin film solar cell based on Cu_2ZnSnS_4 thin films, Sol. Energy Mater. Sol. Cells 65(2001)141-148.

[2] H. Katagiri, K. Jimbo, W. Shwe Maw, K. Oishi, M. Yamazaki, H. Araki, A. Takeuchi, Development of CZTS-based thin film solar cells, Thin Solid Films 517(2009)2455-2460.

[3] A. Weber, H. Krauth, S. Perlt, B. Schubert, I. Kötschau, S. Schorr, H. W. Schock, Multi-stage evaporation of Cu_2ZnSnS_4 thin films, Thin Solid Films 517(2009)2524-2526.

[4] T. M. Friedlmeier, N. Wieser, T. Walter, H. Dittrich, H. W. Schock, Hetero-junctions based on Cu_2ZnSnS_4 and $Cu_2ZnSnSe_4$ thin films, in: Proceedings of the 14th European PVSEC and Exhibition, 1997, P4B. 10.

[5] T. Tanaka, D. Kawasaki, M. Nishio, Q. Guo, H. Ogawa, Fabrication of Cu_2ZnSnS_4 thin films by co-evaporation, Phys. Status Solidi C 3(2006)2844-2847.

[6] K. Ito, T. Nakazawa, Electrical and optical properties of stannite-type quaternary semiconductor thin films, Jpn. J. Appl. Phys. 27(1988)2094-2097.

[7] H. Katagiri, N. Ishigaki, T. Ishida, K. Saito, Characterization of Cu_2ZnSnS_4 thin films prepared by vapor phase sulfurization, Jpn. J. Appl. Phys. 40(2001)500-504.

[8] T. Kobayashi, K. Jimbo, K. Tsuchida, S. Shinoda, T. Oyanagi, H. Katagiri, Investigation of Cu_2ZnSnS_4-based thin film solar cells using abundant materials, Jpn. J. Appl. Phys. 44(2005)783-787.

[9] K. Jimbo, R. Kimura, T. Kamimura, S. Yamada, W. Maw, H. Araki, K. Oishi, H. Katagiri, Cu_2ZnSnS_4-type thin film solar cells using abundant materials, Thin Solid Films 515(2007)5997-5999.

[10] J. Seol, S. Lee, J. Lee, H. Nam, K. Kim, Electrical and optical properties of Cu_2ZnSnS_4 thin films prepared by magnetron sputtering process, Sol. Energy Mater. Sol. Cells 75(2003)155-162.

[11] T. Tanaka, T. Nagatomo, D. Kawasaki, M. Nishio, Q. Guo, A. Wakahara, A. Yoshida, H. Ogawa, Preparation of Cu_2ZnSnS_4 thin films by hybrid sputtering, J. Phys. Chem. Solids 66(2005)1978-1981.

[12] K. Sekiguchi, K. Tanaka, K. Moriya, H. Uchiki, Epitaxial growth of Cu_2ZnSnS_4 thin films by pulsed laser deposition, Phys. Status Solidi(C)3(2006)2618-2621.

[13] K. Moriya, J. Watabe, K. Tanaka, H. Uchiki, Characterization of Cu_2ZnSnS_4 thin films prepared by photo-chemical deposition, Phys. Status Solidi C 3(2006)2848-2852.

[14] K. Tanaka, N. Moritake, H. Uchiki, Preparation of Cu_2ZnSnS_4 thin films by sulfurizing sol-gel deposited precursors, Sol. Energy Mater. Sol. Cells 91(2007)1199-1201.

[15] N. Nakayama, K. Ito, Sprayed films of stannite Cu_2ZnSnS_4, Appl. Surf. Sci. 92(1996)171-175.

[16] N. Kamoun, H. Bouzouita, B. Rezig, Fabrication and characterization of Cu_2ZnSnS_4 thin films deposited by spray pyrolysis technique, Thin Solid Films 515(2007)5949-5952.

[17] Y. B. Kishore Kumar, G. Suresh Babu, P. Uday Bhaskar, V. Sundara Raja, Preparation and characterization of spray-deposited Cu_2ZnSnS_4 thin films, Sol. Energy Mater. Sol. Cells 93(2009)1230-1237.

[18] J. J. Scragg, P. J. Dale, L. M. Peter, Towards sustainable materials for solar energy conversion: Preparation and photoelectrochemical characterization of Cu_2ZnSnS_4, Electrochem. Commun. 10(2008)639-642.

[19] C. P. Chan, H. Lam, C. Surya, Preparation of Cu_2ZnSnS_4 films by electrodeposition using ionic liquids, Sol. Energy Mater. Sol. Cells 94(2010)207-211.

[20] T. Todorov, M. Kita, J. Carda, P. Escribano, Cu_2ZnSnS_4 films deposited by a soft-chemistry method, Thin Solid Films 517(2009)2541-2544.

[21] T. Unold, I. Sieber, K. Ellmer, Efficient $CuInS_2$ solar cells by reactive magnetron sputtering, Appl. Phys. Lett. 88(2006)213502.

[22] Kikuo Tominaga, Masahiro Kataoka, Haruhiko Manabe, Tetsuya Ueda, Ichiro Mori, Transparent ZnO: Al films prepared by co-sputtering of ZnO: Al with either a Zn or an Al target, Thin Solid Films, 290-291

(1996) 84-87.

[23] H. Katagiri, Cu_2ZnSnS_4 thin film solar cells, Thin Solid Films 480-481 (2005) 426-432.

[24] M. Altosaar, J. Raudoja, K. Timmo, M. Danilson, M. Grossberg, J. Krustok, E. Mellikov, $Cu_2Zn_{1-x}Cd_xSn(Se_{1-y}S_y)_4$ solid solutions as absorber materials for solar cells, Phys Stat Sol(a) 205 (2008) 167-170.

[25] P. A. Fernandes, P. M. P. Salomé, A. F. da Cunha, Growth and Raman scattering characterization of Cu_2ZnSnS_4 thin films Thin Solid Films 517 (2009) 2519-2523.

[26] Qijie Guo, Hugh W. Hillhouse, and Rakesh Agrawal, Synthesis of Cu_2ZnSnS_4 Nanocrystal Ink and Its Use for Solar Cells, J. AM. CHEM. SOC. 131 (2009) 11672-11673.

[27] Ingrid Repins, Miguel A. Contreras, Brian Egaas, Clay DeHart, John Scharf, Craig L. Perkins, Bobby To, Rommel Noufi, 19.9%-efficient $ZnO/CdS/CuInGaSe_2$ Solar Cell with 81.2% Fill Factor, Prog. Photovolt: Res. Appl. 16 (2008) 235-239.

[28] K. Tanaka, M. Oonuki, N. Moritake, H. Uchiki, Cu_2ZnSnS_4 thin film solar cells prepared by non-vacuum processing, Sol. Energy Mater. Sol. Cells 93 (2009) 583-587.

[29] S. M. Pawar, A. V. Moholkar, I. K. Kim, S. W. Shin, J. H. Moon, J. I. Rhee, J. H. Kim, Effect of laser incident energy on the structural, morphological and optical properties of Cu_2ZnSnS_4 (CZTS) thin films Current Applied Physics 10 (2009) 565-569.

[30] J. J. Scragg, P. J. Dale, L. M. Peter, Synthesis and characterization of Cu_2ZnSnS_4 absorber layers by an electrodeposition-annealing route, Thin Solid Films 517 (2009) 2481-2484.

[31] D. Lincot, H. G. Meier, J. Kessler, J. Vedel, B. Dimmler, H. W. Schock, Photoelectrochemical study of p-type copper indium diselenide thin films for photovoltaic applications, Sol. Energy Mater. 20 (1990) 67-79.

[32] C. Guillen, J. Herrero, D. Lincot, Photovoltaic activity of electrodeposited p-$CuInSe_2$/electrolyte junction, J. Appl. Phys. 76 (1994) 359-362.

[33] J. Zhang, L. X. Shao, Cu_2ZnSnS_4 thin films prepared by sulfurizing different multilayer metal precursors, Sci China Ser E-Tech Sci 52 (2009) 269-272.

[34] H. Katagiri, N. Sasaguchi, S. Hando, S. Hoshino, J. Ohashi, T. Yokota, Preparation Cu_2ZnSnS_4 thin films by and evaluation of sulfurization of E-B evaporated precursors, Sol. Energy Mater. Sol. Cells 49 (1997) 407-414.

[35] Y. B. Kishore Kumar, G. Suresh Babu, P. Uday Bhaskar, V. Sundara Raja, Effect of starting-solution pH on the growth of Cu_2ZnSnS_4 thin films deposited by spray pyrolysis, Phys. Status Solidi A 206 (2009) 1525-1530.

Electrodeposition of Cobalt Selenide Thin Films*

Abstract Cobalt selenide thin films have been prepared onto tin oxide glass substrates by electrodeposition potentiostatically from an aqueous acid bath containing H_2SeO_3 and $Co(CH_3COO)_2$ at a temperature of 50℃. The electrodeposition mechanism was investigated by cyclic voltammetry. The morphological, compositional, structural and optical properties of the deposited films have been studied using scanning electron microscopy(SEM), energy dispersive X-ray spectroscopy(EDS), X-ray diffraction(XRD) and optical absorption techniques, respectively. The formation of cobalt selenide was confirmed to proceed via under potential deposition mechanism. Se-rich CoSe thin films with compact and homogeneous morphology and hexagonal crystal structure were obtained at a deposition potential of −0.5V vs. saturated calomel electrode(SCE). The electrodeposited CoSe film exhibits an optical absorption coefficient of higher than $10^5 cm^{-1}$ and an optical band gap of $(1.53±0.01)$ eV.

Key words Cobalt selenide; Electrodeposition; Cyclic voltammetry; Under potential deposition; Solar cells

1 Introduction

Late transition metal selenides have received considerable attention in the last few years due to their unusual structures and electronic properties[1,2]. These materials, in thin film form, have found many applications such as in solar cells[3], light emitting devices[4], catalysts[5], superionic conductor[6], etc. Several methods have been attempted to prepare these selenides thin films: molecular beam epitaxy[7], metal organic chemical vapor deposition[8], evaporation[9], chemical bath deposition[10], electrodeposition[3,11], and spray pyrolysis[12]. Compared with the other methods, electrodeposition has numerous advantages[3,13], including: (a) a low-cost, high-rate process involving very simple and inexpensive equipment; (b) a large-area, continuous, multicomponent, low-temperature deposition method; (c) deposition of films on a variety of shapes and forms; (d) no use of toxic gases, effective material use and minimum waste generation(solution can be recycled). Numbers of attempts have been made for the electrodeposition of late transition metal selenides thin films such as Ni_xSe[11], Fe_xSe[3,14], $ZnSe$[15] and Cu_xSe[16]. However, the preparation of cobalt selenide by this effective, low-cost technique has not been reported.

In this work, cobalt selenide thin films were prepared by electrodeposition from aqueous solution. The results of investigation of the deposition mechanism by cyclic voltammetry, as well as film composition, morphology, structure and optical properties were presented.

2 Experimental

The electrochemical experiments, including cyclic voltammetric (CV) and electrodeposition,

---

* Copartner: Fangyang Liu, Wangbo, Yanqing Lai, Jie Li, Zhian Zhang. Reprinted from Journal of Electrochemical Society. 2012, 157(10): D523-D527.

were carried out in a three-electrode cell configuration with a SnO$_2$-coated glass substrate (20Ω/sq) as working electrode, a purity graphite plate as counter electrode, and a saturated calomel electrode (SCE) as reference. All potentials are reported with respect to this reference. All substrates were ultrasonically cleaned with acetone and rinsed with deionized water (18.2MΩ·cm^{-1}), and subsequently dried. The electrolyte solution consists of 2mM H$_2$SeO$_3$, 2mM Co(CH$_3$COO)$_2$ and 100mM LiCl. The pH of the solution was adjusted to 2.0 using concentrated HCl. Residual oxygen in the bath was removed by bubbling N$_2$ for 20min prior to each experiment. A Princeton Applied Research 2273A potentiostat was used for all electrochemical experiments. The cyclic voltammograms were measured at a scan rate of 10mV/s and were first scanned in the negative direction. All experiments were performed in a stagnant bath at 50℃.

The chemical composition, surface morphology and crystalline properties of the prepared films were characterized by energy dispersive X-ray spectroscopy (EDS, EDAX-GENSIS60S), scanning electron microscopy (SEM, JSM-6360LV) and X-ray diffraction (XRD, Rigaku3014), respectively. The optical properties of the films were determined by Shimadzu UV-2450 spectrophotometer.

3 Results and Discussion

Fig. 1 shows the cyclic voltammograms for SnO$_2$ electrode in 100mM LiCl, pH 2.0 solution in presence of 2mM Co(CH$_3$COO)$_2$ and that corresponding to the blank solution. For blank solution, it is observed that there are one cathodic peak at about −0.76V and one anodic peak at about −0.41V, which can be assigned to a diffusion limited reduction of protons and hydrogen oxidation (Eq. (1)) respectively[17,18]. For Co(CH$_3$COO)$_2$ solution, the appreciably negative shift of the cathodic peak, suggesting that hydrogen evolution is somewhat inhibited, may be due to the decrease of H$^+$ concentration near cathode surface caused by competitive adsorption of Co^{2+} and H$^+$ ion. After this cathodic peak, there is a gradual increase in current with further shifting potential negatively up to a particular potential (−0.82V in our experiment) at which the current shows a much steeper increase abruptly than that in blank solution. Therefore, this particular potential is considered as the deposition potential of cobalt according to Eq. (2)[19]. At potentials equal to, or below this deposition potential, both deposition of cobalt and evolution of hydrogen occur simultaneously. Anodic branch of cyclic voltammogram also can provide information about the deposit formed. The anodic peak located at about −0.48V relates to hydrogen oxidation, as already shown in cyclic voltammogram of the blank solution. Another anodic peak at −0.10V with a shoulder at about −0.38V which has been frequently observed by others[20-23], can be associated either with the dissolution of cobalt from electrode to the solution in form of different ionic species or with the dissolution of different cobalt phases, previously formed during the cathodic scan[20,21,24]. It is important to remark that the current remains cathodic upon the sweep reversal, crossing over the current recorded during the negative sweep at −0.17V. This current loop is attributed to the fact that the deposition of metals onto its own, in this case cobalt on cobalt, occurs at lower overpotentials that the deposition of metal onto different nature substrates, in this case cobalt on SnO$_2$[25].

$$2H^+ + 2e \longrightarrow H_2 \qquad (1)$$
$$Co^{2+} + 2e \longrightarrow Co \qquad (2)$$

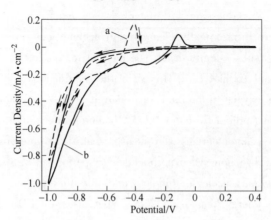

Fig. 1 Cyclic voltammograms for SnO_2 electrode at 10mV/s scan rate and 50℃ in 100mM LiCl, pH2. 0 solution with(curve a) and without(curve b)2 mM $Co(CH_3COO)_2$

Fig. 2 illustrates the cyclic voltammograms for SnO_2 electrode in 100mM LiCl, pH 2. 0 solution in presence of 2mM H_2SeO_3 and that corresponding to the blank solution. For H_2SeO_3 solution, the curve displays two cathodic peaks at −0. 36V and −0. 65V respectively. In combination with previous studies[26-32], the assignment of these two peaks is as follows: the weak cathodic peak at potential of about −0. 36V corresponds to bulk selenium deposition through four-electron reduction of Se(Ⅳ) to Se(0) proceeded by Eq. (3). Another cathodic peak at about −0. 65V corresponds to six-electron reduction of Se(Ⅳ) to Se(−Ⅱ) according to Eq. (4). The charge involved in peak at −0. 36V much smaller than that involved in peak at −0. 65, also contrasts with a four-electron process and a six-electron process respectively. The product Se(−Ⅱ) then undergoes a conproportionation reaction with Se(Ⅳ) in solution(Eq. (5)), leading to the chemical formation of Se(0)[29,32]. This process of selenium deposition according to the reactions Eq. (4) and Eq. (5) looks like the four-electron reduction(Eq. (3)) when the concentration of H_2SeO_3 is high enough. It is also observed from the inset in Fig. 2 that, there is an initial reductive feature before peak at −0. 36V in the potential range from 0. 15V to −0. 05V identified from a very weak reduction current, similar to that reported by other groups[31,32] and us[33]. This weak reduction current may correspond to the four-electron predeposition of selenium on SnO_2 substrate before the over potential deposition of bulk selenium caused by deposit-substrate interaction according to Eq. (3)[29-33]. No anodic peaks are seen although the anodic current is detected obviously because the potential is not positive enough for development into an anodic peak for selenium oxidation.

$$H_2SeO_3 + 4H^+ + 4e^- \longrightarrow Se + 3H_2O \qquad (3)$$
$$H_2SeO_3 + 6H^+ + 6e^- \longrightarrow H_2Se + 3H_2O \qquad (4)$$
$$H_2SeO_3 + 2H_2Se \longrightarrow Se + 3H_2O \qquad (5)$$

Fig. 3 presents the cyclic voltammograms for SnO_2 electrode in 100mM LiCl+2mM $Co(CH_3COO)_2$+2mM H_2SeO_3, pH 2. 0 solution. Additionally, a voltammogram corresponding

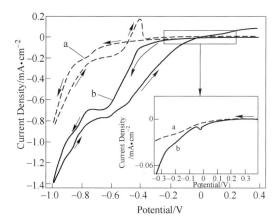

Fig. 2 Cyclic voltammograms for SnO$_2$ electrode at 10mV/s scan rate and 50℃ in 100mM LiCl, pH2.0 solution with(curve a) and without(curve b) 2mM H$_2$SeO$_3$

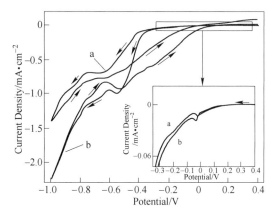

Fig. 3 Cyclic voltammograms for SnO$_2$ electrode at 10mV/s scan rate and 50℃ in 100mM LiCl+2mM H$_2$SeO$_3$, pH 2.0 solution with(curve a) and without(curve b) 2mM Co(CH$_3$COO)$_2$

to 100mM LiCl+2mM H$_2$SeO$_3$, pH 2.0 solution is shown for the sake of comparison. The initial reductive feature beginning from 0.15V and corresponding to the predeposition of selenium is observed again from the inset in Fig. 3. However, the current of this reductive feature in binary Co-Se system is slightly larger than that in unitary 2mM H$_2$SeO$_3$ solution, and this increase in current continues to extend to four-electron bulk Se deposition potential region. The small increase in current can be reasonably attributed to the contribution of additional reduction reaction with very slow deposition kinetics, which involves cobalt forming cobalt selenide proceeded by Eq. (6). The deposition of Co thus begins much earlier on a Se surface than on the SnO$_2$ surface. Whatever the cobalt selenide stoichiometry is, Co$_x$Se is presented here for simplicity. This is characteristic of the induced underpotential deposition mechanism, known as Kröger's mechanism[34], which is caused by the large energy release in the formation of cobalt selenides. For instance, the value of the Gibbs free energy of the formation of CoSe is −40.74kJ/mol[35], and therefore the redox potential of the reaction Eq. (6) is shifted by an amount of $-\Delta G/2F = +0.211$V theoretically with respect to the standard deposition potential of metallic cobalt. This mechanism also has been employed for electrodeposition of many other compound semiconductors[36-39]. It is also obvious that the peak for six electrons reduction reaction shows a significant positive shift, revealing that Co^{2+} helps promoting the six-electrons reduction of H_2SeO_3. This is because the generated H_2Se immediately reacts with Co^{2+} in the solution according to Eq. (7) due to the large free energy of formation of cobalt selenide again, which can facilitate the reaction (Eq. (4)) in forward direction[40]. By combining Eq. (4) and Eq. (7) to form Eq. (8), it is indicative once more that Co^{2+} promotes the six-electrons reduction of H_2SeO_3. It should be noted that this positive shift leads to overlapping of four-electrons and six-electrons Se(IV) reduction potential regions. Therefore, the much greater increase in current when potential reaches about −0.35V can be due to the contribution of two reduction reations proceeded by Eq. (6) and Eq. (8) together. Moreover, the increase in current becomes more and more significant with further shifting potential towards negative direction, which can be due to the growing reaction rates of both reactions Eq. (6) and Eq. (8) under the enlarged reaction driving force with the negative shift of cathodic potential.

$$Se + xCo^{2+} + 2xe \longrightarrow Co_xSe \qquad (6)$$

$$H_2Se + xCo^{2+} \longrightarrow Co_xSe + 2H^+ \qquad (7)$$

$$H_2SeO_3 + xCo^{2+} + 4H^+ + (4+2x)e \longrightarrow Co_xSe + 3H_2O \qquad (8)$$

All above results allow us to conclude that the deposition of Co into Co_xSe solid phase can proceed through two different routes: surface induced reduction by Se and reaction with H_2Se. In any case, the underpotential deposition of cobalt as cobalt selenide is apparent.

Fig. 4 shows the EDS composition of cobalt selenide films deposited at different potentials between −0.3 and −0.7V from 2mM $Co(CH_3COO)_2$+2mM H_2SeO_3, pH 2.0 solution. The films electrodeposited at potentials more negative than −0.7V show lots of pinholes and poor adherence because of hydrogen evolution and have not been considered. For films deposited at −0.3V or more positive potentials, the content of Co is very low making it difficult to confirm the incorporation of Co into films by EDS(the EDS measurement error is within 3at%~5at%), which is attributed to the extremely slow kinetics of underpotential deposition of cobalt selenide when deposition potential is not negative enough based to above CV studies. It is observed that,

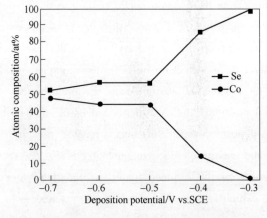

Fig. 4 The EDS composition of electrodeposited cobalt selenide films at 50℃ and different potentials between −0.3 and −0.7V from 2mM $Co(CH_3COO)_2$+2mM H_2SeO_3, pH2.0 solution

as the deposition potential shifts negatively, the cobalt content increases rapidly and the selenium content decreases in the film, leading to a composition of 56.6at% Se and 43.4at% Co obtained at -0.5V. Further negative shift of deposition potential, however, does not change the composition of the films significantly, which indicates that the electrodeposition process is mainly controlled by diffusion independent on deposition potential. This diffusion-controlled process has been maintained until onset of the evolution of hydrogen which disturbs the double layer forming on the electrode surface leading to the instability of electrodeposition and accordingly film composition fluctuation[41].

Fig. 5a-e show the dramatic difference of surface morphologies of electrodeposited Co-Se films at varies deposition potentials from -0.3V to -0.7V. Film deposited at -0.3V consists mainly of selenium and shows some clusters with sizes between 0.2 and 1μm and some agglomerates with size larger than 2μm (Fig. 5a). Similar morphologies of Se films were observed by M. O. Solaliendres et. al[42]. The Se clusters are eliminated significantly with incorporation of a certain amount of Co with deposition potential shifting negatively to -0.4V (Fig. 5b). When the deposition potential reaches -0.5V, the film shows very compact and homogeneous surface morphology having isolated grains with uniform size and well-defined boundaries (Fig. 5c). With further negative shift of deposition potential, the films became rough and porous (Fig. 5d), or even loose structure with flocculent in appearance (Fig. 5e). This is due to the high electrode reaction rate and excessive concentration polarization[43]. Generally, compact and smooth films are needed for application. Therefore, the deposition potential of -0.5V was considered to be optimum for cobalt selenide thin film electrodeposition. In order to get insight into the microstructure profile of the electrodeposited cobalt selenide thin film at -0.5V, the SEM image of cross section is also present, as shown in Fig. 5f. It is seen that the compact and uniform cobalt selenide thin film consisting of large grains extending from the bottom to the top of the film is obtained, and the thickness of the film is about 400nm.

Fig. 6 displays the XRD patterns measured for electrodeposited cobalt senelide thin film at -0.5V and SnO_2/glass substrate. The results clearly indicate that all diffraction peaks, except peaks of SnO_2 (JCPDS card No. 41-1445) which come from the substrate, perfectly match with the hexagonal phase of CoSe (JCPDS card No. 89-2004). No characteristic peaks were observed for other impurities. The excess Se according to the EDS composition can not be indexed by XRD due to its amorphous nature. Therefore, it is demonstrated that the deposites on surface of SnO_2 film consist of compact film of Se-rich CoSe sample.

Fig. 7 shows the optical absorption coefficient (α) of the electrodeposited CoSe thin film as a function of photon energy ($h\nu$), converted from the transmission spectra recorded in the range of 300-900nm. The sharp line near 300nm (4.13eV) is due to the change of source and the lower wave length absorption by the substrate[44]. The absorption coefficient is larger than $10^{-5} cm^{-1}$ in the visible region, which supports the direct band gap nature of material[3] and reveals that the CoSe film can be considered to be a suitable material for photovoltaic solar energy conversion. Based on the allowed direct interband transition, the band gap is determined to be (1.53 ± 0.01)eV by extrapolating the linear $(\alpha h\nu)^2$ vs. $h\nu$ plots to $(\alpha h\nu)^2 = 0$, as depicted from the

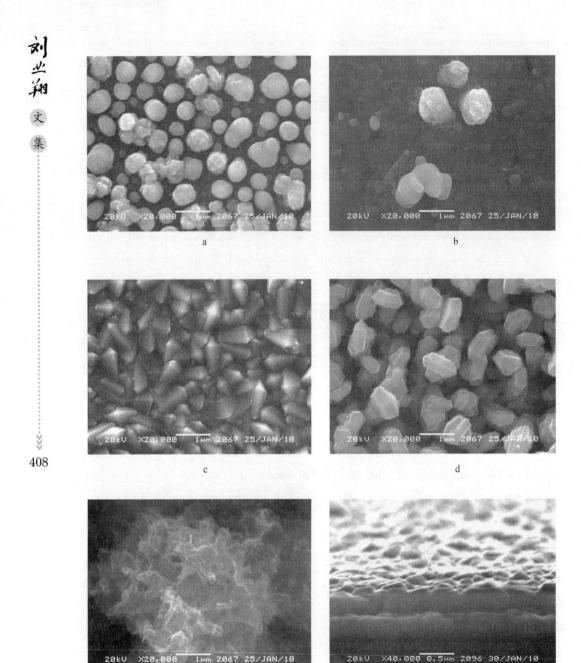

Fig. 5 The SEM micrograph of electrodeposited cobalt selenide thin films at 50℃:
a) Deposition potential = −0.3V; b) deposition potential = −0.4V; c) deposition potential = −0.5V;
d) deposition potential = −0.6V; e) deposition potential = −0.7V; f) a crosssectional
view of cobalt selenide film electrodeposited at −0.5V

inset of Fig. 7. This value is quite close to the theoretical optimal value of light absorber for a single-junction solar cell[45]. These optical characteristics indicate that CoSe is a very promising material for thin film solar cells.

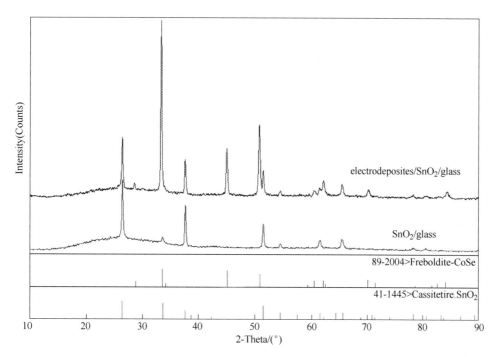

Fig. 6 X-ray diffraction pattern of cobalt selenide thin film electrodeposited at −0.5V and 50℃ on SnO$_2$ substrate

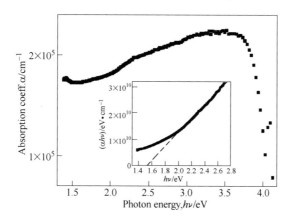

Fig. 7 Optical absorption coefficient (α) of the electrodeposited CoSe thin film at −0.5V and 50℃. The inset shows ($\alpha h\nu$)^2 vs $h\nu$ for CoSe film; the estimated bandgap is 1.52eV

4 Conclusions

Cobalt selenide thin films have been prepared onto tin oxide glass substrates by electrodeposition potentiostatically from an aqueous acid bath containing H$_2$SeO$_3$ and Co(CH$_3$COO)$_2$ at a temperature of 50℃. The formation of cobalt selenide was inferred to proceed via an under potential deposition mechanism from cyclic voltammetry study, which may involve two routes: Co^{2+} reduction by a surface-induced effect from Se and/or reaction with H$_2$Se. Se-rich CoSe thin films with compact and homogeneous morphology, hexagonal crystal structure and a direct optical band gap of (1.53 ± 0.01) eV were obtained at a deposition potential of −0.5V vs. SCE.

References

[1] W. S. Sheldrich and M. Wachhold, Angew. Chem. Int. Ed. , 36, 206(1997).

[2] W. Tremel, H. Kleinke, V. Derstroff and C. Reisner, J. Alloys. Compd. , 219, 73(1995).

[3] S. M. Pawar, A. V. Moholkar, U. B. Suryavanshi, K. Y. Rajpure and C. H. Bhosale, Sol. Energy. Mater. Sol. Cells. , 91, 560(2007).

[4] I. M. Dharmadasa, A. P. Samantilleke, J. Young, M. H. Boyle, R. Bacewicz and A. Wolska, J. Mater. Sci. -Mater. Electron. , 10, 441(1999).

[5] Yongjun Feng, Ting He and Nicolas Alonso-Vante, Electrochim. Acta, 54, 5252(2009).

[6] S. Kashida and J. Akai, J. Phys. C: Solid State Phys. , 21, 5329(1988).

[7] Y. Takemura, H. Suto, N. Honda, K. Kakuno and K. Saito, J. Appl. Phys. , 81, 5177(1997).

[8] X. J. Wu, Z. Z. Zhang, J. Y. Zhang, Z. G. Ju, D. Z. Shen, B. H. Li, C. X. Shan and Y. M. Lu, J. Cryst. Growth, 300, 483(2007).

[9] N. Hamdadou, J. C. Bernède and A. Khelil, J. Cryst. Growth, 241, 313(2002).

[10] P. P. Hankare, B. V. Jadhav, K. M. Garadkar, P. A. Chate, I. S. Mulla and S. D. Delekar, J. Alloys. Compd. , 490, (2010)228.

[11] Z. Zainal, N. Saravanan and H. L. Mien, J. Mater. Sci. -Mater. Electron. , 16, 111(2005).

[12] B. Ouertani, J. Ouerfelli, M. Saadoun, B. Bessaïs, H. Ezzaouia and J. C. Bernède, Sol. Energy Mater. Sol. Cells, 87, 501(2005).

[13] Raghu N. Bhattacharya and Arturo M. Fernandez, Sol. Energy Mater. Sol. Cells, 76, 331(2003).

[14] S. Thanikaikarasan, T. Mahalingam, K. Sundaram, A. Kathalingam, Yong Deak Kim and Taekyu Kim, Vacuum, 83, 1066(2009).

[15] Remigiusz Kowalik, Piotr Żabiński and Krzyszt of Fitzner, Electrochim. Acta, 53, 6184(2008).

[16] Laurent Thouin, Sylvie Rouquette-Sanchez and Jacques Vedel, Electrochim. Acta, 38, 2387(1993).

[17] E. Gómez, R. Pollina and E. Vallés, J. Electroanal. Chem. , 386, 45(1995).

[18] R. Oriňáková, M. Streĉková, L. Trnková, R. Rozik and M. Gálová, J. Electroanal. Chem. , 594, 152 (2006).

[19] S. S. Abd El Rehim, Magdy A. M. Ibrahim and M. M. Dankeria, J. Appl. Electrochem. , 32, 1019(2002).

[20] A. B. Soto, E. M. Arce, M. Palomar-Pardavé, I. González, Electrochim. Acta, 41, 2647(1996).

[21] Manuel Palomar-Pardavé, Ignacio González, Ana B. Soto and Elsa M. Arce, J. Electroanal. Chem. , 443, 125(1998).

[22] Darko Grujicic and Batric Pesic, Electrochim. Acta, 49, 4719(2004).

[23] L. H. Mendoza-Huizar, J. Robles and M. Palomar-Pardave, J. Electrochem. Soc. , 152, C265(2005).

[24] C. H. Rios-Reyes, L. H. Mendoza-Huizar and M. Rivera, J. Solid State Electrochem. , 14, 1432(2010).

[25] M. Palomar-Pardavé, B. R. Scharifker, E. M. Arce and M. Romero-Romo, Electrochim. Acta, 50, 4736 (2005).

[26] R. W. Andrews and D. C. Johnson, Anal. Chem. , 47, 294(1975).

[27] Chang Wei, Noseung Myung and Krishnan Rajeshwar, J. Electroanal. Chem. , 375, 109(1994).

[28] Baoming M. Huang, Tedd E. Lister and John L. Stickney, Surf. Sci. , 392, 27(1997).

[29] Murat Alanyalioglu, Umit Demir and Curtis Shannon. J. Electroanal. Chem. , 561, 21(2004).

[30] Mauro C. Santos and Sergio A. S. Machado, J. Electroanal. Chem. , 567, 203(2004).

[31] Thomas A. Sorenson, Tedd E. Lister, Boaming M. Huang and John L. Stickney, J. Electrochem. Soc. , 146, 1019(1999).

[32] Marianna Kemell, Heini Saloniemi, Mikko Ritala and Markku Leskela, Electrochim. Acta, 45, 3737 (2000).

[33] Yanqing Lai, Fangyang Liu, Jie Li, Zhian Zhang and Yexiang Liu, J. Electroanal. Chem. , 639, 187

(2010).
[34] F. A. Kröger, J. Electrochem. Soc. ,125,2028(1978).
[35] H. Jelinek and K. L. Komarek, Monatshefte für Chemie,105,689(1974).
[36] K. Rajeshwar, Adv. Mater. ,4,23(1992).
[37] H. Saloniemi, T. Kanniainen, M. Ritala, M. Leskelä and R. Lappalainen, J. Mater. Chem. ,8,651(1998).
[38] Doriane Del Frari, Sébastien Diliberto, Nicolas Stein, Clotilde Boulanger and Jean-Marie Lecuire, Thin Solid Films,483,44(2005).
[39] M. Kemmell, M. Ritala, H. Saloniemi, M. Leskela, T. Sajavaara and E. Rauhala, J. Electrochem. Soc. ,147,1080(2000).
[40] P. W. Atkins, Physical Chemistry, third edition, Oxford University Press,1985.
[41] M. Paunovic, M. Schlesinger, Fundamentals of Electrochemical Deposition, Wiley-Interscience, New Jersey,1998.
[42] M. O. Solaliendres, A. Manzoli, G. R. Salazar-Banda, K. I. B. Eguiluz, S. T. Tanimoto and S. A. S. Machado, J. Solid State Electrochem. ,12,679(2008).
[43] Yanqing Lai, Fangyang Liu, Zhian Zhang, Jun Liu, Yi Li, Sanshuang Kuang, Jie Li and Yexiang Liu, Electrochim. Acta,54,3004(2009).
[44] R. Chandramohan, T. Mahalingam, J. P. Chu and P. J. Sebastian, Sol. Energy Mater. Sol. Cells, 81, 371 (2004).
[45] Shannon C. Riha, Bruce A. Parkinson and Amy L. Prieto, J. Am. Chem. Soc. ,131,12054(2009).

Fabrication of Ternary Cu-Sn-S Sulfides by Modified Successive Ionic Layer Adsorption and Reaction (SILAR) Method[*]

Abstract Ternary Cu-Sn-S chalcogenides, Cu_2SnS_3, $Cu_5Sn_2S_7$ and Cu_3SnS_4, have been successfully synthesized by annealing three precursor film samples deposited via modified successive ionic layer adsorption and reaction (SILAR) method. The mechanism of ion-exchange and improvement of rinsing procedure were introduced to the process of SILAR for the purpose of achieving the codeposition of different metal sulfides and increasing the growth rate of thin films. The crystal structure, composition, surface morphology, optical and electrical properties for three ternary sulfides samples have been characterized. Besides the temperature dependence of Seebeck coefficient and electrical conductivity of Cu_3SnS_4 thin film sample have been measured between 294K and 573K. Owing to the intrinsic advantages of SILAR method and the improvement of SILAR process, ternary Cu-Sn-S thin films can be deposited on glass substrates at a speed of 400nm per hour and the surface morphologies of thin films can be comparable with that of thin films prepared by vacuum based methods, which can satisfy the requirement for large-scale industrial production. With the appropriate band gap energies (1.0eV, 1.45eV and 1.47eV for Cu_2SnS_3, $Cu_5Sn_2S_7$ and Cu_3SnS_4 respectively) and considerable absorption coefficients ($\alpha > 10^4 cm^{-1}$), most importantly with Earth-abundant elements, Cu-Sn-S thin films can be used as alternative absorber layer materials in thin film solar cells. Additionally for $Cu_5Sn_2S_7$ and Cu_3SnS_4 film samples, some novel properties (such as strong optical absorption in NIR band, excellent conductivity, suitable carrier concentration and good Seebeck coefficient) make them attract new potential researching interest in lithium battery and thermoelectric materials.

1 Introduction

Compounds in the Cu-Sn-S system belong to the Ⅰ-Ⅳ-Ⅵ ternary chalcogenides including many kinds of important sulfides such as Cu_2SnS_3, Cu_3SnS_4 and Cu_4SnS_4.[1,2] These ternary chalcogenides show various phase structures (tetragonal, cubic, triclinic, hexagonal, metastable zincblenvarousde and wurtzite structure) with outstanding optical, electrical and optical-thermal-mechanical properties.[3-5] Therefore it is entirely natural for Cu-Sn-X (X=S, Se) compounds to attract increasing research interests.[6-9] Especially the p-type semiconductors, Cu_2SnS_3 and Cu_3SnS_4 have become another focus of photoelectronic field because of having the suitable band-gaps and optical absorption coefficients similar to the earth-abundant CZTS (Cu_2ZnSnS_4) materials which are being comprehensively studied currently.[8,10-18] Moreover, the control of composition and phase structure on Cu_2SnS_3 and Cu_3SnS_4 compounds is more convenient due to

---

[*] Copartner: Zhenghua Su, Kaiwen Sun, Zili Han, Fangyang Liu, Yanqing Lai, Jie Li. Reprinted from Journal of Materials Chemistry, 2012, 22: 16346-16352.

their fewer elements compared with CZTS. In addition, with excellent conductivity and supercell crystal structure, Cu-Sn-S compounds have been regarded as another conducting sulfide and have attracted researches on the application of lithium battery.[19]

As for the synthesization methods of Cu-Sn-S system compounds, a few works have been reported on sequential chemical deposition, solvothermal processing, solid-state reaction, spray-pyrolysis and sputtering.[4,6,12,15,16,20] However, all these reported methods involved in at least one of those limitations, such as the difficulty of controlling chemical compositions, bad surface morphology, complex synthesis process with high temperature or high vacuum and so on. In this report we introduced the successive ionic layer adsorption and reaction(SILAR) method to synthesize three Cu-Sn-S system compounds. Combining both advantages of chemical bath deposition(CBD) and atomic layer epitaxy deposition(ALD), SILAR is simple, inexpensive and convenient for large area deposition on various substrates without special restrictions so it is widely used to deposit various thin films and core/shell nanostructure compounds.[21-28] Thin films prepared by SILAR method generally have compact mirror-like surface morphology and the thickness and compositions of thin film can be controlled easily and accurately by changing the cycle numbers and the immersing process. More importantly, the metal sulfides prepared by SILAR method contain enough sulfur and do not contain any other toxic and organic additives, which avoids the deterioration of surface morphology(such as cracks and holes in thin films) aroused by the volatilization of organic compounds and volume expansion during the sulfurization process.[29] Nevertheless, for the reason of competitive adsorption between different metal ions caused by significant difference in solubility products, it is hard to deposit ternary or quaternary metal sulfides in single cation solution by SILAR method. Furthermore, the domination of ion-by-ion growth mechanism and avoiding homogeneous precipitation limited the deposition rate and throughput of SILAR method. For the purpose of the codeposition of different metal sulfides and improving throughput, in this work we modified the process of SILAR and introduced a new growth mechanism then synthesized three Cu-Sn-S compounds which have been studied and assessed preliminarily for the possible applications in photovoltaic and thermoelectric materials fields.

2 Experimental Section

Fig. 1 shows the main experiment processes for the synthesis of ternary Cu-Sn-S thin films. In fact, the precursor thin films of ternary metal sulfides prepared by SILAR method are made up of two different binary metal sulfides rather than single ternary metal sulfides.[24,25] At first if we synthesize precursor Cu-Sn-S compounds(Cu_xS+Sn_xS) using Cu^{2+} and Sn^{2+} solutions as a mixed cation solution, it is unavoidable to meet with the problem of hydrolysis of Sn^{2+}. In addition, owing to weak adsorption ability of Cu^{2+}, it is difficult to deposit high quality Cu_xS thin films by SILAR under alkaline condition, let alone acid condition.[30] Finally we choose ammonium fluoride as the complexing agent of Sn^{2+}. Thus the pH of mixed cation solution can be increased to 5 without the hydrolysis of Sn^{2+} but Cu_xS thin films still cannot be deposited easily on glass substrate under this condition by SILAR. In order to overcome this dilemma, we proposed an approach of ion exchange combining with ion-by-ion growth mechanism to deposit Cu_xS and

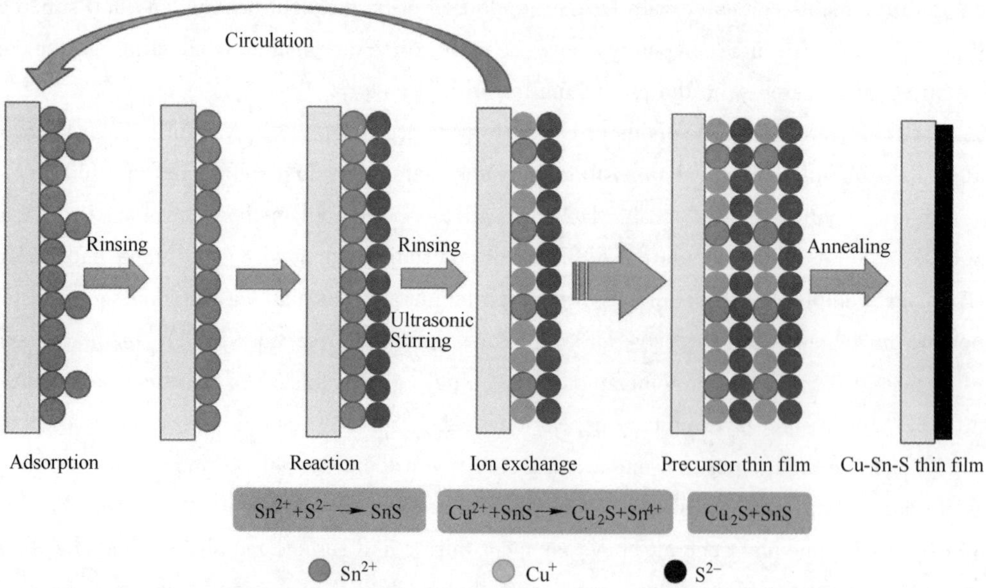

Fig. 1 Schematic illustration for the formation of Cu-Sn-S thin films

SnS thin films in single mixed caion solution.[31] Under weak acid condition, the deposition of SnS thin film on glass substrate can be easily carried out in SILAR process. As the solubility product of Cu_2S($K_{sp} = 10^{-48}$) is far lower than that of SnS($K_{sp} = 10^{-26}$), partial SnS thin films would be transformed to Cu_2S thin films in the process of SILAR so that the codeposition of Cu_2S and SnS thin films can be achieved in single caion solution.

To verify the feasibility of codeposition, we synthesized preliminarily Cu-Sn-S(Cu : Sn : S = 1.9 : 1.0 : 2.7) precursor thin film via SILAR method using a mixture of $CuSO_4$ and $SnSO_4$ aqueous solution as cation solution. The in situ precursor thin film was analyzed by x-ray photoelectron spectroscopy(XPS) to determine the element valence states and existing forms of metal sulfides in Fig. 2. From the XPS of in situ precursor thin film, it can be seen that the binding energy of Cu $2p_{3/2}$ and Cu $2p_{1/2}$ are 931.4eV and 951.3eV respectively corresponding to the binding energy of Cu^+ while the characteristic peak of Cu^{2+}(942 eV) cannot be observed, which implies only the valence state of Cu^+ exists in in-situ thin films.[32,33] Moreover, the binding energy of Sn $3d_{5/2}$ and Sn $3d_{3/2}$ of in-situ thin film peak at 486.0eV and 494.4eV corresponding to the value of Sn^{2+}.[34] The XPS of S(2p) core level spectrum peaks at 161.1eV illustrating the existence of S^{2-}.[35] From the analysis of XPS it reveals that the two possible forms of metal sulfides in precursor thin film prepared by SILAR are Cu_2S and SnS, and explicitly the mechanism of ion exchange can work and play an important role in implementing the codeposition of different metal sulfides via SILAR method.

In order to increase the growth efficiency of thin films prepared by SILAR, we modified the procedure of rinsing and decreased the rinsing and immersing time when preparing ternary Cu-Sn-S compounds. Commonly, each immersing time or rinsing time will last 20 sec at least so that occupy most of the cycle time, which is the primary cause of low deposition efficiency. In order to decrease the time of rinsing process, we introduced ultrasonic and stirring washing simultane-

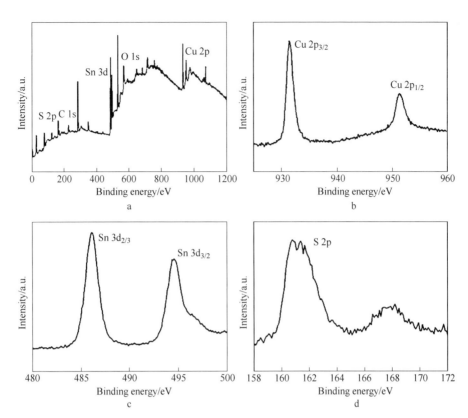

Fig. 2 XPS spectra of precursor thin film
a—Typical survey spectrum; b—Cu 2p core level; c—Sn 3d core level; d—S 2p core level

ously when rinsing and the deionized water in rinsing vessel was replaced automatically after ultrasonic and stirring washing to avoid homogeneous precipitation and the cross-contamination of cation and anion. Thus the rinsing time was decreased to 5 sec and the rinsing effect was improved. Moreover, if we decrease the adsorption and reaction time from 20 sec to 5 sec, we found that the growth rate of thin film only reduced by 10%. Therefore, we decreased one cycle time of SILAR from 80 sec to 20 sec and the growth rate of one cycle was improved significantly.

By preliminary experiments and analysis mentioned above, we determined the mixture aqueous solution of $CuSO_4$, $SnSO_4$ and NH_4F as the cation solution, and $Na_2S \cdot 9H_2O$ aqueous solution as anion solution. According to the designed SILAR procedure, we have synthesized three different Cu-Sn-S compounds: cubic Cu_2SnS_3 (Sample A), $Cu_5Sn_2S_7$ (Sample B) and tetragonal Cu_3SnS_4 (Sample C). When preparing Sample A, B and C, the molar concentration of $CuSO_4$ was 0.005M, 0.01M and 0.015M respectively. The molar concentration of $SnSO_4$, NH_4F and $Na_2S \cdot 9H_2O$ was kept at 0.02M, 0.01M and 0.01M respectively for three Samples. Microscope glass slides were used as substrates after ultrasonically cleaned in turn with 50% ammonia, acetone and deionized water for 10min. The SILAR equipment was made up of one SIEMENS PLC unit and driver devices. In a cycle of SILAR, the adsorption time, reaction time and rinsing time are 5 s respectively. After 100 cycles of SILAR, the precursor thin films of Cu-Sn-S compounds can be obtained with the thickness of about 500nm. Adding the time intervals between every immersion and rinsing, the whole growth rate of thin film by SILAR is about 400nm per hour, which is much larger than previously reported growth rates.[21] The precursor thin films of Cu-Sn-S compounds synthesized at room temperature have poor crystallization and contain the mixture of Cu_2S and SnS so it is necessary to anneal Cu-Sn-S precursor thin films by sulfurization. Then three Cu-Sn-S precursor thin films were sulfurized in a tubular furnance in N_2 and sulfur vapour atmosphere at 400℃, 450℃ and 500℃ respectively for an hour to get Cu_2SnS_3(Sample A), $Cu_5Sn_2S_7$(Sample B) and Cu_3SnS_4(Sample C) thin film compounds.

The crystallinity, Raman spectra, chemical composition and surface morphology of sample films were characterized by X-ray diffraction (XRD, Rigaku3014), Jobin-Yvon LabRAM HR800-Horiba spectrometer, energy dispersive X-ray spectroscopy (EDS, EDAX-GENSIS60S) and scanning electron microscopy (SEM, FEI Quanta-200) respectively. The optical properties of the films were determined by Shimadzu UV-3600 spectrophotometer. The carrier concentration and mobility of thin films were obtained by Hall Effect measurement (HMS-3000/0.55T) at room temperature. A hot point probe and a four-point probe were used to determine conductivity type and the resistivity for samples. The temperature dependence of Seebeck coefficient and electrical conductivity were detected by thermoelectric instrument designed and fabricated by authors.

3 Results and Discussion

Table 1 shows the results of EDS and the thickness of the precursor and annealed thin films for Sample A, B and C. It is clear that the composition of metals in precursor thin film can be easily achieved to desired ratio only by changing the concentrations of metal ions in precursor cation solution. In addition, for three annealed samples the contents of Cu increase slightly compared with the precursor thin films indicating the fact of the loss of Sn during sulfurization process. Moreover, the contents of sulfur in three precursor samples can also illustrate the existence of Cu_2S and SnS coinciding with the XPS analysis. It is worth mentioning that the thickness of three film samples did not change significantly before and after annealing. This phenomenon suggests that the precursor thin films contain considerable sulfur and that sufficient sulfur in precursor films will obviously reduce the deterioration of loose, porous and poor adhesion caused by excess interdiffusion of sulfur after sulfurization.

Table 1 EDS and thickness for three film samples before and after annealing

Sample		EDS			Thickness/nm
		Cu	Sn	S	
A	Precursor	1.9	1.0	2.7	520
	Sulfurized	2.1	1.0	3.3	610
B	Precursor	2.4	1.0	2.9	540
	Sulfurized	2.5	1.0	3.6	640
C	Precursor	2.9	1.0	3.1	550
	Sulfurized	3.1	1.0	4.1	640

The XRD patterns of three Samples after annealing at different temperature are shown in Fig. 3 and the detailed XRD spectrums at diffraction angles, 2θ, around 28.5°, 47.5° and 56.5° are demonstrated in Fig. 4. The XRD of Sample A has three strong peaks at (111), (220) and (311), which can be assigned to JCPDS(89-2877) with cubic crystal structure of Cu_2SnS_3.[14] However, the main peaks (2θ = 28.447°, 47.312° and 56.133°) of tetragonal Cu_2SnS_3 (JCPDS 089-2877) are very close to that of JCPDS(89-4714) (2θ = 28.541°, 47.474° and 56.326°), therefore from the detailed XRD patterns in Fig. 3 the Sample A may contain weak etragonal Cu_2SnS_3. It can also be found that the three peaks at (112), (220) and (312) of Sample C correspond with JCPDS (33-0501) well.[7,11,20] Generally the crystal structures of Cu_2SnS_3 and Cu_3SnS_4 varies with different synthesizing conditions such as temperature or atmosphere so it is acceptable that various crystal structures about Cu_2SnS_3 and Cu_3SnS_4 have been reported.[14,15,19] Although almost no detailed XRD data for $Cu_5Sn_2S_7$ has been published to the best of our knowledge, the XRD of Sample B is still found corresponding with the JCPDS(40-0924)

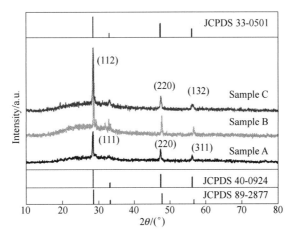

Fig. 3 The XRD patterns for three annealed Samples and the standards for cubic Cu_2SnS_3, $Cu_5Sn_2S_7$ and tetragonal Cu_3SnS_4

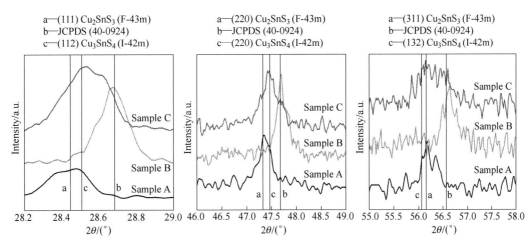

Fig. 4 Detailed XRD spectrums for three samples at around 28.5°, 47.5° and 56.5° with the standards of JCPDS

in addition to weak $Cu_{2-x}S$ phase and the detailed lattice parameters for Sample B should be further estimated.[2]

In order to further study and confirm the structure types of three samples, the Raman spectroscopies for three samples have been characterized in Fig. 5. Sample A has two strong peaks at 302cm^{-1} and 352cm^{-1} which can be assigned to cubic (F-43m) Cu_2SnS_3. However a weak peak at 337cm^{-1} belonging to the Raman spectroscopy of tetragonal (I-43m) Cu_2SnS_3 was observed. For Sample A, the results of Raman spectroscopy and XRD confirm the coexistence of major cubic and weak tetragonal phase of Cu_2SnS_3.[14] The peaks of Sample B are found at 332cm^{-1} and 296cm^{-1} but until now we have not found the reported Raman spectroscopies of $Cu_5Sn_2S_7$. As for tetragonal (I-42m) Cu_3SnS_4, the peaks at 330cm^{-1} and 295cm^{-1} are close to that of CZTS (Cu_2ZnSnS_4) thin film but not similar to that of reported Cu_3SnS_4 nanorods.[7] Perhaps due to different structures, the Raman values of Cu_3SnS_4 thin films in this report are more rational because of similar Sn-S vibration mode with CZTS thin films. The result of Raman data for Sample B also implies the presence of $Cu_{2-x}S$ at the peak of 475cm^{-1} in accordance with the result of XRD.

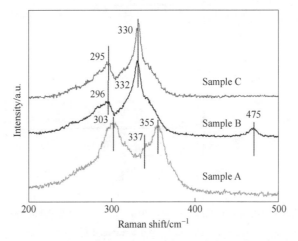

Fig. 5 Raman spectrums for annealed samples

The surface morphologies of three film samples through SEM before and after annealing are shown in Fig. 6. It is clear that all three precursor film samples have smooth and compact surface morphologies and no any holes or obvious grain boundaries can be observed, which indicates that the quality of precursor thin films via SILAR can be comparable with vacuum based methods such as sputtering and evaporation. Besides the natural advantages of SILAR, the improved washing procedure with ultrasonic and stirring also contributes much to the high quality of thin films. Additionally the surfaces of annealed films are still considerably smooth with uniform grains and are significantly better than that of thin films prepared by sulfuring stacked metallic precursors via sputtering or evaporation. Through the above composition analysis it can be concluded that the precursor films contain considerable sulphur, which can avoid volume expansion of film caused by excess interdiffusion of sulfur during sulfurization, and as a result smooth and compact thin films can be obtained. In addition, almost no any $Cu_{2-x}S$ and other particles exist on the surface of annealed sample films. In thin film solar cell structure, as far as absorber layer, the relatively smooth surface and no impurity phases can decrease the defect states and carrier recombination, which is very beneficial to improve the photovoltaic conversion efficiency.

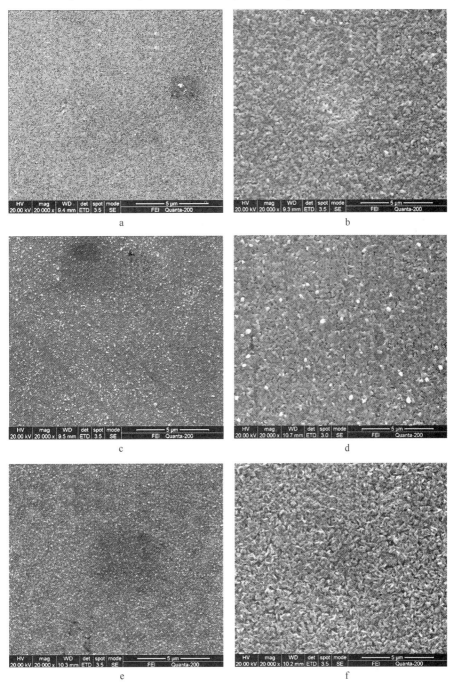

Fig. 6 Surface morphologies for three samples before and after annealing
a—Precursor film for Sample A; b—Annealed film for Sample A; c—Precursor film for Sample B;
d—Annealed film for Sample B; e—Precursor film for Sample C; f—Annealed film for Sample C

As shown in Fig. 7, the optical transmission spectrums of three samples after annealing were obtained in the wavelength of 400-2000nm. It is interesting that all three samples have a strongest peak at 1380nm, 932nm and 888nm respectively and have much intense absorption less than 500nm. As for Sample B and C, strong optical absorption can also be observed at wavelength greater than 1500nm of near-infrared band. Fig. 8 presents the optical absorption spectrums and estimations of band gap energy for three Samples. The absorption coefficients for three Samples achieve the order of $10^4 cm^{-1}$ from 400nm to 2000nm and the order of $10^5 cm^{-1}$ in the UV band. The excellent optical absorption coefficient is becoming one of the indispensable properties of good absorber layer materials in next generation thin film solar cells for the reason of the limitation of using materials. From the transmission spectrums and absorption spectrums, the Sample B and C have considerable absorption coefficients at the band of NIR. Although it has been reported that Cu_xS has a characteristic broad absorption band in the NIR region, from the XRD and Raman results there is only slight Cu_xS phase existed in Sample B so it could not be responsible for the intensive NIR absorption.[36] Nevertheless it is not certain that the optical NIR absorption is aroused by molecular vibration absorption or electronic absorption. On the one hand the electronic absorption in NIR band can bring the possibility of fabricating full

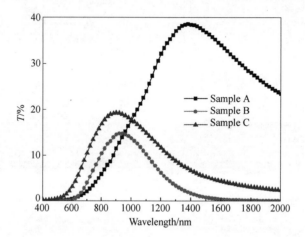

Fig. 7 The optical transmission spectrums for three samples

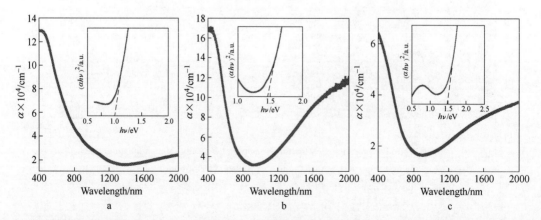

Fig. 8 The optical absorption spectrums and estimations of band gap energy for three annealed samples
a—Sample A; b—Sample B; c—Sample C

spectrum thin film solar cells. On the other hand, the absorption of molecular vibration also supports the feasibility of manufacturing semiconductors with infrared absorption such as thermoelectric materials. As a result, the photovoltaic and thermoelectric properties can be integrated and used in single absorber material for photo-thermoelectric conversion.

The optical band gaps of three samples are estimated in Fig. 8 by extrapolating the linear region of a plot of $(\alpha h\nu)^2$ vs. $(h\nu)$ to zero absorption coefficient ($\alpha = 0$). The band gap energy of Sample A is estimated to be 1.02eV close to the reported values.[12,14] The gap energy value of Sample B and C is 1.45eV and 1.47eV respectively but the band gaps of $Cu_5Sn_2S_7$ and tetragonal(I-42m)c thin films have not yet been reported before. According to the requirement of band gap in thin film solar cell, three Cu-Sn-S thin films prepared by SILAR can be used as absorber layer materials.

Table 2 shows the electrical properties of three samples including conductive type, carrier concentration, mobility and resistivity. All three samples are the type of p-type semiconductor with carrier concentration and mobility in the order of $10^{19} cm^{-3}$ and $10 cm^2 \cdot V^{-1} \cdot s^{-1}$ respectively. In addition, according to the classic Coulomb's low, the relatively larger radius of Cu^+ atomic nucleus contributed to the conductivity of the compounds.[19] So it is rational for the resistivity of three samples decreased with the increase of Cu content while Sample C achieved in the order of $10^{-4} \Omega \cdot cm$. Owing to the excellent conductivity and appropriate crystal structure, ternary Cu-Sn-S compounds currently have attracted increasing researches on the application in electrode materials for Li-ion battery. Furthermore, in addition to suitable conductivity and much intensive infrared absorption discussed above, the carrier concentrations (10^{19}-$10^{20} cm^{-3}$) for Sample B and C also meet with the requirement for good thermoelectric materials with high zT.[37] Recently increasing works have been attracted on quaternary Cu_2ZnSnX_4 (X = S, Se) for thermoelectric application.[38-41] To investigate potential thermoelectric performance of Cu-Sn-S compounds, the electrical conductivity and Seebeck coefficient for Sample C (tetragonal Cu_3SnS_4) have been measured between 293K and 573K. From Fig. 9 it is apparent that the positive Seebeck coefficient increases from $60\mu V/K$ (20K) to $98\mu V/K$ (573K), which is comparable with reported CZTS nanocrystals.[40,41] Additionally, the electrical conductivity of tetragonal Cu_3SnS_4 shows relatively high values from room temperature to 573K (400-700S/m) comparing with stoichiometric CZTS nanocrystals. Due to very thin thickness of Cu_3SnS_4 thin film prepared by SILAR in this work, it is difficult to find a suitable method to measure therm-

Table 2 Electrical properties for three Samples

Sample	Conductive type	Carrier concentration/cm^{-3}	Mobility /$cm^2 \cdot V^{-1} \cdot s^{-1}$	Resistivity /$\Omega \cdot cm$
A	p	3.1×10^{19}	11.0	2.4×10^{-2}
B	p	5.3×10^{20}	22.3	8.7×10^{-3}
C	p	9.1×10^{20}	16.7	1.5×10^{-4}

al conductivity according to our lab situations. Thus in order to assess comprehensively the application in thermoelectric materials for Cu_3SnS_4 thin film, the measurements on thermal conductivity and calculated zT value for Cu_3SnS_4 thin film are expected to carry out in future and the feasibility of the Cu-Sn-S compounds can be applied in thermoelectric materials will be studied more.

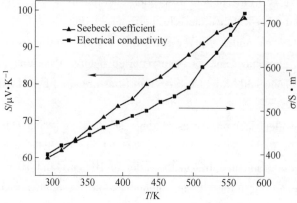

Fig. 9 Temperature dependence of Seebeck coefficient and electrical conductivity for Cu_3SnS_4 thin film

4 Conclusions

This work presents the synthesization of ternary sulfides, Cu_2SnS_3, $Cu_5Sn_2S_7$ and Cu_3SnS_4 compounds, via sulfurizing precursor thin films prepared by a modified SILAR method. The Cu-Sn-S precursor film compounds deposited by SILAR method can be produced easily and the growth rate of thin films was improved remarkably through the introduction of ion-exchange and the improvement of SILAR procedure. Ternary sulfides, Cu_2SnS_3, $Cu_5Sn_2S_7$ and Cu_3SnS_4 thin film compounds can be obtained by sulfuring precursor thin films at different temperatures. For three ternary sulfides samples, the crystal structure, composition, surface morphology, optical, electrical and thermoelectric properties (for Cu_3SnS_4) have been characterized. The prepared thin films have high quality for surface morphology even after sulfurization owing to the intrinsic advantages of SILAR method and the improvement of rinsing process. For Sample A sulfurized at 400℃, the crystal structure is cubic (F-43m) Cu_2SnS_3 with a band gap energy of 1.02eV. For Sample B, the annealing temperature is increased to 450℃ and the crystal structure can be arranged to $Cu_5Sn_2S_7$ (JCPDS 40-0924) with band gap energy of 1.45eV. After annealed at 500℃, tetragonal (I-42m) Cu_3SnS_4 (JCPDS 33-0501) can be obtained with band gap energy of 1.47eV for Sample C. Three Cu-Sn-S thin films compounds have high optical absorption coefficients from UV band to NIR band in the order of $10^4 cm^{-1}$. From the electrical characterizations it can be seen that all three samples are p-type semiconductors with carrier concentrations of $3.1×10^{19} cm^{-3}$, $5.3×10^{20} cm^{-3}$ and $9.1×10^{20} cm^{-3}$ respectively. The resistivity for three samples is $2.4×10^{-2} \Omega \cdot cm$, $8.7×10^{-3} \Omega \cdot cm$ and $1.5×10^{-4} \Omega \cdot cm$ respectively.

With appropriate optical band gaps and outstanding absorption coefficients, most importantly containing earth-abundant chemical elements and relatively less compositions for three Cu-Sn-S thin films, it is extremely possible to be alternative absorber layer materials in next generation thin film solar cells. In addition, the improved SILAR method can also benefit to the industrial application in large-scale for high quality, high efficiency and low-cost production of thin films. Moreover, some novel properties found in Cu-Sn-S thin films may be applied in researching on thermoelectric and lithium battery materials. Nevertheless, much works should be performed further, such as the real assessment of performance in complete solar cell as well as thermoelectric cell devices.

Acknowledgements

This work was supported by the National High Technology Research and Development Program of China(863 Program, Grant No. 2011aa050529), the Fundamental Research Funds for the Central Universities(Grant No. 201021100029) and the Research Fund of Young Scholars for the Doctoral Program of Higher Education in China(Grant No. 200805331121).

Notes and References

To whom correspondence should be addressed:

School of Metallurgical Science and Engineering, Central South University, Changsha 410083 China. Fax:86 731 88876454;Tel:86 731 88830474;

Fangyang Liu, E-mail:liufangyang@ csu. edu. cn

Yanqing Lai, E-mail:csulightmetals03@ 163. com

References

[1] S. Fiechter, M. Martinez, G. Schmidt, W. Henrion and Y. Tomm, J Phys Chem Solids, 2003, 64, 1859-1862.
[2] Daqing Wu, Charles R. Knowles and Lure L. Y. Chang, Mineralogical Magazine, 1986, 50, 323-325.
[3] X. Chen, H. Wada, A. Sato and M. Mieno, J Solid State Chem, 1998, 139, 144-151.
[4] M. Onoda, X. A. Chen, A. Sato and H. Wada, Mater Res Bull, 2000, 35, 1563-1570.
[5] Q. H. Liu, Z. C. Zhao, Y. H. Lin, P. Guo, S. J. Li, D. C. Pan and X. L. Ji, Chem Commun, 2011, 47, 964-966.
[6] B. Li, Y. Xie, J. X. Huang and Y. T. Qian, J Solid State Chem, 2000, 153, 170-173.
[7] Y. J. Xiong, Y. Xie, G. A. Du and H. L. Su, Inorg Chem, 2002, 41, 2953-2959.
[8] G. Marcano, C. Rincon, G. Marin, R. Tovar and G. Delgado, J Appl Phys, 2002, 92, 1811-1815.
[9] X. Y. Chen, X. Wang, C. H. An, J. W. Liu and Y. T. Qian, J Cryst Growth, 2003, 256, 368-376.
[10] MTS Nair, C. Lopez-Mata, O. GomezDaza and P. K. Nair, Semicond Sci Tech, 2003, 18, 755-759.
[11] H. M. Hu, Z. P. Liu, B. J. Yang, X. Y. Chen and Y. T. Qian, J Cryst Growth, 2005, 284, 226-234.
[12] M. Bouaziz, M. Amlouk and S. Belgacem, Thin Solid Films, 2009, 517, 2527-2530.
[13] M. Bouaziz, K. Boubaker, M. Amlouk and S. Belgacem, Journal of Phase Equilibria and Diffusion, 2010, 31, 498-503.
[14] P. A. Fernandes, PMP Salome and A. F. da Cunha, J Phys D Appl Phys, 2010, 43, 215403.
[15] D. Avellaneda, MTS Nair and P. K. Nair, J Electrochem Soc, 2010, 157, D346-D352.
[16] M. Bouaziz, J. Ouerfelli, S. K. Srivastava, J. C. Bernede and M. Amlouk, Vacuum, 2011, 85, 783-786.
[17] T. K. Todorov, K. B. Reuter and D. B. Mitzi, Adv Mater, 2010, 22, E156.
[18] D. B. Mitzi, O. Gunawan, T. K. Todorov, K. Wang and S. Guha, Solar Energy Materials and Solar Cells, 2011, 95, 1421-1436.

[19] C. Z. Wu, Z. P. Hu, C. L. Wang, H. Sheng, J. L. Yang and Y. Xie, Appl Phys Lett, 2007, 91, 143104.

[20] M. Bouaziz, J. Ouerfelli, M. Amlouk and S. Belgacem, Physica Status Solidi a-Applications and Materials Science, 2007, 204, 3354-3360.

[21] Y. F. Nicolau, Applications of Surface Science, 1985, 22/23, 1061-1074.

[22] J. J. Li, Y. A. Wang, W. Z. Guo, J. C. Keay, T. D. Mishima, M. B. Johnson and X. G. Peng, J Am Chem Soc, 2003, 125, 12567-12575.

[23] H. M. Pathan and C. D. Lokhande, B Mater Sci, 2004, 27, 85-111.

[24] J. X. Yang, Z. G. Jin, T. J. Liu, C. J. Li and Y. Shi, Sol Energ Mat Sol C, 2008, 92, 621-627.

[25] Y. Shi, Z. G. Jin, C. Y. Li, H. S. An and J. J. Qiu, Appl Surf Sci, 2006, 252, 3737-3743.

[26] H. Lee, M. K. Wang, P. Chen, D. R. Gamelin, S. M. Zakeeruddin, M. Gratzel and M. K. Nazeeruddin, Nano Lett, 2009, 9, 4221-4227.

[27] J. J. Wu, W. T. Jiang and W. P. Liao, Chem Commun, 2010, 46, 5885-5887.

[28] Xiangdong Gao, Xiaomin Li and Weidong Yu, J Inorg Mater, 2005, 20, 965-970.

[29] H. Katagiri, Thin Solid Films, 2005, 480, 426-432.

[30] S. Lindroos, A. Arnold and M. Leskela, Appl Surf Sci, 2000, 158, 75-80.

[31] A. Wangperawong, J. S. King, S. M. Herron, B. P. Tran, K. Pangan-Okimoto and S. F. Bent, Thin Solid Films, 2011, 519, 2488-2492.

[32] David Cahen, P. J. Ireland, L. L. Kazmerski and F. A. Thiel, J. Appl. Phys., 1985, 57, 4761.

[33] L. D. Partain, R. A. Schneider, L. F. Donaghey and P. S. McLeod, J. Appl. Phys., 1985, 57, 5056-5065.

[34] A. R. H. F. Ettema and C. Haas, J. Phys.: Condens. Matter, 1993, 5, 3817-3826.

[35] L. Hernan, J. Morales, L. Sanchez, J. L. Tirado, J. P. Espinos and A. R. Gonzalez Elipe, Chem. Mater., 1995, 7(8), 1576-1582.

[36] Ivan Grozdanov and Metodija Najdoski, J Solid State Chem, 1995, 114, 469-475.

[37] G. J. Snyder and E. S. Toberer, Nature Materials, 2008, 7, 105-114.

[38] Min-Ling Liu, Fu-Qiang Huang, Li-Dong Chen and I. Wei Chen, Appl Phys Lett, 2009, 94, 202103.

[39] C. Sevik and T. Çağın, Appl Phys Lett, 2009, 95, 112105.

[40] Alexey Shavel, Doris Cadavid, Maria Ibanez, Alex Carrete and Andreu Cabot, J Am Chem Soc, 2012, 134, 1438-1441.

[41] Haoran Yang, Luis A. Jauregui, Genqiang Zhang, Yong P. Chen and Yue Wu, Nano Lett, 2012, 12, 540-545.

Colloidal Synthesis and Characterisation of Cu_3SbSe_3 Nanocrystals[*]

Abstract Cu_3SbSe_3 nanocrystals have been synthesized using hot-injection method. The Cu_3SbSe_3 nanocrystals possess a band gap of 1.31eV and the corresponding nanocrystals-electrode shows an incident photon to current efficiency (IPCE) of 10%-35% in the visible region. Our work demonstrates that Cu_3SbSe_3 nanocrystals have the potential in photo-electric conversion devices application.

Semiconductor nanocrystals are promising candidates in the fields of photovoltaics, thermoelectric, Li-ion batteries and light emitting diodes. Among them, the Cu-Ⅴ-Ⅵ (Ⅴ = Sb, Bi; Ⅵ = S, Se) nano-materials, composed of environmentally benign elements, were considered to be the potential candidates for next-generation solar energy conversion materials and thus this class of the ternary chalcogenides including $CuSbS_2$,[1-4] $Cu_{12}Sb_4S_{13}$,[4,5] Cu_3SbS_4,[6] Cu_3SbS_3,[4,7] $CuSbSe_2$[8-10] and Cu_3BiS_3[11] have received much attention due to their novel properties, earth abundant nature and low cost potential.

Ternary Cu_3SbSe_3 (CASe) is an emerging semiconductor.[12] Like other copper-based ternary and quaternary selenides, CASe has shown excellent thermoelectric properties, which possesses anomalously low thermal conductivity induced by the presence of two additional nonbonding electrons of the Sb^{3+} ions.[13-15] Recently, theoretical calculations[16,17] have demonstrated that the CASe owns good photovoltaic properties. The CASe thin film, which displays p-type conductivity behaviour and a direct optical band gap of 1.68eV, has been synthesized using electrodeposition method by Fernandez and Turner.[18] Maiello et al. used $Cu_3Sb(Se_xS_{1-x})_3$ (CASSe) film to fabricate solar cell devices by two steps synthesis route, obtaining open circuit voltage (V_{oc}) and short circuit current density (I_{sc}) of 3.5mV and 1.6mA·cm^{-2}, respectively.[19] Owing to its strengths of low cost and non-toxicity, CASe holds the promise for realising low cost fabrication of thin film solar cells and thermoelectric devices. There are few works on the synthesis of CASe. CASe films or bulk materials were reported to be prepared via Cu-Sb or Cu-Sb-Se precursor annealing methods at present.[18,19] However, to the best of authors' knowledge, there has been no previous report on the synthesis of CASe nano-materials until now. In order to expand CASe's application and realize high output roll to roll process, it is of great importance to explore new synthetic methods enabling low cost, large scale and high quality CASe production. Nanocrystal-ink painting approach is such a technology possessing up-scaling potential for thin film fabrication, which has been extensively used, especially in the photovoltaic

---

[*] Copartner: Yike liu, Jia Yang, Ening Gu, Tiantian Cao, Zhenghua Su, Liangxing Jiang, Chang Yan, Xiaojing Hao, Fangyang Liu. Reprinted from Journal of Materials Chemistry A. 2014, 2: 6363-6367.

field.[20-25] In this approach, the semiconductor nanocrystals were dispersed in solvents to form a stable nano-ink solution. The nano-ink could then be coated on a substrate using a roll-to-roll method and sintered to form film materials.

In this communication, we firstly reported the synthesis of high quality CASe nanocrystals by a hot injection method in oleylamine(OLA). A novel highly reactive selenium(Se) precursor was used in the all experiments for the purpose of phosphine-free synthesis of CASe nanocrystals. It is well known that Se powder is difficult to dissolve in OLA to form a reactive precursor on its own. But it has recently been reported that Se powder can be dissolved in the presence of OLA via phase transfer protocol.[26,27] In this protocol, Se was first reduced by sodium borohydride($NaBH_4$) or dodecanethiol(DT) into low valence Se, which further associates with OLA to form OLA_mSe_n, thus making the element Se soluble. However $NaBH_4$ is easily oxidized in the open air, and its reduction by $NaBH_4$ must be assisted with ultrasonication. Besides, DT is unstable at high synthesis temperature, which would introduce elemental sulphur due to DT decomposition during the synthesis process. In contrast, dimethylamine borane(DMAB) is a more stable reducing reagent. Herein DMAB was employed to facilitate the dissolution of Se powder in OLA. Se was first reduced by DMAB to generate an alkylammonium selenide at 110℃, which could then be dissolved in OLA to form a bright-yellow color solution. This Se precursor is highly reactive and suitable for the synthesis of CASe nanocrystals. The necessary conditions to synthesize phase pure nanocrystals with narrow size distribution are discussed and corresponding growth process is preliminarily investigated. The synthesized nanocrystals have the diameters in the range from about 13nm to 18nm and an optical band gap of 1.31eV. These nanocrystals can be easily dispersed in a comparably low-toxic solvent to form a stable ink, and then the CASe film can be readily fabricated using the developed ink by drop casting method. The obtained films demonstrate a clear photoresponse, showing that CASe thin films synthesized from its nanocrystals have the potential in the application of photo-electric conversion devices.

The syntheses of CASe nanocrystals were carried out utilizing a hot-injection method via a standard air-inert Schlenk line. In a typical synthesis, Se powder was dissolved in OLA completely with the assistance of DMAB as co-solvent at 110℃, obtaining a Se/DMAB/OLA solution with bright-yellow color. Then the Se/DMAB/OLA solution was quickly injected into the hot solution of OLA containing stoichiometric amount of copper(Ⅱ)acetylacetonate and antimony(Ⅲ)acetate[atom ratio Cu∶Sb=3∶1]at 190℃ for 30min under Ar atmosphere. Full experimental details can be found in the Supporting Information. Fig. 1 demonstrates basic structural characterization of synthesized nanocrystals. They are round shaped nanocrystals with an average diameter of (15.7±2.7) nm as indicated in the low-magnification TEM image (Fig. 1a). The corresponding size distribution chart is given as an inset in Fig. 1a(additional TEM images can be found in the Fig. 1). A high-resolution TEM(HRTEM)image(Fig. 1b) shows a clear crystalline surface with the interplanar spacing of 1.95, 2.00 and 2.32Å corresponding to the(322),(400)and(222)planes of CASe, respectively, demonstrating the nano-scale evidence that the synthesized nanocrystals possess the structure of CASe. The lattice data calculated from selected area electron diffraction(SAED)pattern of randomly chosen region of CASe nanocrystals agrees well with the lattice parameters of CASe(Fig. 1c).

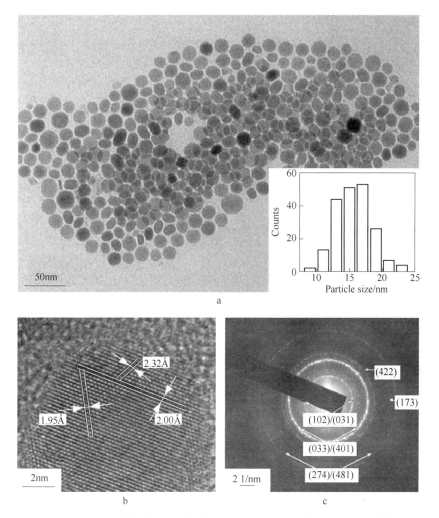

Fig. 1 Low resolution TEM image of CASe nanocrystals(a); High resolution TEM image showing interplanar spacing of 1.95Å, 2.00Å and 2.32Å(b); The SAED pattern indexed to CASe(c)

Fig. 2 illustrates the X-ray diffraction(XRD) pattern of the prepared nanocrystal. All the diffraction peaks can be readily indexed to the orthorhombic phase (Pnma) of CASe (JCPDS no. 86-1751). No evidence of other impurities such as Cu_2Se (JCPDS no. 65-2982), $CuSbSe_2$ (JCPDS no. 75-0992), etc., can be found from the XRD pattern. The element composition of synthesized CASe nanocrystals estimated by Energy Dispersive X-ray spectroscopy (EDS) is $Cu_{2.98}SbSe_{3.09}$ (See Fig. 2), which is close to the stoichiometric composition of CASe.

The composition and valence states of the CASe nanocrystals were further investigated by XPS. As shown in Fig. 3, the binding energies of Cu 2p, Sb 3d, and Se 3d of the nanocrystals in the XPS are consistent with those reported in literatures.[7,28,29] The valence states of Cu, Sb, and Se ions are +1, +3, and -2, respectively. The composition determined from XPS is also close to the stoichiometric composition of CASe, agreeing well with the EDS result.

Besides, we investigated the effect of precursor ratios and found they play a key role in the synthesis process (XRD patterns for synthesized product at different ratios of precursors were given in the Fig. 3). Considering elemental ratio Cu : Sb of 3 : 1 in CASe, a stoichiometric of

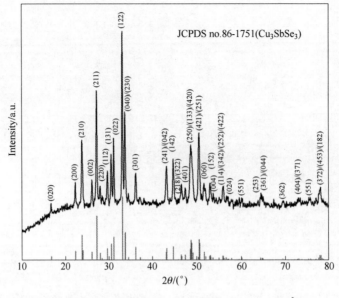

Fig. 2 XRD pattern of CASe nanocrystals [Cu Kα radiation (λ = 1.54Å)]. The reference pattern is standard tetragonal CASe (JCPDS no. 86-1751)

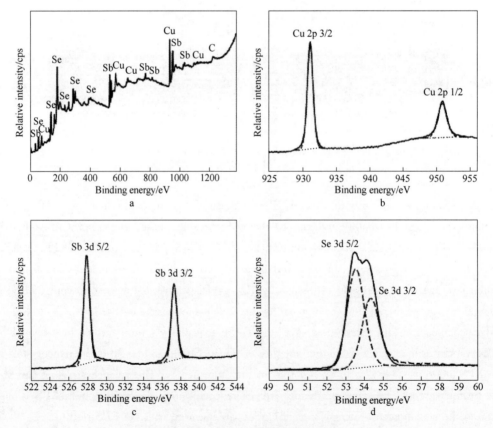

Fig. 3 XPS spectra of CASe nanocrystals
a—The full Scan spectrum; b—Cu 2p spectrum; c—Sb 3d spectrum; d—Se 3d spectrum
The fits are also shown. (dotted lines)

3-fold excess of copper over antimony(Cu : Sb = 3 : 1) was applied in all experiments. The formation of pure CASe was found to be highly dependent on the content of Se precursor. During a series of trial reactions, the amount of injection Se was varied. For mole ratios of metal(Cu+Sb) to Se within the range of 4 : 3 to 4 : 4, the CASe phase was produced while both $CuSbSe_2$ (JCPDS no. 75-0992) and Cu_2Se (JCPDS no. 65-2982) were detected simultaneously. Interestingly, with the increase of content of Se, peaks assigned to $CuSbSe_2$ and Cu_2Se gradually disappeared. Pure CASe can be obtained when (Cu+Sb) : Se ratio was 4 : 4. In addition, the use of excess Se was found to be necessary to balance relativity of cationic precursors to avoid the side reactions and the formation of impurities. A similar synthetic strategy has also been previously reported and used for synthesis of I-III-VI semiconductor nanocrystals via hot-injection method.[28,30,31]

To explore the possible pathway for the formation of CASe nanocrystals, the phase data of the reaction products upon different reaction time were studied by XRD(See Fig. 4). In the initial stage(1min), CASe phase has already been formed, indicating that the Se precursor has high reactivity, and thereby facilitates the reaction between Cu, Sb and Se simultaneously. No other impurities phase, such as CuSe, Cu_2Se, and $CuSbSe_2$ were detected in the XRD pattern, which suggests that the synthesis of CASe nanocrystals should be one-step process. With the reaction time proceeded to 4min, 8min, 16min and 32min, according to XRD, the main phase of CASe remained unchanged. However, when the reaction time was reached 64min, peaks assigned to Cu_3SbSe_4(JCPDS no. 25-0263) were detected, which might result from the further reaction between CASe nanocrystals and excess Se.

In addition, experiments at different reaction (injection) temperatures have been trialled. As shown in Fig. 5-Fig. 6, at the relative low temperature, such as 160℃, 190℃ and 220℃, the products are pure CASe. The optimum reaction temperature of the reaction is found to be around 190℃. When the temperate was elevated to 220℃, NCs showed slight aggregation. While at 160℃, the rate of the reaction decreased and nonuniform nanocrystals were obtained. When the synthesis temperatures are raised above 250℃, $CuSbSe_2$ was detected in addition to the desired phase CASe nanocrystals.

The absorption spectrum of as-synthesized CASe nanocrystals was measured using UV-vis-NIR absorbance spectroscopy(See Fig. 4). The band gap of the CASe nanocrystals is estimated to be 1.31eV by extrapolating the linear region of the plot of the absorbance squared versus energy, as shown in the inset of Fig. 4. The value is close to the optimal value for solar cell application.[32] In addition, a broad band that extends from around 700nm to the NIR region is observed. The origin of this band needs to be further explored in the future work.

To test the photoresponse of CASe nanocrystals, thin film of the nanocrystals was drop-casted onto ITO substrate. This film was mounted to a custom-built three-electrode photoelectrochemical cell with a graphite counter electrode and saturated calomel reference electrode(SCE) containing 0.5M H_2SO_4. In general, photoelectrochemical (PEC) is an excellent tool to assess the photovoltaic performance of new light-absorber materials. This method is popular because it could reduce the recombination probability of minority carriers with photogenerated holes and provide nearly ideal contact to the half cell, avoiding problems such as interface contact and lattice matc-

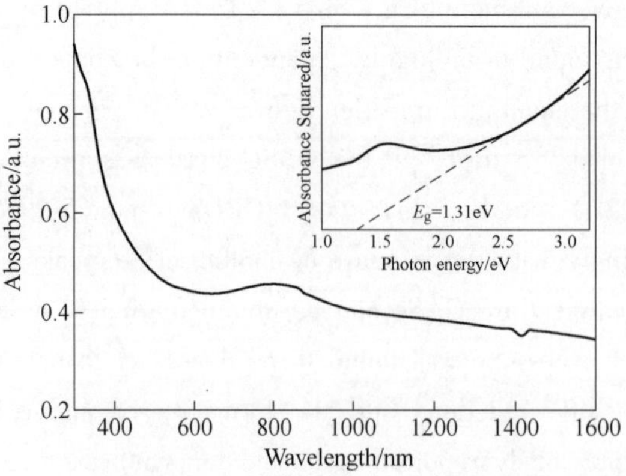

Fig. 4 UV-vis-NIR absorption spectrum of the as-synthesized nanocrystals.
Inset shows the abs^2 vs. eV for the CASe nanocrystals; the estimated band energy is 1.31eV

hing with the full device architecture.[11,33]

Fig. 5a displays the current density versus potential (vs SCE) plots for the prepared CASe films utilizing chopping method (40s light on, 40s light off). The photocurrent density increases with negative shift of the cathodic potential, which is the typical character of p-type semiconductor for CASe nanocrystals film.[27] In this case, electrons are transferred from the conduction band to the oxidant in solution, and then from the back contact (here ITO) into the semiconductor, leading to the observed reductive current. The photocurrent density reaches a saturated value of about 0.22mA · cm^{-2} at −0.65V vs. SCE. The transient photocurrent at this voltage is obtained (see Fig. 7), showing excellent photoresponse characteristics and good photostability of nanoink casting CASe film over many cycles (12 cycles demonstrated here). The magnitude of the gained photoresponse is high enough to acquire the IPCE spectrum, as show in Fig. 5b. Approximately 10%-35% of the incident photons can be readily converted to electron and hole pairs in the wavelength region of 400-980nm that covers most of the visible spectrum. The IPCE of CASe is higher than that of the others Cu-Ⅴ-Ⅵ semiconductor materials[1,10] in the entire absorption region. The IPCE value reaches zero when the photon energy

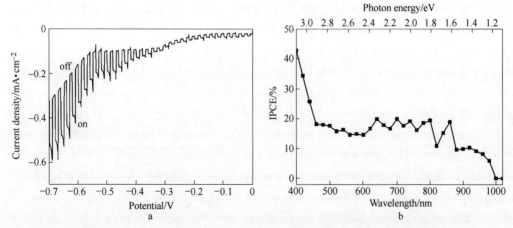

Fig. 5 Photocurrent-potential plot of the CASe films on an ITO substrate in 0.5M H$_2$SO$_4$ under 100mW · cm^{-2} illumination. The inset shows the transient photocurrent spectrum at −0.65V vs. SCE

goes below 1.24eV, which corresponds to the band gap of synthesized CASe and agrees well with the value estimated from the UV-vis-NIR data.

Conclusions

In summary, a colloidal synthesis of CASe nanocrystals via the hot-injection method is presented by utilizing a high reactivity phosphine-free Se precursor. The structure of the synthesized CASe nanocrytals is confirmed simultaneously by high-resolution TEM image, SAED pattern and XRD. The results of EDS and XPS confirmed the composition of the CASe nanocrystals. Both the ratios of precursors and temperature play a crucial role in determining the phase and shape of the synthesised nanocrystals. CASe nanocrystals could be formed in a short period of time owing to the high reactivity of Se precursor. A band gap of 1.31eV for CASe nanocrystals is estimated by UV-vis-NIR data. The CASe nanocrystals film yields an excellent photoresponse in PEC tests and an IPCE of 10%-35% in the visible region. Our work demonstrates that the synthesized CASe nanocrystals have potential in the field of photo-electric conversion devices application. Our future work will focus on the utilization of CASe nanocrystals for the development of full photovoltaic devices.

Acknowledgements

This work was supported by the National Natural Science Foundation of China (Grant No. 51222403) and China Postdoctoral Science Foundation (Grant No. 2012M511403 and 2013T60777).

References

[1] C. Yan, Z. Su, E. Gu, T. Cao, J. Yang, J. Liu, F. Liu, Y. Lai, J. Li and Y. Liu, RSC Advances, 2012, 2, 10481-10484.
[2] A. Rabhi and M. Kanzari, Chalcogenide Letters, 2011, 8, 255-262.
[3] W. Septina, S. Ikeda, Y. Iga, T. Harada and M. Matsumura, Thin Solid Films, 2014, 550, 700-704.
[4] D. Xu, S. Shen, Y. Zhang, H. Gu and Q. Wang, Inorganic Chemistry, 2013, 52, 12958-12962.
[5] J. van Embden, K. Latham, N. W. Duffy and Y. Tachibana, Journal of the American Chemical Society, 2013, 135, 11562-11571.
[6] J. van Embden and Y. Tachibana, Journal of Materials Chemistry, 2012, 22, 11466-11469.
[7] X. Qiu, S. Ji, C. Chen, G. Liu and C. Ye, CrystEngComm, 2013, 15, 10431-10434.
[8] D. Tang, J. Yang, F. Liu, Y. Lai, M. Jia, J. Li and Y. Liu, Electrochemical and Solid-State Letters, 2011, 15, D11-D13.
[9] D. Colombara, L. M. Peter, K. Rogers, J. Painter and S. Roncallo, Thin Solid Films, 2011, 519, 7438-7443.
[10] D. Tang, J. Yang, F. Liu, Y. Lai, J. Li and Y. Liu, Electrochimica Acta, 2012, 76, 480-486.
[11] C. Yan, E. Gu, F. Liu, Y. Lai, J. Li and Y. Liu, Nanoscale, 2013, 5, 1789-1792.
[12] A. Pfitzner, Zeitschrift für anorganische und allgemeine Chemie, 1995, 621, 685-688.
[13] C. Sevik and T. Cagın, Journal of Applied Physics, 2011, 109, 123712.
[14] E. J. Skoug, J. D. Cain and D. T. Morelli, Applied Physics Letters, 2010, 96, 181905.
[15] Y. Zhang, E. Skoug, J. Cain, V. Ozoliņš, D. Morelli and C. Wolverton, Physical Review B, 2012, 85, 054306.

[16] L. Yu, R. S. Kokenyesi, D. A. Keszler and A. Zunger, Advanced Energy Materials, 2013, 3, 43-48.

[17] A. B. Kehoe, D. J. Temple, G. W. Watson and D. O. Scanlon, Physical Chemistry Chemical Physics, 2013, DOI:10. 1039/C3CP52482E, 15477-15484.

[18] A. M. Fernández and J. A. Turner, Solar Energy Materials and Solar Cells, 2003, 79, 391-399.

[19] Pietro Maiello, Guillaume Zoppi, Robert W. Miles, Nicola Pearsall and I. Forbes, Solar Energy Materials and Solar Cells, 2013, 113, 186-194.

[20] Q. Guo, G. M. Ford, W. C. Yang, B. C. Walker, E. A. Stach, H. W. Hillhouse and R. Agrawal, Journal of the American Chemical Society, 2010, 132, 17384-17386.

[21] Q. Guo, H. W. Hillhouse and R. Agrawal, Journal of the American Chemical Society, 2009, 131, 11672-11673.

[22] Q. Guo, S. J. Kim, M. Kar, W. N. Shafarman, R. W. Birkmire, E. A. Stach, R. Agrawal and H. W. Hillhouse, Nano Letters, 2008, 8, 2982-2987.

[23] S. Peng, S. Zhang, S. G. Mhaisalkar and S. Ramakrishna, Physical Chemistry Chemical Physics, 2012, 14, 8523-8529.

[24] M. G. Panthani, V. Akhavan, B. Goodfellow, J. P. Schmidtke, L. Dunn, A. Dodabalapur, P. F. Barbara and B. A. Korgel, Journal of the American Chemical Society, 2008, 130, 16770-16777.

[25] J. Puthussery, S. Seefeld, N. Berry, M. Gibbs and M. Law, Journal of the American Chemical Society, 2010, 133, 716-719.

[26] Y. Wei, J. Yang, A. W. H. Lin and J. Y. Ying, Chemistry of Materials, 2010, 22, 5672-5677.

[27] Y. Liu, D. Yao, L. Shen, H. Zhang, X. Zhang and B. Yang, Journal of the American Chemical Society, 2012, 134, 7207-7210.

[28] H. Zhong, Y. Li, M. Ye, Z. Zhu, Y. Zhou, C. Yang and Y. Li, Nanotechnology, 2007, 18, 1-6.

[29] H. Chen, S. M. Yu, D. W. Shin and J. B. Yoo, Nanoscale research letters, 2010, 5, 217-223.

[30] R. Xie, M. Rutherford and X. Peng, Journal of the American Chemical Society, 2009, 131, 5691-5697.

[31] M. J. Thompson, T. P. A. Ruberu, K. J. Blakeney, K. V. Torres, P. S. Dilsaver and J. Vela, The Journal of Physical Chemistry Letters, 2013, DOI:10. 1021/jz402048p, 3918-3923.

[32] A. Goetzberger, C. Hebling and H. W. Schock, Mat Sci Eng R, 2003, 40, 1-46.

[33] H. Ye, H. S. Park, V. A. Akhavan, B. W. Goodfellow, M. G. Panthani, B. A. Korgel and A. J. Bard, The Journal of Physical Chemistry C, 2010, 115, 234-240.

Kesterite Cu$_2$ZnSn(S,Se)$_4$ Solar Cells with Beyond 8% Efficiency by A Sol-gel and Selenization Process[*]

Abstract A facile sol-gel and selenization process has been demonstrated to fabricate high-quality single-phase earth abundant kesterite Cu$_2$ZnSn(S,Se)$_4$(CZTSSe) photovoltaic absorbers. The structure and band gap of the fabricated CZTSSe absorbers can be readily tuned via varying the [S]/([S]+[Se]) ratio by selenization condition control. The effects of [S]/([S]+[Se]) ratio on device performance have been presented. The best device shows 8.25% total area efficiency without antireflection coating. Low fill factor is the main limitation for the current device efficiency compared to record efficiency device due to high series resistance and interface recombination. By improving film uniformity, eliminating voids and reducing Mo(S,Se)$_2$ interfacial layer, a further boost of the device efficiency is expected, enabling the proposed process the most promising candidate for kesterite solar cells.

Key words kesterite, Cu$_2$ZnSn(S,Se)$_4$, thin fim solar cell, sol-gel, interface recombination

1 Introduction

Kesterite Cu$_2$ZnSn(S,Se)$_4$(CZTSSe) semiconductor has attracted world-wide attention due to its excellent optical and electronic properties comparable to traditional Cu(In,Ga)Se$_2$(CIGS) and CdTe materials for thin film solar cells while consisting of earth-abundant and low-toxic constituent elements. Remarkable progresses have been made in CZTSSe solar cells over the past few years and the highest efficiency(PCE) of 12.6% has been achieved by IBM group[1] showing substantial commercial promise.

Similar to CIGS, CZTSSe thin films can be fabricated by both vacuum and solution processes. For vacuum based processes, over 9% efficiencies of Cu$_2$ZnSnSe$_4$ (CZTSe) devices have been realized by co-evaporation[2,3] which is an in-situ growth process allowing real-time control in chemical composition and reaction path. Cu$_2$ZnSnS$_4$(CZTS) devices from evaporated precursors and sulfurization have yielded efficiencies above 8%.[4,5] The annealing of sputtered precursors under reactive atmosphere containing S$_2$[6] or H$_2$Se[7] has also created devices with efficiencies beyond 9%. For solution based processes, they can be sub-divided into electrodeposition, nanoparticle ink coating and pure solution coating approaches. Devices from electrodeposited metallic precursors followed by sulfurization and selenization have achieved 8%[8] and 7%[9] efficiencies, respectively. Several successful investigations based on nanoparticle ink coating followed by chalcogenization annealing have been reported reaching 8.5% from mixed binary and

---

[*] Copartner: Fangyang Liu, Fangqin Zeng, Ning Song, Liangxing Jiang, Zili Han, Zhenghua Su, Chang Yan, Xiaoming Wen, Xiaojing Hao. Reprinted from ACS applied material & interface, 2015, 7(26): 14376-14383.

ternary nanoparticles,[10] 9.0% from quaternary CZTS nanocrystals,[11] and 8.4% from Ge incorporated CZTS nanocrystals.[12] It is worthy to note that a series of record-setting devices have been developed by IBM group using a hydrazine based hybrid solution-particle slurry process.[1,13-15] In contrast with all aforementioned technologies, pure solution approach offers precursors homogeneity at a molecular scale and accordingly enables precise stoichiometric control and excellent film consistency, which are necessary for low-cost and large-scale production. A number of attractive efficiencies for CZTSSe cells have been achieved based on various solvents: 8.08% by hydrazine-hydrazinocarboxylic acid dissolution of Zn metal,[16] 6.13% by water-ethanol,[17] 6.03% by ethanol-CS_2-1-butyamine-thioglycolic acid,[18] 8.32% by dmethyl sulfoxide(DMSO),[19] and 10.6% by hydrazine(highest efficiency for pure solution approach).[20] Our group has previously reported a facile molecular pure solution route to fabricate sulfide CZTS thin films by a sol-gel and sulfurization process, producing CZTS solar cell with efficiency of 5.1%[21] which was then improved to 5.7% by low pressure sulfurization.[22] Our sol solution was made by dissolving Cu, Zn and Sn salts and excess thiourea in 2-methoxyethanol. Thiourea was added into the solvent to form metal-thiourea-oxygen complex with metal ions, then the complexes were thermally decomposed into CZTS xerogel precursor films during air annealing[21]. This is different from conventional sol-gel processes which entail forming metal-oxygen-metal bonds in sol solution and metal oxides in gel film[23,24] with the need of high reactive but toxic H_2S in post-sulfurization process for sulfides formation.[25,26] Here, employing the same CZTS xerogel precursor materials, we present an approach to fabricate sulfoselenide CZTSSe absorbers by post-selenization and report the solar cell device with a total-area efficiency of 8.2% without antireflection coating. The capability to control[S]/([S]+[Se])ratio and resulting structural and optoelectronic properties of the absorbers by adjusting the selenium partial pressure during selenization has also been demonstrated.

2 Experiment

Precursor sol solution preparation: The precursor solution was prepared by dissolving $Cu(CH_3COO)_2 \cdot H_2O$(0.46mol/L, AR), $Zn(CH_3COO)_2 \cdot 2H_2O$(0.27mol/L, AR), $SnCl_2 \cdot 2H_2O$(0.27mol/L, AR) and $SC(NH_2)_2$(2mol/L, AR) into 2-methoxyethanol(AR) while stirring at 50℃ for 1hour to get dark yellow solution. After aging at room temperature in air for 24h, the prepared precursor solution was converted into sol solution and then proper monoethanolamine was added and stirred to avoid cracks during spin coating. All chemical reagent were purchased from Sinopharm Chemical Reagent Co., Ltd.

CZTSSe thin film preparation: The precursor sol solution was spin coated on molybdenum coated soda-lime glass(SLG) substrate at 4500r/min for 30s followed by annealing at 270℃ for 10min on a hot plate in air. This coating step was repeated 12times to get thick CZTS xerogel precursor. After that, the prepared xerogel precursor was annealed at 560℃ in selenium/N_2 atmosphere for 40mins with controlled selenium partial pressure to obtain desired[S]/([S]+[Se])ratio and crystallinity.

CZTSSe solar cell device fabrication: The solar cell device was completed using a chemical bath deposited 70nm CdS buffer, RF magnetron sputtered 50nm intrinsic ZnO, and DC magnetron sputtered 400nm ITO window layer sequentially. Finally Al was thermally evaporated on ITO layer to form top contact fingers via shadow mask. Each device had a total area of approximately 0.45cm² defined by mechanical scribing.

Characterization and analysis: The surface and cross-section morphology of thin films were characterized by SEM (FEI Quanta-200 and NOVA NanoSEM 230). The X-ray diffraction (XRD) patterns and Raman spectra were collected by using Rigaku-TTR Ⅲ X and Jobin-Yvon LabRAM HR-800, respectively. Energy Dispersive Spectrometer (EDS, EDAX-GENSIS60S in NOVA NanoSEM 230) was used to check the elemental composition and its distribution. A FEI Tecnai G2 equipped with energy dispersive spectroscopy (EDS) detector was used for the transmission electron microscopy (TEM) analyses. Current density-voltage (J-V) characterization for solar cells were performed using Xe-based light source solar simulator (Newport, 91160 and KEITHLEY 2400) to provide simulated 1 sun AM 1.5G illumination which was calibrated with a standard Si reference cell, traceable to the National Renewable Energy Laboratory. The external quantum efficiency (EQE) was measured using chopped monochromator beam and lock-in amplifier, with calibration into the NIR by Si and Ge diodes measurements (QEX10 spectral response system from PV measurements, Inc). Time-resolved photoluminescence (TRPL) measurement was performed on device using the time-correlated single photon counting (TCSPC) technique (Microtime200, Picoquant) at a wavelength near band gap. The excitation is a 467nm laser with tunable repetition.

3 Results and Disussion

Top view SEM image of a typical CZTSSe absorber with atomic ratio of Cu/(Zn+Sn) of 0.85, Zn/Sn of 1.00 and [S]/([S]+[Se]) of 0.25 determined by EDS analysis is depicted in Fig. 1a, and low magnification SEM images are presented in Fig. 1 as well. The CZTSSe film has a relative dense structure consisting of large grains with size in the range of 1-2μm. Grain sizes up to 3μm can also be detected. It should be remarked that between the large grains, smaller grains are observable. EDS mapping measurements exhibited in Fig. 2 show that some small grain regions seems have slightly more sulfur signal than their surroundings, which may be associated with smaller grain size due to less Se incorporation and thus lower volume expansion during seleniation. All metal elements are homogeneously distributed which benefits from molecular level mix of them as well as S in sol-gel process and suggests inhomogeneous substitution Se for S in selenization process probably. As the nonuniformity of chalcolgen elements will have great impacts on resulting V_{oc} and efficiency, Raman mapping technology representing [S]/([S]+[Se]) ratio distribution was employed in order to more precisely assess the degree of this fluctuation of chalcolgen elements (i.e. [S]/([S]+[Se])). Before mapping, a representative Raman spectrum of the CZTSSe absorber was measured for phase identification and purity check first, as shown in Fig. 1b. For this sample with intermediate [S]/([S]+[Se]) value, the main peaks from A1 mode of anion vibrations in both selenide CZTSe and sulfide CZTS[27] are presented but deviate toward one other. This coexistence of both groups of bands, i.e. a bimodal behavior, has also been observed for chalcopyrite Cu(In,Ga)(S,Se)₂ material system[28,29] and

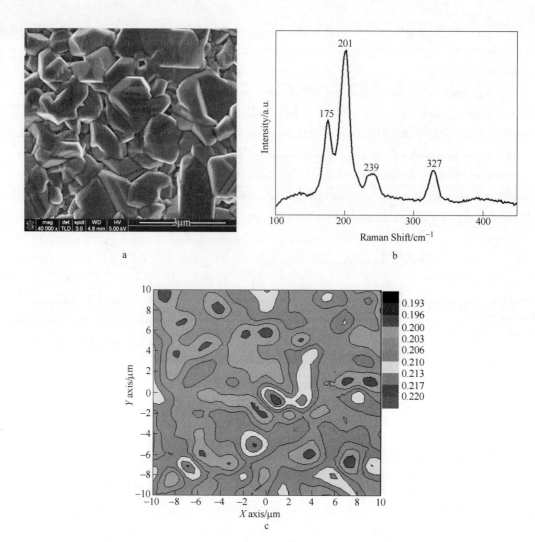

Fig. 1 Typical SEM top view(a) and Raman spectrum of the typical CZTSSe absorber(b) and Raman mapping representing the ratio of the peak intensity for sulfide and selenide peaks after background subtracting(c)

can be explained in terms of the large mass difference between these two types of anion and the resulting substantial difference between the frequencies of the respective phonons. [S]/([S] + [Se]) ratio calculated from the ratio of the peak intensity for sulfide(~327 cm^{-1}) and selenide (~201 cm^{-1}) peaks after background subtracting is 0.21, correlating well with that determined by EDS. Moreover, no evidence of other possible binary (Cu_2S, CuS_2, ZnS, $ZnSe$, SnS, $SnSe$ etc.) or ternary(Cu_2SnS_3 and Cu_2SnSe_3) secondary phases[30-32] can be observed from the prepared CZTSSe absorber. Then the distribution of [S]/([S] + [Se]) ratio is characterized by Raman mapping measurement at a $20 \times 20 \mu m^2$ area (441 points) as illustrated in Fig. 1c. Although there are some small regions appearing to have higher [S]/([S] + [Se]) ratio, all [S]/([S] + [Se]) ratios are concentrated in a very narrow range from 0.193 to 0.220 and the variation of [S]/([S] + [Se]) ratio distribution by a statistical analysis of the data is only 2% (mean value 0.205 with standard deviation 0.004), suggesting an acceptable distribution homogeneity. Even though, this slight lateral fluctuations in [S]/([S]+[Se]) ratio and resulting band gap may still considerably degrade the achievable efficiency of solar cells[33,34]. Therefore, further optimization on selenization process to improve lateral uniformity is needed.

Cross sectional SEM (CZTSSe device) and TEM (absorber, bright field) images are shown in Fig. 2a and b, respectively. It is shown that the absorber composed of mainly large grains spanning the entre layer with thickness of approximately 1μm. Again, few smaller grains between large grains are observed, leading to the uneven thickness. The carbon-rich and/ or fine grain bottom layer frequently observed in absorbers synthesized by solution processes using organic solvent cannot be found here, while a lot of isolated voids are visible at the bottom part of the

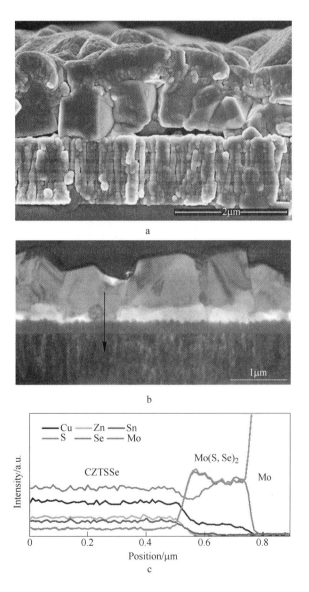

Fig. 2 SEM cross section of the typical CZTSSe device(a), Bright field TEM cross section of typical CZTSSe absorber(b) and Elemental profiles determined by EDS line scan along the blue arrow from absorber to Mo in panel b(c)

absorber. Compositional profiling by EDS line scan in Fig. 2c demonstrates uniform elemental distribution along the thickness direction. Both of the SEM and TEM data highlight a $Mo(S,Se)_2$ interfacial layer with a thickness of nearly 300nm between CZTSSe absorber layer and Mo, as is more evident in EDS composition profiling. The $Mo(S,Se)_2$ interfacial layer may facilitate an electrical quasi-ohmic contact and improve the adhesion of CZTS(e) to Mo back contact[35], but leads to high series resistance and accordingly deteriorates the fill factor and the resulting device efficiency if not thin enough,[36,37] similar to the case of CIGSe solar cells.[38] The 300nm $Mo(S,Se)_2$ layer is considerably thick compared with those in some devices with efficiency higher than 9%.[1-3,7]

As an absorber material, CZTSSe possesses a direct band gap in the range of ~1.0eV(CZTSe) to ~1.5eV(CZTS) depending upon the [S]/([S]+[Se]) ratio. The tailoring of band gap and its effects on device performance are of great interest. Based on the theoretical calculations, the optimum band gap for terrestrial single-junction solar cell is around 1.5eV. One would expect that the larger band gap absorbers of the kesterite system would result in higher efficiency, but this is not the case-most efficiencies achieved by CZTS or high sulfur-containing CZTSSe absorber are considerably lower than those achieved by high selenium-containing absorber. For instance, the band gap of kesterite absorber in record efficiency is 1.13eV.[1] Experimentally, Duan et al[39] reported that when enlarging the band gap of the absorber by increasing its sulfur content, the V_{OC} increases but the overall efficiency is reduced reflected by large V_{OC} deficit and poor carrier collection due to deeper defect energy level and higher bulk defect density. The capability to adjust the [S]/([S]+[Se]) ratio and resulting band gap is crucial to achieve high efficiency solar cells and has been demonstrated here by controlling the temperature of solid selenium source and accordingly the partial pressure of selenium vapor. Fig. 3a shows the XRD patterns for several CZTSSe absorbers with varying [S]/([S]+[Se]) ratios. All samples display the presence of several peaks between the ones for the CZTSe(PDF #52-0868) and CZTS(PDF #26-0575) phases, which allows us to believe the formation of kesterite CZTSSe. From the zoom of (112) peaks in the inset of Fig. 3a, the peak position shifts to smaller 2θ values linearly (Fig. 3a) as the [S]/([S]+[Se]) ratio decreases due to the increase in lattice parameters when larger Se atoms (1.98Å) replace the smaller S atoms (1.84Å), in accordance with other reports.[40,41] Raman spectra of these absorbers in Fig. 3b show that the A1 mode peaks shift toward each other linearly (Fig. 3b) with some noticeable peak broadening with [S]/([S]+[Se]) ratio away from both pure sulfide (338cm^{-1}) and selenide (197cm^{-1}) side, following the similar trend in other reports[16,42]. Moreover, combining XRD and Raman data, no secondary phases can be identified for all absorbers. The band gap values calculated from external quantum efficiency (EQE) of the solar cell devices using absorbers with the [S]/([S]+[Se]) ratio of 0.26, 0.35 and 0.68 are 1.10eV, 1.16eV and 1.40eV, respectively, as seen in Fig. 3c. The band gap of sulfide CZTS by sol-gel and sulfurization process[22] is also given here for a reference. These results exhibit the ability and flexibility of our process to tune the band gap of CZTSSe absorber.

Fig. 4 illuminates the current density-voltage (J-V) characteristic curves for devices fabricated using absorbers with different [S]/([S]+[Se]) ratios from 0.1 to 1. The performance parame-

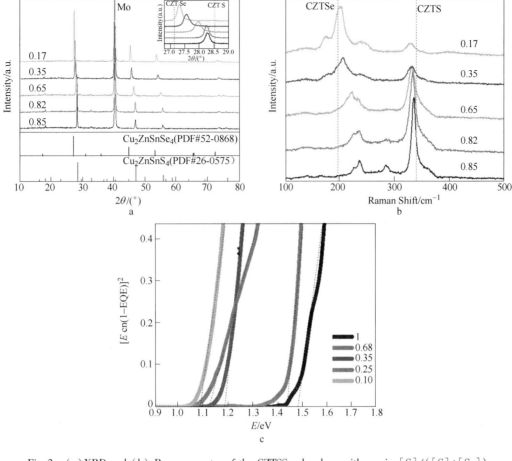

Fig. 3 (a) XRD and (b) Raman spectra of the CZTSSe absorbers with varying [S]/([S]+[Se]) ratios (determined by EDS); (a, inset) magnified view of (112) peaks. (c) Band gap calculated from EQE measurements of the device with varying [S]/([S]+[Se]) ratios

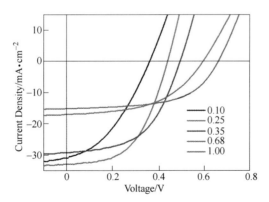

Fig. 4 Current density-voltage (J-V) data of the devices using CZTSSe aborbers with various [S]/([S]+[Se]) ratios (determined oy EDS)

ters deduced from J-V measurements are summarized in Table 1. As the [S]/([S]+[Se]) ratio increases, the open-circuit voltage (V_{OC}) rises, consistent with its enlarged band gap but the Voc deficit shows an increase trend, revealing that higher [S]/([S]+[Se]) ratio not only enlarges band gap but also leads to more severe recombination. The devices fabricated using absorbers with low [S]/([S]+[Se]) ratios have much larger short-circuit current density (J_{sc}) owing to low band gap having enhanced light harvesting in longer wavelength region. High efficiencies above 7% can be achieved for device using absorber with low [S]/([S]+[Se]) ratios (0.25 and 0.35, for example). Note that when [S]/([S]+[Se]) of the absorber is <0.1 the fabricated device efficiency is also decreased. This is because to obtain such high Se content usually requires extremely high Se partial pressure in the selenization process, during which absorber morphology has been deteriorated (very low shunt resistance) and very thicker Mo(S,Se)$_2$ interfacial layer was formed (~500nm, high series resistance) as seen from the cross-sectional SEM image in Fig. 4 Good reproducibility is demonstrated from several batches of devices fabricated under slightly varying conditions with efficiencies above 7%. In order to check large area film uniformity, a typical set of J-V curves from 9 devices on the same substrate are exhibited in Fig. 5. Efficiencies in the range of 7% ~ 8% with an average of 7.47% and standard deviation of 0.31 are achieved, revealing good large area uniformity. These results show that the sol-gel based route is very promising for large-scale production of kesterite CZTSSe solar cells.

Table 1 Performance parameters for device using CZTSSe absorbers with various [S]/([S]+[Se]) ratios

[S]/([S]+[Se])	η /%	V_{OC} /mV	J_{SC}/mA·cm^{-2}	FF /%	E_G /eV	E_G/q-V_{OC} /mV	R_s /Ω·cm^2	R_{sh} /Ω·cm^2
x=1.00	5.73	664	15.15	57.0	1.43	766	2.6	545
x=0.68	5.06	596	17.03	50.0	1.40	804	3.8	320
x=0.35	7.53	493	29.20	52.3	1.16	667	2.1	254
x=0.25	7.45	439	32.83	51.6	1.10	661	1.7	306
x=0.10	4.23	358	30.85	38.3	—	—	2.4	90

Fig. 5 depicts the J-V characteristics of the highest efficiency device achieved up to date by sol-gel solution method under dark and simulated AM 1.5 illumination. The cross-sectional SEM image has been shown in Fig. 2a. This device shows a total area efficiency of 8.25% with an open-circuit voltage of 451mV, a short-circuit current density of 31.7mA/cm^2 and fill factor (FF) of 57.7%. Compared to the record CZTSSe device[1], the efficiency in current device is limited mainly by low FF. The series resistance (R_s), shunt resistance (R_{sh}), diode ideality factor (A), and reverse saturation current density (J_0) extracted from the light J-V curves using the Sites' method[43] are 1.43Ω·cm^2, 476Ω·cm^2, 2.23 and 3.2×10^{-6} A/cm^2, respectively. The R_s, A and J_0 values are significantly higher than those in record device,[1] which accounts for the low FF. The high series resistance can be reasonably attributed to the presence of lots of voids at the bottom part of the absorber and thick Mo(S,Se)$_2$ interfacial layer between absorber and Mo back contact. By eliminating the undesirable voids and reducing the thickness of Mo(S,Se)$_2$ layer, and thereby reducing the R_s, a further boost in CZTSSe device efficiency is highly possible. The A value larger than 2 is the characteristic of strong space-charge region (SCR) recombination and/or the possibility of additional contributions to SCR recombination from trap assisted tunneling,[44,45] which both can lead to a high J_0.[45]

To gain further insight in device performance, the external quantum efficiency (EQE) of the device was measured and shown in Fig. 6. The integrated short-circuit current density extracted from the EQE data agrees well with that obtained under simulated sunlight, and the band gap

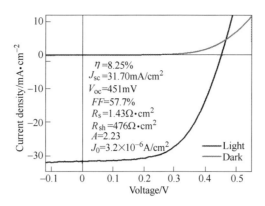

Fig. 5 Current density-voltage (J-V) characteristics of the best CZTSSe device with [S]/([S]+[Se]) of 0.25(determined by EDS) measured in dark and under AM 1.5 simulated sunlight

estimated from the EQE is 1.09eV. The EQE exceeds 80% in the visible light range, however decays at longer wavelength. The lack of long wavelength response reveals a loss of deeply absorbed photons and thereby low carrier collection efficiency, which is examined by the measurement of the ratio of EQE at reverse -1V and 0V bias, as shown on the top panel of Fig. 6. The ratio increases toward long wavelength direction confirms the recombination for minority carriers occurring in neutral region, i.e. the minority carriers generated deep in the absorber by long wavelength photons are difficult to be collected effectively, and a larger depletion width extended into absorber by reverse bias improve the collection efficiency, with a resultant increase in EQE. All EQE(-1V)/EQE(0V) ratios are above one in the whole wavelength range suggesting strong SCR recombination(interface or near the junction).[19,46] These behaviors are most likely caused by short carrier lifetime.[47] In addition, from the reflectivity spectrum of the device with an average reflectivity of 9.2% in the range of 300-1200nm, it is expected that a notable improvement in EQE and resulting J_{sc} should be achieved by the addition of a MgF_2 antireflection coating.

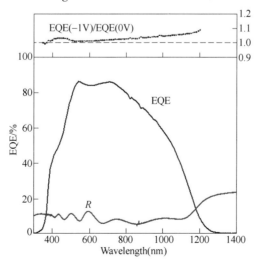

Fig. 6 (Bottom)External quantum efficiency (EQE) at 0V bias and redectance spectrum of the best CZTSSe device. (Top)EQE ratio at -1 and 0V bias

The time-resolved photoluminescence (TR-PL) measurement was performed to check the minority carrier lifetime of the device, as shown in Fig. 7. We use bi-exponential fitting to analyse the curve as described by B. Ohnesorge et al.[48] The initial fast decay with a time constant τ_1 of 2.3 ns is ascribed to Auger recombination in a high injection regime due to much higher carrier density than the normal steady state excitation induced by an ultrashort laser pulse in our measurement. The minority carrier lifetime characterized by the time constant τ_2 is 16.9 ns. This is a relatively long lifetime in CZTSSe system, but device performance is far below the expected level by comparing with the experimental reports[15] and model predictions.[49] This shortfall may be due to some other physical mechanisms that does not affect the measured lifetime.[49] One possible mechanism is the slight lateral nonuniformity in [S]/([S]+[Se]) ratio and resulting band gap fluctuations as revealed by EDS and Raman mapping, which allows significant interface recombination,[34] and may partly contribute to the nonlinear behavior the TRPL data, with the TRPL signal having individual contributions from different region with varying lifetimes.[50] On the other hand, the underperformance of the device also implies the necessity to optimize the device fabrication conditions, especially for CdS buffer, contacts and TCO to push efficiency up to 10% level. Anyway, the minority carrier lifetime is still much shorter than typical tens to hundreds of ns for high efficiency CIGS devices,[51,52] and need further improvement.

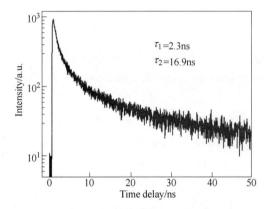

Fig. 7 Time-resolved photoluminescence trace carried out on the finished device

4 Conclusions

In summary, we present a simple and facile syntheses of earth abundant kesterite $Cu_2ZnSn(S,Se)_4$ (CZTSSe) absorbers with large grain, single phase and good uniformity using a sol-gel and selenization process. A slight fluctuation with acceptable deviation in [S]/([S]+[Se]) ratio has been confirmed by EDS and Raman mapping. The capacity for tuning the structural and optoelectronic properties through the adjustment of [S]/([S]+[Se]) composition by controlling the selenium partial pressure has been demonstrated. The maximum total area efficiency without antireflection coating of 8.25% has already been achieved based on this process. Relatively low *FF* is the main limitation for current device performance compared to the record CZTSSe device, caused by high series resistance and significant interface recombination. Optimization to improve film uniformity, eliminate the voids at the bottom part of the absorber, reduce Mo(S,Se)$_2$ interfacial layer and improve device fabrication process is expected to yield further enhancement in efficiency, enabling this process the most promising candidate for fabrication of low-cost, larger-area and high-efficiency kestertite solar cells.

Acknowledgements

The authors thank Professor Martin Green at University of New South Wales for helpful discussions on this work. The authors acknowledge Australian Government for financial support through the Australian Research Council(ARC). The authors also acknowledge the facilities, and the scientific and technical assistance of The Mark Wainwright Analytical Centre(MWAC) at The University of New South Wales.

References

[1] Wang W, Winkler M T, Gunawan O, Gokmen T, Todorov T K, Zhu Y, Mitzi D B. Device Characteristics of CZTSSe Thin-Film Solar Cells with 12.6% Efficiency. Advanced Energy Materials. 2014; 4: 1301465. DOI:10.1002/aenm.201301465.

[2] Repins I, Beall C, Vora N, DeHart C, Kuciauskas D, Dippo P, To B, Mann J, Hsu W-C, Goodrich A, Noufi R. Co-evaporated Cu$_2$ZnSnSe$_4$ films and devices. Solar Energy Materials and Solar Cells. 2012;101:154-159. DOI:10.1016/j.solmat.2012.01.008.

[3] Hsu W C, Repins I, Beall C, DeHart C, To B, Yang W B, Yang Y, Noufi R. Growth mechanisms of co-evaporated kesterite: a comparison of Cu-rich and Zn-rich composition paths. Progress in Photovoltaics: Research and Applications. 2014;22;35-43.

[4] Shin B, Gunawan O, Zhu Y, Bojarczuk N A, Chey S J, Guha S. Thin film solar cell with 8.4% power conversion efficiency using an earth-abundant Cu$_2$ZnSnS$_4$ absorber. Progress in Photovoltaics: Research and Applications. 2013;21:72-76.

[5] Hiroki Sugimoto H H, Noriyuki Sakai, Satoshi Muraoka and Takuya Katou. Over 8% Efficiency Cu$_2$ZnSnS$_4$ Submodules with Ultra-Thin Absorber. 2012 38th IEEE Photovoltaic Specialists Conference(PVSC). 2012: 002997-003000. DOI:10.1109/PVSC.2012.6318214.

[6] Chawla V, Clemens B,. Effect of composition on high efficiency CZTSSe devices fabricated using co-sputtering of compound targets. 2012 38th IEEE Photovoltaic Specialists Conference (PVSC). 2012: 002990-002992. DOI:10.1109/PVSC.2012.6318212.

[7] Brammertz G, Buffière M, Oueslati S, Elanzeery H, Ben Messaoud K, Sahayaraj S, Köble C, Meuris M, Poortmans J. Characterization of defects in 9.7% efficient Cu$_2$ZnSnSe$_4$-CdS-ZnO solar cells. Applied Physics Letters. 2013;103:163904.

[8] Jiang F, Ikeda S, Harada T, Matsumura M. Pure Sulfide Cu$_2$ZnSnS$_4$ Thin Film Solar Cells Fabricated by Preheating an Electrodeposited Metallic Stack. Advanced Energy Materials. 2013;4:1301381.

[9] Guo L, Zhu Y, Gunawan O, Gokmen T, Deline V R, Ahmed S, Romankiw L T, Deligianni H. Electrodeposited Cu$_2$ZnSnSe$_4$ thin film solar cell with 7% power conversion efficiency. Progress in Photovoltaics: Research and Applications. 2014;22:58-68.

[10] Cao Y, Denny M S, Caspar J V, Farneth W E, Guo Q, Ionkin A S, Johnson L K, Lu M, Malajovich I, Radu D, Rosenfeld H D, Choudhury K R, Wu W. High-Efficiency Solution-Processed Cu$_2$ZnSn(S,Se)$_4$ Thin-Film Solar Cells Prepared from Binary and Ternary Nanoparticles. Journal of the American Chemical Society. 2012;134:15644-15647.

[11] Miskin C K, Yang W C, Hages C J, Carter N J, Joglekar C S, Stach E A, Agrawal R. 9.0% efficient

[11] ... Cu$_2$ZnSn(S,Se)$_4$ solar cells from selenized nanoparticle inks. Progress in Photovoltaics: Research and Applications. 2014;DOI:10.1002/pip.2472.

[12] Guo Q, Ford G M, Yang W C, Hages C J, Hillhouse H W, Agrawal R. Enhancing the performance of CZTSSe solar cells with Ge alloying. Solar Energy Materials and Solar Cells. 2012;105:132-136.

[13] Todorov T K, Reuter K B, Mitzi D B. High-Efficiency Solar Cell with Earth-Abundant Liquid-Processed Absorber. Advanced Materials. 2010;22:E156-E159.

[14] Barkhouse D A R, Gunawan O, Gokmen T, Todorov T K, Mitzi D B. Device characteristics of a 10.1% hydrazine-processed Cu$_2$ZnSn(Se,S)$_4$ solar cell. Progress in Photovoltaics: Research and Applications. 2012;20:6-11.

[15] Todorov T K, Tang J, Bag S, Gunawan O, Gokmen T, Zhu Y, Mitzi D B. Beyond 11% Efficiency:Characteristics of State-of-the-Art Cu$_2$ZnSn(S,Se)$_4$ Solar Cells. Advanced Energy Materials. 2013;3:34-38.

[16] Yang W, Duan H S, Bob B, Zhou H, Lei B, Chung C H, Li S H, Hou W W, Yang Y. Novel Solution Processing of High-Efficiency Earth-Abundant Cu$_2$ZnSn(S,Se)$_4$ Solar Cells. Advanced Materials. 2012;24:6323-6329.

[17] Jiang M L, Lan F, Yan X Z, Li G Y. Cu$_2$ZnSn(S$_{1-x}$Se$_x$)$_4$ thin film solar cells prepared by water-based solution process. Physica Status Solidi-Rapid Research Letters. 2014;8:223-227.

[18] Wang G, Zhao W, Cui Y, Tian Q, Gao S, Huang L, Pan D. Fabrication of a Cu$_2$ZnSn(S,Se)$_4$ Photovoltaic Device by a Low-Toxicity Ethanol Solution Process. Acs Applied Materials & Interfaces. 2013;5:10042-10047.

[19] Xin H, Katahara J K, Braly I L, Hillhouse H W. 8% Efficient Cu$_2$ZnSn(S,Se)$_4$ Solar Cells from Redox Equilibrated Simple Precursors in DMSO. Advanced Energy Materials. 2014; DOI: 10.1002/aenm.201301823.

[20] Todorov T, Sugimoto H, Gunawan O, Gokmen T, Mitzi D B. High-Efficiency Devices With Pure Solution-Processed Cu$_2$ZnSn(S,Se)$_4$ Absorbers. IEEE Journal of Photovoltaics. 2014;4:483-485.

[21] Su Z, Sun K, Han Z, Cui H, Liu F, Lai Y, Li J, Hao X, Liu Y, Green M A. Fabrication of Cu$_2$ZnSnS$_4$ solar cells with 5.1% efficiency via thermal decomposition and reaction using a non-toxic sol-gel route. Journal of Materials Chemistry A. 2014;2:500-509.

[22] Zhang K, Su Z, Zhao L, Yan C, Liu F, Cui H, Hao X, Liu Y. Improving the conversion efficiency of Cu$_2$ZnSnS$_4$ solar cell by low pressure sulfurization. Applied Physics Letters. 2014;104:141101.

[23] Kwon S G, Hyeon T. Colloidal Chemical Synthesis and Formation Kinetics of Uniformly Sized Nanocrystals of Metals, Oxides, and Chalcogenides. Accounts of Chemical Research. 2008;41:1696-1709.

[24] Niederberger M. Nonaqueous Sol-Gel Routes to Metal Oxide Nanoparticles. Accounts of Chemical Research. 2007;40:793-800.

[25] Maeda K, Tanaka K, Fukui Y, Uchiki H. Influence of H$_2$S concentration on the properties of Cu$_2$ZnSnS$_4$ thin films and solar cells prepared by sol-gel sulfurization. Solar Energy Materials and Solar Cells. 2011;95:2855-2860.

[26] Tanaka K, Fukui Y, Moritake N, Uchiki H. Chemical composition dependence of morphological and optical properties of Cu$_2$ZnSnS$_4$ thin films deposited by sol-gel sulfurization and Cu$_2$ZnSnS$_4$ thin film solar cell efficiency. Solar Energy Materials and Solar Cells. 2011;95:838-842.

[27] Altosaar M, Raudoja J, Timmo K, Danilson M, Grossberg M, Krustok J, Mellikov E. Cu$_2$Zn$_{1-x}$Cd$_x$Sn(Se$_{1-y}$S$_y$)$_4$ solid solutions as absorber materials for solar cells. Physica Status Solidi a-Applications and Materials Science. 2008;205:167-170.

[28] Palm J, Jost S, Hock R, Probst V. Raman spectroscopy for quality control and process optimization of chalcopyrite thin films and devices. Thin Solid Films. 2007;515:5913-5916.

[29] Bacewicz R, Gebicki W, Filipowicz J. Raman scattering in CuInS$_{2x}$Se$_{2(1-x)}$ mixed crystals. Journal of Phys-

ics:Condensed Matter. 1994;6:L777.

[30] Vigil-Galán O, Espíndola-Rodríguez M, Courel M, Fontané X, Sylla D, Izquierdo-Roca V, Fairbrother A, Saucedo E, Pérez-Rodríguez A. Secondary phases dependence on composition ratio in sprayed Cu_2ZnSnS_4 thin films and its impact on the high power conversion efficiency. Solar Energy Materials and Solar Cells. 2013;117:246-250.

[31] Ahmadi M, Pramana S S, Batabyal S K, Boothroyd C, Mhaisalkar S G, Lam Y M. Synthesis of Cu_2SnSe_3 Nanocrystals for Solution Processable Photovoltaic Cells. Inorganic Chemistry. 2013;52:1722-1728.

[32] Fernandes P A, Sousa M G, Salome P M P, Leitao J P, da Cunha A F. Thermodynamic pathway for the formation of SnSe and $SnSe_2$ polycrystalline thin films by selenization of metal precursors. Crystengcomm. 2013;15:10278-10286.

[33] Rau U, Werner J H. Radiative efficiency limits of solar cells with lateral band-gap fluctuations. Applied Physics Letters. 2004;84:3735-3737.

[34] Werner J H, Mattheis J, Rau U. Efficiency limitations of polycrystalline thin film solar cells: case of $Cu(In,Ga)Se_2$. Thin Solid Films. 2005;480-481:399-409.

[35] Liu F, Sun K, Li W, Yan C, Cui H, Jiang L, Hao X, Green M A. Enhancing the Cu_2ZnSnS_4 solar cell efficiency by back contact modification: Inserting a thin TiB_2 intermediate layer at Cu_2ZnSnS_4/Mo interface. Applied Physics Letters. 2014;104:051105.

[36] Shin B, Bojarczuk N A, Guha S. On the kinetics of $MoSe_2$ interfacial layer formation in chalcogen-based thin film solar cells with a molybdenum back contact. Applied Physics Letters. 2013;102:091907-091904.

[37] Scragg J J, Kubart T, Wätjen J T, Ericson T, Linnarsson M K, Platzer-Björkman C. Effects of Back Contact Instability on Cu_2ZnSnS_4 Devices and Processes. Chemistry of Materials. 2013;25:3162-3171.

[38] Zhu X, Zhou Z, Wang Y, Zhang L, Li A, Huang F. Determining factor of $MoSe_2$ formation in $Cu(In,Ga)Se_2$ solar Cells. Solar Energy Materials and Solar Cells. 2012;101:57-61.

[39] Duan H S, Yang W, Bob B, Hsu C J, Lei B, Yang Y. The Role of Sulfur in Solution-Processed $Cu_2ZnSn(S,Se)_4$ and its Effect on Defect Properties. Advanced Functional Materials. 2013;23:1466-1471.

[40] Salome P M P, Malaquias J, Fernandes P A, Ferreira M S, da Cunha A F, Leitao J P, Gonzalez J C, Matinaga F M. Growth and characterization of $Cu_2ZnSn(S,Se)_4$ thin films for solar cells. Solar Energy Materials and Solar Cells. 2012;101:147-153.

[41] Woo K, Kim Y, Yang W, Kim K, Kim I, Oh Y, Kim J Y, Moon J. Band-gap-graded $Cu_2ZnSn(S_{1-x},Se_x)_4$ Solar Cells Fabricated by an Ethanol-based Particulate Precursor Ink Route. Scientific Reports 2013;3:3069.

[42] Mitzi D B, Todorov T K, Gunawan O, Min Y, Qing C, Wei L, Reuter K B, Kuwahara M, Misumi K, Kellock A J, Chey S J, de Monsabert TG, Prabhakar A, Deline V, Fogel KE. Towards marketable efficiency solution-processed kesterite and chalcopyrite photovoltaic devices. 2010 35th IEEE Photovoltaic Specialists Conference(PVSC),2010;p. 000640-000645. DOI:10.1109/PVSC.2010.5616865.

[43] Sites J R, Mauk P H. Diode quality factor determination for thin-film solar cells. Solar Cells. 1989;27:411-417.

[44] Rau U. Tunneling-enhanced recombination in $Cu(In,Ga)Se_2$ heterojunction solar cells. Applied Physics Letters. 1999;74:111-113.

[45] Contreras M A, Ramanathan K, AbuShama J, Hasoon F, Young D L, Egaas B, Noufi R. Diode characteristics in state-of-the-art $ZnO/CdS/Cu(In_{1-x}Ga_x)Se_2$ solar cells. Progress in Photovoltaics: Research and Applications. 2005;13:209-216.

[46] Hegedus S S, Shafarman W N. Thin-film solar cells: device measurements and analysis. Progress in Photovoltaics: Research and Applications. 2004;12:155-176.

[47] Mitzi D B, Gunawan O, Todorov T K, Wang K, Guha S. The path towards a high-performance solution-processed kesterite solar cell. Solar Energy Materials and Solar Cells. 2011;95:1421-1436.

[48] Ohnesorge B, Weigand R, Bacher G, Forchel A, Riedl W, Karg F H. Minority-carrier lifetime and efficiency of Cu(In,Ga)Se$_2$ solar cells. Applied Physics Letters. 1998;73:1224-1226.

[49] Repins I L, Moutinho H, Choi S G, Kanevce A, Kuciauskas D, Dippo P, Beall C L, Carapella J, DeHart C, Huang B, Wei SH. Indications of short minority-carrier lifetime in kesterite solar cells. Journal of Applied Physics. 2013;114:084507.

[50] Bag S, Gunawan O, Gokmen T, Zhu Y, Mitzi D B. Hydrazine-Processed Ge-Substituted CZTSe Solar Cells. Chemistry of Materials. 2012;24:4588-4593.

[51] Metzger W K, Repins I L, Contreras M A. Long lifetimes in high-efficiency Cu(In,Ga)Se$_2$ solar cells. Applied Physics Letters. 2008;93:022110.

[52] Repins I L, Metzger W K, Perkins C L, Li J V, Contreras M A. Correlation Between Measured Minority-Carrier Lifetime and Cu(In,Ga)Se$_2$ Device Performance. IEEE Transactions on Electron Devices. 2010;57:2957-2963.

七、学术报告与科技评述

我国铝电解技术今后的研究与开发课题[*]

摘　要　本文介绍了我国铝电解技术现状，分析了大多数中小规模铝厂存在的电解槽寿命不长，能耗较高，传感器和精密仪器控制检测技术缺乏，环境污染问题有待解决等问题。在此基础上，从提高铝电解槽寿命、降低电能消耗、传感器和传感技术、惰性电极、环保技术、模拟与设计技术、自动控制技术等方面提出了我国铝电解技术今后的研究与开发课题。

1　技术现状概况

我国第一家铝电解厂（抚顺铝厂）是1954年投产的。48年来铝电解生产技术已取得巨大成就。目前我国铝的年产量已超过400万吨，近几年以23%~25%的年增长速度增长。据专家预测，2005年铝产量可能达到630万吨[1]。截至2002年6月底，我国已有铝电解厂122家，其中年产量大于10万吨的只有10家，5万~10万吨的有16家，其余为规模较小的铝厂，正在改造、扩建和新建的生产能力仍很大。现已能设计、制造、装备160kA、180kA、200kA、280kA、320kA等容量的预焙阳极铝电解槽以及相应的配套工程设施，包括炭素厂、原料运送、干法净化与环保工程等。

在电解槽设计中，已掌握"三场"仿真技术，在模拟与优化设计方面已采用了ANSYS和MHD等软件；能较好地处理电解槽的磁场、热-电平衡等问题，为大型和特大型预焙槽的设计和制造奠定了基础。

我国近几年开发应用的200kA及其以上容量的大型预焙铝电解槽均取得了较好的技术经济指标，以目前已开发应用的最大容量铝电解槽（320kA预焙槽）为例，主要技术经济指标如下。

（1）32台槽生产技术指标[2]：

CE：94.43%；

电耗：13323kW·h/t；

阳极净耗：397kg/t。

（2）环保：采用干法净化后，厂区周边环境大气中氟化物的含量没有增加，烟囱与工作地带氟化物排放浓度分别为2.44mg/m^3（国家标准为15mg/m^3）、0.34mg/m^3（国家标准为15.1mg/m^3）。

（3）劳动生产率为376吨/(人·年)。

以上数据表明，我国铝电解技术已达到国际先进水平，但是要看到我国多数中小规模铝厂离此水平还有相当大的差距。

---

[*]　原发表于《有色金属学会轻金属年会》，2002年。

2 现存的主要问题

2.1 铝电解槽的使用期（寿命）不长

我国自采用预焙槽以来，电解槽的平均寿命不超过1200天，对铝电解厂造成的损失很大。除了产量因停槽损失之外，还增加了大修费用，增加了废旧内衬材料的处理费用，因而增加了制造成本和环境负担。另外，除少数先进国家的铝厂外，槽寿命问题几乎是一个全球性行业中的技术问题，因此要求解决的呼声很高，需求迫切。

2.2 降低能耗

众所周知，铝电解是耗能大户，吨铝消耗电能为 14000~15000kW·h。我国自20世纪90年代开发应用大型预焙铝电解槽，并开发应用新型工艺技术条件和先进控制技术以来，电耗指标有了显著改进，目前一些大型预焙槽生产系列的电耗指标能达到 13300~13500kW·h/t-Al，属于国际先进水平。然而多数铝厂的电耗仍很高。按照国家能源节约的政策和铝厂降低成本的要求来看，当前迫切的任务是大面积降低电耗，达到国际平均先进水平。

铝电解生产的能量效率至今不及50%，还要进一步挖掘潜力降低能耗。从中长期来看，除熔盐电解方法外，能否寻求到更省能、无公害的炼铝新方法，仍可进行理论和实践方面的探索。

2.3 传感器和精密仪器控制检测技术

由于电解过程的问题以及熔盐电解质极具腐蚀性，至今还没有找到在这种严酷环境下能进行工作的敏感元件和保护材料。因此许多现有的传感技术和传感器无法在铝电解中使用。加之电解生产在线监测必需的装置按单个电解槽来配备，投入费用过高，将增加吨铝的制造成本，也是这方面发展缓慢的原因之一。

2.4 环保问题

环境保护条例对铝电解厂的要求日益严格，这是发展趋势。我国仅对大气中含氟量有明确规定，而对 CO_2 排放量没有明确要求。因此控制铝厂及配套产业的 CO_2 排放至今尚未受到重视。应当看到控制温室气体的"京都协议"即将被我国政府批准，今后对 CO_2 的排放必有明确要求。对此要及早采取措施，减少排放。

我国对铝电解厂废旧槽内衬材料的处理还没有环保条例的规定。这类材料未经处理，随意堆放，引起土壤、地下水的污染十分严重，也是必须尽早解决的。

3 今后技术发展待研究的课题

立足我国国情，参考国际电解铝技术的发展方向，今后铝电解技术发展应该以达到提高效率（高产优质、节能降耗）、改善环保、降低成本为总目标。为此要加速铝电解的技术进步，建议以下课题为今后10年研究和开发的主要任务。

3.1 提高铝电解槽的寿命

电解槽（特别是大型槽）使用寿命不长，带来的经济损失是很严重的。除了产量减少、成本增加之外，更增多了废旧炭块的处理量，增加了环境的负担，这一问题几乎是全球性的。我国电解槽寿命普遍较短，平均在1200~1300天。因此提高槽寿命应作为今后的首要目标，具体目标为：平均寿命2500天，少数电解槽达到3000天。为此应加强对以下项目的研究和开发。

3.1.1 有关槽寿命的基础性研究

（1）熔盐中阳离子在阴极电位下与碳界面的交互作用。现已查明，铝电解过程中Na^+对碳阴极的渗透是电解槽破损的重要原因，但对此过程的微（介）观起因并不了解。为了获得"治本"的认识，需进行以下研究：

1）碱金属离子在带电碳表面的优先吸附及其热力学；
2）乱层碳与石墨晶体结构对Na^+嵌入的影响及其过程动力学与机理；
3）Na_xC_y的结构与性质等。

（2）对阴极炭块、侧部材料以及捣固糊等材料的物性参数进行系统和准确的测定，为电解槽阴极与内衬设计（包括计算机仿真设计）提供更准确的数据资料。

（3）开发更先进和准确的电-热-应力模型及仿真软件，以改进电解槽阴极与内衬的设计技术，并用于对阴极与内衬的动态变化过程进行研究。

3.1.2 电解槽破损预报

根据阴极炭块的热膨胀、铝液和电解质渗透以及铝液接触钢棒等情况，采用先进手段提取阴极电流分布、底层温度分布和电阻率等信息，通过槽控机监控，达到预防的目的。

3.1.3 阴极抗破损技术的集成

将现有技术加以提高、优化和集成，分别是：

（1）硼化钛涂层和TiB_2-炭复合成型块；
（2）双阴极棒或圆阴极棒；
（3）底部防渗漏耐火材料层；
（4）新型扎缝材料。

3.1.4 导流槽

由于导流槽的炭阴极呈斜面，其上覆有硼化钛层，生产出的铝能被其润湿呈一薄层覆盖其上，多余的铝流入下方，因此阴极上不积留铝液，不会产生铝的渗漏，可望有较长的使用年限。要继续解决导流槽阴极坡度和硼化钛厚度的最优关系，导流槽的焙烧、启动和正常操作工艺问题，导流槽的能量平衡问题等。

3.1.5 阴极新材料

（1）长寿垫：这是一种表面含有硼化钛材料，中部含有抗钠侵蚀成分的炭素垫子。用沥青粘结在常规炭块底上，形成抗腐蚀、抗钠渗透和易于筑炉施工的技术。

（2）自组装、多功能复合阴极：这是一种新型复合材料。通过高温焙烧后，形成硼化钛表层，具有抗钠、抗渗透、导电率高的中间层，自催化生成的石墨化基体和底表面的防渗层。

3.1.6 研究相适应的焙烧、启动工艺

着重研究焦粒（或石墨粉）焙烧启动、燃气（油）焙烧启动工艺制度与电解槽阴极及内衬温度分布、热应力分布、变形状态、裂纹发展动态过程的关系，为最佳工艺制度的选择提供依据。

3.2 进一步降低电能消耗

3.2.1 减小炭素阳极和炭素阴极电阻率

根据当今先进预焙槽的电压平衡测定，槽电压的主要项目数据见表1。

由表1的比较可见，炭素阳极和炭素阴极电压降这两项同国外的差距较大，今后要着力解决如下问题。即：提高现有炭素阳极的制备工艺，改进配方和改进焙烧制度，提高磷生铁的导电率，减少爪头与炭块的接触电阻以及加强换极前阳极的预热措施，增加阴极石墨化的程度，改进棒和炭块的接触质量等。

表1 Hall-Héroult 铝电解槽的槽电压分布

电压类型	Haupin[3]	320kA 槽[2]	160kA 槽[2]
分解电压和过电压/V	1.779	1.700	1.700
阳极/V	0.235	0.3852	0.439
电解质/V	1.640	1.579	1.403
阳极气泡/V	0.259	—	—
阴极/V	0.300	0.3179	0.360
外部电路/V	0.160	0.225	0.228
阳极效应/V	—	—	—
总压降/V	4.373	4.207	4.20
电流效率/%	93	94.4	93
能耗/kW·h·(kg-Al)$^{-1}$	14.02	13.33	13.60
槽电流/kA	—	—	—
阳极电流密度/A·cm^{-2}			

3.2.2 降低极距

根据 Haupin 的数据[3,4]，极距每降 1cm 可以减少电解质电压降 360mV，或者节电 1180.5kW·h/t-Al。其前提条件是电解槽自控良好，工况稳定，保温性能好。

3.2.3 研究新型结构的阳极

在阳极结构上要求易于排出气体，以减少气泡引起的电压降，同时增大阳极面积，有利于降低阳极电流密度和减小阳极过电压。这种新结构可能是多片状的或者带有排气孔和排气沟道的炭素阳极。

3.2.4 控制合适的氧化铝浓度和保持合适的电解质导电率

目前较流行的低分子比、低温操作技术获得了较高的电流效率，但因电解质导电率降低而升高了槽电压。采用先进的下料控制技术，保持合适的氧化铝浓度范围，对于稳定地控制槽电压在理想范围非常重要，因为既要防止氧化铝浓度偏高而降低电解质导电率并可能因沉淀而增大槽底电压降，又要防止氧化铝浓度偏低而升高了过电压。此外，可考虑应用能降低电解液比电阻的添加剂，例如在电解质中添加 LiF 可以获得更好的降低比电阻的效果。

3.2.5 降低电解质温度，实行低温电解

根据 $\Delta t = t_{体系} - t_{环境}$，电解系统的温度降低，则它与环境的温差 Δt 减小，所以向环境散放的热量损失降低，只要能相应地降低槽电压便可以节能。低温也就是改进电解

质的系统，其关键在于利用新的电解质系统以后，要保持或者增大 Al_2O_3 的溶解度。通常由于电解质温度降低造成 Al_2O_3 的溶解度降低，引起电解槽内沉淀增多，造成工况紊乱。采取适当分子比和添加 LiF 等添加剂，可以作为研究的重点。

3.2.6 钾冰晶石的应用

冰晶石（60wt%）+钾冰晶石（40wt%）混合物的熔点为 827℃。钠盐的使用可以增加电导率，增大 Al_2O_3 的溶解度，但是对炭阴极的渗透和破坏作用比钠盐更为严重。如果在采用非炭阴极和内衬的新型槽中使用，这种可能性是值得探索的。

3.3 传感器和传感技术

铝电解生产过程更实时、更精确的控制技术，需要先进的传感技术和传感器的支持。目前传感器及相关的传感技术远远不能适应这一需要。因此在控制上显得灵敏度差、反应迟钝以及重要的监控参数不足。这是需要大力研究和开发的领域。

急需的、能在极端条件下工作的传感器如下：

(1) Al_2O_3 浓度的在线测定；
(2) 电解质温度和过热度的在线测定；
(3) 电解质分子比或过剩 AlF_3 含量的在线或快速测定；
(4) 阳极和阴极间距离的测定；
(5) 各个阳极和各个阴极棒的电流分配；
(6) 阳极和阴极的电位变化；
(7) HF、CF_4 浓度传感器；
(8) CO_2/CO 比值的在线测定。

除了常规的接触式传感技术之外，还需要引入和研究非破坏式、非接触式传感技术。

3.4 惰性电极

(1) 继续研究新型复合材料，达到如下要求：抗熔融冰晶石的腐蚀和新生态氧的腐蚀；具有良好的导电性和抗热震性；连续工作的使用寿命达到初期 6 个月、中期 1 年、远期 3 年。

(2) 研究开发加速寿命试验的研究方法，建立电极耐腐蚀性（寿命）预测模型。

(3) 解决大型化制备中的工程问题，如复杂形状成型与热处理问题，电极与导杆连接问题等。

(4) 研究开发金属和合金为基体的可润湿性阴极，阴极的薄形化和异型成型。

(5) 新型槽：对阴、阳极竖式排列及电解时防止短路进行模拟和仿真。

(6) 研究和开发双极性电极。

3.5 环保技术

(1) 严格控制原料和环境引入水分及炭素材料引入含硫元素，减少电解生产过程有害物的排放。进一步减少阳极效应和改进干法净化效率的研究。

(2) 开发废旧炭块及内衬材料的处理和回收技术，可以借鉴国外利用这些废弃物加工高档铺路材料的技术。

3.6 模拟与设计技术

铝电解槽全息仿真系统是一项创新性的项目,目前国际和国内铝业界还没有系统地建立,有必要加强研究,完善我国已有的仿真软件,建立一个大平台,以此为基础,研制建立"数字铝电解槽",具体如下:

(1) 建立有关参数的数据库,包括电解槽结构参数、操作参数、物性参数等;

(2) 研究和完善如下计算模块,它们是电场、磁场、流场、热场、热应力分布场和物料平衡、电压平衡、能量平衡;

(3) 与相关传感器配合进行在线测定、仿真测定和仿真显示,实现电解过程的定时显示,包括电解温度与过热度、电解质中 Al_2O_3 浓度、分子比、阳极电流分布、熔体水平及界面形状、多阳极极距分布以及产铝盘存,进行包括阳极效应、电解槽的破损、电流效率和吨铝能耗的早期或快速预报。

3.7 自动控制技术

(1) 在传感器技术和模拟技术改进的同时,相应地改进铝电解过程控制技术;在传感器无突破或使用成本过高的情况下,可综合应用多种参数估计方法和计算机动态仿真技术建立"软测量"算法,以解决铝电解槽等复杂被控对象的重要状态信息的获取问题,并继续应用模糊控制等智能控制技术解决铝电解过程控制问题[5,6]。

(2) 研究开发新型分布式或全分布式计算机控制系统。其特点是:1) 基于新型的现场总线和工业以太网技术;2) 硬件和软件均可以组态方式实现;3) 可与企业内部 Intranet 或 Internet 相联接;4) 可在恶劣环境下稳定可靠运行。这样,可为企业综合解决生产过程的控制问题和实现企业的综合自动化建设目标提供高性能、低价格的先进控制系统。

(3) 围绕企业的综合自动化建设,开展技术基础研究[7]。按过程控制与资源管理一体化网络的建设原则进行全厂计算机网络的整体规划与建设,解决原来的自动化"孤岛"问题,并避免产生新的自动化"孤岛";注重信息资源管理系统的开发与应用,提高各类信息收集、处理、共享及应用水平,并在此基础上建立实用的企业信息及决策系统,提高企业决策和管理现代化水平。对于大型铝电解企业,应按照 CIMS(流程工业现代集成制造系统)的思想,并结合铝电解工业自身的特点,进行基于 PCS(以产品质量和工艺要求为指标的先进控制技术)/MES(以经济指标为目标的生产过程优化运行、优化控制与优化管理技术)/ERP(以财务分析决策为核心的整体资源优化技术)三级结构的 CIMS 建设,并且将 CIMS 的控制、优化与决策功能与办公自动化、电子商务、远程监控等功能集成于一体,构成企业的综合自动化网络体系。

参 考 文 献

[1] 朱研. 安泰科发布对未来中国铝市场的供需平衡预测结果 [J] 中国铝业, 2002 (7): 36.
[2] 姚世焕: 贵阳铝镁设计研究院,资料交流.
[3] W. Haupin. Interpreting the Components of Cell Voltage [J] Light Metals. 1998, ed. B. J. Welch, Warrendale, PA: TMS, 1998: 531.

[4] H. Kvande and W. Haupin. Cell voltage in Aluminum Electrolysis: Practical Approach. J. Metals, 2000, 52 (2): 31.

[5] 李劼,丁凤其,邹忠,李民军,等. 铝电解模糊专家控制器的开发与应用[J]. 有色金属, 2000, 52 (4): 558.

[6] 李劼,王前普,肖劲,等. 铝电解槽智能模糊控制系统[J]. 中国有色金属学报, 1998, 8 (3): 557.

[7] B. J. Welch. Advancing the Hall-Heroult Electrolytic Process (Overview)[J]. J. of Metals, 2000, 52 (8).

论我国铝企业的产品结构调整*

摘　要　针对我国电解铝厂和氧化铝厂的产品结构现状，指出产品品种单一影响盈利和竞争，介绍了电解铝厂和氧化铝厂有开发价值的若干产品领域，以及热点行业所需的各种铝合金与氧化铝产品，提出了结构调整的建议。

1　我国铝厂现行的产品品种单一

自新世纪以来，我国电解铝厂的改建和新建发展迅猛，预计 2003 年底产能可达到 500 万吨，经营这些企业的铝厂很多，他们在原料、电力、技术、资金以及营销方面都有相当的实力，很多方面有挑战国营中铝公司的表现，使我国除氧化铝生产以外的原铝产、供、销产生新的变化。在众多的铝企业中，不难发现电解铝厂的产品单一很突出，主要为原铝的普通铝锭。氧化铝厂产品的品种也有待增加。

电解铝厂产品单一，其附加值少，因此工厂盈利率较低。近年来，有一些铝厂开始出口铝锭，业内人士认为这是发展的很好路径。不足的是出口盈利仍然有限。然而，出口无异于出口能源而留下污染，因此不是最佳的出路。

造成以上产品品种单一的情况主要是没有摆脱计划经济的生产经营模式。目前，已经是按市场经济原则逐步合理的调整产品结构的时候了。

国际上大铝业公司十分注意产品的结构配置，发展产品链，以最大限度获得盈利。例如，美国铝业公司（Alcoa），以铝为核心的产品已经扩大到 8 个领域，紧密结合市场需求，各领域产品品种繁多。这 8 个领域为：（1）氧化铝及化学制品；（2）铝锭；（3）轿车及轻型卡车；（4）商业运输；（5）家庭住宅及商业建筑；（6）工业产品及服务；（7）包装及消费品；（8）航空航天工业。

根据该公司 2001 年度和 2002 年度的财务报告[1,2]，当年的总销售额分别为 229 亿美元和 203 亿美元，其中，氧化铝分别为 19 亿和 18 亿美元，电解铝分别为 37 亿和 32 亿美元，此两项均约占总销售额的 25%，其余的 75% 为各式各样铝的加工产品和制成品。需知，美铝公司的氧化铝产量（含 Alcoa 全球各地氧化铝厂）和金属铝产量（含二次铝）均为全球第一。但美铝公司不仅仅依靠这两类原材料性质的产品盈利，而是更依靠发达的中、下游产品，后者占了营销总额的大头。

加拿大铝业公司（Alcan），也是以铝为核心产品，沿着产业链发展了数千种产品，主要产品有：道路和轨道运输的产品，汽车类产品，电器，电缆，包装和医药的特殊包装，食物的包装产品，铝塑及铝-高分子材料，复合材料产品，以及炭素制品。

正在改革和发展的俄罗斯铝业公司（RusAl）[3,4]拥有世界上最大的电解铝厂，如布拉茨克铝厂（Bratsk）年产铝锭 91.5 万吨，其产品中有各种合金锭 11.1 万吨，该公司的克拉斯诺雅尔斯克铝厂（Krasnoyarsk）为世界第二大铝电解厂，年产铝锭 86.5 万吨，其产品中有十分之一为合金产品和高纯铝。上述这些公司都十分注意开发新产品，

---

* 原发表于《中国有色金属学会第五届学术年会》，2003 年 8 月。

在激烈的国际竞争中都能保持良好的盈利能力和持久的竞争力。

2 关于铝电解厂的产品结构调整

在加入 WTO 以后，铝的国际、国内两个市场已经融通，铝的价格基本上受伦敦金属交易所（LME）价格的左右。为了保持我国铝企业的国际竞争力，根据以上分析，我国电解铝厂的产品结构调整已是十分紧迫。

由于各厂和各铝企业的情况不同。需要根据市场调查和本企业的情况，因地制宜地确定产品结构调整的方向和产品品种。笔者以为，近期应以主产品链的最邻近延伸为宜，瞄准市场空间大，产品期效长的领域进行开发和建设，如汽车、建材、电工和包装等领域。特别要重视开发和生产专用铝合金。为此，要加强本企业新产品开发的规划和预研，加强铸造部的更新和改造，与此同时要重视与大用户的联手，合作开发。

3 铝合金

铝合金是电解铝厂发展新产品、增加产品品种的重要领域，它所需要的技术和装备不很复杂，投资也较节省，制造成本远比用户自行配制合金要低，可以在较短的时期内获得产品品种的改进和增值。以下提出几种合金供发展和建设的参考。

3.1 汽车用铝合金系列

（1）汽车用铝是今后发展趋势。当今，我国的汽车用户急剧增加，而且发展势头相当好，但是在节油和减少污染方面社会的呼声很高。对生产厂家而言则要求汽车的轻量化，有资料表明，车重减少 10%可节油 8%，由此，轻量化已成为各种汽车制造厂商关注的焦点。

（2）发展电动汽车是降低大气污染的根本措施，电动车本身也要求轻量化。

（3）军用车辆也把轻量化提到重要位置，以提高机动性。

因此，努力提高汽车的用铝量，实现轻量化，已成为汽车制造业技术进步的重要标志。

迄今为止，全球已发展数千种铝合金，用于汽车的举例如下：

1000 系列，以 99%的铝为主，耐腐蚀，高导电导热性，机械强度低，有很好加工性能，主要有：1100、1200 合金，用于冷凝管或散热片。

2000 系列，以铜为主要合金元素，当进行热处理时，机械性能相当于甚至超过软钢，人工时效可以用来增加其强度。其中 2008、2010、2011、2017、2024 合金，用于外部和内部支撑板（还适用于建筑材料），负载板，座位外壳，机件的螺钉螺帽，机械带扣等。

3000 系列，以锰为主要合金元素，这类合金不能进行热处理，但具有很强的抗蚀性和成型性。其中 3002、3003、3004、3005、3102 合金，用于冷却器，加热器，蒸发器片，加热器进口和出口管，油冷却器，空调液体流动管道，伸展冷凝管等。

4000 系列，以硅为主要合金元素，有 4004、4032、4043、4045、4104、4343 合金。用于黄铜片的保护层，铸造活塞等。

5000 系列，是铝的一种最有效和应用最广的材料，镁是主要的合金元素。这个系列的合金具有可焊接性和良好的抗蚀性，还可以应用于航海。有合金 5005、5052、5182、5252、5454、5457、5657、5754。用于内部板材和构架，卡车缓冲器，结构支撑板，挡溅板，隔热板，空气清洁器，部件，轮子，引擎附件的支架和边框，翻斗车车体，油罐车，拖车等结构和可焊接部分，装饰、名牌和嵌花等。

6000系列，这类合金使用不同比例的镁硅比，具有很好的成型性和抗蚀性，并且具有很高的强度。有6009、6010、6022、6053、6061、6063、6082、6111、6262、6463合金。用于外部和内部支撑板，缓冲器表面板及支撑结构，座位外壳，其中6061合金用于：主体结构、支架、悬挂结构（锻造的）、操纵轴、操纵杆、刹车筒、轮子、燃料输送系统，6262合金用于：刹车机架，刹车活塞，普通的机器螺钉部件等。

7000系列，以锌为主要合金元素，加入少量的镁或者铜，使合金具有可热处理性和很高的强度。有7003、7004、7021、7072、7116、7129合金。用于座椅框架，缓冲器加固结构，冷凝器和冷却器片，枕头板等。

3.2 铸造合金

铝合金铸造件可以通过所有的铸造过程制备，具有不同的成分和工程性能。汽车应用有：319.0，332.0，356.0，A356.0，A380.0，383.0，B390.0，用于支管，汽缸头，引擎内部结构，活塞，轮子，传动机械，操纵齿轮。B390.0用于高磨蚀部分，例如：环形齿轮，内部传动部件等。

3.3 其他多种产品

不同的锻压类铝合金还可发展为包装用铝产品，建筑材料，家用电器，耐用消费品，铝合金家具，电线电缆等，可根据市场需求，作出近、中、长期发展规划，分步建设，以取得更大的经济效益。

4 氧化铝厂的新产品开发

我国氧化铝厂在新产品的开发方面有较好的发展，但是有两方面的情况值得注意：

（1）氧化铝和氢氧化铝的化学制品从品种数量到质量与国外该领域的情况相比还有很大差距。

（2）国内近年来以氢氧化铝为原料发展起来的新产品品种和数量逐渐增多，形成市场分享的局面。因此，我国氧化铝厂要拓宽视野，大力开发和建立新产品，使产品结构更为全面，以赢得更大的国际和国内市场份额。

4.1 以氢氧化铝、煅烧氧化铝为原料的产品

我国氧化铝厂除冶金级氧化铝外，研发和批量生产了若干种多品种氧化铝，但仍然还有许多空白。这方面的产品简介如下。

以氢氧化铝为原料，可制成多种产品，主要有：

（1）超微细氢氧化铝——是目前世界上用量最大的无机阻燃剂，它以无毒、无卤、低烟的阻燃特点，在塑料、建材、高分子材料、电子等许多行业中得到广泛的应用；

（2）聚合氯化铝——用于水处理及个人卫生用品；

（3）铝酸钠——水处理，钛白颜料涂料；

（4）沸石——分子筛，催化剂载体，洁净剂等。

以煅烧氧化铝为原料，可制成的产品有：熔融刚玉，高铝陶瓷，高压电绝缘器，火花塞，特种玻璃，人造大理石，人造宝石，研磨材料等。

4.2 分子筛类产品

这类产品种类很多,主要有:

(1) KTK-3——石油化工中用于硫回收装置中 H_2S 和 CS_2 的排除;

(2) 3A,4A,5A,13X 等分子筛以不同功能用于石化工业。

4.3 活性氧化铝类产品

主要制备各种催化剂,例如:Actiguard Cl,Actiguard S,Actiguard 450S,Actiguard HF,Actiguard 600PC,Actiguard TBC 和 Claus catalyst(脱硫用)等。

4.4 Si 与 Al_2O_3 的复合胶

用于气体脱水,压缩气体中脱除油雾,液体空气过滤及催化剂。

4.5 美国铝业公司的相关产品

国际上以美铝公司的氧化铝及其化学制品的品种最为齐全[5]。其产品系列举例如下:

(1) 片状氧化铝;

(2) 水泥与黏结剂;

(3) 尖晶石;

(4) 水合氧化铝;

(5) 拜尔体氢氧化铝;

(6) 白色氢氧化铝;

(7) 磨细氢氧化铝;

(8) 种分细微氢氧化铝;

(9) 抛光用氧化铝;

(10) 煅烧反应活性氧化铝;

(11) 吸附剂与催化剂;

(12) 氟化铝;

(13) 特种氧化铝。

下面以吸附剂和催化剂系列为例,作简要介绍。

此系列共有 9 个子系列,50 个分子系列,仅举各分子系列于后:

(1) 催化剂活性床支撑体;(2) 催化剂球体(CSS);(3) 色谱用铝氧 CG-20;…;(22) 低密度 G-250 铝氧胶;…;(45) 脱硫剂(Claus catalysts);(46) 一般吸附剂;(47) 选择性吸附剂;…;(50) HF 去除剂。

其中催化剂活性床支撑体有 3 种产品,即高抗压强度,高表面积和高化学纯度的支撑体。

又如低密度 G-250 铝氧胶,这是美铝公司 1997 年的新产品,它的特性为高纯、高反应活性、高黏结性。500℃时转变成高孔率,高表面积的 gama-Al_2O_3,可制成 4 方面的产品:催化剂,化学品(用于提纯、分离),研磨剂(用于精细抛光)和高温黏结剂等。

总之,要充分掌握国际、国内技术与产品信息,对相关产品的市场情况作详细调查，根据的实际情况，作好规划，逐步把产品结构作出合理的调整，以获得持久的盈利能力和企业的核心竞争力。

参 考 文 献

[1] Alcoa 2001 Annual Report. Alcoa pub. 2002.
[2] Alcoa 2002 Annual Report. Alcoa pub. 2003.
[3] http：// English. Pravda. ru/comp /.
[4] http：// www. rusal. com /.
[5] Alcoa World Chemicals product data, USA/3680/R02/0601；USA/6090/R02/0601；USA/0098 - R01/1101.

加强有色冶金基础研究的建议[*]

1 我国冶金基础研究现状

我国 10 种有色金属产量 2001 年达到 883.7 万吨，居世界第二位，产业的发展离不开科技，科技发展是这个领域中广大科技人员积极努力的结果。尽管有色金属冶金的基础研究投入的经费很少，一直比较薄弱，但是在生产技术需求的带动下仍然在基础研究方面做了大量的工作，并取得了显著的成就，这主要表现在如下几个方面。

1.1 矿产资源的分离提取

（1）硫化矿原生电位浮选技术[❶]；
（2）浮选剂分子设计理论及应用；
（3）生物浸矿理论；
（4）选矿—拜尔法新技术及基础理论。
其中许多已获得国家级的奖励，居国际领先水平。

1.2 冶金过程化学平衡、相平衡及热化学

主要的工作集中在：
（1）基本热力学数据的测定与数据库的建立；
（2）溶液热力学及活度计算模型的建立与应用；
（3）相平衡与界面过程热力学；
（4）溶剂提取与离子交换热力学；
（5）极端条件下的多相平衡问题等。

1.3 冶金过程动力学和机理

目前的主要工作集中在：
（1）有色金属体系基本传输性质的测定；
（2）生物催化浸出过程；
（3）电化学冶金电积过程动力学；
（4）溶剂提取与离子交换机理与过程动力学；
（5）外场作用下的动力学等。

1.4 膜领域的基础研究

（1）成膜机理；

---

[*] 原发表于《中国工程院化工、冶金和材料学部第四届学术年会》，2003 年 10 月，长沙。
[❶] 代表获国家科技进步奖二等奖及以上奖励。

（2）膜材料性能；
（3）膜受污染机理与防治技术；
（4）膜中的传递现象及原理；
（5）膜过程设计的理论、模拟及新膜过程开发。

1.5 冶金电化学

（1）湿法炼锑和镍电解的基础理论；
（2）功能电极材料及节能电极的理论基础和应用研究；
（3）矿浆电解原理；
（4）熔盐电解炭阳极上的电催化❶；
（5）铝电解惰性阳极和硼化钛阴极应用基础。

1.6 冶金新技术的基础理论

（1）机械活化冶金❶；
（2）增值冶金（高附加值产品的冶金技术）；
（3）串级萃取理论；
（4）纳米-精细化工材料的电化学制备基础。

1.7 冶金过程和设备的数字仿真和优化

（1）应用电磁场，热-电场仿真软件优化大型铝电解槽"三场"设计，在仿真功能和精度上都进入了世界同行业的前沿；
（2）铝电解槽槽膛内形与闪速熔炼炉反应塔炉膛内形动态仿真与在线显示；
（3）闪速熔炼炉多参数场耦合仿真和优化；
（4）矿热电炉、贫化电炉节能降耗智能决策技术；
（5）铜镍锍吹炼过程智能优化与吹炼终点预报技术。

1.8 冶金过程的自动控制

（1）开发和应用特种检测技术，解决冶金领域的多种特种变量的检测问题；
（2）开发和应用针对冶金领域的专用控制软件、优化软件及专用控制装置；
（3）采用新一代主控系统（包括集散控制系统、可编程控制器、仪表、器件、网络、传感器等）构造过程控制系统；
（4）综合应用多种建模手段及现代先进控制方法与控制策略，解决复杂冶炼过程的建模与控制问题，例如铝电解过程的智能模糊控制。

2 拟定发展战略的几点原则

（1）立足本国的需要，发挥已有科技成果的基础研究优势，引导原创性研究的发展，以推动有色金属科技与生产的跨越式发展为目标；
（2）体现高新技术或先进实用技术改造传统产业的要求，进行学科交叉的研究；
（3）充分利用和借鉴当今相邻学科的最新科学技术成就；

---

❶代表获国家科技进步奖二等奖及以上奖励。

（4）重视相关技术基础的研究工作。

3 建议开展如下重要问题研究

3.1 复杂界面交互作用的研究

有色金属提取冶金是在一定条件下实现有色金属与伴生元素的分离提取。其基本过程可分为：矿物的溶浸与分解，伴生元素的分离，金属与化合物的析出等。这些基本过程涉及各种不同条件下液-液、固-液、固-气等复杂界面的交互作用。通过系统研究复杂界面作用的基础理论问题，可以为复杂和低品位矿的处理、冶金过程的强化、节能降耗、减少污染、新工艺及新技术奠定理论基础。

3.2 矿物的生物加工理论与技术

主要包括：

（1）生物溶出—萃取—电积的理论及技术。
（2）硫化矿与非硫化矿浮选新理论。
（3）微生物选矿药剂结构与性能。
（4）选矿—拜尔法基础理论，铝土矿正、反浮选铝硅分离：
1）矿物晶体结构表面性质与浮选性能；
2）铝、硅矿物浮选体系，溶液化学行为及复杂界面作用；
3）矿物界面极性相互作用及流体动力学；
（5）矿物材料，针对各种资源的处理，研究直接从各种资源中加工制备各种材料的新技术与基础理论：
1）无机非金属矿物超细粉体材料；
2）高纯净化矿物；
3）直接从矿石中生产精细矿物化学品；
4）矿物功能性原状结构的物理加工技术及基础；
5）矿物材料化学加工的技术、新工艺及基础理论；
6）低维纳米矿物的结构与应用性能。

3.3 我国丰产元素资源的深加工

重点应放在稀土、钨、钛等稀有金属，镓、铟、锗等稀散金属，锌、铅、锑、铋、镍等重金属，铝、镁、锂等轻金属及锰、铁等黑色金属上。产品的重点应该是信息材料、新能源材料及阻燃剂、催化剂、高比强结构材料等。

3.4 再生金属资源化和"三废"资源化

再生金属包括废电池、废旧家电、废旧通讯设备、废旧电脑、废旧汽车、废旧仪器设备的金属资源回收处理和再生利用。"三废"包括火法冶金炉渣、精炼渣、烟尘、湿法冶金残渣、赤泥、净化渣、铝电解废旧炭块等资源再生回收的基础理论问题。

3.5 无污染冶金的基础研究

研究无SO_2排放的一步炼铅、一步炼锑、一步炼铋冶金新技术新工艺，用以改造现在的鼓风炉炼铅和炼铅锑技术，减少 SO_2 排放 60 万吨/年，具有重大的社会环境效益；与此相关的许多基础理论问题，如高铁锍与金属炉渣的相平衡，富铁锍中硫氧化成元素硫然后回收等问题亟待解决。

3.6 冶金过程热力学和动力学

（1）冶金过程相关相平衡与界面过程热力学；
（2）溶剂提取与离子交换热力学；
（3）极端条件下的多相平衡问题；
（4）有色金属体系基本传输性质的测定；
（5）生物催化溶出过程；
（6）外场（如电场、磁场、声场、微波场等）作用下的过程动力学。

3.7 机械活化冶金

（1）力场作用下对有色金属功能材料性能的影响。运用力化学原理合成并改善功能材料性能的可能性及其条件；
（2）机械活化与固体界面性能和结构的关系；
（3）力场的作用方式、强度及机械活化效果的影响；
（4）机械-化学作用场强的测量与表征。

3.8 膜科学与技术在冶金中的应用基础研究

（1）无机膜的制备与成膜机理；
（2）膜材料的性能和表征；
（3）无机膜的传质过程机理及传递阻滞；
（4）膜过程设计的理论、模拟及新膜过程的研究；
（5）离子膜的电化学行为；
（6）膜的受污染机理与预防和再生。

3.9 冶金电化学新前沿基础研究

（1）电化学分离、提取与功能增进（如：增氧、增氢、脱硫、多层化、磁性化、多功能修饰化等）；
（2）纳米尺度合金、线晶、多层膜等的电化学制备；
（3）电积用于电子、生物医药、环境及能源转换；
（4）电极过程动力学及电催化（包括界面电子迁移、成核及长大、薄膜中的电迁移与化学反应等）。

3.10 离子液体—新型清洁介质的基础与应用研究

离子液体是绿色化学的、环境友好的反应介质，应用前景极为广阔。用作萃取分离、化学反应、催化反应的介质，燃料电池、太阳能电池、锂电池等的制备等，需要进一步研究以下问题：

（1）研究离子液体的性质；
（2）离子液体的合成，特别是降低其制造成本；

（3）离子液体在萃取、提取与分离方面的应用；

（4）离子液体的催化性质、催化作用及其工业应用；

（5）研究其电化学、光电化学等扩大离子液体的新应用。

3.11　熔盐电解

（1）熔盐电解需要研究的问题：

1）熔盐应用于有色金属"绿色"化学反应及过程的基础研究；

2）熔盐选择提取，熔盐电化学气体分离，熔盐电化学能源等。

因为我国铝的产量已居世界第一位，也是我国有色金属中产量的第一位。考虑特色和重点，开展熔盐电解炼铝的基础研究具有战略意义。

（2）新颖电极材料：

1）极端环境下（高温、氟化物熔盐腐蚀、电化学腐蚀、新生态氧的腐蚀）材料性能的快速表征；

2）极端环境下多功能阳极和阴极材料的界面反应与抗腐蚀性；

3）极端环境下气体电极的氧化、还原功能。

（3）数字铝电解槽系统：

1）建立多参数数据库（结构参数、操作参数、物性参数等）；

2）研究下列计算模块：电场、磁场、流场、热场、物料平衡、电压平衡、能量平衡等；

3）多参数仿真与显示，包括电流分布、熔体/铝界面形状、极距、Al_2O_3 浓度、阳极效应预警、电流效率预测、吨铝能耗预测；

4）实现以上 1）、2）、3）的全息仿真。

3.12　冶金过程和设备的数字仿真和优化

（1）温载流体系中颗粒群行为动力学及其相互作用模型（应用领域：闪速熔炼、闪速吹炼、稀相焙烧、煤粉燃烧、载流干燥等的强化与优化）；

（2）高温熔体与浸没射流间气-固-液三相传输动力模型与反应速率模型（应用领域：熔池熔炼、吹炼、烟化过程等的强化与优化）；

（3）流态化床（浓相）中流固两相的多场耦合仿真模型（应用领域：流态化焙烧炉、煅烧炉、流态化浸出与流态化电积的强化与优化）；

（4）移动床中气固两相的流场、温度场与浓度场的耦合仿真模型（应用领域：竖式焙烧炉、竖式煅烧炉与鼓风熔炼炉的强化与优化）；

（5）高温熔炼炉操作优化智能决策模型（应用领域：燃烧或电热熔炼炉的强化与优化）。

3.13　传感器与自动控制

（1）传感器与传感技术：有色冶金过程的更实时、更精确的控制技术，需要先进的传感技术和传感器的支持。目前传感器及相关的传感技术远远不能适应这一需要。因此在控制上需监控的参数缺乏、灵敏度差、反应迟钝。这是需要大力研究和开发的领域。

有色冶金过程所需的传感器是多种多样的，以铝电解为例，急需在恶劣环境下使

用的传感器,需要开展相关的基础研究:

1) Al_2O_3 浓度的测定;
2) 电解质温度和过热度;
3) 电解质分子比或者是过剩 AlF_3 含量;
4) 阳极和阴极距离;
5) 阳极和阴极的电流分配;
6) 阳极和阴极电位的变化;
7) HF、CF_4 传感器;
8) CO_2/CO 比值的实时测定。

除了常规的接触式传感技术之外,需要引入和研究非破坏式、非接触式传感技术。

(2) 自动控制技术:

1) 智能集成建模方法研究:针对复杂冶炼过程存在物料多变、机理复杂、参数直接在线检测困难、难以得到精确机理模型的问题,研究综合应用机理分析、模糊、知识、逻辑、神经网络、专家经验等建模方法,建立集成型模型,为复杂有色冶炼过程的软测量与控制问题的综合解决提供新的方法;

2) 复杂有色冶炼过程的智能集成控制技术的研究开发:针对大型复杂有色冶炼设备一般具有多变量、非线性、大滞后、时变等特性,重要状态参数与被控参数难以直接检测,具有模型的不确定性等特点,应用专家系统、模糊控制、预测控制、神经网络以及多种智能控制技术的集成,构成控制算法模块库,并使之与"软测量"算法模块有效联接,构成丰富的智能测控算法模块库,满足复杂有色冶炼过程对控制算法的多种需求;

3) 先进控制系统的研究开发:研究开发基于新型的现场总线和工业以太网技术的、硬件和软件均可以组态方式实现的、可与企业内部 Intranet 或 Internet 相联接的、可在恶劣环境下稳定可靠运行的新型分布式或全分布式计算机控制系统,为有色冶炼企业综合解决复杂有色冶炼过程的控制问题和实现企业的综合自动化建设目标提供高性能、低价格的先进控制系统;

4) 有色冶炼企业综合自动化建设的技术基础研究:针对有色冶炼过程的特点,研究基于知识链的 ERP/MES/PCS 三级结构流程工业现代集成制造系统集成技术,主要包括:以知识链为内涵的过程知识集成技术、以信息化的财务结构体系为核心的应用集成技术、以合同实时跟踪为主线的功能集成技术、以产品全息索引为基础的数据集成技术。

论铝电解的节能潜力*

摘　要　针对铝电解用电矛盾突出，节电问题需多加讨论。在介绍了铝电解的理论最低能耗为 6.24kW·h/kg-Al 后提出，要以降低过电压，降低极距和减小阴极电压降为重点挖掘潜力，并引证了若干新思路。对减少电解槽的热损失与废热利用，低温电解与新电解质组成也作了简要讨论。

近年，我国铝电解工业超乎常规的发展引起了诸多问题，其中供电紧张是很尖锐的问题，加之国家对铝电解取消了优惠电价，时下又处于全国短期内电力供应偏紧的局面，使铝电解用电矛盾更为突出，业界自然而然地考虑到铝电解的节能问题。节能虽是个老课题，从降低成本、适应新形势发展的需要，仍需多加讨论、更多地投入研究与开发。所幸，当今的科技进步和学科交叉，为铝电解节能提供了新思路，使业者们的探索有了更加广阔的空间。

我们对铝电解节能应更具信心。自 1954 年我国第一家铝电解厂（抚顺铝厂）投产以来，铝电解生产技术已接近国际先进水平。有资料记载[1]，1975 年全国铝电解平均直流电耗为 16720kW·h/t-Al，至 2002 年，平均直流电耗降为 14200kW·h/t-Al，经 27 年努力，吨铝电耗降低 2500 余度，节能 15% 左右。今后 5~10 年，继续坚持技术进步，努力挖掘铝电解的节能潜力，是可以再创佳绩的。

1　铝电解的理论最低能耗

铝电解生产熔融金属铝理论最低能耗数据是通过热力学计算获得的。计算中，设定进入电解槽的反应物和离开电解槽的产品和副产品都是在室温，而熔融铝离开电解槽时是 960℃。并确定了计算系统（电解槽）的理论边界。计算表明，电解槽的工作温度变化对理论能耗只有很微弱的影响。例如，电解槽的工作温度在 700~1100℃ 范围内发生 100℃ 的变化时，它对理论最低能耗的影响小于 1%[2]。对理论分解电压的影响仅 0.05V。如果假定离开电解槽的气体为 960℃，此时计算所得理论最低能耗比上述条件计算的结果要高 2.5%~3.5%。理论上来说，收集这些排放气体所携带的能量是可能的。然而实际上比较困难，因为气体是从数百个电解槽的集气罩中捕集来的，在排放到大气之前经过了处理。仅有非常小的一部分热被吸附返回到系统中。因此，以设定离开电解槽的气体为 25℃ 为宜。

采用炭阳极进行铝电解时，设定进入电解槽系统的 Al_2O_3 为室温，排放的 CO_2 也为室温，生产出的熔融铝为 960℃，又假定反应是在完美的条件下进行的，也就是说，没有可逆反应和副反应额外地消耗炭阳极，反应物离子到达电极没有浓度梯度，以及没有热损失。那么，根据热力学计算，铝电解的理论最低能耗为 5.99kW·h/kg。其中，推动反应进行所需的能量（ΔG）为 5.11kW·h/kg，保持平衡所需的热能为

---

* 原发表于《轻金属》40 周年专刊约稿，2004 年。

0.49kW·h/kg。如果排放的 CO_2 为 960℃，则总的理论最低能耗为 6.16kW·h/kg-Al[2]。

现今，国际先进铝电解槽的电流效率已经达到 96%，吨铝电能消耗 13kW·h。若按 96% 电流效率算，则理论上吨铝最低能耗为 6.24kW·h。由此可算出现代铝电解槽的能量效率为 46%~48%。

2 铝电解的节能潜力

现行铝电解每吨铝的能耗为 13000~14000kW·h，与理论能耗 6240kW·h 相比，除不可避免的能耗（加热与熔化物料等）之外，节能的潜力是很大的。我们应从两方面来探讨今后的节能途径。当务之急是改进目前的生产状况来挖掘节能潜力，其次是从电解槽革新方面寻求节能潜力。至于远景方面，还要从炼铝新方法及非冰晶石-氧化铝熔盐电解法进行研究和开发，以期原铝生产能耗比现时生产方法有大幅度降低。本文仅针对目前状况作一粗浅讨论。

2.1 目前工艺状况下可挖掘的潜力

众所周知，铝电解的电能消耗与槽平均电压成正比，与电流效率成反比。提高电流效率很重要，但本文不在此讨论，只探讨在槽电压方面可能的节能潜力。

2.1.1 槽电压分配

槽电压的组成可用下式表示：

$$E_{槽} = E_{分解} + E_{过} + E_{电解质} + E_{阳极} + E_{阴极} + E_{连接} + E_{AE}$$

式中，$E_{槽}$ 为槽电压；$E_{分解}$ 为极化电压；$E_{过}$ 为过电压；$E_{电解质}$ 为电解质电压降；$E_{阳极}$ 为阳极电压降；$E_{阴极}$ 为阴极电压降；$E_{连接}$ 为连接母线电压降分摊；E_{AE} 为效应电压分摊。

不同类型的铝电解槽有不同的槽电压分配，但各基本项目电压降的出入不大，以下举出典型的槽电压分配为例。

典型的槽电压分配（V）：

(1) 极化电压（即 Al_2O_3 理论分解电压+过电压）1.70V

（其中：Al_2O_3 理论分解电压 1.2V；阳极过电压 0.4V；阴极过电压 0.1V）；

(2) 电解质电压降 1.60V；

(3) 阳极电压降 0.30V；

(4) 阴极电压降 0.40V；

(5) 效应分摊 0.05V；

(6) 导线连接分摊 0.15V；

(7) 槽平均电压 4.20V。

上列各项都需斤斤计较地加以减小，但限于篇幅，本文不打算逐项分析节能的潜力，只想讨论以下几个重点，即：(1) 降低过电压；(2) 降低极距；(3) 减小阴极电压降。

2.1.2 几个挖潜重点

(1) 降低过电压。过电压为 0.5~0.7V，为阳极过电压与阴极过电压之和，其中阴极过电压仅为 0.1V 左右，主要是阳极过电压，为 0.4~0.6V 占槽电压之 10%~15%。可见这部分对节电影响很大。通常降低阳极过电压可通过适当减小阳极电流密度，增大氧化铝浓度，适当增加冰晶石比等措施来达到。然而，这些措施和当今行之有效的生产工艺条件如低氧化铝浓度，低摩尔比及较高阳极电流密度是有矛盾的。因此，降低过电压的方向在于采用电催化剂以及其他新技术。电催化剂乃是一种阳极添加剂，其作用是在阳极表面上增加反应的活性中心，以加速阳极电化学反应速率。我们对降低阳极过电压的研究表明[3]，往阳极中掺入微量锂盐，即"锂盐糊"，对自焙阳极具有电催化活性，能够降低自焙阳极过电压100~150mV。但是，对于预焙阳极加入锂盐却无效。多年的研究包括同挪威科技大学的合作，已获得实验室成果，即在预焙阳极的配料中掺入少量铝-镁尖晶石及氟化铝，经1200℃下焙烧后阳极仍具有催化活性，铝电解时预焙阳极上可降低阳极过电压40~60mV[4]。数字虽不大，但令人鼓舞，也引起国际同行的重视，这方面的研究工作有待从理论到技术更深入地开展下去，以继续挖掘其潜力。

除此之外，还要减少阳极上气泡膜引起的过电压，以及减小因Al_2O_3分布不均匀引起的浓差极化，特别是大型槽（>300kA），这点不可忽视。

（2）降低极距。降低极距的意义在于从电解质电压降中挖掘节能潜力。电解质电压降的计算简式为

$$E_{电解质} = \rho DL$$

式中，ρ为电解质的比电阻；D为平均电流密度；L为两极间距离。

通常，D为设计所定，生产中一般不改变。至于电解质的导电率（比电阻的倒数），主要受添加剂影响。有关研究指出，大型槽采用低摩尔比后，再添加锂盐其效果并不大。在现代大型槽生产工艺条件下尚未有采用其他添加剂的新举措。因此，降低电解质电压降主要应从减小极距着手。要注意的是，减小极距应以保证电解槽的热平衡为前提，也应考虑铝液在磁场作用下产生的波动有可能在低极距下接触阳极发生短路等，故不能盲目行动。

电解质电压降一般在1.5~1.6V。当极距为4.5cm，电解质电压降为1.6V时，降低极距1cm可降低电解质电压降336mV，也就是说每降低极距1mm可减小电解质电压降33.6mV，相当于100kW·h/t-Al。可见极距的变化对节能影响之显著。笔者的意见是，要重视这部分的潜力，为此，要研究解决现行工艺条件下能耗最小的最佳极距，此外还要研究开发：

1）阳极水平的定位装置，以期阳极电流分布均匀，减少局部电流密度过大引起的短时电压降增高；

2）阳极上开沟或开孔，以利气泡排出，减少电解质电压降；

3）改变两极的排布方式，由电极面水平相对排列改成电极面竖式相对排列。电极的竖式排列可完全消除磁场的负面影响并有利于气体排出和电解质循环。目前虽不能做到，但这是方向。

（3）降低阴极电压降。阴极电压降由铝液层、阴极炭块、阴极导电钢棒三部分的电压降总和组成。我国多数铝厂的阴极电压降数据为0.45V，它占有槽电压的11%以上，因此，努力提高阴极炭块的导电率是节电的重要方向。

石墨炭块是阴极炭块中电阻率最小的产品，但是它的耐磨损能力较差，影响阴极寿命，因而影响槽寿命。采用硼化钛涂层的石墨炭块是可取的，这样既能降低阴极的电阻率，又能延长电解槽寿命。在提高阴极炭块导电率及其寿命方面，还有一些新思路有待研究实现，例如：

1）在阴极炭块中原位生成TiB_2[5]，即在炭块焙烧过程或炭块使用过程中，由于配

料的相互作用而就地生成 TiB_2 及相关化合物；

2）制造嵌入型炭块。在炭块中预先掺入或嵌入高导电化合物，以提高阴极炭块导电率并抑制以后 Na 的渗入[6]；

3）服役中阴极炭块的自动催化石墨化，等等。

另外，在阴极导电钢棒方面，增大导电面积和改变棒的形状（例如圆形双棒）有利增加导电面积，减少接触电阻和减少棒和炭块间的应力损坏，有利于延长槽寿命。生产操作中要严防槽底底部结壳的生成，因为它不仅增加了槽底电阻，而且由于槽底电流分布不均匀，造成局部磁场波动，使电解槽工况不稳定。

2.2 减少电解槽热损失与废热回收利用

减少电解槽的热损失就可节能。根据国内外多种电解槽能量平衡测算，以某一组数据为例：大型槽总供入能量之53%左右变成了热损失，相当于每小时损失能量 300~340kW。可见能量损失之大。其中，从电解槽上表面（包括阳极表面和 Al_2O_3 结壳表面）散热约占总热损失之50%；从槽壳侧部散热约占总热损失之20%；从阴极钢棒约占10%等。在能量平衡基础上，一方面要采取措施减少热损失，如适当加厚阳极表面 Al_2O_3 保温层厚度，形成并保护侧部炉帮，提高槽侧壁与槽底的绝热层效果等，另一方面要研究废热的回收利用。由热转变为电的热-电转换材料在研究开发方面已获得重要进展，此前这类新颖半导体材料由于价格昂贵，仅限于航天应用，如今已转向汽车工程应用，旨在把汽缸的废热转变为电能重新使用[7]，近年已在提高转变效率方面取得成果。交叉学科的技术进步值得我们密切注意。我们期待这种技术日后也能对电解槽的废热回收有用。

2.3 关于低温电解

低温电解是铝业界长期努力的目标。许多学者包括我国邱竹贤院士，在这方面做了大量工作。低温电解的最大优点在于使电解槽系统传递到环境的热损失大为减少，也使加热物料的能耗减少，因此，低温电解在节能方面的主要表现是减少了电解槽的热损失。

实现低温电解主要靠调低电解质的熔度，获得 700~850℃ 的电解温度。低温电解质运行的最大障碍是 Al_2O_3 的溶解度和溶解速度大为减小，使操作困难。在已研究过的低温电解质中，以含锂冰晶石的体系较好，此体系的组成（wt%）为 Na_3AlF_6，53.7%；Li_3AlF_6，33.8%；CaF_5，5%；Al_2O_3，7.5%，熔度为 800℃[8]。此体系的主要缺点是成本太高。

最新进展是探索室温铝电解[9]。铝以固态形式析出，电解在 200℃ 下进行，电解质采用离子液体 C_6mimPF_6 和 C_8mimPF_6，以氧化铝为原料。可惜目前氧化铝在这类离子液体中的溶解度仅为 0.5%~1.0%，无法工业应用。如果改为易溶解入该类离子液体的含铝原料，也可能会有应用前景。

至于电解槽的改进革新，已有许多论述。总的说来，采用导流槽，采用可湿润性阴极与惰性阳极联合使用的竖式电解槽，或双极性电极的多室电解槽，有可能把现有电解方法的能耗再减 1/3，达到 10000~11000kW·h/t-Al。

参 考 文 献

[1] 陈万坤. 有色金属进展：轻金属. 中南工业大学出版社，1995：117-122.

[2] William T. Choate and John A. S. Green, U. S. Energy Requirements for Aluminum Production: Historical Perspective, Theoretical Limits and New Opportunities January 2003, Prepared under Contract to BCS, Incorporated, Columbia, MD. for the U. S. Department of Energy Efficiency and Renewable Energy, Washington, D. C.

[3] Yexiang Liu, Yanqing Lai and Jomar Thonstad, On the Electrocatalysis of the Carbon Anode. in Aluminium Electrolysis. 第六届国际熔盐会议论文集, 2001: 16-27.

[4] Lai Yanqing, Liu Yexiang, Yang Jianhong, and Thonstad Jomar: On Electrocatalysis of carbon anode in aluminium electrolysis RARE METALS, 2002, 21 (2): 117-122.

[5] US. Pat. 5728466; PN: 0586811.

[6] Metals. 1986, 40 (6): 582-584.

[7] David Voss, Thermoelectric materials, Electricity from waste heat is just one of the potential uses. , MIT Tech Rev. April 2002, Upstream.

[8] A. Sterten, et al. , Some Aspects of Low-melting Baths in Aluminum Electrolysis, Light Metals, 1988: 663-670.

[9] Mingming Zhang, Ramana G. Reddy: Alumina Solubility Study in Ionic Liquids with PF_6 Anion, Light Metals, 2004: 315-319.

电化学新材料的若干发展前沿[*]

摘 要 对电化学与电化工领域中应用和制备的若干新材料的主要进展、应用领域和急需研究解决的问题作了简要介绍。这些新材料包括：高性能电池及其电极材料、超级电容器、电化学法制备的纳米材料、富勒烯与碳纳米管、导电聚合物、光伏电池与光催化剂、传感器、柔性电子材料。

当今科技发展突飞猛进，全球电化学与电化工新材料领域的专家们正致力于解决该领域内的若干新问题，他们的研究方向和成果值得了解和借鉴。本文对与笔者研究工作相关的一些重要前沿课题作一个简单介绍。

1 高性能电池及电极材料

市场和环保急迫需求的新能源，主要是指电动汽车和便携式电器用的能源。

在任何国家和地区的电动汽车排放物（NO，CO 等）都是一个严重的污染源。为了满足环保的要求，汽车工业的进一步发展急需解决不用或少用矿物燃料作其动力的问题。现在解决此问题有两种方案：一是采用混合能源，即部分用电，部分用其他能源；二是全部用电。显然，全部用电达到零排放，为最佳方案。对汽车动力用电池的要求是高能量密度、高效率、长寿命和低成本。目前成为研究与开发热点的有燃料电池、高性能蓄电池和超级电容器等。

便携式电源近十年来，由于微电子技术的迅猛发展，推动了 4C（即电脑 Computer、移动电话 Cellular Phone、摄录机 Camcorder 及无绳工具 Cordless Tool）的急剧发展，它们需要高性能便携式电源。仅以手机为例，2000 年全球用户为 4.0 亿；现我国月增长速度达 500 万户，2002 年 6 月底已达到 1.7 亿部，已超过美国（1.26 亿部）。由此可见电源市场需求之广大。

1.1 燃料电池

燃料电池是将燃料的化学能经电化学反应直接转变为电能的一种装置，它是能量转换装置中效率最高的一种。通常将燃料电池按其所使用的电解质分为：碱性、质子交换膜（即固体聚合物）、磷酸、熔融碳酸盐和固体氧化物等多种类型。

阳极和阴极浸入电解质中，阳极用金属或金属氧化物陶瓷等制成多孔状，燃料由此处供入，阴极也用金属或金属氧化物陶瓷构成，引入空气，利用其中之氧与燃料经电化学反应产生电能。燃料一般用 H_2、甲醇或天然气。以 H_2 为例，燃料燃烧热值高，产物为 H_2O，没有公害。

1.1.1 融碳酸盐燃料电池（MCFC）

这种燃料电池已研究多年，它以熔融 Na_2CO_3 为主要电解质成分，工作温度为 650

---

[*] 原发表于《中国工程院化工、冶金和材料学部第五届学术年会》，2005 年 12 月，博鳌。

~700℃，以浸有（K,Li）CO$_3$ 的 LiAlO$_2$ 隔膜为电解质，电极材料以雷尼镍和氧化镍为主，可用净化煤气或天然气为原料。已有 100~1000kW 电厂试验。面临大规模商业化，亟待研究解决如下问题：

(1) 高能量密度方面，研究能够运转 2 万小时以上的装置，以及相关材料；
(2) 含有 MCFC 的电力系统，即 MCFC 与其他燃料电池及燃油发电系统的串联；
(3) 高能量密度的块状电池和电极的制备及其性能，在压力下的运转问题；
(4) 预测电池的腐蚀，电解质的损失，基体材料变化及其他影响电池寿命的问题；
(5) 研究电解质组成和添加剂对高能量性能的影响及其对能量衰减之影响；
(6) 污染物对 MCFC 性能之影响和限制方法。

1.1.2 直接甲醇燃料电池（DMFC）

已成为很具吸引力和应用前景的电源，因为电池较易更换，便于交通及加油站设置。在这方面的研究中，主要的问题有：

(1) 新型聚合物电解质；
(2) DMFC 电池的电催化剂；
(3) 新型电极结构和组成（向电极注入甲醇）；
(4) 单个电池及其系统的数学模型；
(5) 开发块状 DMFC 及其应用（如手机）。

1.1.3 聚合物电解质燃料电池

聚合物电解质是由导电高分子材料构成，它在燃料电池中既可以作为电解质又可作为电极，是汽车用燃料电池的首选对象，目前已制成 10kW 以天然气为燃料的质子交换膜燃料电池。主要待研究的有：

(1) 性能更优导电聚合物膜和复合材料的制备和性能表征；
(2) 研究聚合物层内电荷的迁移特征；
(3) 导电聚合物材料在燃料电池及电极上的电催化作用，及它们在充电电池及超级电容器方面的应用；
(4) 电池应用的稳定性，包括衰退过程，电解质/电极界面产物与机理；
(5) 聚合物电解质电池的制造方法。

1.2 高能锂离子电池

锂离子电池是当今最佳便携式电源，它既可以满足便携式电子通讯器具的需要，也可以为汽车提供新型能源。

锂离子电池在许多性能上远优于 Ni-Cd、Ni-MH 电池，它们的比较见图 1。

因此，锂离子电池已成为当今国际上研究与开发的热点。目前取得的主要进展如下：

(1) 在性能上，锂离子电池已有更大的提高。一个锂离子电池的工作电压相当于三个 Ni-Cd 或 Ni-MH 电池串联。手机电池通话时间达 285min，待机时间可大于 300h，反复充放电次数 1000 次以上。

(2) 正极材料由 LiCoO$_2$ 发展到 LiMn$_2$O$_4$。LiMn$_2$O$_4$ 虽然在重量/能量比上略逊于 LiCoO$_2$，但 Mn 的价格仅为 Co 的 1/40，因而电池成本可大为降低。另外，Mn 比 Co、Ni 好，无毒性，更安全；其他的正极材料还有 LiNiO$_2$，LiCo$_x$Ni$_{1-x}$O$_2$，LiFePO$_4$，

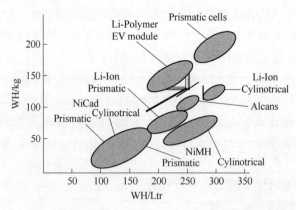

图1　锂离子电池和其他各种电池的性能比较

$Li_xV_2O_5$等。

(3) 薄膜型锂离子电池。由美国橡树岭国家实验室开发出来的固体膜状电池,正极材料为 $LiCoO_2$ 或 $LiMn_2O_4$,负极为 $Sn-SiN_x$,厚度仅 $15\mu m$,可做在塑料膜上,柔软可折叠,或加工成任意形状。

(4) 新型负极材料。由碳负极材料向金属氧化物,金属间化合物,纳米合金复合材料发展,插锂的性质更趋优良,如纳米 SnO_x、$Li_{4.4}Sn$、$LiTi_xO_y$ 和 $AlSi_{0.1}$ 等。

需要继续研究的课题是:

(1) 性能更佳的新型正极材料、负极材料、电解质和隔膜材料。

(2) 电池的充电。使用状况监控技术及相关的电子器件。

(3) 研究和制造用于航天器的大型锂离子电池,要求深度放电达 50%,充放电达 5 万次,工作时间超过 15 年。

1.3 超级电容器

1.3.1 超级电容器的特点及应用

超级电容器的容量比传统电容器大 20~200 倍,通常是由活性炭或高比表面积的 RuO_2 等作为电极材料,由硫酸或含有离子键的聚合物作电解质。

电化学超级电容器综合了电容器和电池的长处,具有脉冲放电能力大(放电电流>200A)、充放电速度快、循环寿命长、维护简单、廉价无污染等优点,广泛应用于能量存储器,尤其适合于短时大电流放电的装置。

电化学电容器既可以单独的使用,也可以与电池等其他能源联合使用。当与电池联合使用时,超级电容器可提供高的功率,而电池可提供高的能量,这样就克服了电容器不能提供高能量的缺点,又保护了电池免受电流浪涌,增加了设备的运行时间及电池的操作寿命。例如,仅以电池作为电动车的能源时,因一般电池的大功率输出能力差,很难满足车辆在启动、爬坡、加速时的动力需求。当电池与电容器组成了混合供电系统之后,电容器弥补了电池的上述不足,电动车的性能也得到了很大的提高。

1.3.2 电化学超级电容器的研究取得了长足进步

电化学超级电容器主要有以下几种类型:

(1) 碳电极电化学电容器。电极材料为活性炭、活性炭纤维和碳纳米材料,电解质为硫酸水溶液、有机电解质和含有四烃基氨盐的固体电解质(该种电解质材料具有极大的比表面,通常达 $1000m^2/g$),另外双电层之间分隔的距离很短,一般只有 2~

3Å，因此贮存容量可达200~500F/g，贮存能量可达3.25kJ/g。

美国Power store company开发了一种电化学双电层超级电容器-空气电容器，可提供高达4000W/kg的比功率密度。主要用于能源转化系统中的负荷平衡等。

这种电容器的生产成本低，为了获得高比表面积和低点阵电阻的活性炭，常采用的方法有：使用碳泡沫和糊状电极；高温分解碳基聚合物；催化热处理；添加贵金属、合金及Ru、Rh、Pd、Os、Ir、Co、Ni、Mn、Pt、Fe等的催化物。

（2）贵金属氧化物电极电化学电容器。主要用RuO_2，IrO_2等贵金属氧化物作为电极材料，以硫酸溶液作为电解液。由于RuO_2电极的导电性比碳好（导电率比碳大两个数量级），电极在硫酸中稳定，可以获得更高的比能量，这种方法制备的电容器比碳电极电容器具有更好的发展前景。

（3）导电聚合物电极电化学电容器。它比贵金属电化学超级电容器性能更加优越。近期研究工作主要集中在寻找具有优良掺杂性能的导电聚合物，提高聚合物电极的导电性能、循环寿命和热稳定性等方面。正、负极的导电聚合物分别进行p型和n型掺杂，充放电后都是掺杂状态，这种聚合物所作的电容器被认为是最有发展前途的电化学电容器。

在材料的选择上，主要集中在可进行n型掺杂的导电聚合物上，如聚乙炔、聚吡咯、聚苯胺、聚噻吩等导电物。

其他的电容器有：（1）孔状的NiO作电极的电化学超级电容器，单极电容达200~256F/g；（2）钴氧化物气凝胶粉末作电极材料的电化学超级电容器，容量可达291F/g。

电化学电容器是一种很有应用前景的新型储能装置，在今后的研究中仍需要通过新材料的研究开发，寻找更为理想的电极体系与材料，深入研究与开发导电聚合物型电化学电容器，另外全固态电化学电容的研制也是今后R&D的重要课题。

2 电化学法制备纳米材料

近年来，纳米材料的研究和开发日新月异，已经成为了材料科学研究的热门课题。

纳米材料在电化学、电化工中获得了广泛的应用。在先进电池工业中，纳米材料可用作锂离子电池的正极材料和负极材料，及其他化学电池、燃料电池、光化学电池的电极，还可以用作电化学贮氢材料、超级电容器材料等。在电催化反应中，纳米Ni-Co合金电极对析氢反应具有良好的催化活性，可以替代昂贵的铂电极，在电池和氯碱工业中有广阔的应用前景。

2.1 纳米材料的电化学制备方法

电化学法制备纳米材料和一般的物理、化学方法相比，其优点十分明显。电化学方法可以在常温、常压下进行，设备简单，易于操作；可以在大面积和复杂形状的零件上（单晶体）获得良好的外延生长层；由于没有高温作业，材料内部几乎没有热应力，并且避免了材料内部的热扩散，可以获得组成一定的单一成分；电沉积速度很快，膜层的厚度可以自由地控制在几个原子层到几万个原子层厚度，整个电积过程可以用计算机控制；电化学法实用范围广，可以制备的纳米材料有颗粒、纯金属、合金、金属陶瓷复合涂层以及块体材料，制备的材料在致密度上要优于其他的方法。

电沉积法主要有单槽沉积和双槽沉积：

单槽沉积法是将两种不同化学活性的金属离子以适当的含量加入统一电解槽中，

控制电极电位在一定范围内周期地变化，可得到组成或结构周期性变化的金属膜。

双槽电沉积法是交替地在含有不同沉积金属的两个电解槽中进行的，按电镀条件也可以分为恒电流法和恒电位法两种，以后者应用得多。

2.2 纳米多层膜的电化学法制备

金属多层膜，又称为组分调制合金，是一种金属或合金沉积在另一种金属或合金上形成的组分或结构周期性变化的材料，多层膜中相邻两层金属或合金的厚度之和等于多层膜的调制波长 L，当 L 小于 100nm 时，这种多层膜就可以称为纳米金属多层膜或纳米组分调制合金。

纳米多层膜的特征主要有：薄膜效应、界面效应、耦合效应、周期性效应，由于这些特征，这种材料表现出各种新奇的特征，引起了人们的极大兴趣。

目前由电沉积法制备的二元纳米多层膜样品已达几十种，对于三元纳米多层膜的研究也有了很多的报道，如 Co-Pt-P 多层膜制备的纳米金属多层膜。已经制备出的纳米金属多层膜样品有：Cu-Ni，Cu-Bi，Cd-Ag，Ag-Pb，Ni-Mo，Cr-Mo，Co-Ni 等。

2.3 纳米晶镍的电化学法制备

采用了喷射沉积法制纳米晶镍，喷射电沉积是利用电解液的冲击对镀层进行机械活化，沉积过程也因此得到了改善；扩散层的厚度大为减小，晶粒细小，镀层也更加致密。

电沉积法制备的纳米 Ni-Co 合金的平均晶粒尺寸为 1.7nm，在 30% 的 KOH 溶液中纳米晶镍-钼合金电极对析氢反应表现出很高的电催化活性。

2.4 高热稳定性锐钛矿型纳米 TiO_2 粉体

在醇中加入少量有机胺导电盐作为电解液，以工业纯钛作"牺牲"阳极，控制电流电解，可分别得到乙醇钛、丙醇钛等相应的醇溶液，然后直接水解，经溶胶-凝胶形成凝胶前驱体，一定温度下煅烧后得到平均粒径 10nm 左右的高热稳定锐钛矿型 TiO_2。

2.5 新型电解法制备纳米粉体

新型电解法制备纳米粉体可以制备出纯度高、粒径分布均匀且表面包覆的纳米金属粉体。

3 富勒烯（Fullerene）和纳米碳管

3.1 富勒烯的结构和制备方法

富勒烯（Fullerene）是由碳原子所形成的系列分子的总称，C_{60} 是其代表物。C_{60} 是由 60 个碳原子所构成的球形 32 面体，即由 12 个五边形和 20 个六边形组成。单个富勒烯分子是由 12 个五边形和不同数目的六边形构成，多个分子成双结对地结合在一起，组成含有 30~600 个碳原子的封闭笼形结构，当这种笼形结构沿某一方向延长，形成具有极大长径比的结构时，因其直径为纳米量级，故称其为碳纳米管。这就是人类所知的第三类碳物质。1985 年发现 Fullerene 的专家们预言，它将为人类文明作出巨大的贡献。进入 90 年代后，对 Fullerene 的研究十分热烈，陆续发现了纳米碳管（管状富勒烯)、碳的纳米簇等许多新颖的品种。并发现这些物质具有很奇特的物理化学性质。人们正在深入研究，不断有新的发现。

碳纳米管的制备方法主要有电弧法、激光法、等离子法及催化法等。催化法是制备多壁碳纳米管的最佳方法，设备简单，产量大（一次可达克级）。

3.2 碳纳米管的潜在用途与发展前景

（1）锂离子电池的负极材料。碳纳米管的层间距离略大于石墨，其充放电容量大于石墨。其首次放电容量达645mA·h/g，可逆容量为700mA·h/g，远大于石墨的理论可逆容量375mA·h/g，是高能量密度电池的理想电极材料，可用来代替石墨作锂离子电池的负极材料。

（2）超级电容器。从结构上看，碳纳米管符合理想电容器电极材料的所有要求（即结晶度高，导电性好，比表面积大，微孔导电性较小），用这种材料制成的超级电容器可获得大于4000W/kg的比功率。

（3）贮氢材料。零排放的氢动力汽车必须解决高性能贮氢技术与贮氢材料的问题，碳纳米管的可逆储/放H_2量在5%左右，大大高于金属氢化物的2%，是迄今最好的贮氢材料。

（4）催化剂组合。碳纳米管具有纳米级的内径、类似石墨的六面体碳原子环网和大量未成键的电子，具有选择吸附和活化一些较惰性的分子（如CO_2、NO等）的能力，曾有实验证实，它的催化活性比铑还好，并很稳定，极可能具有一些与贵金属类似的催化功能，可在一大批贵金属催化反应上得到应用，碳纳米管将给石油化工产业带来一场新的技术革命，其应用前景不可估量。

（5）传感器。利用碳纳米管对气体吸附的选择性和由此引起的导电性变化可做成气体传感器，在碳纳米管内填充碱金属可形成P-N结，填充了湿敏、压敏、光敏等材料后可制成各种功能的纳米级传感器，可预见碳纳米管传感器将是一个很大的产业。碳纳米管还可以制成场致发光的大型平面显示器，新型电子探针，微电子器件，防弹保暖隐身纺织纤维等，在未来可望形成一个庞大的产业群。

3.3 富勒烯的研究现状

目前人们在富勒烯的基础研究、基础特性和潜在的应用领域等方面取得了很大的进展，国际上化学、物理、材料、生物等方面的专家对富勒烯（Fullerene）研究的热点问题如下：

（1）富勒烯和纳米碳管的电化学、电子学、光电化学、催化、电子器件与传感器研究等；

（2）进一步研究各种富勒烯及新出现的相关化合物的性质和应用，它们是纳米碳管，有机富勒烯，电活性富勒烯，超分子富勒烯，有机金属富勒烯，包合物富勒烯，富勒烯薄膜及复合材料等；

（3）富勒烯的光感应过程，有关富勒烯的光物理、光化学情况，富勒烯的功能化问题，以及以富勒烯为基础的材料对光的激发、传输、光能量储存、光能发射以及光电化学电池应用；

（4）纳米碳管材料和新颖纳米级器件，研究富勒烯纳米管、纳米簇及相关材料的制作，微型（纳米级）器件的制造技术，新颖纳米器件的建筑拼块等技术；

（5）碳纳米管的化学功能化就是通过亲核添加、功能团添加、增氢、生成过渡金属络合物、氧化及亲电子反应、多纳米碳管合成、纳米碳管总合成等实现各种功能化；

（6）纳米碳管的固体物理：有关纳米碳管化合物的物理结构与性质，包括化学反应特性、超导、表面研究、衍射研究及热、电性质；

（7）包合物富勒烯和碳缓择体：研究各种包合物富勒烯如金属包合物，稀有气体包合物，碳缓择体和金属缓择体等的合成，性质及应用等；

（8）富勒烯材料在电池，燃料电池方面的应用：研究其电化学和电催化性能，电池特性和有关器件性能及能量储存问题等；

（9）富勒烯材料在生物工程，医学和药物学上的应用。

4 导电聚合物

导电聚合物是 20 世纪 70 年代初的一次偶然机会发现的，是白川美树（日），A. G. MacDiarmid（美）和 A. J. Heeger（美）三位科学家努力探索的结果，他们共同获得了 2000 年度诺贝尔化学奖。

导电聚合物兼有金属和塑料的多种性质，与金属和无机半导体相比，它有许多优点。它有比铜更加优良的导电性，具有塑性和延展性，质量密度低、导热性低、膨胀系数低，抗化学腐蚀性好，此外它还具有多向异性的分子结构和导电性、可调的光学性质（取决于分子结构）、电色性、很强的能量贮存能力。但是导电聚合物很难兼备导电性、可加工性和稳定性，为此进行了大量的研究开发工作。

起初，以聚乙炔为主的导电聚合物的应用集中于电池、非线性光学材料（用于通信）、半导体制造、传感器、抗静电、无线电干扰屏蔽及电磁干扰屏蔽等方面。

后来出现了一系列的导电聚合物，如聚丙烯、聚苯胺、聚吡咯、聚噻吩等，尤其是把导电聚合物的可加工性与比重小的优点和金属导电率相结合的思想极大的推动了其研究和应用的进展，具体应用有：

（1）发光二极管（LED）：根据聚苯乙炔（PPV）及其衍生物的非掺杂导电聚合物的半导体态在外电场激发下可以发光，从而形成发光二极管的原理，已开发出的电子产品有广告牌显示屏、手机显示屏、聚合物电子显示器和塑料发光大屏幕电视等，LED 有可能代替灯泡作为新型照明光源；

（2）塑料芯片：已有报道称，以廉价、柔韧的聚合物为材料，已制出有 864 个晶体管的电路板和其他的电脑芯片元件，塑料芯片的应用前景将相当诱人；

（3）聚合物激光器：用电子而不是光子激发导电聚合物可以使之产生激光，2000 年在美国《科学》杂志上报道了在晶体管中镶入一块能发光的四并苯（tetracence）器件，接通电源后，四并苯就能发出激光；

（4）导电聚合物传感器：利用聚丙烯等导电聚合物现已开发出多种功能的传感器，如便携式数字成像传感器、聚合物传感器的彩色显示器，Akhter 公司用聚烷基噻吩制造出了多种类型的显示器、智能 DNA 传感器以及用于生物学方面的传感器等；

（5）人工肌肉：用电活性聚合物，如聚丙烯腈等制造航天机器人的人工肌肉。

此外，还有导电纺织物、新型光存储器、微型可移动像素等方面的应用。总之，导电聚合物的基础与应用研究已成为热点，在 21 世纪它将在材料科学领域发挥重要作用，并取得丰硕成果。

5 光伏电池、光催化剂

5.1 光电化学、光伏电池

当物质（例如金属板）吸收辐射能（电磁波、日光）之后释放出电荷的现象被称为光电效应，利用光能（日光）产生电能的装置被称为光伏电池或太阳能电池。1887年赫兹发现了光电效应，50年代，美国的 Bell 实验室人员开发了 Si 太阳能电池。如今，地球上各种各样光电池技术已有三四十年的发展历史了，全球已建立了一个年产值数十亿美元且日益增长的市场。

现有三种类型的光伏电池：单晶硅，多晶硅，薄膜无定形硅及密集光伏电池组。

密集光伏电池组是用棱镜把太阳光聚焦于小面积的硅太阳能电池上。一个太阳能光电池板通常由 36 块单晶硅太阳能电池组成，板的光-电转换效率可达 15%，最高可达 18%。多晶硅太阳能电池的转换效率较低，为 13% 左右，无定形硅则为 4%～7%（最先用于计算器上）。

光伏电池最早用于海上航标灯和人造地球卫星上，还用于边远地区广播站、农村电话网和交通不便的林区哨所等。现已发展到抽水灌溉、草原放牧、乡村或山区医院及山区别墅照明、空调供电等方面。

为了今后的更大发展，关键是降低成本，此外，尚有以下热点问题正在开展研究：

(1) 光伏电池新器件和新结构，如薄膜硅，微晶硅或无定形硅，纳米晶材料，生物态材料，有机材料，光电化学电池以及这类材料或器件组合而成的新器件和新结构；

(2) 超高效率的外延太阳能电池，包括Ⅲ-Ⅴ族材料和器件，长寿命技术，单晶薄膜材料与低成本材料的复合；

(3) 新材料的高效、低成本制造技术；

(4) 太阳能制氢气，太阳能产生电色及新颖的太阳能发电与储存方法；

(5) 光导薄膜及光电致色材料；

(6) 半导体电极的改性、光刻及光催化；

(7) 采用纳米级材料对光电化学废料处理的技术。

5.2 光催化剂

自从 1972 年 Fujishima 和 Honda 首次发现光催化现象以来，人们围绕太阳能的转化与贮存，在探索多相光催化过程原理及其开发应用方面已进行了大量的工作，光催化在水和空气的净化、太阳能的贮存和利用方面显示了相当诱人的商业前景。

当半导体受到光激发时，本身会产生电子-空穴对，在此过程中，电子居于较高的能态，可作为一个还原剂，而价带中的空穴具有较高的氧化电势，当它们被吸附的反应物所捕获时能分别进行氧化或还原作用（而不是复合），这样就有可能被用来做光催化反应的催化剂，这种太阳光激发的氧化与还原作用可以用于去除污染、杀菌消毒等。

目前已研究开发出多种光催化剂，它们多为大禁带 n 型半导体氧化物，如 TiO_2、CdS、SiC、WO_3、In_2O_3、SnO_2、SiO_2、Fe_2O_3 等，其中 TiO_2 表现最优。它不仅无毒无害，非常稳定，且在水相、气相和非水相溶液中都十分有效。已开发的光催化应用有：

(1) 降解水中有机污染物：如含有卤衍生物，有机氯化物，DDT，染料废水，农药废水，表面活性剂，含油废水等；

（2）水中无机污染物的光催化处理：如 Hg^{2+}，Ag^+，Cr^{6+}，CN^-，SO_2，H_2S，NO，NO_2 等的还原式氯化物；

（3）灭菌消毒：对细菌细胞产生光化学氧化作用，降低生物体中辅酶的活性而导致细胞死亡，从而使水中的酵母菌、大肠杆菌、轧制杆菌和葡萄球菌等不能存活，可用于水处理，食品工业等；

（4）气相污染物的光催化降解：主要对挥发性有毒有机物有光催化降解作用，如氟利昂、三氯乙烯等，用于自净化、玻璃态和建筑材料等；

（5）CO_2 的光催化还原：CO_2 是主要温室气体成分，去除 CO_2 并将其转化为有机物具有十分重要的意义。利用光敏半导体颗粒，如 TiO_2、WO_3、ZnO 等可将水中的 CO_2 还原为甲酸、甲醛和甲醇等。

由此可见，这是一个造福人类的新技术领域，但还有许多基础与实用工作有待完成，主要有：

（1）寻求对可见光敏感、抗腐蚀、寿命长、选择性好的高效催化剂材料；

（2）研究与开发新的制备光催化剂方法，尤其是电化学制备法和可视化流程与设备，减少污染，提高效率，降低成本；

（3）寻求合适的光催化剂载体，固定化技术及高效、多功能的水处理装置；

（4）研究固液、固气界面光催化氧化作用机理，光载流子传输机理以及光催化剂中毒等问题。

6 传感器

能够感知各种物理、化学、生物量，并由此响应而产生出某种可检测信号的器件称为传感器。第二次世界大战以来，它已得到十分广泛的发展，特别是在信息系统和流程自动控制系统中，已成为关键基础的器件。在航空航天中，如宇宙飞船、航天飞机、人造地球卫星等需用成千上万个传感器，宇航员穿着的宇航服中也有数十个之多。军事应用中功能齐全、种类繁多，如火箭、导弹制导等仍在不断研制和开发新型传感器。飞机、汽车用于自动驾驶的传感器研制已甚成熟，它们可测温度、速度、高度、距离、压力、风速、油耗等，家用电器、家庭安全报警系统等已甚普及，在医疗保健方面已广泛使用的心电图、B超、脑血流图、验血、验尿、体内外专项检测等都有专用的传感器。特别是在工业过程控制中，更需要各种各样的传感器与计算机配合实现自动监测和控制的目的，由于多方面的需要，传感器的研究与开发一直受到重视，这是一个十分活跃的领域，目前世界传感器的种类约有2万种，市场前景广阔。

根据传感器的工作原理，可以分为光学、红外、热学、超声、化学、电化学、电磁学及生物传感器等。它们的单独、联合、成阵列使用，可用来检测温度、湿度、压力、流量、速度、水平（深浅）、气体组成（H_2、O_2、CO_2、CO、H_2S、CH_4、NO）、地声、气味、生理生化专项、食物新鲜程度及衣着舒适程度等，它们所涉及的敏感材料也是多种多样的。

目前，受到重视和研究开发的有以下几方面：

（1）野外用的各类型传感器；

（2）环境检测用的传感器，人们特别重视对有机污染物的分析检测及电化学传感

器。污染防治方面还需要许多新型高效传感器,此外,腐蚀传感器也在开发中;

(3) 声波传感器(可作非接触测定),用于检测各种物理、化学、生物的参数,并可以监测正在进行的过程;

(4) DNA(脱氧核糖核酸)传感器,这是生物传感器中最受重视的课题,因需要迫切而发展很快,这种传感器可用于鉴别细菌组织,检测并抑制核糖核酸以追踪疾病的进展,鉴定多形核白细胞的发生以预示人的疾病,以及用此种传感器来检测和发现毒品,还有与此相关的酶、葡萄糖和氮氧化物传感器等;

(5) 超微型传感器,它们与微型机器人、微型机械电机系统,植入体内的微型传感器配套使用。

7 柔性电子材料

7.1 柔性电子材料 FE(Flexible Electronics)

(1) 柔性电子材料(FE)是既柔软又有电子器件功能的材料。简单地说,就是把有机半导体材料以薄层形状沉积在塑料或者金属箔上,形成一种具有电子器件功能的柔软材料(例如:柔性显示屏等)。

(2) 与传统的多晶硅半导体材料制成的电子器件完全不同,FE 制成的平板显示屏柔软而轻巧,更省电,也更便宜,同时还可以生物降解,不会构成对环境的污染。

(3) 由于可以利用印刷或喷涂技术来制造,因此,可以像印刷报纸一样进行大规模连续生产,目前已经出现所谓卷-卷(Roll to Roll)生产技术。因此,FE 产品成本可以大幅度降低。

(4) IBM 的技术有机半导体并五苯(碳和氢的化合物)在经过专门处理后,能够得到比较粗大的晶体,具有良好的半导体性质。

7.2 制备方法

柔性电子材料是美国 Lawence Livemore National Lab 在 1997 年发明的。目前已吸引许多大、小公司进行研究和开发,并加速商品化。

当前,制造柔性电子材料主要有两种方法:

(1) 把有机半导体材料(如并五苯)薄膜沉积在塑性物质上(这是 IBM 公司的制造技术);

(2) 把无定形硅半导体沉积在塑性材料上。

7.3 实际应用

作为一种新兴的显示材料,柔性电子材料获得了国际电子行业的极大重视,预计可以形成数十亿美元的产业。当前,已经开发和正在开发的产品有数十种。举例如下:

(1) 计算机显示屏。由于它的柔软和可折叠性,这种产品可容易地放进口袋甚至穿在身上或戴在手上。

(2) 智能卡(Smart-Card)。以前,芯片是粘贴在塑料卡上的(智能卡制造费用最高的工序)。现在,由于柔性电子材料的出现,电子器件可制备在塑料表面上,而且不需要电极。这种智能卡可以用来作身份证、驾驶证、病历卡和各种管理卡等。

（3）无线电报纸。由 FE 制成的这种报纸，与蓝牙技术结合，可以时时更新，只需购买一次便可长久使用。

（4）大容量存储器件。可以用 FE 生产存贮容量达 5TB 的记忆元件。

另外，它还可以用来生产各种类型的显示器件：如有机光发光二极管、移动电话的信息屏等，其特点是工作电压低、质量轻、结构单一、视角广、反差高且可获得全色的图像。Philips 公司将其应用在剃须刀上，通电时可显示使用时间和剃须效率，断电时可作为镜子；Sony 公司将其用在大显示屏上，亮度可达 300 烛光/m^2；Tonshiba 公司也制造了一台 17 英寸的全色彩色屏。

7.4 未来展望

自 FE 出现以来，好评如潮，举例如下：

（1）"FE（Flexible Electronics）的出现是电子技术革命的曙光，用柔性的膜替代玻璃制造显示屏幕，而且是连续的、印刷式的制造过程，将使整个电子技术发生一场新的革命"（国际半导体杂志，2001 年 11 月）；

（2）"利用这种技术可在单面上制造出上十亿个晶体管，而且是环境友好的连续式工艺，所用材料可重新使用"（MIT<技术评论>，2002 年元月）；

（3）"把晶体管印刷在有机物上（如塑料）是电子技术革命的一大飞跃，它可以使消费者在日常生活中获得更多的电子服务功能"（Intel 元件研究实验室）；

（4）"卷对卷的生产技术可以做到中间环节少、收率高、成本低并实现大规模生产"（美国 DuPont 公司）；

（5）"用这种技术来生产塑料太阳能电池，要比传统的硅太阳能电池成本更低，而且可以在曲面上工作"（欧洲经济共同体）。

7.5 若干技术研究问题

尽管当前该领域的研究开发工作非常活跃，但是还有大量的科学技术问题亟待解决，有关问题如下：

（1）柔性电子材料：硅沉积在柔性物质上的方法，有机半导体，可印刷的电子材料，柔性线路；

（2）宏观电子学：舒适的大面积电子器件、电子器件新概念以及超大型电子线路；

（3）柔性显示：有机光发光二极管，柔性液晶，柔性电色显示和电泳显示；

（4）相关技术：新材料，柔性物质，制造系统和技术，新的制造过程。

参 考 文 献

[1] D. P. Wilkinson. Fuel Cells [C]. The Electrochem. Soc. Interface, Spring, 2001：22-25.

[2] G. Halpert, et al. Batteries and Fuel Cells in Space [C]. The Electrochem. Soc. Interface, Fall, 1999：25-30.

[3] 高利珍，等. 纳米碳管的生产及其应用 [J]. 科技导报，2001（6）：16-19.

[4] 万梅香. 2001 科学发展报告 [M]. 北京：科学出版社，2001：42.

[5] Y. Mastai, D. Gal, G. Hodes. Nanocrystal-Sized Control of Electrodeposited Nanocrystalline semiconductor Film by Surface Capping [J]. J. Electrochem. Soc. 2000, 147（4）：1435-1439.

［6］E. Chasseing, et al. Nanometric Cu/Co Meltilayers Electrodeposited on Indium-Tin Oxide Glass ［J］. J. Electrochem. Soc. 1999, 146（4）: 1794-1797.

［7］Jeff Dorsh. Flexible Electronics: On the Brink of a Tech Revolution ［J］. Semiconductor Magazine, 2001, 2（11）.

［8］Wade Roush. Flexible Transistors on a Roll ［J］. MIT's Technolgy Review, 14, Jan. , 2002.

［9］11th International Meeting on Lithium Batteries, Program Information, June 23-28, 2002.

能源新材料的若干发展前沿*

摘　要　对新能源领域中的若干新材料的主要进展、应用领域和需研究解决的问题作了简要介绍，包括：高性能锂离子电池、超级电容器、光伏电池、柔性电子材料与可印刷电源。

引　言

在科学发展观的指引下，我国电化学与新能源领域的专家们正致力于研究该领域内的若干新问题，开发新技术，提供新产品，积极为经济繁荣与社会进步服务。参照国内国际的发展，结合我们的研究工作，本文对相关领域的一些重要前沿作一简单介绍。

1　高性能锂离子电池

1.1　锂离子动力电池

锂离子电池是目前比能量最大的二次电源，在许多性能上远优于 Ni-Cd、Ni-MH 电池。相对于小型的单体电池，大型动力电池组是锂离子电池发展的更重要的方向。电动车能源的应用前景使得研究人员将制备材料与电池的标准定位于高容量高功率和低价格上。近两年主要研发进展如下。

第二代锂离子电池首见于美国能源部车用技术办公室支持的"先进运输用电池"项目（BATT），旨在开发纯电动车（EVS）和混合动力车（HEVS）用可充电电池[1-3]。项目要求达到：10年内容量衰退小于20%，充放电1000次；工作温度为-40~65℃；比能量大于170W·h/kg；比功率大于300W/kg；成本小于150美元/（kW·h）。BATT 项目由美国劳伦兹-伯克利国家实验室（Lawrence Berkeley National Laboratory，LBNL）牵头，组织了全美6个国家级实验室共同研发，2006~2007年着重研究了：

（1）三类新正极系统的性能和局限，包括 $LiFePO_4$、Mn 尖晶石及尖晶石复合系统、镍酸盐系统；

（2）新型负极系统，包括中间金属化合物（Cu-基、Li-Cu-M-Sn、InSb 等）、纳米级 Sn 基负极材料、树枝状结构材料等；

（3）新颖电解质及其表征，研究内容有杂质对电解质界面行为的影响、聚合物电解质的纳米结构、氟化聚醚与离子液体、模拟新颖离子液体及聚合物电解质、离子液体表征、胶体电解质。

美国先进电池协会（U. S. Advanced Battery Consortium，USABC）也在从事高性能锂离子电池研究。有美国三大汽车公司（DaimlerChrysler, Ford and General Motors Corpo-

---

* 原发表于《中国工程院化工、冶金与材料学部第六次学术年会论文集》，2007年10月，济南。

ration）加盟，成立协会的目的是：

（1）开发电动车，混合动力车及燃料电池车用电池技术；

（2）继续开发高功率电池技术，以降低成本至20美元/kW并延长电池寿命至15年；

（3）开发用于混合电动车的超级电容器技术；

（4）对高功率和高能量电池及超级电容器的应用进行技术规范化工作。

中南大学针对锂离子电池组建了湖南业翔晶科新能源有限公司，主要瞄准动力电池领域，开发了：1）电动摩托车使用的36V/15A·h系列电池组（如图1所示），已用于法国Solex公司电动摩托车上；2）高安全性矿灯电池；3）电动工具用的复合型电池。

电池性能提高的关键取决于电极材料和电解液性能的提高。下面重点介绍锂离子电池电极材料和电解液近两年的研发进展。

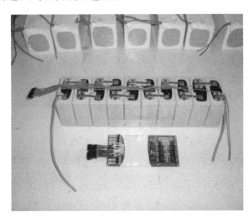

图1　36V/15A·h电池组

1.2　锂离子电池电极材料

1.2.1　锂离子电池正极材料

（1）橄榄石型材料。

$LiFePO_4$被认为是最有希望应用于电动车的锂离子电池正极材料，关于提高它的大电流放电问题仍然是研究者们关注的焦点。目前采用的改进方法主要分为：1）表面炭包覆；2）高价金属离子取代Li^+；3）合成纳米级的细小颗粒。$LiFePO_4$是电子/离子混合导体，充放电过程中，Li^+和e的输运都将影响材料的输出电流。J. B. Goodenough[4]采用电沉积的方法将$C/LiFePO_4$与导电聚吡咯复合在不锈钢网集流体上，获得的$(C-LiFePO_4)_{0.84}(PPy)_{0.16}$在进行高达10C的充放电时，仍有80~90mA·h/g的比容量。L. V. Thorat[5]等认为以长径比比较大的炭纤维为导电剂更有利于电导率的提高，可使$LiFePO_4$的放电功率达到USABC提出的车用锂离子电池的功率目标。M. S. Song等[6]采用微波技术合成经过球磨的原料，仅用2min就获得了性能优异的$LiFePO_4$。D. H. Kim等[7]采用醋酸锂、醋酸铁和磷酸二氢铵为原料，在四甘醇中，335℃反应回流16h，就获得了长方体型的纳米级$LiFePO_4$，0.94C放电电流时的比容量为166mA·h/g，而在30C放电时仍有58mA·h/g。A123systems公司[8]则宣称其$LiFePO_4$电池可以进行100C的脉冲放电和35C 100%深度放电，输出功率高达3000W/kg和5800W/L。

硅酸盐因具有比磷酸铁锂储量更丰富和环境更友好的特性而更具有吸引力，有人预计 Li-M(Fe,Mn)-Si-O 聚离子系统将是下一代安全、廉价的锂离子电池正极材料。2006 年，Zaghib 等[9]通过真空烧结的工艺合成了 Li_2FeSiO_4，该材料的充放电平台分别为 2.80V 和 2.74V。杨勇[10]等采用液相合成前驱体的方法，制备了掺杂化合物 $Li_2Mn_xFe_{1-x}SiO_4$，发现利用 Mn 从+2 价氧化到+4 价，$Li_2Mn_{1-x}Fe_xSiO_4$ 可以进行多于 1 个电子的反应，掺杂 20% 的 Mn 就会导致理论容量超过 200mA·h/g，这将是这一材料发展的新方向。

（2）层状化合物。

各类用以取代 $LiCoO_2$ 的层状化合物始终是研究人员感兴趣的课题，从 Co/Ni，到 Co/Ni/Al，再到 Co/Ni/Mn 和 Ni/Mn，性能、价格和安全性的综合考评很难全面满足。

早期 $LiCoO_2$-$LiNiO_2$ 被普遍看好，但其不佳的热稳定性和上涨的 Ni 价格阻碍了其在电动车上的应用。$LiCoO_2$-$LiMnO_2$ 吸引了很多研究者的关注。Rossen 和 Dahn 首先对 $LiNiO_2$ 和 $LiMnO_2$ 的固溶体进行了研究，并得到了 $x>0.4$ 的单相层状 $LiNi_xMn_{1-x}O_2$ 材料。低放电倍率时，$LiNi_{0.5}Mn_{0.5}O_2$ 的可逆容量大约为 150mA·h/g。D.K. Lee 等[11]采用共沉淀方法合成的 $Li[Li_{0.2}Ni_{0.2}Mn_{0.6}]O_2$ 材料在 2~4.6V 之间可达到 270mA·h/g 的比容量，并且在循环 50 次后，容量保持率为 92%。

$LiNiO_2$ 和 $LiCoO_2$ 也能很容易与 $LiMnO_2$ 合成三元固溶体材料。其中最典型的材料是 $LiNi_{1/3}Mn_{1/3}Co_{1/3}O_2$，近两年有大量关于这种三元体系的合成条件、不同成分配比、包覆处理和掺杂处理等方面的研究报道。其中 N. Yabuuchi 等[12]制备的 $LiCo_{1/3}Ni_{1/3}Mn_{1/3}O_2$ 在 55℃，20C 的放电容量达到 160mA·h/g。

（3）尖晶石型化合物。

尖晶石 $LiMn_2O_4$ 由于其三维结构具有比 $LiCoO_2$ 更大的 Li^+ 扩散速度而成为电动车锂离子电池的备选材料之一，为克服其稳定性差的问题，各类掺杂和改性的研究还在不断的进行。

1.2.2 锂离子电池负极材料

炭材料仍是负极材料的主角，其中 SiC 复合物是研究者关心的课题。合金材料因嵌脱锂引起的粉化现象至今尚无法很好的解决，但并不阻碍一些公司对其大容量的兴趣，如松下电池产业公司研发新型锂电池即采用合金负极。同时锂金属聚合物电池也引起了关注。

1.3 电解液

双乙二酸硼酸锂（LiBOB）在 1999 年由 Lischka 等首次报道，近两年对其的研究主要集中于 LiBOB 及其他新盐化学结构的优化设计以及与其他类型锂盐搭配使用的研究；不同溶剂组分配比筛选，新型电解质体系设计以及凝胶电解质的研究；使用硼酸酯类新盐及 LiBOB 基电解质盐的新电池体系的优化设计等，阻碍该材料大规模应用的仍是其较高的成本，同时其较低的电导率和低温特性也需要提高[13]。而在电解液中添加特殊的物质来防止过充和过放，将使得锂离子电池组的应用更加方便[14]。

离子液体具有热稳定性好、不挥发、低毒性、电化学窗口宽等优点，其作为电解质是电解液发展的另一个可能方向。

1.4 热点和难点

今后需要继续研究的主要课题有：

（1）性能更佳的新型正极材料、负极材料、电解液和隔膜材料以及它们之间的良好匹配；

（2）电池的充电、使用和实况监控技术及配套的电源管理系统；

（3）研究和制造用于航空航天器的大型锂离子电池组，要求深度放电达50%，充放电达5万次，工作时间超过15年；

（4）加紧开发生物医药用高性能锂离子电池，要求电池长寿命、可体外充电，可驱动植入体内的微型起搏装置，如Bion®。加紧开发硅氧烷基聚合物+$LiFePO_4$正极+C负极材料。

2 超级电容器

超级电容器是介于传统电容器和电池之间的一种新型储能器件。超级电容器一问世便受到人们的重视，已成功地应用到仪器仪表、电动工具、消费电子、玩具等领域[15]。下面列举几个正在蓬勃发展的超级电容器的应用领域：

（1）军事方面。新一代激光武器、粒子束武器、潜艇、导弹以及航天飞行器等高军事装备在发射阶段除装备常规高比能量电池外，还须与超级电容器组合才能构成"致密型超高功率脉冲电源"。

（2）无线通信领域。超级电容器适合在大功率的脉冲电源上应用，特别是那些使用无线技术的便携装置。

（3）电动汽车和混合电动汽车。开发混合电动汽车（HEV），用电池为电动汽车的正常运行提供能量，而加速和爬坡时可以由超级电容器来补充能量。而完全用超级电容器作为主电源的电动汽车，目前正成为各国科学家积极追求的目标。

2.1 超级电容器的产业现状

美国、日本、俄罗斯和欧洲正积极地进行超级电容器的研究和开发，下面列举主要国家的超级电容器产业状况。

（1）俄罗斯。研发机构包括ELIT、ESMA等。主要用于军事和电动车辆以及内燃机启动等领域。

（2）美国。主要由Army Research Laboratory，Lawrence Livermore National Lab，Maxwell Tech，Evans，Copper等进行研究。其产品系列全面，涉及消费电子、通信领域，电动汽车和军事领域。

（3）日本。主要有Panasonic，NEC，ELNA，Tokin等公司，开发的双电层电容器占据了世界大部分市场。

（4）欧盟。欧盟实施的Joule Ⅲ项目，其目标是满足电动汽车的要求，为工业开发作准备。

（5）韩国和澳大利亚。韩国NESS公司和澳大利亚的Cap-xx公司的超级电容器在GSM/GPRS通信、GPS定位仪、数码相机和便携式数据终端等设备中获得了成功的应用。

（6）中国。我国已有超级电容器生产厂家的产品多为小型钮扣式和卷绕式单体，且与美国、日本、韩国相比，相关技术仍有较大差距，主要表现在：1）产品体积能量密度低；2）内阻大，功率密度低；3）漏电流大。这些问题严重制约了在高端领域的推广应用。

中南大学针对超级电容器发展的新动向，主要开发大型、组合型有机系超级电容器单体和模块，已经开发出 2.7V/4000F 单体和 500V/30F 高压大功率超级电容脉冲电源系统（见图2）。

图 2　基于超级电容器的高压大功率超级电容脉冲电源系统

2.2　超级电容器的关键材料

2.2.1　电极材料

电极材料是决定超级电容器性能与成本的关键因素[16]，主要有碳材料、金属氧化物和导电聚合物等。

（1）炭电极材料。除已产业化应用的活性炭材料外，国内外还针对碳气凝胶和碳纳米管等新型碳材料开展了大量研究。保持高比表面积的同时，进一步优化其孔径结构，同时降低其成本是活性炭发展的方向。我们课题组正在优选新的前驱体，新的炭化和活化工艺，新的除杂工艺制备可工业化的多孔炭材料[17]。

（2）金属氧化物电极材料。主要有贵金属氧化物 RuO_2、IrO_2 和贱金属氧化物 MnO_2、NiO、Co_3O_4、V_2O_5 等。目前的发展趋势是在制备上采取掺杂、纳米化等特殊处理来改善其性能。

（3）聚合物电极材料。聚合物电极材料存储电能的机理主要是赝电容。近期研究工作主要集中在寻找具有优良掺杂性能的导电聚合物，提高聚合物电极的导电性能、循环寿命和稳定性等方面[18]。

2.2.2　电解液

离子液体作为一种新的电解液，也成为超级电容器的一个研究的方向。针对耐高温、耐燃、宽电位窗口的离子液体成为研究的热点[19,20]。

为满足薄膜化和全固态化，高离子导电率、良好的机械性能、稳定的电化学性能，以及能够实现与电极材料的良好匹配的新型聚合物电解质等成为研究的热点。

2.3　热点、难点

（1）高性能的超级电容器器件及其组件技术。超高能量的电容器单体的制备技术成为发展的难点。另外，由于超级电容器的单体电压较低，需要对多个单体进行串并联，从而形成高电压超级电容器组件，因此要求单体的一致性好。另外需要高性能电源管理电路系统与之配套，从而实现高电压、高容量的要求。

（2）通过采用新型高性能电极材料、功能化的电解液，开发导电性好、高比容量、低杂质含量和低成本的电极材料，以及高安全、长寿命、低漏电流、低内阻、宽的使用温度范围使未来超级电容器能够满足多种场合需要，成为拓宽市场需求的一个主要发展方向。

（3）超级电容器的检测和评估手段需要进一步完善，提高电容器的可靠性、安全性，延长循环寿命也是发展的重点。

3 太阳电池

太阳能是取之不尽、用之不竭、无污染的廉价可再生能源，在太阳能利用中，太阳电池成为发展最快、最具活力和最受瞩目的领域。

3.1 太阳电池的发展状况

图 3 为全球太阳电池产量的变化，可以看出全球太阳电池产业呈快速发展态势。2000 年以来，发达国家大力开发太阳能：美国推出了国家光伏计划，在每个州都建有一座太阳能住宅作为示范，还推出了"太阳能路灯计划"[21]；日本推出了"新阳光计划"，实施"太阳能住宅 7 万套工程计划"；欧洲则推出著名的"尤里卡"高科技计划。这些进展促进了 PV 电池的研发、生产和应用的迅猛发展。

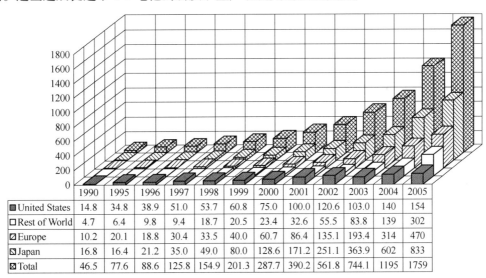

图 3　1990~2005 年全球太阳电池/组件产量变化

3.2 太阳电池的发展现状

根据太阳电池材料的发展历程，可以将太阳电池的发展划分为 3 个阶段[22]。

（1）第一代太阳电池。

第一代太阳电池基于晶体硅（单晶硅和多晶硅），光电转化效率高（分别为 20% 和 15% 左右），技术也比较成熟，目前在工业生产和市场上处于主导地位，产量占整个太阳电池的90%以上。但其价格高，材料成本占总成本60%~80%，成为目前推广的主要障碍。

（2）第二代太阳电池。

为了节省材料，有效降低太阳电池的成本，基于薄膜材料的第二代太阳电池逐渐显示出巨大优势和发展潜力，成为近些年来太阳电池研究的热点和重点。常见的薄膜太阳电池有多晶硅、非晶硅、铜铟硒和碲化镉薄膜电池等[23,24]。此外，目前研究十分活跃的染料敏化纳米晶太阳电池和有机聚合物太阳电池也被划归为第二代太阳电池的范畴[25]。目前第二代太阳电池的效率仍较低，商用薄膜电池的光电转换效率只有10%左右。

（3）第三代太阳电池。

澳大利亚、欧美各国研究者在研究了太阳电池的能量损失机理和极限转化效率的基础上，提出了第三代太阳电池的概念[26,27]，可使电池转化效率提高近2~3倍。图4描述了三代太阳电池的光电转化效率与其成本之间的关系[28]。

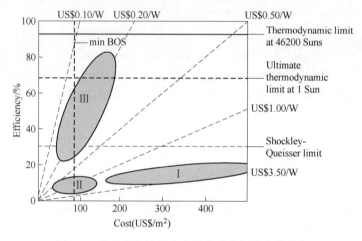

图4 三代太阳电池的光电转化效率与其成本变化

第三代太阳电池要求具有薄膜化结构、更高的光电转化效率、原料丰富且无毒等。目前第三代太阳电池还在进行概念、试验研究和前期产品开发，已经提出的第三代太阳电池有叠层太阳电池、多带隙太阳电池、热载流子太阳电池、热光伏电池和量子点太阳电池等。

叠层太阳电池概念是采用多个不同带隙的单结太阳电池来分别吸收能量稍微大于其带隙能量的光子，从而大大的消除热损失。叠层太阳电池概念可以通过光谱分裂或堆叠电池来实现，如图5所示。按照带隙宽度递减的方式堆叠电池（图5a）能够实现自动滤光作用，效率也随着堆叠电池数目的增加而增加，这种方式的理论转化效率为86.8%，而且能够避免光谱分裂方式需对每一单个电池进行复杂的独立操作（图5b）[29,30]。叠层太阳电池目前已经有商业化产品，如三结GaInP/GaAs/Ge太阳电池等。

多带隙太阳电池则是通过掺杂等方式在半导体材料中引入若干中间带隙，使不同能量的光子能够将电子激发至不同能级，从而有效利用各个波段的太阳光。这种多带隙太阳电池的极限转化效率也是86.8%。

热载流子太阳电池是利用一个特殊的能带结构，在光生载流子与晶格中的原子进行非弹性碰撞之前，即载流子冷却之前，将光生载流子捕获，从而减少热损失；而普通电池则只是在光生载流子复合之前将其收集。

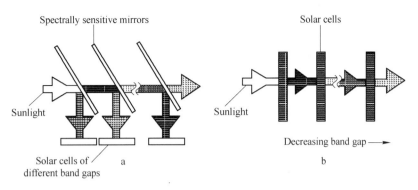

图 5 叠层太阳电池概念
a—光谱分裂；b—堆叠电池

热光伏电池是利用一个加热体而不是太阳作为光发生器。热光伏电池中包含两个二极管，分别作为太阳电池和光发射器，两个二极管之间是绝热和光学耦合的。光发射器被加热后向太阳电池发射光子能量为其带隙能量、接近单色的光，太阳电池能将光发射器发出的光高效地转化为电能。

利用量子点材料构造第三代太阳电池是当前研究的热点之一。量子点是尺寸约为10nm的纳米结晶。量子点材料能增大光吸收系数，通过调整尺寸提高与太阳光匹配程度，还能进行量子点间的结合而形成中间带隙，以利用更宽频谱的太阳光。此外，最近美国科学家利用光束照射直径仅为几个纳米的量子点时发现，一个光子能够在量子点中产生多个电子，这一成果有望大大提高太阳电池的转化效率。据报道，量子点太阳电池可获得的理论转化效率可达66%[31]，但量子点对光生载流子的逃逸、捕获与复合机理等还有待于进一步探讨。

3.3 太阳电池发展的热点和难点

今后除降低成本是关键外，尚有以下热点问题：
（1）以有机分子，聚合物，染料及光色电泳为基础的 PV 电池。
（2）光伏电池新器件和新结构，如薄膜硅、微晶硅或无定形硅，纳米晶材料，生物态材料，有机材料，新颖 PV 化合物；光电化学电池以及这类材料或器件组合而成的新器件和新结构。
（3）超高效率的外延太阳能电池，包括Ⅲ-Ⅴ族材料和器件、长寿命技术、单晶薄膜材料与低成本材料的复合。
（4）新 PV 过程，如整流天线器件，光子管理及载流子控制等。

4 有机电子（印刷电子）材料与印刷电源

4.1 有机电子材料

4.1.1 有机电子材料

有机电子，又称塑料电子、柔性电子、印刷电子等，是电子学的一个分支，涉及导电聚合物或导电塑料。所谓"有机"是因为聚合物中的分子都是碳-基的，这与传统的电子学相反，后者依赖于铜或硅那样的无机导体。导电聚合物更轻更柔软，也比无机导体便宜，这使得它有希望替代无机导体得到许多应用[32,33]。

聚合物的导电性质是 20 世纪 70 年代后期发现的。三位开拓者 Alan J. Heeger、Alan G. MacDiarmid 和 Hideki Shirakawa 于 2000 年获得了诺贝尔奖。其实，早在 1963 年澳大利亚的 D. E. Weiss 等就发现了掺碘经氧化处理的聚吡咯黑（即黑色素 Melanin）具有高达 1S/cm 的导电率。

4.1.2 有机半导体

有机半导体是一种具有半导体性质并具有异常高导电率的有机材料。1974 年，发现具有半导体性质的 Melanin Switch 是第一个有机半导体。

目前，有机半导体分为短链和长链两种。短链有机半导体有并五苯（Pentacene）$C_{22}H_{11}$、蒽（Anthracene）$C_{14}H_{10}$、红荧烯（Rubrene）$C_{42}H_{28}$ 等。长链有机半导体有 poly(3-hexylthiophene)、poly(p-phenylene vinylene)、F8BT、聚乙炔及其衍生物[34,35]。

4.1.3 印刷电子

印刷电子是把电子线路用标准印刷方法印刷在纸、塑料和纺织物上的一种新技术。它采用常规印刷设备，但用的是一系列的有电子功能的电子墨水。

印刷电子同有机电子紧密相关，因为许多功能墨水是由碳基化合物构成的。印刷电子可用任何一种溶解材料，包括有机半导体，无机半导体，金属导体，纳米颗粒，纳米管等以沉积方法印刷在媒体上。

印刷电子材料是美国 Lawence Livemore 国家实验室在 1997 年发明的。目前已吸引许多大、小公司进行研究和开发，并加速商品化[36,37]。

4.2 有机电子材料的实际应用

应用范围已涉及：集成电路、柔性显示、照明及信号、光伏电池、有机材料传感器、智能包装、无线电频率识别（RFID）（电子标签）、智能纺织物、燃料电池及一、二次电池等。

作为一种新兴的显示材料，有机电子材料获得了国际电子行业的极大重视，预计可以形成数十亿美元的产业。当前，已经开发和正在开发的产品有数十种。举例如下：（1）计算机显示屏。由于它的柔软和可折叠性，这种产品可容易地放进口袋甚至穿在身上或戴在手上。（2）智能卡。以前，芯片是粘贴在塑料卡上的（智能卡制造费用最高的工序）。现在，由于柔性电子材料的出现，电子器件可印制在塑料表面上，而且不需要电极。这种智能卡可以用作身份证、驾驶证、病历卡和各种管理卡等。（3）无线电报纸。由柔性电子材料制成的这种报纸，与蓝牙技术结合，可以时时更新，只需购买一次便可长久使用。（4）大容量存储器件。

目前已成功印刷了导线、电池、指示器、天线、电容器和电活性物质等。紧接着是研究和开发薄膜燃料电池和太阳能电池的印刷制造。今后还有一些更奇妙的印刷制品，如可印刷的微处理器，它能根据日期自行调节使用等。

4.3 可印刷电池

可印刷电池的典型产品以柔性薄膜电池、薄膜 Li 离子电池为例。

4.3.1 柔性薄膜电池

薄膜微电子器件用电池（Graphic Solutions International and Power Paper 公司创造的）是一种环境友好的、可生物降解、无毒和无腐蚀作用的电池（但不可燃烧、过热）。此种电池无外壳封装，可以印刷、涂抹和多层制作在纸、塑料，或其他媒质上，可以连续印刷且成本较低。该种电池材料与碱性电池相同,即以 Zn-MnO_2(C) 为电极材料,但分别制成墨水状;该种薄膜电池能集成到各种表面上,制成不同的形状和尺寸,用丝网印刷法能大批量低成本地生产。也可作为产品的集成部分或配件,成本低、超薄、柔软,能经受机械力弯折。适用于需要柔性和易一次成型的场合[38,39]。

柔性薄膜电池的结构如图 6 所示。柔性薄膜电池与标准 Zn-C 电池的放电曲线比较如图 7 所示。

柔性薄膜电池具有以下特性:电压 1.5V,厚 0.5mm,质轻,标准容量 2.5mA·h/cm^2,标准连续电流密度 0.1mA/cm^2,峰值脉冲放电电流(0.1ms)5mA/cm^2,最大内阻 15Ohms,储存寿命 2.5 年,工作温度 -20~60℃,可用于高湿度范围,在人体体温及湿度下性能最佳。

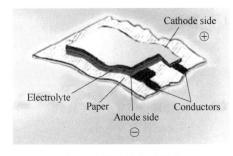

图 6 柔性薄膜电池结构

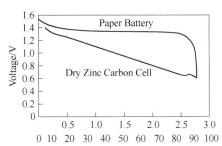

图 7 柔性薄膜电池与标准 Zn-C 电池性能比较

4.3.2 薄膜 Li 离子电池

美国橡树岭国家实验室发明的薄膜 Li 离子电池有以下特点:超薄(4~15μm),面积可大可小;正极为 $Li_4Mn_5O_{12}$,负极用 Li、Sn_3N_4 等;电解质为 LiPON;在 4.5~2.5V 放电时容量约 145mA·h/g,在 1.5V 放电时约为 270mA·h/g;当正极为结晶 $LiCoO_2$,厚度 4μm 时,最优应用电流密度约为 50μA/cm^2;在连续放电超过 3mA/cm^2 时,正极的利用率大于 50%。

采用 Li 负极和不同正极时的薄膜 Li 电池性能如图 8 所示,其中 c 为晶态,a 为无定形,n 为纳米晶态。薄膜 Li 离子电池可用于以下领域:微电子器件用电源,如植入人体的医药器件用电源;遥感电源;微型转换器,智能卡;MEMS 器件用电源等,以及许多方面的后备电源应用(如 PCMCIA 卡和其他类型的 CMOS-SRAM)。图 9 是集成在多芯片模块上的薄膜电池,在陶瓷封装体的背面制作了一个 3.8~4.2V/150μA·h 的 Li-$LiCoO_2$ 薄膜电池,在前端与线路相接触的是沉积上去的正、负极导线,后者穿过镀金的开孔。

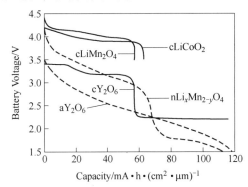

图 8 薄膜 Li 电池在低电流下放电曲线

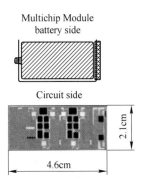

图 9 集成在多芯片模块上的薄膜 Li 电池

4.4 无线电频率识别（RFID）-电子标签

4.4.1 射频识别技术

射频识别技术是 20 世纪 90 年代兴起的一项非接触式自动识别技术。它是利用射频方式进行非接触双向通信，以达到自动识别目标对象并获取相关数据[40-43]。已有专家预言，它有可能成为继移动通讯技术、互联网技术之后又一项影响全球经济与人们生活的新技术。电子标签是新一代有电池驱动的标签应答系统，如图 10 所示。

图 10 RFID 卡及其中的薄膜电池

4.4.2 用途

有源（带印刷电池）电子标签（RFID）的用途举例：

（1）商品零售。可高度可视化地跟踪供应链上的货物箱，监测货箱装卸、返回情况，有 100%的可靠性，提高配送效率。

（2）运输与后勤保障。可为运输业快速提供货物，沿供应链跟踪货柜，重新配发货物箱。减少了费用，提高了效率。

（3）军用。用于部队后勤供应系统，全程跟踪供应链上的贵重军事装备，部队服装、设备、工具，武器零配件等。美国国防部称：电子标签为其全球后勤供应链节省开支 3 亿~5 亿美元，减少货柜近 90%，在伊拉克战争中把供应循环时间由数天缩短至数小时。此外，美国国防部 RFID 最终政策的进度表中，RFID 无源标签的实施定在 2006 年 1 月份。DoD 将要求供应商在货箱、汽油包装箱、润滑油箱、油箱、化学药品箱等包装箱上使用 RFID 无源标签。DoD 的供应商们可以使用任何标签供应商所提供的标签，只要其满足 64 或 96 字节的 EPC Class0 或 Class1 标签的要求即可。自 2007 年 1 月开始，所有运往 DoD 的商品货物的货箱和货盘都要求贴上 RFID 标签才可放行[44,45]。

（4）工业制造。在工业制造业如汽车、航天、电子和化工产业方面。

（5）药物安全与防伪。主要用于药物的安全配供和防止假冒。

4.5 热点和难点

印刷电子若干紧迫科学技术问题有：
（1）深入开展有机电子学、有机半导体及印刷电子材料的基础研究；
（2）探讨、开发、制造低成本的柔性印刷电子材料技术与表征；
（3）电子器件制造中的新颖印刷技术，新型纳米印刷术方法；
（4）太阳能电池、柔性计算机的印制；生物材料与生物传感器的印刷，等等。
下一步的另一热点和难点就是尽快推进印刷电子的商业化生产与应用。

4.6 问题和展望

柔性电子材料，可印刷电子与材料的研究与开发正在迅猛发展，许多新产品刚刚面世，其制造方法和设备等的开发还需数年时间，可印刷电池的研究与开发及商业化还有大量的工作要做。

参 考 文 献

[1] V. Battaglia, T. Duong, et al. An Overview of the DOE's Advanced Technology Development Program [C]. 24th Meeting of The Electrochemical Society, 2003.
[2] http://www.uscar.org/.
[3] Tien Q. Duong, K. Striebel. Batteries for Advanced Transportation Technologies (BATT) Program for Electrochemical Energy Storage, 1st Quarter Report 2004, 3: 5.
[4] Y. H. Huang, K. S. Park, J. B. Goodenough. Improving Lithium Batteries by Tethering Carbon-Coated $LiFePO_4$ to Polypyrrole [J]. J. Electrochem. Society, 2006, 153: A2282-A2286.
[5] I. V. Thorat, V. Mathur, J. N. Harb, et al. Performance of carbon-fiber-containing $LiFePO_4$ cathodes for high-power applications [J]. J. Electrochem. Society, 2006, 162: 673-678.
[6] M. S. Song, Y. M. Kangb, J. H. Kim, et al. Simple and fast synthesis of $LiFePO_4$-C composite for lithium rechargeable batteries by ball-milling and microwave heating [J]. Journal of Power Sources 2007, 166: 260-265.
[7] D. H. Kim, J. Kim. Synthesis of $LiFePO_4$ Nanoparticles in Polyol Medium and Their Electrochemical Properties Electrochem [J]. Solid-State Lett., 2006, 9: A439-A442.
[8] http://www.a123systems.com/.
[9] K. Zaghib, A. A. Salah, N. Rav, et al. Structural, magnetic and electrochemical properties of lithium iron orthosilicate [J]. J. Power Sources, 2006, 160: 1381-1386.
[10] Z. L. Gong, Y. X. Li, Y. Yang. Synthesis and Characterization of $Li_2Mn_xFe_{1-x}SiO_4$ as a Cathode Material for Lithium-Ion Batteries Electrochem [J]. Solid-State Lett., 2006, 9: A542-A544.
[11] D. K. Lee, S. H. Park, K. Amine. High capacity Li[$Li_{0.2}Ni_{0.2}Mn_{0.6}$]O_2 cathode materials via a carbonate co-precipitation method [J]. J. Power Sources, 2006, 162: 1346-1350.
[12] N. Yabuuchi, T. Ohzuku. Electrochemical behaviors of $LiCo_{1/3}Ni_{1/3}Mn_{1/3}O_2$ in lithium batteries at elevated temperatures [J]. J. Power Sources, 2005, 146: 636-639.
[13] 郑洪河,等. 锂离子电池电解质 [M]. 北京:化学工业出版社, 2007.
[14] D Aurbach, B Markovsky, G Salitra, et al. Review on electrode-electrolyte solution interactions, related to cathode materials for Li-ion batteries [J]. J Power Sources, 2007, 165: 491-499.
[15] Conway B E. Electrochemical supercapacitor: scientific fundamentals and technological applications [M]. New York Kluwer Academic/Plenum Publishers, 1999.
[16] Frackowiak E, Beguin F. Carbon materials for the electrochemical storage of energy in capacitors [J]. Carbon, 2001, 39: 937-950.
[17] Li Jing, Li Jie, Lai Yanqing, et al. Influence of KOH activation techniques on pore structure and elctrochemical property of caron electrode materials [J]. Journal of Central South University of Technology, 2006, 13: 360-366.
[18] Lai Yanqing, Li Jing, Li Jie, et al. Preparation and electrochemical characterization of C/PANI composite electrode materials [J]. Journal of Central South University of Technology, 2006: 353-359.
[19] A Bakdyccui, F Soavi, M Mastragostino. The use of ionic liquids as solvent-free green electrolytes for hybrid supercapacitors [J]. Applied Physics A, 2005, 339: 3402-3407.
[20] A Fernicola, B Scrosati, H Ohno. Potentialities of ionic liquids as new electrolyte meda in advanced electrochemical devices [J]. Ionics, 2006, 12: 95-102.
[21] M. John. Solar battery roofing for a solar house [J]. Renewable Energy, 1997, 11: 395.
[22] Martin A. G. Third generation photovoltaics: solar cells for 2020 and beyond [J]. Physica E, 2002, 14: 65-70.
[23] Vasilis M. Fthenakis. Life cycle impact analysis of cadmium in CdTe PV production [J]. Renewable and

Sustainable Energy Reviews, 2004, 8: 303-334.

[24] K. Ramanathan, R. N. Bhattacharya, M. A. Contreras, et al. CIGS Thin-Film Solar Cell Research at NREL: FY04 Results and Accomplishments [C]. 2004 DOE Solar Energy Technologies Program Review Meeting. Denver, Colorado, October 25-28, 2004.

[25] Christoph Lungenschmied, Gilles Dennlera, Helmut Neugebauer, et al. Flexible, long-lived, large-area, organic solar cells [J]. Solar Energy Materials and Solar Cells, 2007, 91: 379-384.

[26] M. A. Green. Third Generation Photovoltaics-Advanced Solar Energy Conversion [M]. Springer series in photonics, 2003.

[27] R. McConnell, R. Matson. Next-Generation Photovoltaic Technologies in the United States [C]. The 19th European PV Solar Energy Conference and Exhibition, Paris, France, June 7-11, 2004.

[28] Nathan S. Lewis. Toward Cost-Effective Solar Energy Use [J]. SCIENCE, 2007, 315: 789-801.

[29] Antonio Marti, Gerardo L. Arafijo. Limiting efficiencies for photovoltaic energy conversion in multigap systems [J]. Solar Energy Materials and Solar Cells, 1996, 43: 203-222.

[30] M. A. Green. Photovoltaics: technology overview [J]. Energy Policy, 2000, 28: 989-998.

[31] A. J. Nozik. Quantum Dot Solar Cells [C]. The NCPV Program Review Meeting. Lakewood, Colorado, 2001: 14-17.

[32] Noel S. Hush. An Overview of the First Half-Century of Molecular Electronics [J]. Ann. N. Y. Acad. Sci., 2003, 1006: 1-20.

[33] D. Gamota, P. Brazis, K. Kalyanasundaram, et al. Printed Organic and Molecular Electronics [M]. New York: Kluwer Academic Publishers, 2004.

[34] Malliaras lab/Cornell University, Organic Semiconductors Make Cheap, Flexible Photovoltaics and LEDs, Science Daily, September 21, 2006.

[35] J. M. Shaw, P. F. Seidler. IBM Journary of Research and Development, 2001, 451.

[36] http: // www. idtechex. com /.

[37] Tucholski, Gary R, Russell, Ed T, McComsey, et al. Thin printable flexible electrochemical cell and method of making the same [P]. US: P20050260492, Nov 2005.

[38] Daniel R, et al. Printed organic and molecular electronics [J]. Materials Today, 2004, 7: 53-60.

[39] W Knight. Printable battery rolls off the presses [J]. New Scientist, 14 October 2005.

[40] Anyonio L, et al. CLELIA: a multi-agent system for publishing printed and electronic media [J]. Expert Systems with Applications, 2005, 28: 725-734.

[41] V Surbamanian, J Frechet, P C Chang, et al. Progress toward development of all-printed RFID tags: Materials, Processes, and Devices [J]. Proceedings of the ieee, 2005, 93: 7.

[42] J Kim, H Kim. Vulnerability and Considerations in Mobile RFID environment [C]. Advanced Communication Technology, the 8th International Conference, 2006.

[43] http: // www. st-andrews. ac. uk/.

[44] http: // rfid. idtechex. com/knowledgebase/en/sectionintro. asp? sectionid = 121.

[45] http: // networks. silicon. com/lans/0, 39024663, 39151548, 00. htm.

战略金属铍材料的可持续发展*

1 Be 的独特性质

铍（Be）具有非常独特的性质，它是最轻的金属结构材料，它的密度比铝小 1/3，它的强度（按强/重比考虑）比钢大 6 倍。因此，作为高可靠性的结构材料在高技术应用范围内很有优势，用于超精密卫星元件、太空望远镜的反射镜，如哈勃太空望远镜的反射镜，军用寻的系统和红外定位系统。在商业应用方面也非常广泛。

铍的某些复合的光学材料具有非常精密的尺寸稳定性，用于超精确的军用寻的系统。铍的光学材料用在天文望远镜方面；很多军用和民用的可回收的卫星和 NASA 多种空间科学探测发射都用了铍的结构元件。铍材料的独特的核性能（反射中子、对 X 射线透明）使其大量用于核实验堆以及 CT 机的 X 射线窗口。

氧化铍类的陶瓷材料，具有高的导热性和高的电阻率，是非常稳定的材料，可用近净成形方法来制备。主要用于高级电子设备。

铍合金：往金属中加入 0.5%~2.75% 的铍可明显提高金属的性质，例如铜和镍。铍铜合金有格外强的强度、导热性和导电性，非常出色的耐磨性。铍铜合金也是一种可加工硬化的合金，可以通过不同的退火和时效过程改变它的机械性质。

铸造铍铜合金，全球各地铸造厂可采用各种方法来生产此种合金。通过融化和铸造制成各种各样的元部件，例如：从微型电子的连接件到大型焊枪的结构体，后者用于汽车装配生产线的焊接机器人上。

变形铍铜合金，可以通过煅造、挤压、拉伸和轧制制成杆、板、管、棒，提供用户。

2 应用的四大领域

2.1 核能和核武器

（1）核聚变反应堆（ITER）。

1）铍作为第一壁板材料。第一壁板是聚变堆的核心部件，铍材料具有优良热电性能、尺寸稳定性好、刚度和比强度大、低的原子和中子吸收截面等特点。铍作为第一壁板材料的理由：铍受等离子污染的影响小，辐射功率损失低，获得氧的能力好，没有化学的分解，容易现场修复，其主要的功能是保护冷却壁的结构不受高热流损害，也不受等离子体的直接接触。各种牌号的材料中，候选材料有 S-65C VHP（真空热压级，由 brush wellman inc.，USA 生产），其优点是产品中的氧化铍和金属杂质含量最低、高温强度和热循环的破损非常低。另一个候选材料是 DShG-200（RF 公司生产）。

2）中子倍增器。几乎各种各样的原子反应堆都要用铍作中子反射体，特别在建造

---

* 原发表于《院士专家宁夏行报告》，2008 年，银川。

用于各种交通工具的小型原子锅炉时更需要。建造一个大型的原子反应堆，往往需要动用2t多金属铍。

(2) 太空反应堆。

早先，美国橡树岭国家实验室的研究指出，铍和氧化铍是太空反应堆中中子反射器非常好的候选材料。因为这种材料密度低，中子反射能力高。

(3) DEMO反应堆（核聚变示范反应堆）。

采用铍的中间金属化合物，如铍12钛（Be12Ti）、铍12钒和铍12钼。这种材料要比金属铍的熔点更高，并具有更高的中子反射能力。

(4) 用于粒子物理研究。

2.2 航空航天（包括太空望远镜等）

铍合金是制造飞机的方向舵、机翼箱和喷气发动机金属构件的好材料。现代战斗机上的许多构件改用铍制造后，由于重量减轻，装配部分减少，使飞机的行动更加迅速灵活。

航天器用铍铝合金。美国空军和NASA在2005年初所发射的航天器中就用到了铍铝合金制备的元件。其中4个元件用在空军的XSS-11实验极地轨道卫星（用于监测美国的其他在轨物体），另外一个用在NASA设计的三个微型航天器之一，在恶劣环境中使用。

为减少铍中毒，已全面开发了含铍新材料替代铍铝合金，例如牌号为AlBeMet 162H，Beryllium SF-220-H，and Beryllium I-70-H。

另外，由于铍铜合金能改善组件服务的寿命，提高耐久性，并降低维护成本以及优越的抗摩擦磨损特性，已应用于商业和军用飞机的起落架衬套和轴承。

2.3 军用（寻的、制导系统）

(1) 军用寻的与制导系统。

(2) 高速旋转应用。

2.4 电工电子

(1) 铍铜合金带材广泛应用于电子的连接件，如通信、计算机和汽车电子应用等。

(2) 铍铜合金棒材和铍铜铸造合金用来制造各式各样的导电连接件，海底光缆的中继器外壳及其相关的组件。数十年来，该器件的服务寿命很长，在严酷的深海环境中表现出色。

(3) 不断发展的电子技术，以及应用在手机，个人数据设备，电脑及周边设备，都要求更高的可靠性，更高性能的材料。铍铜合金特别是混合铍铜/铍镍合金，具有高强度和导电性，优良的耐疲劳和磨损性能能满足这些应用要求。

(4) 当今各种各样的现有材料的复合和制造技术，使含铍合金成为今天电子应用方面强大的支撑材料。

(5) 电阻焊。高导电性能的铍铜合金棒（C17510，17500）用来制造电阻焊的设备元件，如图1、图2所示。

铍铜合金棒、板特别适合于焊工模具、焊接轮和有关的电阻焊接组件、汽车装置流水线焊枪机器人，使得数百万计的个别焊件能利用C17510合金制成的焊枪臂焊接。采用铍铜合金是由于其具有极强的高温疲劳强度、高强度/硬度和高热/电传导性能。

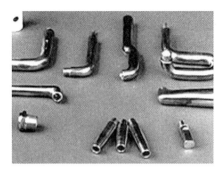

图 1　Holders, Shanks and Adaptors

图 2　Portable Barrel Type Welding Gun

3　制造与加工

下面介绍几个铍制品加工制造公司。

3.1　美国布劳希工程材料公司（Brush Engineered Materials Inc.）

http：// www. beminc. com/（2008-08-17）

图 3　美国布劳希工程材料公司的主页

3.1.1　公司概况

美国布劳希工程材料公司是著名的国际高性能工程材料的生产和供应商。属下有四家子公司，即（1）Brush Wellman 公司，简称 BWI，是全球著名的铍、含铍合金和氧化铍陶瓷材料公司；（2）TMI 主要是生产铍制品；（3）WAM 主要生产贵金属和特殊金属产品；（4）它们的国际通用名称为 Brush international（BI）。

该公司经营四个方面的业务：

（1）特殊工程合金。公司通过冶金制备方法来解决特殊用户的合金产品各种性能要求。主要产品为铜铍合金条、块。主要应用在计算机、通信、汽车电子、航空、采油、水下光缆和塑料模具等方面。

（2）铍和铍复合材料。铍的质量很轻，具有独特的机械和热学性质，它的比刚度要远比铝、钛和钢这类工程结构材料高得多。铍产品有：AlBeMet 和 E-materials，用在非常多的高性能场合以及国防、航空、X 射线和声学市场。

图 4　特殊工程合金

图 5　铍及铍合金材料

（3）工程材料系统。工程材料系统是由 TMI 公司来制造的，是把贵金属和非贵金属以连续铸轧的形式制成产品。用于通信系统、汽车和计算机中复杂的电子和电器元件。

（4）先进材料技术与服务。WAM 公司的产品主要由纯金属和特殊金属合金构成，要满足高标准和高可靠性的要求，所制备的有特别价值的各类材料，广泛用于混合微电子、半导体、光通信、电子管、磁头材料等方面。

图 6　工程材料系统中的应用

图 7　先进材料方面的应用

3.1.2　计算机和通信设备

铜铍合金用于整个的通信网络和线路，对其要求是强度、柔性、可靠性和省钱。在台式机和笔记本电脑中，此种材料能提供理想的电连接材料，使得互联网工作更有效、更可靠。

在通信工程中，铜铍合金提供可靠的电池连接和移动电话的电连接。

此外，还用于无线基础设施的电子元件的封装、深海光缆的中继器以及开关、导线框架、屏蔽设施等。

该公司的 Zentrix 材料可以使通信和信息传输设备做得更小、更可靠。另外，陶瓷产品是无线通信设备当中高频元件的心脏，这些元件在无线通信的微型化方面起着集成化的作用，工程材料公司的产品能够适应通信基础设施不断变化的更高的需求。

图 8　计算机和通用设备中的应用 Ⅰ　　图 9　计算机和通用设备中的应用 Ⅱ　　图 10　计算机和通用设备中的应用 Ⅲ

3.1.3　汽车电子

Brush 工程材料公司为汽车工业提供重要的技术和产品。在车辆的制造当中，几乎各个方面都可以找到该公司生产的元件——电子信号发生器、动力元件、传感器的开关、延迟和敏感元件、热管理、点火元件的电绝缘、安全器件元件等。

铜铍合金材料可以在恶劣的汽车环境下（靠近引擎附近高温 400°F、震动、腐蚀材料、海边气候、电池漏液）提供无损伤的工作。另外，铜铍合金元器件在小型化时不会降低其功能。

氧化铍陶瓷可制成电子点火系统当中的散热元件。

图 11　汽车电子中的应用 Ⅰ　　　　　　　图 12　汽车电子中的应用 Ⅱ

3.1.4　工业元器件

Brush 工程材料公司还为飞机着落系统、油汽开采、塑料工业模具提供优质材料。

图 13　作为工业元器件应用 Ⅰ　　图 14　作为工业元器件应用 Ⅱ　　图 15　作为工业元器件应用 Ⅲ

金属铍和 AlBeMet 用来制造低密度、高刚度的精密机械元件，用在需要高速度、精细运动的场合，如机器人、电子装配设备、光学扫描设备。

3.1.5 光学媒体

铍和铍合金的溅射靶材，用于制造永久性的或可读写的储存光盘，因而广泛用于游戏、音乐、电影工业。

铍和铍合金的溅射靶材还大量用于光通信的制造业，要求这种材料有反射性、半反射性、介电性，或者使光碟上具有相变层。这种材料为光通信应用提供优良的反射性、抗腐蚀性、长寿命和读/写的可靠性。

3.1.6 航天与国防

铍材料广泛用于航天和国防工业，在保护生命、捍卫国土、保障国家安全、洲际运输和外层空间的探索方面起着重要作用。

图16 光学媒体中的应用

（1）国防军工。铍的制品用于喷气式战斗机、轰炸机、运输机、救生机、直升机、坦克等武器中。在很多场合下要经受极端的震动、失重和热压的考验，铍制元件是不可替代的。

图17 航天方面应用

图18 国防军工方面应用

（2）航天探测。NASA最先使用铍防热罩来保护航天飞机，这些材料由该公司生产，当时为全美第一把交椅。到目前为止，NASA仍采用该公司很多的铍材料制防热罩；NASA还选用了该公司光学级铍作为太空望远镜的主镜片，投资8.25亿美元制造了James Webb太空望远镜。如今，NASA的Spitzer空间观测站已收到经由铍镜获取的宇宙图像。

2003年元月，NASA的"精神号"和"机会号"先后登陆火星，使用了350多个以AlBeMet为基材的复合材料制成的结构件。

（3）飞机工业。铍合金材料用于制造飞机起落架的活动门、机翼传感器、燃料泵和油泵以及屏蔽设备，为飞机的减轻重量、增加载荷、减少维修和节约燃料起了重要作用。

3.2 日本碍子公司（NGK INSULATORS，LTD.）

铍制品属该公司的 Specialty Metals Business 部门。

http：//www.ngk.co.jp/cn/（2008-08-17）

该公司的产品有铍铜延伸材。延伸材主要是以板、带、棒材形式提供产品，提供了完整的各种合金规格和硬度，能根据用途不同进行选择。

可提供完整的各种合金规格和硬度，能根据用途不同进行选择。最薄的厚度可以提供到0.045mm。

图 19　日本碍子公司主页

图 20　铍铜延伸材

（1）铍铜细丝。细丝具有和板带材一样的特征，在微型化的电子部件的实际应用正在增加。可以提供直径 0.05mm 以上。

（2）激冷排气口。是在制造铝和镁制品时，通过高压排气法的方法，在铝气等的排放过程中产生冷却作用的模具。

图 21　铍铜细丝

图 22　激冷排气口

（3）MC 合金。是应用于模具新开发的高性能低价格的铜合金。具有优秀的热传导性和耐高温的硬度。

（4）微软连接器。应用密集型，高速化的 BGA（Ball Grid Array）和 LGA（Land Grid Array）类型的半导体插座。

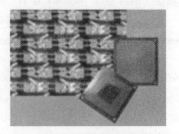

图 23　MC 合金　　　　图 24　微软连接器 I　　　　图 25　微软连接器 II

（5）铍铜加工品。根据客户要求通过铸造，锻造，机械加工等制造最终部件。

（6）铍铜安全工具。作为防爆工具在各种作业场合都有广泛应用，具有不着火，无磁性，耐腐蚀，高强度等特点。

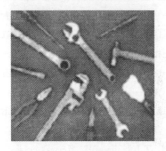

图 26　铍铜加工品　　　　　　　　图 27　铍铜安全工具

（7）其他的铍铜关联制品。在海底光缆的中转器的外壳和铍铝合金等方面也有产品应用。

3.3　美国 NASA

制造太空望远镜 Be 镜片的过程。

http：//www.jwst.nasa.gov/mirror.html（2008-08-16）

NASA 需要建造一个新型的望远镜，性能比哈勃望远镜更好，取名为 James Webb 太空望远镜（简称 JWST），可以观察到远离地球数十亿光年的太空情况。

望远镜的核心部件为光学镜面和镜头。如何制造一个又大又轻的镜子是一巨大挑战。下面是制备过程。

（1）又大又轻的镜片。

JWST 需要一个巨大的镜面来收集距离地球 130 亿光年远的宇宙传来的光线。专家组决定主镜面必须有 6.5m 的跨度才能测量到那么远的光线。迄今，地球上还没有做过这么大的镜面，且从来没有把这么大的镜子送到外层空间。另外，还没有一种火箭能够大到足以装载直径为 6.5m 的镜子。于是，专家组决定先建造若干小块的镜面，可以拼成大镜子，以折叠的形式装进火箭，在着落后能够打开成大镜子。

除了做得小以外，还要做得轻。如果 JWST 的大小同原哈勃太空望远镜（2.4m）

一样的话，那么发射进入轨道又嫌太重。专家组必须寻求新材料以建造足够轻的镜片，其单位面积重量为哈勃镜片的 1/10。而且强度必须非常高，才能保持镜片的形状。

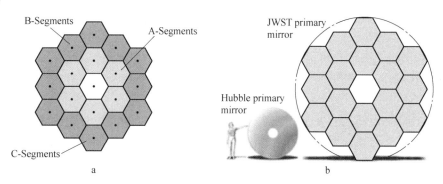

图 28　镜片的示意图
a—哈勃原镜片；b—JWST 原镜片

（2）低温镜子。

为了观察到宇宙间的红外光，JWST 就不能像哈勃望远镜那样有本身发出的红外光。因此，JWST 就必须保持非常冷，使它的镜子能在 $-220\ ℃$ 下工作，这样才能捕捉到远距离的红外光。镜子需要这样低的温度且不变形。要保持 JWST 的低温，需把 JWST 送入深空，远离地球，而且要为镜子遮住太阳的热。

（3）选定镜子材料。

为了满足 JWST 的这些要求，NASA 要找出新的方法来建造望远镜的镜片。为此，设立了一个镜子系统专项，进行了四年的研究。该课题完成了两种镜片的建造和全面的实验，一种是铍制镜片（Ball Aerospace 制造），另一种由特种玻璃制成（Kodak 制造）。经过专家试验和选择，确定用铍镜，理由是在深度低温下，铍能保持其形状不变。由 Northrop Grumman Space Technology 来主持铍镜的制造，由 NASA 的 Goddard Space Flight Center 管理。

（4）铍镜的制造。

铍镜的原料来自美国 Utah 州，在 Brush Wellman 公司提纯。专用于 JWST 铍镜的铍粉为"O-30"。

把铍粉放入不锈钢的罐筒中，压成平板状。取出后，把铍块割成两半，作为两个镜片块，每个镜片块跨度约 1.3m。JWST 的镜面由 18 个六边形的镜片块组成。

图 29　制作中铍镜

Beryllium Mirror Segments，Axsys Technologies

图 30　JWST 铍镜

各个镜块经过检测后，送往 Axsys Technologies（in Cullman，Alabama）加工。最初两个镜块花了很长时间才加工成最后的形状（2004 年 3 月完成）。因为需要仔细解决铍尘的毒害问题。后来，该公司建立了一个能同时处理 8 个镜块的工厂。

加工成形先是把铍镜块背面多余的部分切割掉，剩下的是一个很薄的结构，仅 1mm 厚。尽管大量的材料被切去了，但铍镜外形很稳定。然后，对其前表面进行抛光再加工，使之能完全适合于拼装入大镜面的最后位置。经 Axsys 完成切削加工之后，再将镜片送往 SSG/Tinsley 进行抛光。

SSG/Tinsley 对镜子的表面进行研磨，直到它接近最后的形状。

在镜片的后面装上支撑结构，其中有小的机构能够帮助最后镜片的聚焦；然后，进一步的抛光，再用激光和专用仪器来分析镜子的精度和质量。

经过多次的重复，直到镜子非常完美。镜片抛光后，再送到 NASA 的 Marshall 航天中心。

在 Marshall 航天中心对镜块的形状进行低温试验，模仿深空-220℃ 环境，并再检查镜面的形状。如果检查有问题的话，要尽快送还 SSG/Tinsley 重新处理。

镜块的形状在深冷下完全正确，然后要在其上镀金，黄金能保证镜子对红外光的反射。镜子镀金以后，再次送回 Marshall 航天中心作最后的形状检测。如果测试合格，再送回 Goddard 航天中心，在此基地由 ITT 公司把 18 个镜片组装成一个大镜子。除了镜片以外，镜子的背面结构，由盐湖城的 ATK 公司建造，由 ITT 公司把镜块装在背面结构里。在镜子的中央部位装 12 块镜片，两翼各有 3 片，两翼可以折叠回来，这样整个镜子就可以装在火箭里面。

整个镜子建成后，再把科学仪器装在里面，全部组装完成后还要经过几轮的测试，例如热试、振动、模拟火箭振荡、外空间的真空和低温。一切检验合格以后，镜子就要放在火箭的顶部。

火箭发射后，JWST 将要花三个月的时间到达它在太空 L2 区域，大约离地球 150 万公里开始工作。

图 31　火箭中的应用

图 32　JWST 的镜子

4　环保与安全（略）

5　建议

对铍材料可持续发展总的看法：加大研发力度，大力提高加工制造水平，掌握关键技术，降低成本，解决环保，扩大应用。

（1）规划与布局；

（2）以企业为核心，整合 R&D 力量，组建研发队伍；

（3）争取与国际上著名制造加工企业合作，提高我们的制造加工水平；

（4）加大投入；

（5）人才队伍培养。

八、其他（氧化铝、TiAl 合金、自蔓延高温合成、传感器、光催化剂、燃料电池电催化、泡沫材料、超细粉体）

强化烧结法生产氧化铝新工艺的研究与实践*

摘　要　针对我国铝土矿资源特点和我国烧结法生产氧化铝的现状，论述了强化烧结法生产氧化铝新工艺。采用铝硅比为 9.62 的矿石，按照钙比为 1.3~2.0，碱比为 1 进行配料，在 1250℃ 以上进行熟料烧成。实验结果显示：烧成的熟料溶出条件宽松，二次反应程度弱；溶出浆液在 170℃ 左右直接进行加压脱硅，脱硅后溶液的硅量指数大于 200，再加入适量的石灰进行深度脱硅，即使溶液中氧化铝浓度超过 200g/L 时，精液的硅量指数也大于 600，且得到的水化石榴石中二氧化硅饱和系数大于 0.28；通过加入晶种和采用新的碳酸化分解工艺制度，产品中的 SiO_2 含量降至 0.025%，Na_2O 含量小于 0.37%；加入表面活性剂，不仅能将碳分母液蒸发至 Na_2O_T 浓度大于 300g/L，而且还可有效减缓表面上结疤的形成速度。部分结果已应用于工业实践中，单台窑产能提高了 19.7%，工艺能耗降低了 24.71%。

1880 年和 1902 年缪列尔和帕卡尔德在萨特里确立的两成分烧结法的基础上提出了三成分（碳酸钠、石灰和铝土矿）烧结法生产氧化铝工艺，该工艺的关键是使含硅化合物转变成 $β-2CaO·SiO_2$，较好地实现了铝硅分离，从而能处理品位较低的铝土矿，奠定了传统烧结法生产氧化铝工艺的基础[1-3]。传统烧结法生产氧化铝时，存在熟料折合比高、技术经济指标欠佳等弊端，因而通过提高矿石品位，降低熟料折合比，提高产出率应是烧结法生产氧化铝工艺发展的主要方向。本文作者提出了强化烧结法氧化铝新工艺，即提高火法系统中氧化铝含量和湿法系统中氧化铝浓度，从而提高氧化铝产能。

基于拜尔法系统补碱的需要，前苏联对高铝硅比矿烧成进行了长期的生产实践，乌拉尔铝厂[4]烧结法系统中矿石的铝硅比在 10.4 左右，博戈斯洛夫铝厂[4]的烧结法系统和第聂泊铝厂[2]烧结系统所用矿石的铝硅比大于 7。我国也做过类似的研究[5]。但这些实践与研究所用矿石的铁含量高（Fe_2O_3 20%以上），熟料容易烧结，且此烧结系统本身不生产氧化铝，工艺简单。因此，这些研究对我国强化烧结法生产氧化铝工艺的研究指导作用不大。

针对强化烧结法，本文主要研究高铝硅比矿的配料和烧成，熟料溶出中二次反应的规律，高浓度粗液的脱硅，高浓度铝酸钠溶液的碳酸化分解以及碳分母液的深度蒸发。

---

* 本文合作者：李小斌、刘祥民、刘桂华、彭志宏。原发表于《中国有色金属学报》，2004，14（6）：1031-1036。

1 实验

1.1 原料

铝土矿取自中州铝厂,主要成分如表 1 所列。

表 1　铝土矿的成分　　　　　　　(质量分数,%)

SiO_2	Al_2O_3	Fe_2O_3	CaO	TiO_2	A/S
7.27	69.91	2.67	1.62	3.18	9.62

配料所用的 $Ca(OH)_2$、Na_2CO_3、Na_2SO_4 和 K_2CO_3 均为分析纯。

1.2 实验步骤

生料的配制是根据碱比 N_R、钙比 C_R(均为摩尔比)进行的,计算公式如下:

$$N_R = ([Na_2O] + [K_2O])/([Fe_2O_3] + [Al_2O_3])$$
$$C_R = [CaO]/[SiO_2]$$

根据计算结果,准确称取各物料,均匀混好生料,然后称取混匀的生料装入刚玉坩埚中,移入马弗炉(SX-8 型,长沙实验电炉厂)中,在设定的温度下烧结一定时间,取出烧结后的样品冷却至室温,磨细后,密封备用。按照熟料溶出时设定的含量称取熟料,在 DY8-群釜低压装置(中南工业大学机械厂,甘油为加热介质)进行溶出实验。在设定时间里取样,测定溶液中各组分浓度,渣相经充分洗涤后烘干,以进行物相分析。

1.3 分析方法

Na_2O_K 和 Al_2O_3 的浓度采用容量法测定,SiO_2 的浓度采用硅钼蓝分光光度法测定(7230G 分光光度计,上海分析仪器总厂)。CO_2 采用 CYES-Ⅱ O_2/CO_2 气体测定仪(上海嘉定学联仪表厂)测定。

2 结果与讨论

2.1 配料与烧成

在烧结法生产氧化铝的过程中,配料与烧成的好坏直接影响整个生产工艺的经济指标。在新工艺中,各因素对熟料烧成的影响结果如表 2 所列。

表 2　不同配料条件下烧成温度对熟料烧成的影响

Sample No.	Charge		Sintering Temperature /℃	Extraction ratio/%	
	Alkali ratio	Calciunr ratio		$\eta(Al_2O_3)$	$\eta(SiO_2)$
1	1.0	1.3	1225	92.19	72.68
			1250	99.60	75.65
			1275	100	66.77
			1300	100	59.25
2	1.0	1.5	1225	95.88	71.88
			1250	96.41	70.10
			1275	100	66.87
			1300	100	54.17

续表2

Sample No.	Charge		Sintering Temperature /℃	Extraction ratio/%	
	Alkali ratio	Calciunr ratio		$\eta(Al_2O_3)$	$\eta(SiO_2)$
3	1.0	1.7	1225	89.93	74.62
			1250	97.81	69.22
			1275	100	61.62
			1300	100	48.48

In addition, there are K_2O 7%, Na_2O 1.5% in charge, sintering time is 20min. Sinter leaching.
Conditions: temperature 90℃; time 20min; amount of sinter 350g/L. Solution for sinter leaching: Na_2O_K 77.12g/L, Na_2O_C 12g/L, Al_2O_3 70.55g/L.

实验结果表明，当碱比为1，钙比小于2，在1250℃以上进行熟料烧成时，熟料中的氧化铝和氧化钠溶出率较高（大于96%），且随着烧结温度的升高，二氧化硅的溶出率逐渐下降，氧化铝溶出率也略有升高。通过改进烧结工艺制度，得到的熟料为具有一定强度的多孔块状物，高铝硅比矿烧结无困难。对熟料的X射线衍射谱进行分析，熟料中的主要物相为：$NaS_2O \cdot Al_2O_3$、$\beta\text{-}2CaO \cdot SiO_2$、$CaO \cdot SiO_2$、$CaO \cdot TiO_2$等。对烧成过程进行的热力学分析结果表明，降低钙比时，熟料中的$CaO \cdot SiO_2$含量增大，且烧成温度的提高有利于$CaO \cdot SiO_2$的生成[6]。

2.2 熟料溶出与分离

熟料成分为：Al_2O_3 43.9%，Na_2O 27.51%，SiO_2 4.56%，CaO 6.39%，K_2O 6.87%，Na_2O_S 1.48%，熟料粒度小于0.38mm。氧化铝溶出率的变化规律如图1所示。

实验结果表明：当溶出液中的Al_2O_3浓度大于200g/L时，在30min以内熟料溶出比较充分（氧化铝溶出率大于96%）；在80℃溶出时，随着时间的延长，氧化铝溶出率变化不明显，二次反应也不明显；而在90℃溶出时，前30min氧化铝浓度变化不明显，但在60min时氧化铝浓度降低。固相成分分析结果表明，其主要原因是在熟料溶出后期发生了生成钠硅渣的脱硅反应，这说明二次反应也不明显。强化烧结法熟料溶出时，二次反应程度弱的主要原因之一是熟料中的$CaO \cdot SiO_2$含量增大。X射线衍射分析结果和热力学计算结果也表明，这种硅酸钙在熟料溶出前后稳定性较好[6,7]。实验中还发现，熟料溶出后得到的赤泥在高压条件下比较稳定，又由于熟料在溶出过程中二次反应不明显，且渣量较少，因此在工艺实施中将溶出浆液直接进行脱硅。实验结果显示浆液沉降性能很好。这不仅改善了浆液的沉降性能，而且省去了一次分离工序。

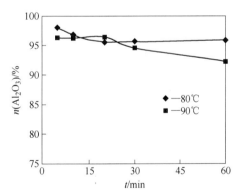

图1 不同温度下熟料溶出时氧化铝的溶出率与时间的关系

(Sinter leaching condition: amount of sinter, 350g/L; Solution for sinter leaching: Na_2O_K 85g/L, Al_2O_3 78.55g/L, Na_2O_C 13g/L, α_K 1.78; Green liqor: Al_2O_3 203~210g/L, SiO_2 10~11g/L)

2.3 粗液脱硅

基于传统烧结法已有的生产设备，粗液脱硅采用两段脱硅工艺，即加压脱硅和常压深度脱硅，以保证精液的硅量指数大于600，满足碳分的要求。

2.3.1 加压脱硅

温度对高浓度粗液加压脱硅的影响结果如表3所示。

表3 温度变化对高压脱硅的影响

Temperature /℃	Content of Aluminate solution/g·L^{-1}				Mass ratio of Al_2O_3 to SiO_2
	Na_2O_K	Na_2O_T	Al_2O_3	SiO_2	
160	195.29	208.71	208.59	0.95	221
164	192.80	210.19	210.23	0.97	217
170	191.56	212.66	212.28	0.90	235

Initial solution: Na_2O_K 197.75g/L, Al_2O_3 219.46g/L, SiO_2 13.81g/L, Na_2O_S 10g/L, K_2O 60g/L; time: 60min, seeds 60g/L.

表3的结果说明：粗液在160~170℃脱硅时，能使93%左右的SiO_2转化成钠硅渣，脱硅后溶液的硅量指数大于200。

2.3.2 常压深度脱硅

虽然加压脱硅后溶液的硅量指数大于200，但仍不能从经济上满足碳分生产氢氧化铝的要求，因而需加入含钙化合物进行深度脱硅，使其中大部分二氧化硅转化成水化石榴石，并将精液的硅量指数提高至600以上。本文列出了加入石灰进行深度脱硅的部分实验结果（如图2所示）。

石灰脱硅的反应式如下：

$$xSiO_2(OH)_2^{2-} + 2Al(OH)_4^- + 3Ca(OH)_2 =\!=\!= $$
$$3CaO·Al_2O_3·xSiO_2·(6-2x)H_2O + (2+2x)OH^- + 2xH_2O$$

根据实验结果和反应方程式，石灰加入量越大，脱硅效果越好，精液的硅量指数也越高；但石灰加入量太多，氧化铝损失量就越多，需处理的硅渣量也越多，因此适宜的石灰加入量为15g/L左右。实验结果表明，随着脱硅时间的延长，溶液中SiO_2浓度渐渐降低，硅量指数升高；但在120min以后溶液中SiO_2浓度变化较小，因而可认为脱硅时间在120min左右为宜。同时，研究中还发现，加入活性更好的含钙化合物进行深度脱硅，能大幅度降低CaO的加入量。

与传统烧结法深度脱硅相比，在适宜条件下加入石灰进行深度脱硅的新工艺中得到的水化石榴石二氧化硅的饱和系数（x值）较大，钙硅渣中的铝硅比较低，实验结果如表4所列。

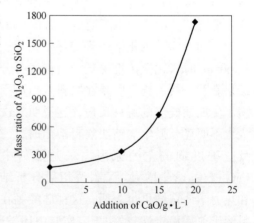

图2 石灰加入量对常压深度脱硅的影响
（Initial solution: Na_2O_K 186.58g/L, Al_2O_3 209.81g/L, SiO_2 1.31g/L, Na_2O_S 10g/L, Na_2O_C 12.88, K_2O 60g/L; time: 90min）

表4 常压深度脱硅的渣相化学成分

Amount of lime/g·L⁻¹	Mass fraction/%				x	A/S
	Na_2O	SiO_2	Al_2O_3	CaO		
10	1.56	4.06	25.44	41.99	0.270	6.27
15	1.08	4.43	25.33	42.53	0.292	5.72
20	0.44	4.29	25.34	42.63	0.282	5.91

2.4 铝酸钠溶液的碳酸化分解

研究强化烧结法氧化铝生产新工艺的目的一方面是为了提高产能，降低生产成本；另一方面也要提高产品的质量，满足现代铝电解对氧化铝物理、化学质量的要求。研究了在碳分过程中分解率 $\eta(Al_2O_3)$ 与氢氧化铝中的 SiO_2 和 Na_2O 含量之间的关系，实验结果如表5所列。

表5 不同情况下碳分氧化铝中 Na_2O 和 SiO_2 的含量

Carbonization conditions				Pregnant solution			$\eta(Al_2O_3)$ /g·L⁻¹	Product	
Seed ratio	Concentration of CO_2/%	Velocity /m³·h⁻¹	Time /min	$\rho(Al_2O_3)$ /g·L⁻¹	α_k	$\rho(Na_2O_C)$ /g·L⁻¹		$w(SiO_2)$ /%	$w(NaO)$ /%
0.15	34	0.26	182	195.85	1.59	14.70	87.70	0.058	
			190				91.03	0.044	0.30
			165	191.02	1.58	16.09	90.05	0.052	
			174				92.42	0.048	0.35
			155	185.93	1.60	16.08	87.78	0.037	0.27
			165				92.13	0.044	0.26
			130	194.50	1.44	8.27	92.70	0.034	0.36
0.15	32	0.25	146	193.04	1.57	17.50	90.70	0.031	
			154				93.06	0.031	
			162	190.42	1.56	9.00	90.05	0.025	0.37
			172				93.85	0.029	
0.25	32	0.25	152	195.62	1.54	7.50	88.73	0.035	
			167				92.37	0.032	0.28

实验结果表明，控制一定的碳分条件（如加入适量的晶种），在碳酸化分解率达到93%左右时，氢氧化铝中的 SiO_2 含量降低至0.025%，Na_2O 的含量降低至0.37%，即化学质量达到了氢氧化铝国家二级品标准。进一步研究的结果表明，在碳酸化分解过程中，产品中 SiO_2 含量主要受钠硅渣平衡溶解度的影响。通过控制工艺条件，降低 HCO_3^- 浓度，阻止丝钠铝石的形成，能明显降低产品中的 Na_2O 含量[8-11]。

2.5 碳分母液的深度蒸发

强化烧结法氧化铝生产新工艺中碳分母液深度蒸发是确保工艺水分平衡的关键。据初步估算，采用铝硅比为8左右的铝土矿进行生产，Na_2O_T 浓度应介于280~320g/L。而目前碳分母液蒸发后 Na_2O_T 浓度在220g/L左右，且结疤严重，水洗和酸洗设备频繁。本文研究了加入表面活性剂进行碳分母液深度蒸发的技术。实验结果如图3和图4

所示。

在类似于强制循环蒸发器的研究实验中，添加适当的表面活性剂，不仅可以将碳分母液蒸发至 Na_2O_T 浓度大于 300g/L，而且还可有效减缓加热表面上结疤的形成速度，大大延长机组的运行周期，降低能耗和清理费用。

图 3　碳分母液未加表面活性剂蒸发 6h 后蒸发器的照片

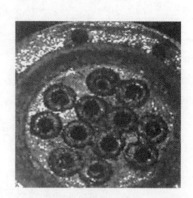

图 4　碳分母液加表面活性剂（0.015g/L）蒸发 16h 后蒸发器的照片

2.6　工业实践

强化烧结法氧化铝生产新工艺[12]从 1999 年 11 月起在中州铝厂逐步实施以来，为保证生产的平稳过渡，首先提高矿石的铝硅比，然后逐渐提高湿法系统中氧化铝浓度。在新工艺逐步实施中已取得很好的成绩，主要指标如表 6 所列。工业运行初步结果表明，采用强化烧结法生产氧化铝新工艺后，熟料中氧化铝含量增加了，熟料的折合比降低了，赤泥产出率降低了，单台窑产能提高了 19.7%，工艺能耗降低了 24.71%。

表 6　逐步消化吸收强化烧结法与传统烧结法生产氧化铝的主要指标对比

Period	A/S of bauxite	A/S of sinter	Al_2O_3 content of sinter	Al_2O_3 net leaching ratio of sinter	Na_2O net leaching ratio of sinter	Output of red mud	Operation ratio of kiln/%	Output of rotary kiln /t·h^{-1}	t-specific sinter per ton Al_2O_3	Energy consumption of per ton Al_2O_3/kg
1998	4.77	3.30	34.79	86.80	95.51	46.70	88.53	52.99	4.09	1821.00
2000	6.48	4.42	37.99	89.97	97.30	38.23	92.41	55.81	3.57	1447.30
Jan.-May of 2001	6.83	4.61	38.81	90.97	97.58	34.71	94.03	56.58	3.49	1371.70

3　结论

（1）采用铝硅比为 9.62 的矿石，按照钙比为 1.3~2.0，碱比为 1.0 配料，在 1250℃以上进行熟料烧成，熟料烧成无困难。

（2）熟料溶出条件宽松，二次反应程度弱，溶出浆液直接进行加压脱硅，浆液沉降性能好。

（3）在适宜的深度脱硅工艺制度下，即使溶液中氧化铝浓度超过 200g/L，精液的硅量指数也能大于 600，得到的水化石榴石中二氧化硅饱和系数大于 0.28。

(4) 采用新的碳酸化分解工艺制度，产品的 SiO_2 含量降低至 0.025%，Na_2O 含量小于 0.37%。

(5) 加入表面活性剂，不仅能将碳分母液蒸发至 Na_2O_T 浓度大于 300g/L，而且还可有效减缓加热表面上结疤的形成速度。

参 考 文 献

[1] 李小斌，张宝琦，程裕国，等. 强化烧结法氧化铝生产工艺 [P]. 中国专利：ZL99109676，2002，12：4.

[2] Lainr A I. Alumina Production [M]. Moskwa：Metallurgy，1962：11-135.

[3] Eremin N I. Process and Equipment in Alumina Production [M]. Moskwa：Metallurgy，1980：12-156.

[4] Arljuk B I. Sinter's Leaching [M]. Moskwa：Metallurgy，1979：10-145.

[5] 中南工业大学氧化铝课题组. 平果铝土矿两成分炉料烧结苛化补碱的试验研究 [A]. 第七届全国氧化铝学术会议论文集 [C]. 峨嵋：中国有色金属学会，1992：90.

[6] LIU Gui-hua, LI Xiao-bin, PENG Zhi-hong, et al. Behavior of calcium silicate in leaching process [J]. Trans Nonferrous Met Soc China, 2003, 13 (1)：213-216.

[7] LIU Gui-hua, LI Xiao-bin, PENG Zhi-hong, et al. Stability of calcium silicate in basic solution [J]. Trans Nonferrous Met Soc China, 2003, 13 (5)：1235-1238.

[8] 李小斌，刘祥民，苟中入，等. 铝酸钠溶液碳酸化分解的热力学 [J]. 中国有色金属学报，2003，13 (4)：1005-1010.

[9] 彭志宏，李小斌，苟中入，等. 铝酸钠溶液碳酸化分解产品中的 Na_2O [J]. 中国有色金属学报，2002，12 (6)：1285-1289.

[10] 苟中入. 高浓度铝酸钠溶液碳酸化分解的机理与工艺 [D]. 长沙：中南大学，2002.

[11] LI Xiao-bin, PENG Zhi-hong, LIU Guihua, et al. Intensifying digestion of diaspore and separation of alumina and silica [J]. Trans Nonferrous Met Soc China, 2003, 13 (3)：671-677.

[12] 赵东峰. 强化烧结法生产氧化铝的技术经济效果 [A]. 第四届全国轻金属冶金学术会议论文集 [C]. 青岛：中国有色金属学会，2001，210.

氢氧化铝晶粒强度的应力状态分析[*]

摘　要　采用材料力学应力状态分析理论，通过计算氢氧化铝晶粒叠合面的正应力与剪应力，研究了叠合面上关键点的应力状态，建立了叠合方式-应力状态-强度的关系，并对二段种分氢氧化铝生产工艺中采用精液分流技术的产品形貌、结构和强度进行了研究。结果表明：氢氧化铝晶粒强度与叠合面承受的应力状态密切相关，正应力成为强度的关键因素；各种不同氢氧化铝晶体间叠合方式中，晶体底面与底面叠合时，氢氧化铝晶粒强度较高并随两个晶体间叠合面积的减小而减小；采用精液分流技术生产出的产品氢氧化铝和氧化铝晶粒，晶体间多为底面与底面叠合且叠合面积较大，氧化铝强度较高。

随着环境保护的日益严格，砂状氧化铝以其粒径粗、强度高等优良性能成为氧化铝生产的趋势。国内外对从过饱和铝酸钠溶液中析出氢氧化铝的过程进行了大量的研究[1-3]，但对晶体强度的研究报道不多，并且这些研究多着重于生产工艺条件对强度的影响[4,5]，未从本质上寻找氧化铝强度的主要因素，因此很难为砂状氧化铝生产提供理论指导。

氧化铝强度由其独特的晶体结构决定。李等[6]发现铝酸钠溶液分解过程氢氧化铝晶体结构是生长基元叠合和化学结晶的直接结果。谭等[7,8]研究了具有不同晶体结构的氧化铝的磨损行为，发现具有镶嵌型结构的氧化铝强度高。吴等[9,10]对氢氧化铝晶体面之间的相互作用能进行了量子化学的理论计算，计算了三水铝石有利生长基元最优结构 $Al_6(OH)_{18}(H_2O)_6$ 的总能量、布居数、原子静电荷等，结果表明，$Al_6(OH)_{18}(H_2O)_6$ 较为有利的成键方位是桥联 OH 基团方位。因此要获得高强度的砂状氧化铝所具有的晶粒结构，必须控制铝酸钠溶液分解过程氢氧化铝晶体的叠合方式。

任何氢氧化铝晶粒的强度与其在外力作用下的力学行为有关。氢氧化铝晶粒中存在晶界，由两个单晶体叠合而成的晶粒的力学行为异于单晶体。根据晶界两侧晶粒取向的不同，晶界对整个晶粒力学行为的影响程度也不同。由晶界面上应变相容性的要求，晶界直接影响晶粒中各组元晶体的力学行为。因此研究晶粒的力学行为，重点在于获得各组元晶体叠合界面的应力状态。任等[11-13]用三维各向异性有限单元法对3种不同的铜双晶体中晶界附近的弹性应力场和各组元晶体主滑移系上分解切应力进行了计算分析，结果表明：同轴双晶体晶界附近的弹性应力场及主滑移系上分解切应力是连续的，非同轴双晶体晶界附近的弹性应力场主滑移系上分解切应力是不连续的。

本文作者拟运用材料力学应力状态分析方法[14-16]，以两块规则氢氧化铝晶体叠合而成的氢氧化铝晶粒为研究对象，通过计算在外力作用下不同叠合方式形成的氢氧化铝晶粒的叠合面应力状态，研究了叠合方式与晶粒强度的关系以及叠合面的破坏方式，

---

[*]　本文合作者：李旺兴、花书贵、尹周澜、叶露升。原发表于《中国有色金属学报》，2005，15（5）：775-781。

以期建立叠合方式—应力状态—强度的关系，为氧化铝强度的研究提供理论基础。同时结合实际生产中的氢氧化铝和氧化铝产品形貌，进一步说明产品显微结构与强度的关系。

1 物理模型

由于规则氢氧化铝晶体一般为六棱柱形，为研究不同叠合方式的晶体在重叠界面上的受力情况，把两个氢氧化铝晶体之间的叠合方式分为底面与顶面叠合，侧面与侧面叠合以及侧面与底面叠合3种方式[17,18]，分别如图1所示。

根据材料力学理论[19]和研究对象的实际情况，为简化计算，作以下假设：

（1）氢氧化铝晶体为连续均匀介质，平整且没有内部结构，具有一定对称性质的正规六棱柱体，在分析其应力状态和应变时，忽略其微小变形，以简化边界条件。

（2）将氢氧化铝晶粒简化为一端固定，另一端受载荷的固定端杆状模型；确定氢氧化铝晶粒杆状模型的约束为：底端3个方向的平动位移和3个方向的转动位移均为零；在上方氢氧化铝晶体顶面一边中点处施加静载荷，根据氢氧化铝晶体的材料性能，忽略应力集中效应。氢氧化铝（氧化铝）属脆性材料，因此分析叠合面的应力状态依照"最大拉应力理论"判断其强度以及破坏方式。

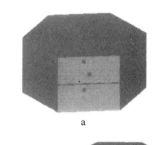
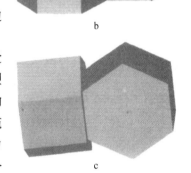
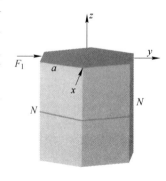

图1 3种不同的叠合方式

在以上约束和载荷条件下，氢氧化铝晶粒将有发生弯曲变形的趋势。本文作者重点研究晶体叠合方式与晶粒整体强度的关系，并分析叠合面的应力状态分布。

2 应力分析方法

现以底面与底面叠合为例来说明在外力作用下，晶体的叠合面所受应力的情况。

假设下方晶体固定不动，规则氢氧化铝晶体正六边形底面的边长为 a，正六棱柱的高为 b。外力 F_1 沿 y 轴方向垂直作用在六棱柱底边的中点处，使氢氧化铝晶粒有发生横力弯曲的趋势，叠合面 $N\text{-}N$ 上产生正应力和剪应力，过叠合面 $N\text{-}N$ 的形心且与 x 轴平行的轴称为中性轴，过叠合面 $N\text{-}N$ 的形心且与 y 轴平行的轴为叠合面沿 y 轴的对称轴，如图2所示。

由力矩平衡条件，有

$$M_{N\text{-}N} = F_1 \times b \tag{1}$$

式中，$M_{N\text{-}N}$ 为横截面 $N\text{-}N$ 上的弯矩；F_1 为外力；b 为外力 F_1 的作用点到横截面的距离。

图2 两晶体完全叠合受力图

由静力平衡条件，晶体在 y 轴方向上不发生相对移动，叠合面上产生剪力 F_2，有

$$F_2 = F_1 \tag{2}$$

3 结果与讨论

3.1 正应力与叠合面积的关系

根据平衡条件，采用截面法，对两块晶体在叠合面上的应力分布进行计算，由于外力 F_1 的作用，在截面上产生顺时针方向的弯距 $M_{N\text{-}N}$。

由力矩平衡方程有

$$F_1 \times b + M_{N\text{-}N} = 0 \tag{3}$$

横力弯曲时的正应力为

$$\sigma = M_{N\text{-}N} \cdot y / I_x \tag{4}$$

式中，I_x 为叠合面对中性轴的惯性矩，根据计算公式 $I_x = \int y^2 \cdot dA$，得到 $I_x = \dfrac{5\sqrt{3}}{16}a^4$，将它代入式（4）并与式（3）联立可得

$$\sigma = \frac{16F_1 b}{5\sqrt{3}\,a^4} y \left(-\frac{\sqrt{3}}{2}a \leqslant y \leqslant \frac{\sqrt{3}}{2}a \right) \tag{5}$$

式中，σ 为距离中性轴 y 处的正应力，σ 对于 y 轴对称分布。

由式（5）可见，叠合面的正应力不仅与 $M_{N\text{-}N}$ 有关，而且与 y/I_x 有关，亦即与叠合面的形状和尺寸有关。最大正应力 $\sigma_{\max}$ 发生于叠合面上离中性轴最远处。于是由式（4）得

$$\sigma_{\max} = \frac{M_{N\text{-}N} \cdot y_{\max}}{I_x} \tag{6}$$

当 $y = \pm\dfrac{\sqrt{3}}{2}a$ 时，$|\sigma_{\max}| = \dfrac{8F_1 b}{5a^3}$。

当两块晶体底面不完全重合时，即沿 y 轴方向发生位移，设位移为 c，受力情况同上，如图 3 所示。

当 $y = \pm\dfrac{\sqrt{3}a - c}{2}$ 时，

$$|\sigma_{\max}| = \frac{F_1 b(\sqrt{3}a - c)}{2\left(\dfrac{4-\sqrt{3}}{3}a - \dfrac{c}{6\sqrt{3}}\right)\left(a - \dfrac{c}{2}\right)^3} \tag{7}$$

由式（7）可知，当两块晶体沿 y 轴方向位移 c 增大时，$|\sigma_{\max}|$ 将随之增大，即在晶界承受同等临界应力的强度条件时，整个晶粒的强度将随界面所发生的位移 c 的增大（即叠合面积的减小）而减小。

实际中，外加载荷为 $F_1 = 10\text{kN}$（25mm 管道中的压强为 0.4MPa[20]），$a = 10\mu\text{m}$[21,22]，代入式（7）计算结果如图 4 所示。

图 3 两晶体不完全叠合

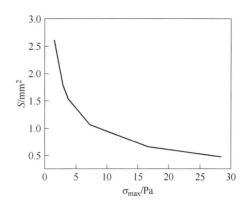

图 4 正应力与叠合面积的关系

3.2 剪应力与叠合面积的关系

根据平衡条件，采用截面法，对两块晶体的叠合面上的剪应力分布进行计算，由于外力 F_1 的作用，在截面上产生水平方向的剪力 F_2。

由静力平衡方程有

$$F_2 - F_1 = 0 \tag{8}$$

弯曲剪应力为

$$\tau = \frac{F_2 \cdot S_x}{I_x \cdot b} \tag{9}$$

将 $S_x^* = \int y dA$ 与 $I_x = \frac{5\sqrt{3}}{16}a^4$ 代入式（9），并与式（8）联立解得

$$\tau = \frac{8F_1}{5\sqrt{3}ab} - \frac{16F_1}{5\sqrt{3}a^3b}y^2 + \frac{32F_1}{45a^4b}y^3 \quad (-\frac{\sqrt{3}}{2}a \leq y \leq \frac{\sqrt{3}}{2}a)$$

当 $y = -\frac{\sqrt{3}}{2}a$ 时，$|\tau_{max}| = \frac{32\sqrt{3}F_1}{45a^2}$；当 $y = \frac{\sqrt{3}}{2}a$ 时，$|\tau| = 0$。

同理，当两块晶体底面不完全重合时，$y = -\frac{\sqrt{3}a - c}{2}$，有

$$|\tau_{max}| = \frac{F_1 \cdot \left\{\left[\left(1 - \frac{2}{3\sqrt{3}}\right)a - \frac{c}{6\sqrt{3}}\right]\left(a - \frac{c}{2}\right)^2\right\}}{\left(\frac{4-\sqrt{3}}{3}a - \frac{c}{6\sqrt{3}}\right)\left(a - \frac{c}{2}\right)^3\left(\frac{3}{2}a - \frac{c}{2\sqrt{3}}\right)} + \frac{f_1\left[\frac{(\sqrt{3}a-c)^2}{4}\left(-\frac{13}{12}a + \frac{7}{12\sqrt{3}}c\right)\right]}{\left(\frac{4-\sqrt{3}}{3}a - \frac{c}{6\sqrt{3}}\right)\left(a - \frac{c}{2}\right)^3\left(\frac{3}{2}a - \frac{c}{2\sqrt{3}}\right)} \tag{10}$$

将实际参数代入式（10）计算，结果如图 5 所示。

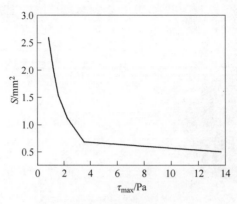

图 5 剪应力与叠合面积的关系

由图 5 可知,当两块晶体沿 y 轴方向位移 c 增大时,将随之增大。图 4、图 5 所示分别为不同方式叠合的晶体,其正应力、剪应力与叠合面积的关系,该关系也适用于晶体侧面与侧面,侧面与底面叠合时的情况。

3.3 叠合面的应力状态分析

两块晶体完全叠合时,在叠合面上顶边中点处、形心处、底边中点处、叠合面沿 y 轴的对称轴上距离形心分别为 $\frac{\sqrt{3}}{4}a$ 和 $-\frac{\sqrt{3}}{4}a$ 处,分别沿 x、y、z 轴方向取正方体微小单元进行应力状态分析。因为在叠合面上 x 轴两侧应力对称分布,而且在 x 轴方向上不产生正应力和剪应力,因此,将六面体微元投影到 yoz 面上,分析其二向应力状态。

由本文 3.1 和 3.2 节分析计算,各处沿 z 轴的正应力 σ_z 和沿 y 轴的剪应力 τ_{zy} 列于表 1。

表 1 不同位置处的正应力和剪应力

Site	σ_z	σ_y	τ_{zy}	τ_{yz}
Mid-point of top margin	$\dfrac{8F_1b}{5a^3}$	0	$\dfrac{32\sqrt{3}}{45a^2}F_1$	0
$y=-\dfrac{\sqrt{3}}{4}a$	$\dfrac{4F_1b}{5a^3}$	0	$\dfrac{3\sqrt{3}}{10ab}F_1$	0
Center	0	0	$\dfrac{8}{5\sqrt{3}ab}F_1$	0
$y=\dfrac{\sqrt{3}}{4}a$	$-\dfrac{4F_1b}{5a^3}$	0	$\dfrac{11\sqrt{3}}{30ab}F_1$	0
Mid-point of bottom margin	$-\dfrac{8F_1b}{5a^3}$	0	0	0

根据上述不同位置微元的应力计算结果,应力状态分析结果列于表 2,其中 α_0、α_1 分别为最大正应力平面法线和最大剪应力平面法线与 z 轴的夹角。

代入相关参数,由计算得知,在叠合面上,y 轴负方向一侧(即靠近受力点一侧),正应力较大,而且距 z 轴距离越大正应力越大(图 2)。因此,随着载荷的增大,靠近受力点一侧将首先发生破坏。

由于氢氧化铝晶体属脆性材料,且叠合面为整块晶粒最为脆弱之处,尽管在叠合面上主应力分布方位较为复杂，但其破坏的发生主要是由沿 z 轴方向上的正应力所决定。

表 2 不同位置的应力状态

Site	Maximum tension stress		Maximum shear stress $(a_1 = a_0 + \frac{\pi}{4})$
	σ_{max}	σ_0	
Mid-point of top margin	$\frac{4F_1 b}{5a^3} + \frac{4F_1}{5a^2}\sqrt{\frac{b^2}{a^2}+\frac{192}{81}}$	$\frac{1}{2}\arctan\left(-\frac{8\sqrt{3}a}{9b}\right)$	$\frac{4F_1}{5a^2}\sqrt{\frac{b^2}{a^2}+\frac{192}{81}}$
$y = -\frac{\sqrt{3}}{4}a$	$\frac{2F_1 b}{5a^3} + \frac{F_1}{5a}\sqrt{\frac{2b^2}{a^4}+\frac{27}{4b^2}}$	$\frac{1}{2}\arctan\left(-\frac{3\sqrt{3}a^2}{4b^2}\right)$	$\frac{F_1}{5a}\sqrt{\frac{2b^2}{a^4}+\frac{27}{4b^2}}$
Center	$\frac{8}{5\sqrt{3}ab}F_1$	$\frac{\pi}{4}$	$\frac{8}{5\sqrt{3}ab}F_1$
$y = \frac{\sqrt{3}}{4}a$	$-\frac{4F_1 b}{5a^3} - \frac{F_1}{5a}\sqrt{\frac{4b^2}{a^4}+\frac{363}{36b^2}}$	$\frac{1}{2}\arctan\left(\frac{11\sqrt{3}a^2}{11b^2}\right)$	$-\frac{F_1}{5a}\sqrt{\frac{4b^2}{a^4}+\frac{363}{36b^2}}$
Mid-point of bottom margin	$-\frac{8F_1 b}{5a^3}$	0	$-\frac{4F_1 b}{5a^3}$

3.4 产品氢氧化铝的形貌、结构与强度

在高浓度二段种分生产氧化铝工艺中，第一段为附聚段，第二段为长大段。当第一段经过附聚后的晶粒随浆液进入第二段后，由于此时溶液过饱和度已经比较低，影响产品氧化铝强度。采用精液分流技术，使部分精液直接进入长大段，提高长大段的初始过饱和度，可以改善产品氧化铝的强度。图 6 所示为精液分流产品氢氧化铝晶粒形貌，图 7 所示为精液不分流产品氢氧化铝晶粒形貌。

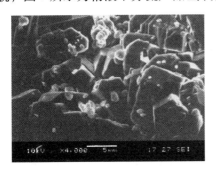
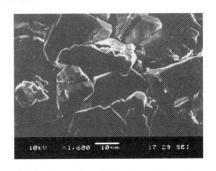

图 6 精液分流产品氢氧化铝晶粒 SEM 的形貌　　图 7 精液不分流产品氢氧化铝晶粒 SEM 的形貌

从图中可看出，精液分流产品晶粒表面氢氧化铝晶体之间的叠合多为底面与底面的相互叠合，叠合面积较大，产品强度高，经小窑焙烧得到的产品氧化铝粒度小于 45μm 和小于 20μm 的晶粒分别为 8%~11% 和 2% 以下，磨损指数可以达到 20% 左右。而精液不分流产品晶粒的晶体之间较为松散，底面与侧面，侧面与侧面的叠合较多，叠合面积较小，氢氧化铝整体晶粒的强度相对较小。

4 结论

（1）通过对晶体叠合方式及受力情况进行分析，初步建立了晶体的受力模型与叠合方式-应力状态-强度的关系。

（2）通过对叠合面进行应力状态分析，研究了晶体叠合面应力状态分布，发现界面破坏主要由于正应力所致，且界面上靠近受力点一侧正应力较大。因此随着载荷的增大，将首先发生破坏，该位置是决定强度的关键部位。

（3）氢氧化铝晶粒强度与叠合面承受的应力状态密切相关。各种不同氢氧化铝晶体间叠合方式中，晶体底面与底面叠合时，氢氧化铝晶粒强度较高并随两个晶体间叠合面积的减小而减小。

（4）采用精液分流种分技术，产品晶粒表面氢氧化铝晶体之间的叠合多为底面与底面的相互叠合，叠合面积较大，氧化铝强度较高。

参 考 文 献

[1] 杨重愚. 氧化铝生产工艺学 [M]. 北京：冶金工业出版社，1993：119-126.

[2] 周辉放，杨重愚，陈光渊，等. 铝酸钠溶液中晶体附聚机理研究 [J]. 有色金属，1994，46（4）：54-58.

[3] Veesler S, Rource S, Boistelle R. Abut supersaturation and growth rates of hydragillite Al(OH)$_3$ in alumina caustic solution [J]. J Crystal Growth, 1993, 130 (3, 4)：411-415.

[4] Kubota N, Yokota M, Mullin J W. The combined influence of supersaturation and impurity concentration rationon crystal growth [J]. Journal of Crystal Growth, 2000, 212 (3, 4)：480-488.

[5] Sang J V. Factors affecting the att rition strength of alumina products [J]. Light Metals, 1987, 121-127.

[6] 李洁. 铝酸钠溶液结构与分解机理的研究 [D]. 长沙：中南大学，2001.

[7] 张樵青. 对拜耳法高浓度铝酸钠溶液两段分解细晶种附聚的研究 [J]. 轻金属，1994（4）：5-9.

[8] 陈国辉，陈启元，尹周澜，等. 铝酸钠溶液种分成核和附聚研究进展 [J]. 湿法冶金，2003，22（1）：14-18.

[9] 吴争平，陈启元，尹周澜，等. Al$_6$(OH)$_{18}$(H$_2$O)$_6$ 的结构及成键方位的从头算法及密度泛函分析（Ⅰ）[J]. 中国稀土学报，2004，22（专辑）：283-288.

[10] 吴争平，陈启元，尹周澜，等. Al$_6$(OH)$_{18}$(H$_2$O)$_6$ 的结构及成键方位的从头算法及密度泛函分析（Ⅱ）[J]. 中国稀土学报，2004，22（专辑）：289-293.

[11] 任德斌，彭兆行，李守新，等. 双晶体弹性应力场和在滑移上分解的切应力分析 [J]. 东北大学学报（自然科学版），2000，21（1）：34-37.

[12] 任德斌，彭兆行，李守新，等. Cu 三晶体中各晶粒滑移系上切应力分析 [J]. 沈阳工学院学报，1999，18（3）：50-54.

[13] 任德斌，李守新，王中光，等. Cu 三晶交点应力集中特征的分析 [J]. 沈阳工学院学报，1998，17（2）：45-49.

[14] 陈肖虎，程立，赵莉. 砂状氧化铝生产分解过程实验研究 [J]. 贵州工学院学报，1994，23（6）：65-70.

[15] 刘宝琛，曹秋良. 大晶粒多晶铝试样细观变形测量 [J]. 清华大学学报（自然科学版），1994，34（5）：26-31.

[16] 张广平，王中光. 晶体取向 Ni$_3$Al 合金单晶体疲劳行为的影响 [J]. 金属学报，1997，33（10）：1009-1013.

[17] Malkin A J, Kuznet sor Y G, Mepherson A. Defect structure of macromolecular crystals [J]. Joural of Structural Biology, 1996, 117 (2)：124-137.

[18] Yasuhisa A. Lowering friction coefficient under low loads by minimizing effects of adhesion force and viscous resistance [J]. Wear, 2003, 254 (10)：965-973.

［19］刘鸿文．材料力学［M］．北京：高等教育出版社，1992：173-191．
［20］Seyssiecq I, Veesler S, Boistelle R, et al. Agglomeration of gibbsite Al(OH)$_3$ crystals in Bayer liquors, influence of the process parameters［J］. Chemical Engineering Science, 1998, 53 (12): 2177-2185.
［21］YS/T 438.2—2001．砂状氧化铝物理性能测定方法——磨损指数的测定［S］．
［22］李旺兴，张江峰，尹周澜，等．铝酸钠溶液晶种分解附聚过程研究［J］．中国稀土学报，2004，22（专辑）：223-226．

A New Heat Treatment Processing for TiAl Based Alloy*

Abstract Owing to the limitation of the normal thermomechanical treatment on TiAl based alloy, a double-temperature heat treatment technology for hot forged TiAl based alloy was suggested. The effect of the new technology on microstructure and tensile properties of the alloy at ambient temperature was also studied. By TEM analysis, the mechanism that the double temperature heat treatment improves the homogeneity of the microstructure in TiAl based alloy was discussed.

1 Introduction

Having the merits of low density and excellent elevated temperature properties, TiAl based alloy is regarded as an ideal high temperature structure material which has the promising application in aerospace. However, owing to poor thermal deformability[1], brittleness at ambient temperature[2] and insufficient oxidation resistance above 850℃ [3] etc, the practical application of this alloy is hindered. Its brittleness at ambient temperature is more severe. Studies[4-6] showed that, the ambient temperature mechanical properties of TiAl based alloy were significantly influenced by its microstructure, and the specimen with fine, homogeneous duplex structure demonstrated excellent ambient temperature ductility[7-9]. However, the coarse cylindrical α_2/γ lamellar colonies in the ascast microstructure of TiAl based alloy can be disintegrated only through hot pressing deformation. The lamellar colonies of TiAl based alloy exhibit anisotropy in mechanical properties[10] and in deformation substructure[11-13], and the zigzagboundaries[14] restrict the rotation of grains, therefore, the deformation substructures of all the α_2/γ lamellar colonies exhibit monotonity during the thermal deformation process. As a result, several coarse lamellae colonies still remain perfectly lamellae after thermal deformation process[15]. Moreover, because of high rheological stress in TiAl based alloy and high friction between the specimen and punch, the distribution of internal stress is heterogeneous in the specimen. Stress may be very low in some area in which thermal deformation characteristics experienced by material are not apparent, and coarse lamellae colonies still exist[14]. Owing to the complexity of the thermal deformation process of TiAl based alloy, several indestructed coarse lamellae colonies always remain in the deformed specimen[17]. Those lamellar colonies exhibit microstructural thermal-stability when annealed at high temperature and can not be fined during subsequent annealing in two-phase field. Therefore, in order to acquire fine, homogeneous thermal-deformation structure for further study, specimens should be selectively cut from the ingot[18]. Moreover, due to low thermoplasticity of TiAl based alloy, a large amount of deep cracks come out on the periphery and upper

---

* Copartner: He Yuehui, Huang Baiyun, Liu Yong. Reprinted from Transactions of the Nonferrous Metals Society of China, 1996, 6(3): 96-102.

and lower surface of the ingot, and homogeneous, thermal-deformation microstructure only partially exist. Therefore the quality of the ingot is very poor, the selected specimens always demonstrate poor mechanical properties because of containing some coarse lamellar colonies.

This article concentrates the investigation on the effects of double-temperature heat treatment process on the microstructure and ambient temperature mechanical properties of randomly selected specimens from TiAl based alloy ingot.

2 Experimental

The alloy with nominal composition (in weight fraction) of Ti-33Al-3Cr-0.5Mo was prepared with consumable arc melting technique in argon atmosphere. In order to reduce composition inhomogeneity, the alloy ingot was remelted. After homogeneously annealing at 1050℃ for 48h, the ingot was spark-eroded into dia. 50mm×100mm cylindrical specimen, which was then HIP-ed (1250℃, 170MPa, 4h). Plastic deformation through hot press treatment was conducted under isothermally forging machine, process parameters were: at 1040℃, $3 \times 10^{-4} s^{-1}$, 82%. After the crack parts on the ingot periphery were cut off, specimens were selected randomly from the remain parts for heat treatment and mechanical property tests. Ambient temperature tensile test was conducted on CSS 112 type electronic versatile test machine. The effective dimensions of cylindrical tensile specimen are dia. 3mm×18mm. Fractograph analysis was conducted under scanning electron microscope. The metallographic specimens after heat treatment were prepared in a standard fashion and etched with the Kroll's solution. Microstructural-metallographic analysis were conducted under Leica-Quantimet 520 type optical microscope and morphology analysis instrument, and scanning electron microscope. The foils for transmission electron microscope were prepared by twinjet technique using a solution consisting of 70mL alcohol, 120mL methanol, 100mL butane-1-01 and 80mL perchloric, at a voltage of ~45V, a current of 7 to 10mA, and temperature of −40℃. Observation of the foils was conducted in a CM-12 transmission electron microscope operated at 120kV.

3 Results

Table 1 describes the mechanical properties of the alloy at ambient temperature which was normally heat treated according to the route shown in Fig. 1. The experiment data show that, Ti-33Al-3Cr-0.5Mo alloy exhibits low ambient temperature mechanical properties, especially poor ambient temperature ductility.

Table 1 The mechanical properties at ambient temperature of Ti-33Al-3Cr-0.5Mo alloy after normal heat treatment

Sample No.	Heat treatment processing	Processing type	$R_{p0.2}$/MPa	R_m/MPa	A/%
1	900℃, 24h, A.C.	I	390	428	1.3
2	1000℃, 24h, A.C.	I	453	480	1.2
3	1180℃, 4h $\xrightarrow{F.C.}$ 950℃, 6h, A.C.	II	460	519	1.2
4	1250℃, 4h $\xrightarrow{F.C.}$ 950℃, 6h, A.C.	II	510	537	1.8
5	1280℃, 4h $\xrightarrow{F.C.}$ 950℃, 6h, A.C.	II	521	570	1.5
6	1380℃, 2h, A.C. +950℃, 6h, A.C.	III	621	746	1.3

Note: A.C.—Air colling; F.C.—Furnace colling.

Fig. 2 illustrates the microstructure of specimens normally heat-treated. Indestructed coarse lamellae colonies still remain when heated in two phase-field as shown in Fig. 2a. Chemical composition EDAX analysis of the microstructure in selected area is shown in Fig. 2b, no apparent difference between the coarse lamellae colony and the lamellar colony in fine duplex microstructure thereabout was found. TEM analysis shows that, coarse lamellae

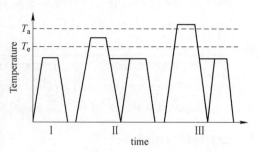

Fig. 1 Scheme of normal heat-treatment processing for TiAl based alloy sample IF'ed

colony still consists of α_2 lathes and γ lathes, strict orientational relation exists between adjacent lathes, as shown in Fig. 3. Fig. 4 describes the SEM fractograph of tensile specimen after annealed in two-phase field. The fracture of coarse colony can be found in the fractograph, and the large amount of secondary cracks also can be detected around these lamellar colonies boundary.

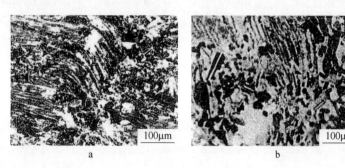

Fig. 2 Optical micrographes of Ti-33Al-3Cr-0.5Mo alloy normally heat treated

a—900℃ ; b—1250℃ $\xrightarrow{F.C.}$ 950℃, 6h

Fig. 3 TEM micrograph and diffractional pattern of coarse lamellae colony of TiAl base alloy after normal heat-treatment

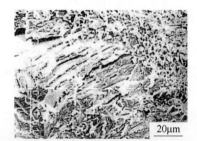

Fig. 4 Fractograph by SEM analysis for Ti-33Al-3Cr-0.5Mo alloy ingot heat-treated at 1250℃, 4h $\xrightarrow{F.C.}$ 950℃, 6h

When randomly selecting specimens from TiAl based alloy ingot, it exhibits inhomogeneous microstructure and poor mechanical properties at ambient temperature. For this reason, a new technique-double-temperature heat treatment is presented, as shown in Fig. 5. Table 2 illustrates the ambient temperature tensile property of Ti-33Al-3Cr-0.5Mo alloy treated by this new technique.

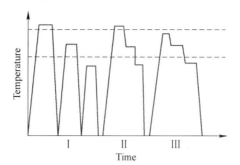

Fig. 5 Scheme of double-temperature heat-treatment processing

Table 2 The effects of double-temperature heat-treatment types on mechanical properties of Ti-33Al-3Cr-0.5Mo

Sample No.	Heat treatment processing	Processing type	$R_{p0.2}$/MPa	R_m/MPa	A/%
1	1310℃,4h,A.C. + 1250℃,4h, A.C. + 950℃,6h,A.C.	I	521	583	2.1
2	1310℃,4h $\xrightarrow{F.C.}$ 1250℃,4h, A.C. + 950℃,4h,A.C.	II	590	673	2.4
3	1280℃,4h $\xrightarrow{F.C.}$ 1250℃,4h, A.C. + 950℃,6h,A.C.	III	543	657	2.8

Compared with the experiment data in Table 1, the ambient temperature heat mechanical properties of TiAl based alloy ingot after double-temperature treatment are improved. Fig. 6 shows two optical metallographs for the microstructure of TiAl based alloy ingot after various double-temperature heat treatment. After I-type double-temperature heat treatment, microstructure of mixed grains is obtained-fine equiaxial γ grains or fine lamellae colonies distributing in coarse equiaxial lamellae colonies, as shown in Fig. 6a; while after II-type double-temperature heat-treatment, fine, homogeneous duplex microstructure is obtained, as shown in Fig. 6b. In this article, the process parameters of II-type double temperature are further studied, and their experiment results are shown in Table 3, from which it can be seen that variation of the heating time at the first annealing temperature in double-temperature heat treatment influences the ambient temperature tensile property, i.e. diminishing the first annealing time will improve the ambient temperature mechanical properties, especially the ductility of TiAl based alloy. In order to investigate how the II-type double-temperature heat treatment influences the ambient temperature mechanical properties of the alloy, their microstructures are analyzed. The results are as follows: when heated at 1310℃ for 4, 3, 2, 1 and 0.5h, the mean grain diameters of duplex structure in the finally heat-treated specimens are respectively 48, 35, 28, 25 and 18μm, part of the optical metallographs are shown in Fig. 7.

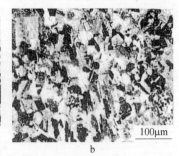

Fig. 6　Micrographes of Ti-33Al-3Cr-0.5Mo alloy heat-treated at double-temperature

a—1310℃,4h,A.C.+1250℃,4h;b—1310℃,4h $\xrightarrow{F.C.}$ 1250℃,4h,A.C.+950℃,6h,A.C.

Table 3　The effects of Ⅱ-type heat-treatment processing on mechanical properties of Ti-33Al-3Cr-0.5Mo alloy at ambient temperature

Sample No.	Heat treatment processing	$R_{p0.2}$/MPa	R_m/MPa	A/%
1	1310℃,3h $\xrightarrow{F.C.}$ 1250℃,4h $\xrightarrow{F.C.}$ 960℃,6h,A.C.	432	577	2.6
2	1310℃,2h $\xrightarrow{F.C.}$ 1250℃,4h $\xrightarrow{F.C.}$ 960℃,6h,A.C.	403	507	2.8
3	1310℃,1h $\xrightarrow{F.C.}$ 1250℃,4h $\xrightarrow{F.C.}$ 960℃,6h,A.C.	426	603	3.3
4	1310℃,0.5h $\xrightarrow{F.C.}$ 1250℃,4h $\xrightarrow{F.C.}$ 960℃,6h,A.C.	429	572	3.2

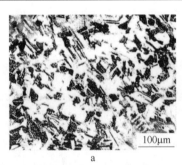

Fig. 7　Micrographes of TiAl basedalloy after double-temperature heat-treatment

a—1310℃,3h $\xrightarrow{F.C.}$ 1250℃,4h,A.C.;b—1310℃,0.5h $\xrightarrow{F.C.}$ 1250℃,4h,A.C.

4　Discussions

One largely hot forging plasticity deformation can not disintegrate all the coarse lamellae colonies in TiAl based alloy. When annealing in two phase field, owing to in-situ order → disorder phase transformation of α_2, lamellar colonies with perfect α_2/γ lamellae structure can transform to α/γ lamellar structure at elevated temperature, which inherit the original α_2/γ or γ/γ twin interface relation. Due to strict orientation relation on the lathes boundaries, it is very difficult for nucleation. Additionally, the strict orientation relation restricts the boundary migration when lathes grow widely and equiaxially(as shown in Fig. 8). Therefore, the coarse α_2/γ lamellae colonies remaining after thermal deformation exhibit high microstructure thermal-stability during subsequent annealing in two-phase field. Poor mechanical properties at ambient temperature are mainly due to coarse colonies in specimens, as shown in Fig. 2 and Table 1. However, the ambient temperature mechanical properties of TiAl based alloy are greatly improved through appropriate double-temperature heat treatment, as illustrated in Table 2 and Table 3, owing to having obtained a fine, homogeneous duplex microstructure (Fig. 6 and Fig. 7). Fig. 9 describes how II-type double-temperature heat treatment can refine the microstructure of TiAl based alloy specimen hot-forged. After thermoplastic deformation, a large amount alloys of lattice defects are generated in TiAl based alloy, keeping high internal energy store. When heating in monophase field, a large number of α nucleus comes into forming. Without the restriction on new crystal boundary migration by lathes boundaries when heating in two-phase field, α phase nucleuses grow equaxially. Because of high nucleation rate, the equiaxial grains are not so large, as shown in Fig. 9a. When cooling in furnace and retaining at constant temperature in two phase field, γ phase will precipitate. The formation of γ nucleuses is homogeneous, and the growth of them is equiaxial, as shown in Fig. 9b. At this time, the microstructure is characterized by dispersion of finely equiaxial γ phase grains in coarsely equiaxial α phase grains. Subsequently, when cooling from α + γ two phase field to room temperature, the transformation of $\alpha \rightarrow \alpha_2 + \gamma(\alpha_2/\gamma)$ occurs. Main steps are $\alpha \rightarrow \alpha + \alpha^{SF} \rightarrow \gamma_2 + \alpha \rightarrow \gamma + \alpha \rightarrow \gamma + \alpha_P(\alpha_2/\gamma)$, i.e. during $\alpha \rightarrow \alpha_2 + \gamma$ transformation process, γ phase acts as an antecede phase, then $\alpha \rightarrow \alpha_2$ ordered transformation occurs at T_e thus, α_2/γ lamellar colony is obtained. In order to decrease nucleation energy, the γ lathes formed in α_2/γ lamellar colony should adhere to initially formed equiaxial γ grains, i.e. nucleation locations are provided on the surface of the initially formed equiaxial γ grains. According to nucleation theory, certain orientation relation should be acquired between γ lathes and boundaries of initial γ grains, herein the twin relation is {111}<110>, which has been proved by experiments, as shown in Fig. 10. In α_2/γ lamellar colony, lath B of γ phase adheres to an equiaxial grain A of γ phase, nucleating and growing up. As also shown in Fig. 10, the selected area electron diffraction pattern of B/A interface is $[011]_A // [0\bar{1}\bar{1}]_B$, twin face is (111). In an original α grain there exist many γ equiaxial grains whose surfaces are composed of many crystal faces with various orientation. Hence, many α/γ colonies with various lamellae orientation come into forming in an original α grain. When cooling to room temperature, ordered transformation of α phase occurs. As a result, through II-type double-temperature heat treatment, the fine, homogeneous duplex microstructure is obtained, as shown in Fig. 9c. The microstructure analyzed by SEM is also shown in Fig. 11 that α_2/γ colonies nucleate and grow up around equiaxially γ phase grains. When diminishing heating time at first temperature, compositional heterogeneity in α grain gets more and more, and the nucleation of γ grains is easily enhanced when cooling from α monophase field to α + γ two-phase field and retaining at a constant temperature, as a result, the number of γ grains increases. Moreover, due to shorter heating time, the original α phase grain is fine, and favorable for acquiring fine duplex microstructure. However, through I-type double-temperature heat treatment, α_2/γ lamellar colonies are firstly obtained, as shown in Fig. 12a. When reheating to α+γ two-phase field, $\alpha_2 \rightarrow \alpha$ in-situ phase transformation occurs and α/γ lamellar colonies thus be generated. Moreover, the renucleation and growth of both α grains and γ grains happen in some local area, the microstructure is shown in Fig. 12b. When cooling to room temperature, $\alpha \rightarrow \alpha_2$ in-situ ordered transformation occurs in α/γ lamellar colonies, hence α_2/γ lamellar colonies are obtained, and finally a microstructure of mixed grains is generated, as shown in Fig. 12c.

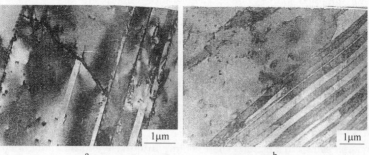

Fig. 8 TEM micrograph of Ti-33Al-3Cr-0.5Mo alloy ingot IF'ed after annealing of heated at $\alpha + \gamma$ two phase field

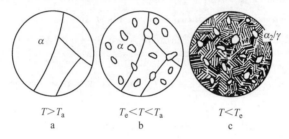

Fig. 9 Scheme of microstructure transformation after annealing at type II processing double-temperature

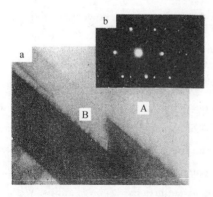

Fig. 10 TEM micrograph and diffraction pattern of sample annealed at double-temperature

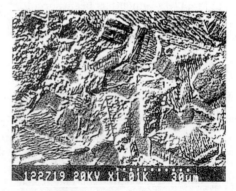

Fig. 11 SEM BS micrograph of TiAl based alloy sample double-temperature heat-treated

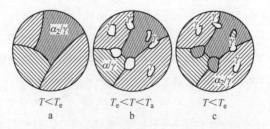

Fig. 12 Scheme of microstructure transformation after annealing at double-temperature in type I processing

5 Conclusions

(1) After normal heat treatment, the specimen, which is randomly selected from TiAl based alloy ingot hot-forged, exhibits inhomogeneous microstructure. In the microstructure, coarse α_2/γ lamellae colonies can be easily found, which has been proved to be inherited from as-cast microstructure. The alloy specimens exhibit poor mechanical properties at the ambient temperature.

(2) Through appropriate double-temperature heat treatment, the fine and homogeneous duplex microstructure can be acquired from TiAl based alloy ingot. The ambient temperature mechanical properties of the specimens are significantly improved.

References

[1] Minoru Nobuki et al. Journal of the Japan Institute of Metals, 1986, 50(9):840.
[2] Shethtman D et al. Metall Trans, 1974, 5(6):1373.
[3] Kazuo Kasahara et al. Journal of the Japan Institute of Metals, 1989, 53(1):58.
[4] Sastry S M L et al. Metall Trans, 1977, 8A:299.
[5] Huang S C et al. In:6th world Conference on Titanium, Cannes, 1988.
[6] Vasudevan et al. Scripta Metall, 1989, 23:467.
[7] Huang S C. Metall Trans, 1992, 23A(1):375.
[8] Pu Denjie et al. Acta Metallurgica Sinica, 1993, 28(8):A365.
[9] Kim Y W. JOM, 1989, 7:24.
[10] Umakoshi Y et al. Scripta Metall, 1991, 25:1525.
[11] Yamaguchi S et al. Trans of Japan Institute of Metals, 1991, 30(1):48.
[12] He Yuehui et al. Hot Working Technology, (in Chinese), 1995, 5:13.
[13] He Yuehui et al. Materials Science and Engineering, (in Chinese), 1996, (1):35.
[14] McQuang P A et al. Scripta Metall, 1991, 25:1689.
[15] He Yuehui et al. Hot Working Technology, (in Chinese), 1995, 1:13.
[16] He Yuehui et al. High Technology Letters, (in Chinese), 1994, 12:26.
[17] He Yuehui et al. MRS. U. S. A., Boston, 1994.
[18] Kim Y W. Acta Metall Mater, 1992, 40(6):1121.

自蔓延高温合成过程中绝热温度的编程计算

摘要 本文研究了自蔓延高温合成过程中绝热温度的计算问题。在建立有关热力学数据库的基础上用 FoxBASE 语言编程计算了自蔓延高温反应过程中的绝热温度,并将所计算的值和资料所报道的绝热温度进行了比较分析,同时示例计算绘制了预热温度和添加稀释剂对反应绝热温度影响的曲线。

1 前言

自蔓延高温合成法(self-propagating High-temp. Synthesis-SHS)是由前苏联学者 Merzhanov 和 Brovinskaya 等于 1967 年首先提出来的[1],由于该方法在材料合成方面有许多独特的优点[1,2],目前引起了世界各国的普遍关注,并迅速波及到了许多科学领域。而在 SHS 研究的过程中,反应体系必须满足一定的热力学条件,反应才能自行蔓延,其基本的热力学参数是反应的绝热温度(adiabatic temperature)。所谓绝热温度是指反应物点燃后所释放的热全部用来给反应体系升温所能达到的最高温,它是一个理论计算值。研究表明,要使自蔓延反应自我维持,必须满足:

绝热温度 $T_{ad} \geq 1800K$[1-4]。同时,对反应物进行预热和添加稀释剂能直接影响 SHS 反应的绝热温度,而且反应温度的高低也直接影响生成物的相态以及 SHS 反应机理。因此,在 SHS 研究的过程中,体系的绝热温度的定量计算是十分必要的。而以往的研究主要集中在由单质元素室温下合成简单化合物时绝热温度的估算。对于初始条件(预热、添加稀释剂)对 T_{ad} 的影响以及较复杂的 SHS 反应(如 $10Al + 3TiO_2 + 3B_2O_3 = 5Al_2O_3 + 3TiB_2$)$T_{ad}$ 计算研究尚无系统的资料可查。由于考虑初始条件对 T_{ad} 的影响,且如果生成物中有相态变化,其中牵涉的热力学数据较多,计算工作量大,本研究在建立数据库的基础上用 FoxBASE 语言编程只需输入 SHS 反应方程式和预热温度,便可迅速准确计算该 SHS 反应体系在一定的初始条件下(稀释剂直接写入反应方程式)最终的反应绝热温度(T_{ad}),并自行判断生成物中是否发生固态相变和熔融,且计算固态相转变率和熔化率,最后运用计算结果示例绘制了 T_0-T_{ad} 以及 $D_u\%$-T_{ad} 的曲线关系图。该研究在 SHS 理论研究和应用研究方面有很大的指导意义。

2 计算方法及编程

设物质恒压克分子热容 C_p(卡/(开·克分子))为:

$$C_p = A_1 + A_2 \times 10^{-3}T + A_3 \times 10^5 T^{-2} + A_4 \times 10^{-6} T^2 + A_5 \times 10^8 T^{-3} \tag{1}$$

按物质名称、该物质不同相态下 C_p 的系数、固相转变点(T_u)、固相转变热(ΔH_{tr})、T_M(熔点)、熔化热(ΔH_M)以及该物质的标准生成焓($H_{298}^{\ominus}$)为字段名建立常用物质的有关热力学数据库。

---

* 本文合作者:胡文彬、郑子樵、刘威威。原发表于《材料工程》,1993(9):36-39。

根据热力学第一定律可以得到：

$$C_p = dH/dT \tag{2}$$

在常温和 T 间积分上式，就有：

$$H_T^\ominus - H_{298}^\ominus = \int_{298}^T C_p dT \tag{3}$$

式（3）即为本研究的基本计算公式，但式（3）只适用于物质在所研究的温度区间没有发生相变的情况。如果在该温度区域内有固态相变，则相对焓的计算公式采用：

$$H_T^\ominus - H_{298}^\ominus = \int_{298}^{T_{tr}} C_p dT + \Delta H_{tr} + \int_{T_{tr}}^T C_p' dT \tag{4}$$

式中，C_p' 为第二种固相变型的恒压克分子热容。

若该物质在所研究的温度区间既发生固相转变同时发生熔化，则按下式计算相对焓：

$$H_T^\ominus - H_{298}^\ominus = \int_{298}^{T_{tr}} C_p dT + \Delta H_{tr} + \int_{T_{tr}}^{T_M} C_p' dT + \Delta H_m + \int_{T_M}^T C_p' dT \tag{5}$$

式中，C_p' 为该物质在液态时的恒压克分子热容。

若该物质在所研究的温度区间不发生固相转变但发生熔化，则：

$$H_T^\ominus - H_{298}^\ominus = \int_{298}^{T_M} C_p dT + \Delta H_M + \int_{T_M}^T C_p' dT \tag{6}$$

最后计算绝热温度 T_{ad} 的公式为：

$$\sum_i m_i H_{T_0}^\ominus(\text{反应物}) + \sum_k f_k H_{T_0}^\ominus(\text{稀释剂}) = \sum_j n_j H_{T_{ad}}^\ominus(\text{反应物}) + \sum_k f_k H_{T_{ad}}^\ominus(\text{稀释剂}) \tag{7}$$

式中，m_i、f_k、n_j 为方程式中物质的摩尔系数。

为了在编程过程中简化程序，上式简化为：

$$\sum_i m_i H_{T_0}^\ominus(\text{反应物} + \text{稀释剂}) = \sum_i n_j H_{T_{ad}}^\ominus(\text{反应物} + \text{稀释剂}) \tag{8}$$

上式是一个关于 T_{ad} 的方程，考虑有关热力学温度精确到1K，在求解 T_{ad} 时采用循环试算法求解。试算函数设为 $f(T_{ad})$：

$$f(T_{ad}) = \sum_i n_j H_{T_{ad}}^\ominus - \sum_i m_i H_{T_0}^\ominus \tag{9}$$

设初值 $T_{ad} = 1000K$，如果 $f(T_{ad}') < 0$，则循环试算步长为+100，$T_{ad}' = T_{ad}' + 100$ 继续试算，直至 $f(T_{ad}') \geq 0$，再将循环步长改为-10，$T_{ad}' = T_{ad}' - 10$，循环试算直至 $f(T_{ad}') \leq 0$；然后再将试算步长改为+1，$T_{ad}' = T_{ad}' + 1$，循环试算直至 $f(T_{ad}') \geq 0$，这样计算的 T_{ad} 的最后值为 $(T_{ad}' - 0.5) \pm 0.5K$。如果在循环试算后最终得到 $T_{ad}' = T_e$ 或 T_M，则需判断该物质相态变化的情况，计算固态相转变率和熔化率。相转变率的计算公式为：

$$U_m = (\sum_i m_i H_{T_0}^\ominus - \sum_j n_j H_{T_{ad}}^\ominus + n_j \Delta H)/n_j \Delta H \tag{10}$$

式中，n_j 为研究物质在方程式中的摩尔系数；ΔH 为研究物质的相转变热。

当 $U_m > 0$ 时相变确实存在，给出 U_m 的值，$T_{ad} = T_{ad}$；

当 $U_m = 0$ 时相变未发生，$T_{ad} = T_{ad}'$；

当 $U_m < 0$ 时未达到相变转变温度，给出 $T_{ad} = (T_{ad}' - 0.5) \pm 0.5K < T_0$（或 T_M）。

程序流程图如图1所示。

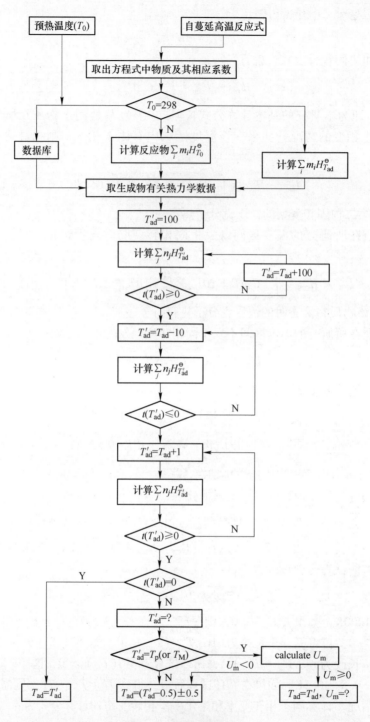

图 1 计算程序流程图

3 常用自蔓延高温反应绝热温度的计算

本程序采用人机汉字对话的形式，只需输入预热温度和反应方程式，自动进入数据库读取有关物质的热力学数据，可方便准确迅速地计算 SHS 反应最终的绝热温度。采用本程序计算常见的 SHS 反应的绝热温度示于表 1。

表1 常见自蔓延高温反应绝热温度的计算值[*]

自蔓延高温反应方程式 SHS reaction equation	T_{ad}/K	相转变情况 Phase transition condition	T_{ad}资料[1]报道值/K T_{ad} (reported by ref. [1])/K
Ti + 2B = TiB$_2$	3193	熔化,熔化率 27.0% melted, melted rate 27.0%	3190
Ti + C = TiC	3290	熔化,熔化率 16.4% melted, melted rate 16.4%	3210
Si + C = SiC	1858.5		1800
4B + C = B$_4$C	1038.5		1000
2Ti + N$_2$ = 2TiN	4908.5	全熔化 all melted	4900
Zr + 2B = ZrB$_2$	3323	熔化,熔化率 78.2% melted, melted rate 78.2%	3310
Nb + 2B = NbB$_2$	2317.5		2400
W + C = WC	1161.5		1000
Ta + C = TaC	1711.5		2700
2Ta + N$_2$ = 2TaN	3363	熔化,熔化率 97.3% melted, melted rate 97.3%	3360
2Nb + N$_2$ = 2NbN	3474.5		3500
Nb + 2Si = NbSi$_2$	2013.5		1900
5Ti + 3Si = Ti$_5$Si$_3$	2547.5		2500
Mo + 2Si = MoSi$_2$	1942.5		1900
V + C = VC	2228.5		
5Zr + 3Si = Zr$_5$Si$_3$	2755.5		2800
Zr + C = ZrC	3757.5		
Ni + Al = NiAl	1912	熔化,熔化率 41.8% melted, melted rate 41.8%	
Ti + 3Al = TiAl$_3$	1516.5		
4Al + 3TiO$_2$ = 3Ti + 2Al$_2$O$_3$	1805.5		
4Al + 3TiO$_2$ + 3C = 3TiC + 2Al$_2$O$_3$	2303	Al$_2$O$_3$ 熔化,熔化率 99.5% Al$_2$O$_3$ melted, melted rate 99.5%	
10Al + 3TiO$_2$ + 3B$_2$O$_3$ = 5Al$_2$O$_3$ + 3TiB$_2$	2531.5	Al$_2$O$_3$ 全部熔化 Al$_2$O$_3$ all melted	
2Al + B$_2$O$_3$ = Al$_2$O$_3$ + 2B	2303	Al$_2$O$_3$ 熔化,熔化率 74.4% Al$_2$O$_3$ melted, melted rate 74.4%	
5Mg + TiO$_2$ + B$_2$O$_3$ = 5MgO + TiB$_2$	3098	MgO 熔化,熔化率 21.5% MgO melted, melted rate 21.5%	
2Al + Fe$_2$O$_3$ = Al$_2$O$_3$ + 2Fe	3148.5	Al$_2$O$_3$ 全熔化 Al$_2$O$_3$ all melted	

[*] 有关热力学数据取自文献 [5]。

由表1可见,运用本计算程序计算所得到的绝热温度基本上和文献 [1] 报道的值相接近,造成计算结果差别的主要原因有两方面:一方面,不同的作者采用的热力学数据有差别;另一方面,迄今为止,有关资料报道的绝热温度都是一个估算值,这一点从表1引用文献 [1] 中报道的绝热温度就可看出,报道值一般精确到100K或10K,

估算结果较粗糙,更不能反映生成物的相转变情况。而本研究设计的计算程序能方便迅速精确地计算 SHS 反应的绝热温度,如果有相转变,能给出有关相转变情况。

运用本程序还可计算一定起始条件(预热和加稀释剂)下 SHS 反应的绝热温度。图 2 示例绘出了预热温度对绝热温度影响的曲线。图 3 绘制了 SHS 反应 $10Al + 3TiO_2 + 3B_2O_3 + xAl_2O_3(or\ TiB_2) = 5Al_2O_3 + 3TiB_2 + xAl_2O_3(or\ TiB_2)$ 在不同的预热条件下($T_0 = 298K、400K、500K、600K$)下的 $X\text{-}T_{ad}$ 曲线图。

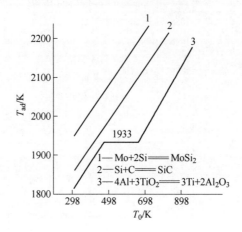

图 2 预热温度(T_0)对绝热温度的影响曲线

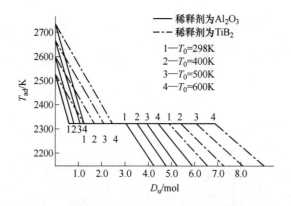

图 3 稀释剂对 SHS 反应绝热温度的影响曲线

从图 2、图 3 可以看出,在一定的初始条件下 SHS 反应的绝热温度可以一目了然。运用 SHS 反应的绝热温度可以帮助分析,探讨反应过程和反应机理、反应产物存在状态等。

4 结论

(1)在建立数据库的基础上,只需输入预热温度和 SHS 反应方程式,便可方便、迅速、准确地计算任一 SHS 反应的绝热温度。

(2)运用计算结果,可以绘制在一定初始条件(预热、加稀释剂)对 SHS 反应绝热温度影响的曲线,绝热度可以一目了然。

参 考 文 献

[1] Z. A. Mnir. Synthesis of High Temperature Materials by Self-propagating Combustion Methods. Am Ceram Soc Bull. , 1988, 67 (2): 342-349.

[2] Z. A. Mnir et al. Self-propagating Exothermic Reaction: The Synthesis of High Temperature Materials by Combustion. Mater. Sci. Rep. , 1989, 3: 277-365.

[3] Frankhouser et al. Advance Processing of Ceramic Compounds. Park Rldge, Noyes Data Corporation, Jerney. U. S. A. 1987.

[4] J. F. Crider. Self-propagating High Temperature Synthesis A Solvent Method for Producing Ceramic Materials. Ceram. Eng. Sci. proc. , 1982, 3 (9-10): 519-528.

[5] I. Barin et al. Thermo Chemical Properties of Inorganic Substances Spinger-Verlag-Berling-Heidelnboy-New York. 1973, 1977 (Supplement).

Study on CuO-BaTiO$_3$ Semiconductor CO$_2$ Sensor*

Abstract The prototype of a CO$_2$ sensor made of CuO-BaTiO$_3$, which has capacitance sensitive effect, is designed based on the pn heterojunctions of CuO and BaTiO$_3$ semiconductors. The preparation of BaTiO$_3$ semiconductor powders is pointed out, using the coprecipitation and semiconducting techniques. The characteristic quantities relating to the capacitance sensitive effect of the sensor are studied systematically with the aid of a gas tester. A reasonable mechanism of the sensor is proposed.

1 Introduction

In recent years, a great attention has been paid to the development and application of environmental gas CO$_2$ sensors[1-3]. The researches on solid electrolyte sensor are rather active[4-6]. CO$_2$ was determined by the voltage difference between CO$_2$ gas and the alkaline carbonate coated on working electrode, but the structure of the sensor is complicated, and the alkaline carbonate is deliquescent and strongly affected by water vapor, so, the requirements of the electrode preparation are very strict. Since the chemical property of CO$_2$ is stable, it was thought that the metal oxide semiconductor-type sensor could not be used for determining CO$_2$, while the research[7] reported SnO$_2$ which adding alkaline oxide is sensitive to CO$_2$ to some degree, but its preference to CO$_2$ is bad and the stability is not ideal. In 1990, Ishihara et al. found that the complex oxide compound of BaTiO$_3$ and PbO is sensitive to CO$_2$, and developed a semiconductor capacitive-type sensor for CO$_2$ gas[8]. From then on, more sensors for CO$_2$ gas have been introduced. They are equimolar mixtures of BaTiO$_3$ and metal oxide[9-11], such as NiO-BaTiO$_3$, PbO-BaTiO$_3$, Y$_2$O$_3$-BaTiO$_3$, CuO-BaTiO$_3$, etc. Studies on their microstructure, technological parameters and physical characteristics are also in progress.

In this paper, in order to reveal the mechanism of the capacitance effect of the CO$_2$ sensor, a special designed technique for preparing the semiconductors of CuO-BaTiO$_3$, which exhibit the capacitance effect for CO$_2$ gas, is introduced. The correlation between the pn heterojunction effect and the capacitance effect is investigated systematically for the CuO-BaTiO$_3$ semiconducting sensors as an example. A reasonable sensing mechanism is proposed.

2 Experimental

2.1 Preparation of gas sensitive elements

Using H$_2$TiO$_3$, Ba(NO$_3$)$_2$, (NH$_4$)$_2$CO$_3$ and NH$_3$·H$_2$O as the raw materials, BaTiO$_3$ powders

---

* Copartner: Liao Bo, Wei Qin, Wang Kaiyi. Reprinted from Sensors and Actuators, B: Chemical, 2001, 80(3): 208-214.

were prepared by the means of coprecipitation. After purification and beating, H_2TiO_3 turned into colloidal state and was fully mixed with $Ba(NO_3)_2$ aqueous solution, then calculated amount of $(NH_4)_2CO_3$ aqueous solution was added to it, using $NH_3 \cdot H_2O$ to adjust the pH value of the mixture. After coprecipitation through filtration, drying, washing and calcinations, high purity $Ba(NO_3)_2$ powders were obtained.

To improve the conductivity of $BaTiO_3$, the following semiconducting procedure is introduced. Nb_2O_5 with 99.99% purity is added as an impurity to $BaTiO_3$ to a concentration of 0.1mol%, and the even mixture of $BaTiO_3$ and Nb_2O_5 is sintered at 1150℃ in air for 2h. The phase analyses by X-ray diffraction confirm that $BaTiO_3$ is of single phase, with Nb atoms substituting some of the Ti atoms. The resistivity of $BaTiO_3$ becomes $10^{4.3} \sim 10^{4.9} \Omega \cdot cm$, which is of the order of semiconductors. With the impurity atoms Nb replacing part of the Ti atoms, the Ti-O bond becomes weak, thus favoring the production of the oxygen holes, which may improve the diffusion of volume and grain boundaries in $BaTiO_3$.

Lastly, the obtained $BaTiO_3$ powders were mechanically mixed with CuO (99.99%) powders in an equimolar amount, then the mixture was pressed into disks(13mm diameter and 0.6mm thickness). After calcination at 773K for 5h, Ag paste was applied on both faces of the disks to make electrodes(6mm in diameter), thus the sensitive element for determination was achieved.

2.2 Sensitive characteristic determinations of elements

A measurement system based on an inductance-resistance-capacitance tester, a device for gas analysis and a personal computer are installed, which can fulfill gas mixing, temperature control and gas analysis dynamically[12]. The working frequency is $f = 20Hz-100kHz$ and that of the temperature is $T = 293-873K(\pm 0.5K)$. With this system, a steady flow of gas can be maintained quantitatively, and the concentration precision of CO_2 gas can reach 200ppm. The sensitivity is defined as the ratio of element capacitance in CO_2 sample gas to that in air stream, namely Cs/Co.

The sensitive properties of the gas sensor such as the response and recovery of the capacitance, characteristics of the temperature, frequency, concentration, repetition and selectivity to CO_2 were investigated.

2.3 Study on sensing mechanism of elements

The microstructure of elements was examined by D-500 X-ray diffractometer with Ni filtered and scanning electron microscope[13]. Influences of the component of elements and semiconductor characteristic of $BaTiO_3$ on the sensitivity to CO_2 was also investigated. The carbonate reaction of copper dioxide was analysed by thermodynamics calculation. The I-V characteristic of elements was studied.

3 Results and Discussions

3.1 Study on sensitivity of elements

3.1.1 Response and recovery characteristic of elements

For electric-ceramic $CuO-BaTiO_3$ to be used as a dielectric, the elements which were fabricated by spreading silver electrode on both sides not only have capacitance but also resistance and impedance. Sensitivity of the three electric parameters to 2% CO_2 were inspected, which were expressed by Cs/Co, Rs/Ro and Zs/Zo, respectively. Results shown in Fig. 1 indicate that the capacitance of elements has a good response and recovery sensitivity to CO_2, which changes rapidly within several seconds and tends to become constant at 90s or above, while it takes more time to recover completely. Meanwhile, compared to the capacitance, resistance and impedance of elements belonging to capacitive-type sensors have little sensitivity to CO_2, which makes it easy to characterize the sensitivity of the elements, simplify sensor electric circuit design and provides more possibilities to realize the practicality of the sensor[14].

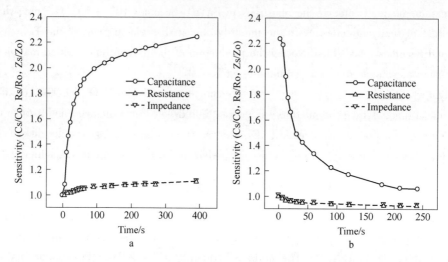

Fig. 1 Response and recovery characteristic of the electric parameters of elements
a—Response; b—Recovery
(f = 100kHz, T = 701K, CO_2 = 2%)

3.1.2 Temperature characteristic of elements

Placing the sensitive elements in air stream under various constant temperature for some time to measure their stable capacitance, then switching the air stream rapidly to 2% CO_2, inspecting the temperature sensitive characteristic of the elements, Fig. 2 shows that the sensitivity of the elements depends strongly on the operating temperature. The sensitivity to CO_2 gas increases with elevation in operating temperature and reaches a maximum of 2.34 at T = 685K, the characteristic is similar to the work done by Ishihara[15], while the maximum value of 2.98 was obtained at higher temperature T = 729K.

3.1.3 Frequency characteristic of elements

The sensitivity of CuO-$BaTiO_3$ element is strongly dependent on the applied ac frequency, as shown in Fig. 3. With the elevation of the operating frequency (20Hz – 100kHz), the element sensitivity increases non-linearly. Complex dielectric constant varies with the frequency and the capacitance of elements decreases with increase in the frequency, which can be affected easily by CO_2 gas and increases the sensitivity. In order to obtain high gas sensitivity, elements should be operated at higher frequency because, if the frequency is too high the capacitance will be very small, which may bring difficulties in measurement and design of electric circuit, so range of frequency is usually chosen between 50–100kHz.

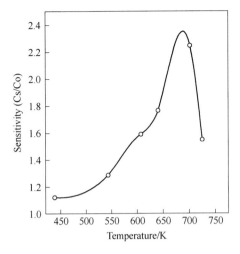

Fig. 2 Temperature characteristic
curve of the element
($f = 50kHz, CO_2 = 2\%$)

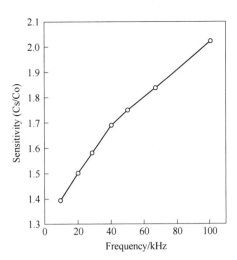

Fig. 3 Frequency characteristic
curve of the element
($T = 606K, CO_2 = 2\%$)

3.1.4 CO_2 concentration characteristic of elements

The sensitivity of CuO-$BaTiO_3$ at 606K is plotted as a function of CO_2 gas in Fig. 4. It increases with the concentration of CO_2, the detection range is rather wide. A linear relationship exists between the sensitivity and the concentration of CO_2 under 5%, which is of advantage to be standardized.

3.1.5 Repetition characteristic of elements

We compared the sensitivity of elements to 2% CO_2 under three different conditions, the result is shown in Fig. 5. The situation is (1) fresh undetected element; (2) after repeated transient exposure to CO_2 during various concentrations, especially endurance to high concentration of CO_2 (>90%); (3) after 48h recovery from operation (2). It is shown that the sensing tendency to CO_2 gas under three conditions are similar, while there exists some difference in the response and recovery time. The fresh element has the best sensitive response to CO_2 gas. After the strong

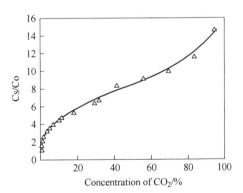

Fig. 4 CO_2 concentration characteristic
curve of the element
($f = 100kHz, T = 606K$)

stimulation of high concentration, the response speed becomes slower, but after 48h recovery, the response characteristic reaches the original level, which is similar to the recovery characteristic. From the result, we can see that the sensitive characteristic of elements can be similar after repeat transient exposure to CO_2, endurance to hot stimulation, electric detection, while only some difference exists between response and recovery time, but the recovery tendency of the elements is good. So, the repetition characteristic of elements is good, while longer and more test experiments need to carry on.

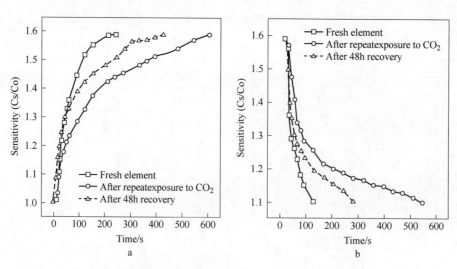

Fig. 5 Repetition characteristic curve of the elements
a—Response; b—Recovery
($f = 50\text{kHz}, T = 606\text{K}$)

3.1.6 Selectivity characteristic of elements

The selectivity of the sensor is an important problem, for the hydrocarbons are expected to interface because they burn to form CO_2, so we investigated the influence of CO, CH_4, and H_2 on CO_2 sensing. Typical responses of 2% CH_4, H_2, CO and CO_2 are shown in Fig. 6. CH_4 and H_2 had no appreciable effect, but the same concentration of CO affected the capacitance. Capacitance changes by CO seem to be caused by the carbon dioxide formed by the element. Since silver and copper dioxide are known to be effective catalysts for the oxidation of CO[16], further studies are required to be carried out. It is clear that although carbon dioxide is the most chemically stable among the gases examined in Fig. 6, the mixed-

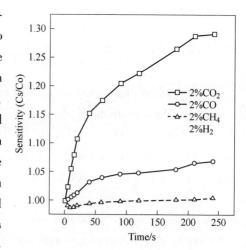

Fig. 6 Selectivity characteristic curve of the element
($f = 1\text{kHz}, T = 606\text{K}$)

oxide capacitor of CuO-$BaTiO_3$ is highly sensitive to CO_2. In practice, the interference by H_2O is anticipated because the dielectric constant of water is large, but the capacitance change caused by CO_2 is larger than that by humidity at the same concentration, and the sensitivity to 2% CO_2 was independent of the existence of humidity[10], this suggests the selectivity of the CO_2 sensor is good.

3.2 Study on sensing mechanism of elements

Though we adopted a new method to prepare the basic sensitive material $BaTiO_3$ powders, while the element made from it has similar sensitive characteristic to CO_2 gas as shown by Ishihara[15], and the characteristics of repetition, selectivity and concentration are also good, the phenomena supported the mechanism we considered, i. e. the surface adsorption mechanism and grain boundary effect of semiconductors mechanism. It was tested and verified by other electrical and microstructure analysis experiments.

3.2.1 Surface adsorption mechanism

From Fig. 2, the temperature characteristic of elements displays a regular bell curve and has the same characteristic as[9], while Ishihara et al. explains it by supposing the carbonate reaction of the copper oxide exists. Since the reaction rate between solid and gas is generally low at lower temperatures, the rate of the reaction would increase with elevation in the operation temperature and the depth of the surface layer participating in carbonation will be increased. On the contrary, the equilibrium favors the left-hand side at the elevated temperature, the authors thought this can account for the shape of the sensitivity curve qualitatively[9], and considered that the carbonation of CuO is a possible mechanism to the sensitive element[9, 10, 17, 18].

However, our calculations of Gibbs free energy based on the thermodynamic theory of chemical reaction processes indicate[19] that there is no chemical carbonation between CuO or $BaTiO_3$ and CO_2 gas, and the XPS for the sensor sample shows no evidence of carbonating valency states. These results make the carbonating mechanism quite questionable. So, we consider that the temperature sensitive characteristic curve just attributed to the typical surface adsorptive principle as most oxide gas sensor. Adsorption of gas sensitive elements can be divided into physical and chemical one. Physical adsorption plays a main role under low temperature, while chemical adsorption increases with temperature. So, the sensitivity increases with the temperature, but when the temperature reaches a constant value, disadsorption of CO_2 gas is aggravated, and physical and chemical adsorption all decrease. Thus, the sensitivity decreases with temperature after it reaches a maximum value. The specific process needs to be studied further. The morphology of elements was observed by JSM-35C SEM, the crystals and gas holes are well-distributed, which is beneficial to the absorption of CO_2 gas[13].

3.2.2 Grain boundary effect of semiconductors mechanism

3.2.2.1 Influence of component on the sensitive characteristic of elements

We inspected the influence of component on the sensitive characteristic of elements, i. e. to fabricate CuO and $BaTiO_3$ samples by following the same process as CuO-$BaTiO_3$ element, then detect the temperature sensitive characteristic to CO_2 by comparing with that of CuO-$BaTiO_3$ element. The result is shown in Fig. 7. The experiment indicates that CuO and $BaTiO_3$ samples have no sensitivity to CO_2. Meanwhile, the XRD analysis shows no diffraction peaks except for those observed from CuO and $BaTiO_3$, and no changes in the diffraction angles nor in the relative intensities of each diffraction peak from CuO and $BaTiO_3$ can be recognized, i. e. no solid reaction took place between CuO and $BaTiO_3$ during the calculation of temperature 0 - 773K. The CuO-$BaTiO_3$ mixed oxide is just a mechanical two-phase mixture[12]. From the experiment referred, the single phase of elements has no influence on the sensitivity to CO_2 gas, while the mechanical mixture of the two single phase can be sensitive to it, so, we think it is reasonable to use the grain boundary effect of semiconductors to explain the sensitivity of elements.

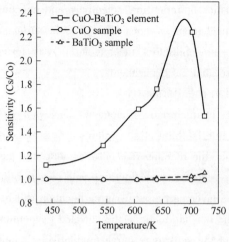

Fig. 7 Influence of component on sensitivity to CO_2
($f = 50kHz, CO_2 = 2\%$)

3.2.2.2 Influence of semiconductor characteristic of $BaTiO_3$ on sensitivity of elements

Copper dioxide is a typical semiconductor, while ordinary $BaTiO_3$ is dielectric material, when we fabricate the sensitive element, we must make $BaTiO_3$ semiconducting. In order to emphasize the necessity of the semiconductor characteristic, we compared the sensitivity of elements made of semiconductor $BaTiO_3$ powders to that of insulating $BaTiO_3$ powders which were obtained through sintering $BaCO_3$ and TiO_2 in oxygen flow stream, the result is shown in Fig. 8.

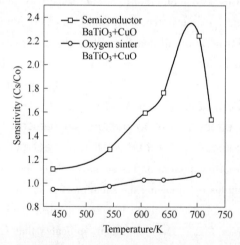

Fig. 8 Influence of $BaTiO_3$ semiconductor characteristic on sensitivity to CO_2
($f = 50kHz, CO_2 = 2\%$)

The element fabricated by the semiconductor $BaTiO_3$ powders displays the typical bell temperature curve, while the sample made from the $BaTiO_3$ which was obtained from sintering in oxygen has no sensitivity to CO_2 during the wide temperature range, so $BaTiO_3$ must be a semiconductor is the key condition to obtain the sensitive characteristic of elements. Only if a mass of p-n interfaces can form within a definite volume a sensitive unit from two different type oxide n-$BaTiO_3$ and p-CuO semiconductor powders, we can fabricate this kind of capacitive-type sensor for CO_2.

3.2.2.3 I-V characteristic of elements

Fig. 9 shows the *I-V* characteristics of the CuO-$BaTiO_3$ element in air stream and 2% CO_2 stream. Under the working conditions, CuO and $BaTiO_3$ are p-and n-type semiconductors, respectively, it is well known that the depletion layer formed at the junction between p-type and n-type semiconductors behaves as a capacitor. A large energy barrier seems to form at the grain junction between CuO and $BaTiO_3$ resulting from the p-n contact, consequently, the *I-V* characteristics of the element were non-linear. On the other hand, the non-linearity in the *I-V* relationship becomes linear characteristic in 2% CO_2, this suggested that the adsorption of CO_2 reduced the height of the energy barrier at the grain junction of CuO and $BaTiO_3$. Therefore, the width of the depletion layer was also reduced, which accords to the capacitance of the binary oxide of CuO and $BaTiO_3$ exhibiting a large increment upon exposure to CO_2. The potential barrier at the boundary between CuO-$BaTiO_3$ is expected to play an important role for the capacitance change by CO_2 gas.

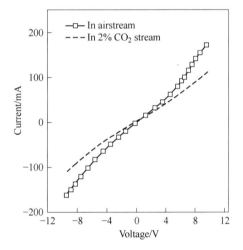

Fig. 9 *I-V* characteristic of the element in air stream and CO_2 stream
(T = 606K)

4 Conclusions

In this paper, CuO-$BaTiO_3$ compound sensitive elements for CO_2 based on the capacitance change were fabricated. Coprecipitation and doping semiconductor technique were adopted to prepare high purity $BaTiO_3$ semiconductor powders, which is different from the sintering method used by the inventor, then adulterated CuO semiconductor powders were made into electrodes by pressing and calcination to obtain the sensitive element.

Sensitive characteristics of elements are detected by a special tester. Experimental results indicate that the capacitance of elements belonging to the capacitive-type gas sensors is sensitive to CO_2 gas. A linear relationship exists between the sensitivity of elements and at the concentration of CO_2 under 5%, the repetition and selectivity of elements is good. In addition, some other experimental results on sensitivity, microstructure, electrical and thermal dynamics analysis of elements are also tested and verified that the main sensing mechanism of the capacitive-type sensor is due to the surface adsorption and grain boundary effect of semiconductors, which negated the mechanism of carbonation of oxide advanced by the inventor.

References

[1] H. Yahiro, S. -I. Katayama. Application of metal ion-exchanged zeolites as materials for carbon dioxide sensor. Denki Kagaku, 1993, 61: 451-452.

[2] D. Mutschall, E. Obermeier. A capacitive CO_2 sensor with on-chip heating. Sens. Actuators B, 1995, 24/25: 412-414.

[3] K. Ogura, H. Shiigi. A CO_2 sensing composite film consisting of base-type polyaniline and poly(vinyl alcohol). Electrochem. Solid State Lett., 1999, 2: 478-480.

[4] W. F. Chu, D. Discher. Thin and thick film electrochemical CO_2 sensors. Solid State Ionics, 1992, 53-56: 80.

[5] L. W. Mason, S. Oh. Development and testing of a solid-state CO_2 gas sensor for use in reduced-pressure environments. Sens. Actuators B, 1995, 24/25: 407-411.

[6] M. Holzinger, J. Maier, W. Sitte. Potentiometric detection of complex gases: application to CO_2. Solid State Ionics, 1997, 94: 217-225.

[7] N. Mizuno, T. Yoshioka. CO_2-sensing characteristics of SnO_2 element modified by La_2O_3. Sens. Actuators B, 1993, 13/14: 473-475.

[8] T. Ishihara, K. Kometani, M. Hashida, Y. Yakita. Mixed oxide capacitor of $BaTiO_3$-PbO as a new type CO_2 gas sensor. Chem. Lett., 1990: 1163-1166.

[9] T. Ishihara, K. Kometani, M. Hashida, Y. Yakita. Application of mixed oxide capacitor to the selective carbon dioxide sensor. J. Electrochem. Soc., 1991, 138: 173.

[10] T. Ishihara, K. Kometani, M. Hashida, Y. Yakita. Mixed oxide capacitor of CuO-$BaTiO_3$ as a new type CO_2 gas sensor. J. Am. Ceram. Soc., 1992, 75: 613.

[11] T. Ishihara, S. Matsubara. Capacitive type gas sensors. J. Electroceram, 1998, 2: 215-228.

[12] L. Bo, L. Hongzhuan, Y. Jianhong. Testing system for calibrating characteristics of carbon dioxide gas sensor. Sensor Tech., 1997, 16: 43-45.

[13] L. Bo, L. Yexiang, W. Qin. Preparation of capacitive-type CO_2 gas sensitive element using nanometer powders. Trans. Nonferrous Met. Soc. China, 1998, 8: 657-660.

[14] Matsubara, et al. Sens. Actuators B, 2000, 65: 128-132.

[15] T. Ishihara. Capacitor types for selective gas sensing in chemical analysis. Selective Electrode Rev., 1992, 14: 1-31.

[16] T. Ishihara, K. Kometani, Y. Mizuhara, Y. Takita. New type of CO_2 gas sensor based on capacitance changes. Sens. Actuators B, 1991, 5: 97-102.

[17] T. Ishihara, K. Kometani, Y. Takita. Impedance studies on the CO_2 detection mechanisms of the mixed oxide. CuO-$BaTiO_3$, as a capacitor type CO_2 sensor, Denki Kagaku, 1993, 61: 911-912.

[18] T. Ishihara, K. Kometani. Improved sensitivity of CuO-$BaTiO_3$ capacitive-type CO_2 sensor by additives. Sens. Actuators B, 1995, 28: 49-54.

[19] B. Liao. Preparation of CuO-$BaTiO_3$ complex oxide electronic ceramic capacitive-type CO_2 gas sensor. PhD thesis, Central South University of Technology, 1997.

纳米 TiO_2 光催化剂的电化学法制备及其表征*

摘 要 采用有机电解-溶胶-凝胶法制备了纳米 TiO_2 光催化剂,通过 XRD、TEM、TG-DTA、激光粒度分析法等对纳米 TiO_2 的结构相变、表面形貌、颗粒大小等进行了表征。实验表明:当热处理温度低于 350℃ 时,TiO_2 的晶相结构均为锐钛矿型,颗粒大小在 25nm 以下,样品的比表面积大于 $65m^2/g$;热处理温度为 400℃ 时,TiO_2 的晶相结构出现锐钛型和金红石型混合相,颗粒大小在 35nm 以下,样品比表面积为 $46.43m^2/g$,实验制备的粉体样品属于纳米级水平;实验测试了各焙烧温度下粉体样品的光催化活性。

1 引言

采用纳米 TiO_2 光催化剂处理环境污染物作为一项新的污染治理技术[1-3],由于具有能耗低,操作简单,反应条件温和与可减少二次污染等突出优点,该技术日益受到重视。近年来关于纳米 TiO_2 粉体制备方法有很大的发展,其制备方法较多,如化学气相沉积法[4]、醇盐水解法[5]、化学沉淀法[6]、水热制备法[7]、固相法[8]、化学氧化法[9]、溶胶-凝胶法[10,11]等。本研究采用有机电解-溶胶-凝胶法,一步法制备纳米 TiO_2 的前驱体钛酸丁酯,通过溶胶-凝胶法制备纳米 TiO_2,该法具有常温常压操作,可通过调节电位,控制电极反应的速度,克服了传统方法的缺点,不需使用昂贵的钛酸丁酯为原料,生产成本低,工作环境好,后处理容易,有潜在的产业化前景。sol-gel 法实验采用电解得到的铁酸丁酯为前驱体,无水乙醇作溶剂,醋酸作催化剂,用氨水作 pH 调节剂,通过 sol-gel 法合成了 TiO_2 凝胶,在不同热处理温度下制备了纳米 TiO_2 光催化剂,并通过 XRD、TEM、TG-DTA、激光粒度分析法等对纳米 TiO_2 的结构相变、表面形貌、颗粒大小及光催化活性等进行了表征。

2 实验

2.1 样品的合成

2.1.1 钛酸丁酯的合成

在自制电解槽内,用处理后的工业纯钛片 TA-1(钛的化学组成参见 GB 3620—83)作电极,电解液为 0.005mol/L 的少量复合导电盐的 200mL 丁醇溶液,控制电压为 30~50V,电解液温度为 20~80℃,丁醇溶液经脱水处理后,在无水条件下电解 10h 后得到淡黄色的钛酸丁酯溶液[12]。

2.1.2 溶胶-凝胶过程

在 20℃,pH=4,钛酸丁酯液:乙醇:水 = 1:3:4(体积比)的实验条件下,将 5mL 的无水乙醇、适量的二次蒸馏水及催化剂(调节 pH 值)配成溶液 A;将上述电解

---

* 本文合作者:陈建军、李荣先、赵方辉。原发表于《功能材料》,2003,34(6):732-734。

得到的钛酸丁酯溶液 5mL 在快速搅拌下缓缓滴入 10mL 无水乙醇和 0.5mL 无水醋酸的混合液中,配成溶液 B;将溶液 A 缓缓滴入溶液 B,得到均匀透明溶胶。在空气中放置,陈化,得固体凝胶。

2.1.3 干燥-焙烧过程

将上述固体凝胶放置于空气中自然干燥 3d 后,用红外干燥箱或恒温干燥箱于 60~90℃下恒温干燥数小时,得固体干凝胶;将干凝胶用玛瑙研钵研磨得初始纳米粉体;将初始粉体置于 250~800℃的马弗炉中焙烧 3.5h 即得纳米粉体。

2.2 纳米 TiO_2 的表征

光催化剂粉体样品的 XRD 分析采用 Cu 靶,Rigaku 衍射仪,比表面积测试采用 ASAP 2010 比表面积测定仪,用 BET 法测定样品的比表面积,TiO_2 光催化剂干凝胶样品的 TG-DTA 分析测试采用 WRT-3P 微量热天平,升温速率为 10K/min,平均晶粒度尺寸采用 XRD 线宽法测定,颗粒度及分布采用 H-800 透射电子显微镜(TEM)和 Zetaplus Potential Analyzer 测试。

2.3 光催化活性测试

采用 500mL 苯酚溶液作为模拟废水,苯酚的初始浓度为 100mg/L,各种温度焙烧后的纳米 TiO_2 粉体的浓度为 200mg/L。将样品置于苯酚溶液中用超声分散器分散 5min,反应液通过冷凝水控制在 35℃左右,光降解光源为 300W 的高压汞灯,紫外灯发光稳定后,溶液预热 4min 后(溶液温度约为 35℃)开始计时,每 10min 取样一次,共取 6 次,将所取试样通过台式离心机分离,取上层清液用 751G 型紫外-可见分光光度计测定其吸光度,根据苯酚的工作曲线求出其光降解后的浓度,从而可比较其光活性的大小。实验所用光反应器(如图 1 所示)由 3 层同心圆筒构成,包括 4 部分:光发生器、气体搅拌器、冷凝系统和反应器。

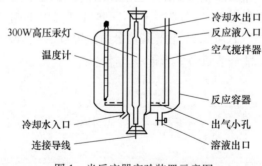

图 1 光反应器实验装置示意图

3 实验结果与讨论

3.1 XRD 物相分析

图 2 给出了试样在 250℃、350℃、400℃、600℃和 800℃热处理温度下的 XRD 衍射谱图。由图 2 可见,当焙烧温度为 250℃和 350℃时,样品均为锐钛相 TiO_2,当焙烧温度为 400℃时,有少部分转化为金红石型,其金红石相 TiO_2 所占比例约为 15%;当焙烧温度为 600℃时,样品大部分转变为金红石型,金红石相占总样品质量的 85%,800℃时样品完全转变为金红石型结构。

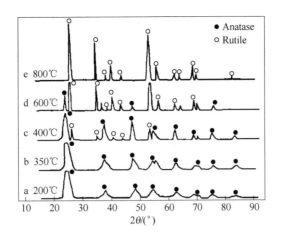

图 2 纳米 TiO_2 粉体样品经不同焙烧温度热处理 3.5h 后的 XRD 图

3.2 TG-DTA 分析

图 3 为干凝胶样品的 TG-DTA 曲线。由图 3 可知在 98.6℃附近有一宽的吸热峰，相应地在 TG 曲线上对应于宽吸热峰区表现为失重较快，这可能是低温区干凝胶粉末中的吸附水、丁醇等有机物的不断挥发或分解所引起的，此温度下对应 TG 曲线上的失重率为 13.3%；在 275~365℃之间，有一小的放热峰，它对应于无定型 TiO_2 向锐钛型转变的晶化热，相应地在 TG 曲线上表现为缓慢失重，在 350.6℃处，其失重率为 34.5%；在 365~550℃之间，DTA 曲线上存在一较大放热峰，它对应于金红石型 TiO_2 晶相的晶化热，这说明在此温度范围内，锐钛型与金红石型两相共存。这一分析结果与 TiO_2 的 XRD 分析结果相吻合。

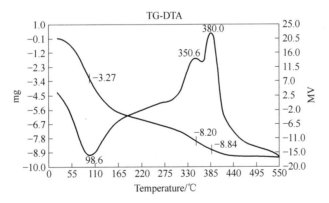

图 3 TiO_2 干凝胶粉末的 TG-DTA 曲线

3.3 平均晶粒度

采用 XRD 线宽法测量了经不同温度热处理后 TiO_2 纳米粉末的平均晶粒尺寸。依据 Scherrer 方程[13]：

$$D = \frac{0.89\lambda}{B\cos\theta} \qquad (1)$$

式中，λ 是波长，nm；B 的单位是 rad；D 的单位是 nm。由此可测得经不同热处理温度后 TiO_2 纳米粉末中锐钛矿晶粒的平均尺寸，结果列于表 1 中，热处理温度低于 600℃ 时，粉末中锐钛矿相的平均晶粒尺寸均 <35nm。经 600℃ 热处理 3.5h 后，粉末样品的晶粒尺寸达到 110nm，经 800℃ 热处理 3.5h 后晶粒尺寸为 146.3nm。

表 1　纳米 TiO_2 平均晶粒及颗粒大小与热处理温度的关系

热处理温度 T_a/℃	晶体大小 D_c/nm	颗粒大小 D_p/nm	直径 D_p/nm
250	8.9	9.0	9.3
300	11.6	11.6	12.2
350	21.2	22.8	22.3
400	31.6	32.3	31.18
600	110	132.0	171.2
800	146.3	397.0	464.6

3.4　TEM 形貌观察和粒径测试

不同温度热处理后的 TiO_2 粉末的 TEM 颗粒形貌照片如图 4 所示，测量结果见表 1 中的第 3 列。实验表明，600℃ 以下热处理粉末其平均颗粒度在 35nm 以下，而且颗粒分布均匀，而当热处理温度高于 600℃ 时，颗粒分布很不均匀，颗粒度分布较宽，这可能与金红石型 TiO_2 的形成有关。

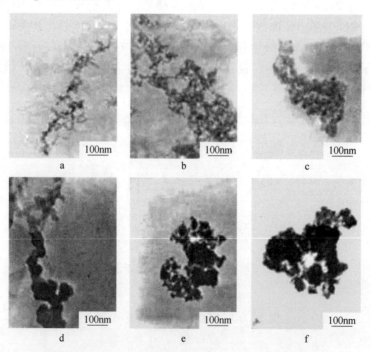

图 4　粉体样品在不同温度下焙烧 3.5h 的 TEM 形貌照片
a—250℃；b—300℃；c—350℃；d—400℃；e—600℃；f—800℃

采用 Zetaplus Potential Analyzer 测定了不同热处理温度下粉体样品的平均粒径尺寸（见表 1 第 3 列）。结果表明 600℃ 以下热处理后的粉末样品其平均颗粒度在 35nm 以下，热处理温度高于 600℃ 时，TiO_2 粉末中颗粒分布较宽，这可能与 TiO_2 晶粒的 A→R

结构相变有关，较高焙烧温度促进了晶粒生长。由于 XRD 线宽法只对 50nm 以下的纳米晶粒（即主要是对锐钛矿相 TiO_2 而言）尺寸准确测量，而 800℃ 的 TiO_2 热处理粉末的金红石晶粒已超过 50nm。由 Zetaplus Potential Analyzer 定量测定的平均颗粒尺寸包括锐钛矿相和金红石相 TiO_2 两种晶粒，因此，表1中列出的 800℃ 粉末的晶粒尺寸与平均颗粒尺寸差别较大。

3.5 比表面积

通过测定粉体单位质量的比表面积 S_w，可计算纳米粉体粒子直径[14]（设颗粒呈球形）：

$$d = 6/\rho S_w \tag{2}$$

式中，ρ 为密度；d 为比表面直径。S_w 的一般测量方法为 BET 多层气体吸附法。不同热处理温度下纳米 TiO_2 粉末样品的比表面积如表2所示。由表可知，通过 BET 求得的固体颗粒粒子直径与其他方法测量的结果相近。

表 2 在不同焙烧温度下粉体样品的比表面积

热处理温度 T_a/℃	比表面积 $S_w/m^2 \cdot g^{-1}$	粒子直径 D_p/nm
250	170.17	9.05
300	124.91	12.33
350	65.21	23.62
400	46.43	33.19
600	10.66	144.54
800	3.69	418.01

3.6 产量及光催化活性表征

将电解得到的 5mL 钛酸丁酯溶液通过溶胶-凝胶法处理后，所制得的 TiO_2 干凝胶粉体分别在 250℃、350℃、400℃、450℃、600℃ 和 800℃ 下进行热处理，得到的纳米 TiO_2 粉体平均质量为 0.3316g，经推算，200mL 电解液中含有钛酸丁酯 56.4317g，可见该法具有一定的实用价值。实验测试了各种样品对苯酚的光降解效率的影响（如图5所示）。

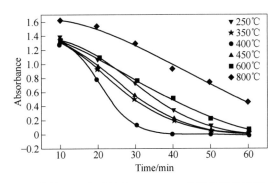

图 5 热处理温度对纳米 nano-TiO_2 光催化活性的影响

由图5可知，400℃ 时样品的光催化活性最好，低于或高于 400℃ 时，其光催化活性均有所降低。当光降解 30min 时，400℃ 热处理温度下的粉体样品对苯酚的光降解率约为92%，250℃、350℃、450℃、600℃和800℃下的试样对苯酚的光降解率分别为55%、65%、62%、55%和10%。由此可见热处理温度对光催化剂的性能影响很大。温度太低，样品颗粒粒径虽小，但可能未形成完整的晶型，光照时，不利于催化剂表面激发光生电子与光生空穴，故活性较低；温度太高，光催化剂内部孔隙减少，粒子团聚度增加，比表面减少，导致光活性中心减少；此外，温度太高，纳米锐钛型 TiO_2 会向金红石型 TiO_2 转变，带隙减少，光生电子与光生空穴复合几率增加，因而活性变差。

4 结论

通过有机电解-溶胶-凝胶法制备了纳米 TiO_2 光催化剂，采用 XRD、TEM、TG-DTA、激光粒度分析法等研究了纳米 TiO_2 的结构相变、表面形貌、颗粒大小等微结构特征，并以苯酚光催化降解作为测试反应，研究了经不同热处理温度焙烧后样品的光催化活性。结果表明：样品的光催化活性和它的微结构特征密切相关。在 250~400℃温度下焙烧，随着样品结晶度的改善，其光催化活性逐渐提高，400℃下焙烧的样品显示最高的光催化活性，其晶相结构出现锐钛型和金红石型混合相；高于400℃焙烧时，晶体尺寸、金红石含量迅速增大，样品的光催化活性逐渐下降。这些结果表明，焙烧温度对样品锐钛矿结晶度、颗粒大小有较大影响，也是影响光催化活性的最重要因素，因此，在电化学法制备和处理纳米 TiO_2 光催化剂时，有效地控制样品的焙烧温度是十分重要的。

参 考 文 献

[1] Mills A, Davies Richard H, Worsley D. [J]. Chem Soc Rev, 1993, 15：417-425.
[2] Fox M A, Dulay M T. [J]. Chem Rev, 1993, 93 (1)：341-357.
[3] Linsebigler Amy L, Lu Guangquan, Yater John T. [J]. Chem Rev, 1995, 95 (3)：735-758.
[4] 朱以华, 陈爱平, 李春忠, 等. [J]. 华东理工大学学报, 1999, 25 (4)：382-385.
[5] 王浚, 高濂, 宋哲. [J]. 无机材料学报, 1999, 14 (1)：170-174.
[6] 赵敬哲, 王子忱, 刘艳华, 等. [J]. 高等学校化学学报, 1999, 20 (3)：467-469.
[7] 汪国忠, 汪春昌, 张立德, 等. [J]. 材料研究学报, 1997, 11 (5)：527-530.
[8] 高荣杰, 史可信, 王之昌. [J]. 金属学报, 1996, 32 (10)：1097-1101.
[9] 陈士仁, 邵艳群, 唐电. [J]. 中国有色金属学报, 1998, 8 (2)：250.
[10] Xu Naping, Shi Zaifeng, Michael Z C, et al. [J]. Ind Eng Chem Res, 1999, 38 (2)：373-379.
[11] 张兆敏, 李斌, 买光昕. [J]. 山东工业大学学报, 1999, 29 (4)：398-401.
[12] 陈建军. 中南大学冶金系博士论文 [D]. 长沙：中南大学, 2001.
[13] 尹元根. 多相催化剂的研究方法 [M]. 北京：化学出版社, 1988.
[14] 张立德, 牟季美. 纳米材料和纳米结构 [M]. 北京：科学出版社, 2001.

Preparation of Precursor for Stainless Steel Foam*

Abstract The effects of polyurethane sponge pretreatment and slurry compositions on the slurry loading in precursor were discussed, and the performances of stainless steel foams prepared from precursors with different slurry loadings and different particle sizes of the stainless steel powder were also investigated. The experimental results show that the pretreatment of sponge with alkaline solution is effective to reduce the jam of cells in precursor and ensure the slurry to uniformly distribute in sponge, and it is also an effective method for increasing the slurry loading in precursor; the mass fraction of additive A and solid content in slurry greatly affect the slurry loading in precursor, when they are kept in 9%–13% and 52%–75%, respectively, the stainless steel foam may hold excellent 3D open-cell network structure and uniform muscles; the particle size of the stainless steel powder and the slurry loading in precursor have great effects on the bending strength, apparent density and open porosity of stainless steel foam; when the stainless steel powder with particle size of 44μm and slurry loading of 0.5g/cm^3 in precursor are used, a stainless steel foam can be obtained, which has open porosity of 81.2%, bending strength of about 51.76MPa and apparent density of about 1.0g/cm^3.

1 Introduction

Porous or cellular metallic materials attract more and more researching interests and their application fields are being developed due to that they can offer many excellent performances such as light quality, high specific strength and stiffness, excellent crash energy and sound absorption abilities[1-4]. Recently, higher melting point porous metallic materials have found an increasing range of applications in many fields. For example, high porosity open-cell Fe-based metal foam can be used as heat exchangers[5], stainless steel foam with 3D open-cell network structure and excellent anti-corrosion property can be used as catalyst carriers in purifying automotive exhaust gases[6], high melting point metal foams can be used as flow field and gas diffusion layer in direct methanol fuel cells[7].

At present, porous metallic materials with higher melting points are usually prepared by powder metallurgy (PM) approaches; sometimes, metal fibre or foaming agents such as $MgCO_3$, $SrCO_3$ or others are used in the preparation process[1,4,8-11]. However, cells existing in these porous metallic materials prepared by the above methods are smaller, sometimes partial cells are closed; in addition, it is difficult to prepare higher melting points porous metallic materials with 3D open-cell network structure by the above methods. So, developing a novel preparation technology of higher melting points porous metallic materials with 3D open-cell network structure is

---

* Copartner: Zhou Xiangyang, Li Shanni, Li Jie. Reprinted from Journal of Central South University of Technology, 2008, 15(2):209-213.

significant.

Recently, interests have been focused on the research of stainless steel foams (SSFs) with 3D open-cell network structure in our group, and a novel preparation technology of SSFs has been developed[12]. The preparation technology of precursor for SSFs was mainly investigated in this work.

2 Experimental

2.1 Preparation of precursors and SSFs

Stainless steel powders with different particle sizes were the main materials used in this study. Their chemical compositions are given in Table 1. Other materials were additive A, polyurethane sponge (0.8 pores/mm) and high-purity argon (99.999%).

Table 1 Chemical compositions of 316L stainless steel powder (mass fraction, %)

C	Mn	Si	Cr	Ni	P	S	Mo	O	Others
≤0.03	0.20	0.80	16.0–18.0	12.0–15.0	≤0.03	≤0.02	2.0–3.0	≤0.30	Bal.

Main apparatuses used in this study were a controlled temperature oven with a maximal temperature of 300℃ and a controlled atmosphere electric furnace with a maximal temperature of 1600℃.

The preparation processes of precursor and SSFs are simply described as follows: (1) confecting slurry containing a given amount of solid mass fraction by mixing the stainless steel powder and additive A; (2) preparing precursor with a certain loading of drying slurry by pouring the slurry into polyurethane sponge and drying at 120℃ for 24h; (3) preparing SSFs samples by sintering the precursors under flowing argon at (1260 ± 2)℃ for 30min. The detailed preparation processes of precursors and SSFs were described in Ref. [12].

2.2 Testing methods

The load refers to the dry mass of slurry in unit volume of polyurethane sponge (g/cm^3).

In addition, digital camera (SONY, DSC-P10) and scanning electron microscope (JEOL, JSM5600) were employed to observe the apparent appearance and section morphologies of samples, respectively. Metallurgical microscope (XJP-6A) was used to investigate the grain sizes and microstructures of samples. Electric omnipotence tester (CSS-44100) was employed to test the bending strength of samples. The apparent density and open porosity of samples were tested according to the usual methods described in Ref. [13].

3 Results and Discussion

3.1 Effects of polyurethane sponge pretreatment on apparent appearance and load in precursors

In this work, pretreatment of sponge refers to immersing sponge in alkaline solution. In order to study the effects of polyurethane sponge pretreatment on precursors, fastening other technologic conditions was necessary. In the experiments, particle size of stainless steel powder was 44μm, the mass fraction of additive A and solid content (i.e. mass fraction of stainless steel powder) in slurry were 10% and 60%, respectively.

Fig. 1 shows the apparent appearance photographs of precursors before and after polyurethane sponge being pretreated with alkaline solution. Obviously, the slurry is more uniformly distributed in pretreated sponge, and the cell-jammed phenomenon is severer in un-pretreated sponge. This may be due to the improvement of wettability between slurry and pretreated sponge.

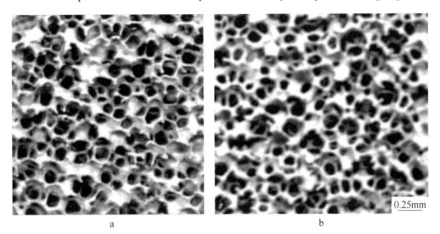

Fig. 1　Apparent photographs of precursors with and without pretreatment
a—Un-pretreated; b—Pretreated

On the other hand, the pretreatment also influences the load of slurry in precursors. Fig. 2 shows the effects of pretreatment on load. Apparently, pretreatment of sponge is an effective method to increase the load. So, all polyurethane sponge used in this work had been pretreated by alkaline solution with the exception of special statement.

Fig. 3 shows the SEM image of SSFs from pretreated sponge. Obviously, the sample exhibits uniform muscles and excellent 3D open-cell network structure.

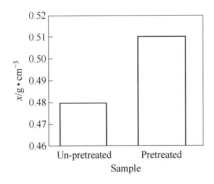

Fig. 2　Effects of pretreatment on load of slurry in precursors

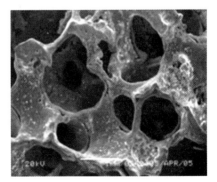

Fig. 3　SEM image of SSFs sample

3.2　Effects of concentration of additive A and solid mass fraction in slurry on load in precursors

In the experiments, the particle size of the stainless steel powder was 44 μm.

Fig. 4 shows the effects of mass fraction of additive A on load in precursors. Apparently, the load gradually increases with rising the mass fraction of additive A in slurry. From experimental phenomena, it is found that the slurry is hardly poured into the sponge when the mass fraction of additive A is less than 9%, and a higher mass fraction of additive A(> 13%) is unsuitable for much more slurry staying in sponge. This may be due to the fact that the thixotropy of slurry is poorer in the case of lower additive mass fraction, and the viscosity of slurry is too larger under higher additive mass fraction[14].

Fig. 5 shows the load changes at different solid mass fractions in slurry. Evidently, the load in precursor increases with rising solid mass fraction. Experimental phenomena also show that the slurry hardly settled in sponge when the solid mass fraction is less than 30% due to the fact that the slurry is too dilute, the cell-jammed phenomenon and bigger hollows co-exist on the surface of samples when the solid mass fraction is less than 52%, and the cell-jammed phenomenon becomes severer when the solid mass fraction is more than 75% due to the worsened fluidity of the slurry.

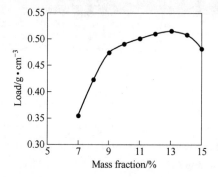
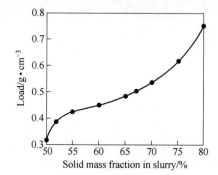

Fig. 4 Effects of mass fraction of additive A on load of slurry in precursor
(Mass fraction of solid in slurry is 66.7%)

Fig. 5 Effects of solid mass fraction in slurry on load
(The mass fraction of additive A in slurry is 11%)

3.3 Effects of load in precursors on apparent density, open porosity and bending strength of SSFs samples

The load range of samples used in the experiment was from 0.42 to 0.69g/cm^3, and the particle size of stainless steel powder was 44μm. The correlations between load in precursors and apparent density of SSFs samples are given in Fig. 6. Fig. 7 presents the effects of load on open porosity and bending strength of SSFs.

It is clearly observed that the apparent density of SSFs gradually increases with rising the load, while the open porosity falls with rising the load, and raising load is beneficial to the enhancement of bending strength. However, the cell-jammed phenomena appear while the load is more than 0.675g/cm^3. A SSFs sample with a bigger open porosity(about 81.2%) and higher bending strength (51.76MPa) can be obtained at the load of 0.5g/cm^3.

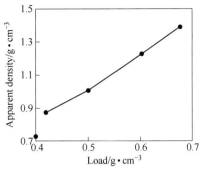

Fig. 6　Effects of load on apparent density of SSFs samples

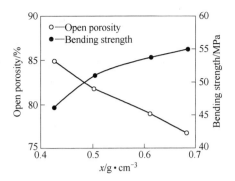

Fig. 7　Effects of load on open porosity and bending strength of SSFs samples

3.4　Effects of particle size of stainless steel powder on open porosity and bending strength of SSFs samples

Four stainless steel powders with different particle sizes were used in the experiments; the particle sizes were 37μm, 44μm, 50μm and 61μm, respectively. Here, a kind of slurry containing additive A with mass fraction of 11% and a solid mass fraction of 67% was used, and the load of precursors was about 0.5g/cm^3.

The effects of particle size are shown in Fig. 8 and Fig. 9. The results indicate that the open porosity of SSFs samples gradually decreases with the decrease of particle size; the bending strength of SSFs from fine particle size is remarkably higher than that from coarse powder, only being 16.33MPa at the particle size of 61μm, but rising to 51.76MPa at 44μm. The reasons why the bending strength of SSFs from coarse powder is inferior to that from fine particle size may be as follows: (1) the intrinsic surface energy driving force of fine particle size powder with larger specific surface area is higher than that of coarse powder, so that the former can be more easily sintered compactly than the latter[15]; (2) the fluidity of slurry from coarse powder is poorer than that from fine powder, which will bring on the un-uniform distribution of slurry in precursor, as a result that the muscles in SSFs are inhomogeneous, which will cause lower bending strength; (3) there are many defects such as lower cells spheroidizing degree, bigger cavities at the grain boundary and bulky crystal grains exist in SSFs sample from coarse powder, which will also lower the bending strength. The defects existing in SSFs samples from coarse powder can be clearly observed in Fig. 10.

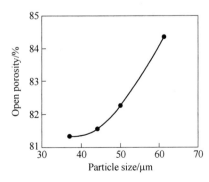

Fig. 8　Effects of particle size on open porosity of SSFs samples

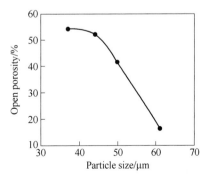

Fig. 9　Effects of particle size on bending strength of SSFs samples

Fig. 10 presents the differences of metallurgical photographs of SSFs samples with size of 61μm and 44μm. The cavities on the grain boundary of sample from coarse powder(61μm) are evidently bigger than that from fine powder(44μm), the cavities are indicated by arrows A and B. In addition, the cells spheroidizing degree of sample from coarse powder is lower; however, for SSFs sample from fine powder, and the crystal grain morphology is near-spherical, and the grain boundary is smooth. The measuring results also show that the crystal grain size of sample from fine powder(44μm) is 1/2 to 1/4 of that from coarse powder(61μm).

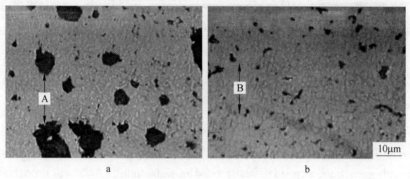

Fig. 10 Metallurgical photographs of SSFs samples with different particle sizes
a—61μm; b—44μm

4 Conclusions

(1) The pretreatment of sponge with alkaline solution is effective to reduce the jam of cells in precursor and ensure the slurry to uniformly distribute in sponge, and pretreatment of sponge is helpful to increase the slurry loading in precursor.

(2) The mass fraction of additive A and solid content in slurry greatly affect the slurry loading in precursor. When the mass fraction of additive A and the solid content in slurry are kept in 9%–13% and 52%–75% respectively, the stainless steel foam may hold excellent 3D open-cell network structure and uniform muscles.

(3) With rising the load, the apparent density and bending strength of SSFs gradually increase, while the open porosity falls. With the decrease of particle size, the open porosity of SSFs gradually decreases and the bending strength rises.

(4) A stainless steel foam sample with open porosity of 81.2% and bending strength of about 51.76MPa and apparent density of about 1.0g/cm^3 can be obtained at stainless steel powder particle size of 44μm and the slurry loading of 0.5g/cm^3 in precursor.

References

[1] ZHAO C Y, LU T J, HODSONA H P, JACKSON J D. The temperature dependence of effective thermal conductivity of open-celled steel alloy foams [J]. Materials Science and Engineering A, 2004, 367(1/2): 123-131.

[2] JOHN B. Manufacture, characterization and application of cellular metals and metal foams [J]. Progress in Materials Science, 2001, 46(6):559-632.

[3] ZHAO C Y, LU T J, HODSONA H P. Thermal radiation in ultralight metal foams with open cells[J]. In-

ternational Journal of Heat and Mass Transfer, 2004, 47(14/16):2927-2939.

[4] PARK C, NUTT S R. PM synthesis and properties of steel foams[J]. Materials Science and Engineering A,2000,288(1):111-118.

[5] LU W, ZHAO C Y, TASSOU S A. Thermal analysis on metal-foam filled heat exchangers (Part Ⅰ):Metal-foam filled pipes [J]. International Journal of Heat and Mass Transfer,2006,49(15/16):2751-2761.

[6] ZHOU X Y, LI J, LONG B, HUO D W. The oxidation resistance performance of stainless steel foam with 3-D open-celled network structure at high temperature [J]. Materials Science and Engineering A,2006,435/436(5):40-45.

[7] ARISETTY S, PRASAD A K, ADVANI S G. Metal foams as flow field and gas diffusion layer in direct methanol fuel cells[J]. Journal of Power Sources, 2007, 165(1):49-57.

[8] LIN Xiao-qing, HE Yue-fei, JIAO Yao, ZHANG Feng-shou. Experimental study on effective parameters for filtration performances of stainless steel porous materials [J]. Materials Science and Engineering of Powder Metallurgy,2005,10(2):128-132(in Chinese).

[9] PARK C, NUTT S R. Effects of process parameters on steel foam synthesis[J]. Materials Science and Engineering A,2001,297(1/2):62-68.

[10] FANG Hui-hui, HUANG Zhao-qiang, XUE Song. SEM analysis of hole structure of metal fibre sintered filtration felt [J]. Physics and Chemistry Test:Physics Fascicule,2000,36(6):258-259(in Chinese).

[11] WANG Tong-qing. Filtration performance and science application of filtration material of metal fibre sintered felt [J]. Filtrating and Separating,2003,13(1):26-28(in Chinese).

[12] ZHOU Xiang-yang, LI Jie, DING Feng-qi, LONG Bo. The preparation means of the porous metal foam with 3-D open cells or portion cells connected with each other:CN,200510032174.7[P]. 2005-09-22 (in Chinese).

[13] LIAO Ji-qiao. Powder metallurgy experiment technology [M]. Changsha:Central South University Press, 2003(in Chinese).

[14] WANG Xiao-gang, WEI Yun-jing, LI Xiao-chi. Major factors influencing the performance of silicon carbide foam ceramic slurry [J]. Journal of Xi'an University of Science and Technology,2004,24(2):190-192(in Chinese).

[15] GUO Shi-ju. Powder metallurgy sintering theory [M]. Beijing:Metallurgical Industry Press, 1998 (in Chinese).

低温热处理以及高能球磨对制备超细氧化铝的影响*

摘 要 以 $Al_2(SO_4)_3$ 与 $(NH_4)_2CO_3$ 为原料，采用液相沉淀法，制备出前驱物 $NH_4Al(OH)_2CO_3$（碱式碳酸铝铵），利用低温热处理以及高能球磨对前驱物进行了处理，并煅烧得到 $\alpha\text{-}Al_2O_3$。运用 XRD、振实密度、SEM 等现代分析检测技术对样品进行表征。结果表明：采用低温热处理能改善 Al_2O_3 粉体的分散性，并能在一定程度上促进 Al_2O_3 的晶型转变；采用高能球磨能在很大程度上改善 Al_2O_3 粉体的分散性，并且能极大地促进 Al_2O_3 的晶型转变，制备出片状 Al_2O_3 粉末。经过低温热处理与高能球磨的前驱物，在 1100℃煅烧 2h，能得到无团聚、粒径为 500nm 左右的片状超细 $\alpha\text{-}Al_2O_3$ 粉末。

作为特种功能材料之一的超细 Al_2O_3 粉，由于具有高强度、高硬度、抗磨损、耐腐蚀、耐高温、抗氧化、绝缘性好、表面积大等优异的特性，目前已在生物陶瓷、精密陶瓷、化工催化剂、稀土三基色荧光粉、集成电路芯片、航空光源器件等方面得到了广泛的应用[1-4]。正是由于超细 Al_2O_3 粉末与常规颗粒相比具有一系列优异的电、磁、光、力学和化学宏观特性，因此近年来世界各地将制备高纯超细 Al_2O_3 粉末作为新材料领域研究的主攻方向之一。而片状 Al_2O_3 具有热容量大、导热率高、密度高、高温性能稳定、活性大、结构有再生性、抗腐蚀和抗破坏性能好等特性，可以用作催化剂基体或载体、环氧树脂和聚酯树脂的填料以及防滑涂层的磨料等，因此对片状 Al_2O_3 的制备技术进行研究非常必要。

超细 Al_2O_3 粉末的制备方法众多[5-11]，大致可以分为液相法、气相法、固相法三大类[12]。液相沉淀法制备的超细 Al_2O_3 粉末具有平均粒径好、粒度分布窄等优点，因此成为了研究最多和最有希望的方法[6-11,13-17]。但是通过液相法制备前驱物，结合低温热处理以及高能球磨来获得高分散片状超细 Al_2O_3 粉末的方法未见报道。本研究采用液相沉淀法制备出前驱物碱式碳酸铝铵，并对前驱物进行了低温热处理、高能球磨，于 1100℃煅烧 2h，得到了无团聚、粒径为 500nm 的片状超细 $\alpha\text{-}Al_2O_3$ 粉末。

1 试验内容

1.1 样品制备

将 0.1mol 的 $Al_2(SO_4)_3\cdot 18H_2O$（分析纯）配置成 0.3mol/L 的溶液，加入一定量的 PEG600 作为分散剂；将 0.4mol 的 $(NH_4)_2CO_3$（分析纯）配制成 2.0mol/L 的溶液，加入一定量的 PEG1000 作为分散剂。将两种溶液均置于恒温水浴预热到 55℃。采用自制的反应器，在高速搅拌（1400r/min）下，将 $Al_2(SO_4)_3$ 溶液加入到 $(NH_4)_2CO_3$ 溶液里面，控制 pH 值为 8~10，反应结束后将体系陈化 1h，采用去离子水洗涤到使用

---

* 本文合作者：肖劲、万烨、邓华、李劼。原发表于《粉末冶金技术》，2006，24（6）：449-452。

$BaCl_2$ 溶液检验滤液无沉淀出现为止。将得到的滤饼使用乙醇洗涤 2 次，80℃真空干燥 12h，稍加研磨便得到前驱物粉末。将前驱物粉末分别进行如下处理：直接放入马弗炉，于 1100℃煅烧 2h，得到产物记为样品 A；300℃低温热处理 3h，移入 1100℃马弗炉煅烧 2h，得到产物记为样品 B；高能球磨 4h，80℃真空干燥 12h，放入 1100℃马弗炉煅烧 2h，得到产物记为样品 C；300℃低温热处理 3h，真空冷却后高能球磨 4h，80℃真空干燥 12h，放入 1100℃马弗炉煅烧 2h，得到产物记为样品 D。

1.2 高能球磨

试验中采用 QM-ISP4 型行星式球磨机，为了避免球磨罐体和球对试验材料的污染，球磨罐和磨球均采用刚玉自制。控制球料比为 10∶1，采用适量无水乙醇作为球磨介质，将需要球磨的物料在 300r/min 下球磨 4h。

1.3 样品表征

采用日本理学 Rigaku D/max 2550VB$^+$ 18kW 转靶 X 衍射仪确定前驱物以及煅烧产物；采用智能拍击（振实）密度测试仪测定 Al_2O_3 粉末的振实密度；采用日本 JSM-5600LV 扫描电镜观察 Al_2O_3 粉末的形貌以及分散情况。

2 结果与讨论

2.1 前驱物 XRD

图 1 为前驱物粉末的 XRD 图谱，由图 1 可知，采用上述工艺条件制备的前驱物为碱式碳酸铝铵（$NH_4Al(OH)_2CO_3$），由此推测，当原料 $Al_2(SO_4)_3$ 和 $(NH_4)_2CO_3$ 的摩尔比为 1∶4 时，反应器中发生了如下反应：

$$Al_2(SO_4)_3 + 4(NH_4)_2CO_3 + 2H_2O = 2NH_4Al(OH)_2CO_3 + 3(NH_4)_2SO_4 + 2CO_2 \quad (1)$$

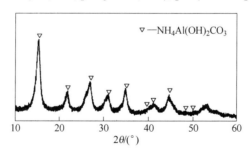

图 1 前驱物的 XRD 图谱

2.2 低温热处理以及高能球磨对前驱物煅烧特性的影响

图 2 是四种样品的 XRD 图谱。通过图 2 我们可以看出，样品 A 为 α-Al_2O_3 与 θ-Al_2O_3 两相混合物，并且主相为 θ-Al_2O_3（见图 2a）；样品 B 同样为 α-Al_2O_3 与 θ-Al_2O_3 两相混合物，但是主相为 α-Al_2O_3（见图 2b）表明低温热处理能促进 Al_2O_3 的晶型转变，原因在于：经过低温热处理之后，样品中含有的水分在低温条件下基本被驱除，水分的驱除能在一定程度上缓解前驱物在煅烧过程中的团聚问题，并能细化晶粒（这

都在后面的 SEM 分析中得到了验证），而分散性的改善以及晶粒的细化都能在一定程度上降低 α-Al_2O_3 的成形温度。

经过高能球磨处理的样品 C 与样品 D 都为纯相 α-Al_2O_3（见图 2c、2d），而没有经过高能球磨的样品 A（见图 2a）、样品 B（见图 2b）均不是纯相 α-Al_2O_3，因此高能球磨能降低 α-Al_2O_3 的成形温度。结合文献资料[18]，笔者认为在采用高能球磨的过程中，前驱物经过长时间的研磨和冲击，会引起结构的紊乱、网络的断裂或错动。经过高能球磨后，粉体的比表面积变大，晶格发生畸变，表面会产生许多破键，粉末内部存储大量的变形能和表面能，因此能显著降低 α-Al_2O_3 的成形温度。

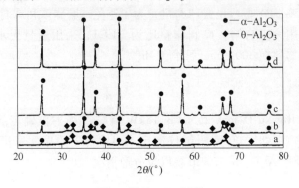

图 2　样品 XRD 图谱

a—样品 A；b—样品 B；c—样品 C；d—样品 D

2.3　低温热处理以及高能球磨对 Al_2O_3 粒子性能的影响

四种样品的粒子性能列于表 1。从表观上观察，四种样品存在非常大的区别：样品 A 为块状团聚体，需要研磨才能进行检测，且研磨后的 Al_2O_3 粉末非常密实，振实密度很高；样品 B 为略带团聚的粉末，需要略加研磨才能进行检测，研磨后的 Al_2O_3 粉末疏松性能较差，振实密度较高；而样品 C 与样品 D 都是疏松的粉末，且两种样品的振实密度都较小。

表 1　样品的部分性能

样品	A	B	C	D
样品的表观特性	严重的块状团聚体	略带团聚的粉末	疏松的粉末	疏松的粉末
振实密度/g·cm^{-3}	0.956	0.715	0.617	0.592

图 3 是四种样品的 SEM 图谱。可以看出：样品 A 颗粒呈纺锤状，大小不均，且相互联结在一起，有明显的烧结颈，说明样品发生了明显的硬团聚（见图 3a）。这是因为采用乙醇置换并不能很彻底地将前驱物中的水分脱除，前驱物在煅烧过程中会出现严重的团聚现象；样品 B 的颗粒也呈纺锤状，粒子较样品 A 细小、均匀，虽然部分区域仍然有颗粒联结成块状，但是与样品 A 相比，样品 B 的分散性能得到了一定程度的改善（见图 3b）。这是由于经过低温热处理之后，前驱物中的大部分水分能在低温条件下被脱除，从而能在很大程度上缓解前驱物在煅烧过程中的团聚问题，从而细化了 Al_2O_3 晶粒、大大改善了样品的分散性；但图中样品 C 颗粒呈片状，粒径较样品 A、B 都要大，为 1~2μm，粒径大小不均匀，但是分散性特别好，基本上不存在颗粒

联结现象（见图3c）。原因在于：外部的机械力量通过球磨，使研磨罐与前驱物粉末之间发生频繁撞击、摩擦和挤压，使粉末受到反复锤锻、撕裂、破碎，促使颗粒表面缺陷密度增加，最终使粉末颗粒形貌由球形变成片状；同时在强烈的球磨作用下，无水乙醇能更好地将前驱物中的水分置换脱除。因此，样品 C 的分散效果得到了更好的改善。样品颗粒变大的机理目前还不清楚，需要在以后进一步深入研究。样品 D 颗粒呈规则的片状，粒径较样品 C 小，为 0.5μm 左右，粒径大小非常均匀，且各单个颗粒松散地靠在一起，基本上没有发生团聚（见图3d）。综合两种工艺条件之后，能很好解决前驱物在煅烧过程中的团聚问题，得到分散性非常好、粒度分布非常均匀、粒径大小为 0.5μm 左右的片状 Al_2O_3 粉末。

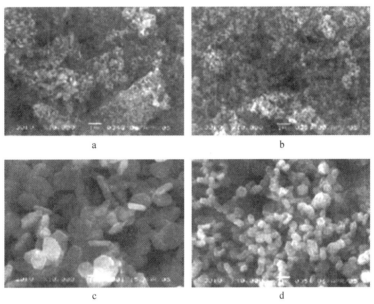

图 3 样品的 SEM 图谱
a—样品 A；b—样品 B；c—样品 C；d—样品 D

3 结论

（1）控制 $Al_2(SO_4)_3$ 与 $(NH_4)_2CO_3$ 的摩尔比为 1∶4，采用液相沉淀法，能制备出前驱物碱式碳酸铝铵（$NH_4Al(OH)_2CO_3$）。

（2）采用低温脱除前驱物中的水分，能在一定程度上解决前驱物煅烧过程中的团聚问题，并能在一定程度上促进 Al_2O_3 的晶型转变。

（3）前驱物经过高能球磨后，煅烧产物 Al_2O_3 的分散性得到了很大改善。同时，高能球磨能显著地降低 α-Al_2O_3 的成形温度。

（4）前驱物经过高能球磨之后，煅烧产物 Al_2O_3 的形貌由球形变为了片状。

（5）将前驱物在 300℃ 低温处理 3h 后，进行高能球磨，于 1100℃ 煅烧 2h，能得到无团聚，粒径为 500nm 左右的片状 α-Al_2O_3 粉末。

参 考 文 献

[1] 付高峰. 超细 Al_2O_3 粉末制备技术. 有色矿冶，2000，16（1）：39-41.

[2] 刘维平. 气流循环粉碎法制备超细 Al_2O_3 粉的正交试验研究. 南方冶金学院学报, 2003, 24 (5): 34-36.

[3] 赵玉成. 液相法纳米 Al_2O_3 粉的制备. 燕山大学学报, 2005, 29 (1): 88-91.

[4] 董海峰. 加热与脱盐对纳米 Al_2O_3 粒度及分布的影响. 功能材料, 2005, 36 (4): 583-588.

[5] 李慧韫, 张天胜, 杨南. 纳米 Al_2O_3 的制备方法及应用. 天津轻工业学院学报, 2003, 18 (4): 34-37.

[6] 王雅娟, 李春喜, 王子镐. 超声波-化学沉淀法制备纳米 Al_2O_3 粒子. 北京化工大学学报, 2002, 29 (4): 8-11.

[7] 谢冰, 章少华. 纳米 Al_2O_3 的制备及应用. 江西化工, 2004, (1): 21-25.

[8] 耿新玲, 袁伟. 纳米 Al_2O_3 的制备与应用进展. 河北化工, 2002, (3): 1-9.

[9] 唐海红. 纳米 Al_2O_3 的制备及应用. 中国粉体技术, 2002, 8 (6): 37-39.

[10] 张永刚, 闫裴. 纳米 Al_2O_3 的制备及应用. 无机盐工业, 2001, 33 (3): 19-22.

[11] 吴志鸿. 纳米 Al_2O_3 的制备及其在催化领域的应用. 工业催化, 2004, 12 (2): 35-39.

[12] Blendell J E, Kent B H, Coble R L. High purity alumina by controlled precipitation from aluminum sulfate solutionsl ceramic bulletin, 1984, 63 (6): 797-802.

[13] 彭天右. 共沸蒸馏法制备超细 Al_2O_3 粉体及其表征. 无机材料学报, 2000, 15 (6): 1097-1101.

[14] Kirk-Othmer. Encyclopedia of chemical technology (Third Edition). New York: Awiley-Interscience Publication, 1992.

[15] 沈志刚. 无机纳米粉体制造技术的现状及展望. 无机盐工业, 2002, 34 (3): 18-21.

[16] 王宝. 纳米材料干燥技术的研究和发展. 通用机械, 2004, (12): 12-13, 72-73.

[17] 仇海波. 纳米氧化锆粉体的共沸蒸馏法制备及研究 1. 无机材料学报, 1994, 9 (3): 365-370.

[18] 韩兵强, 李楠. 高能球磨法在纳米材料研究中的应用. 耐火材料, 2002, 36 (4): 240-242.